Climate Change Management

Series editor

Walter Leal Filho, Hamburg, Germany

More information about this series at http://www.springer.com/series/8740

Walter Leal Filho
Editor

Climate Change Impacts and Adaptation Strategies for Coastal Communities

Springer

Editor
Walter Leal Filho
Faculty of Life Science
HAW Hamburg
Hamburg
Germany

ISSN 1610-2010 ISSN 1610-2002 (electronic)
Climate Change Management
ISBN 978-3-319-70702-0 ISBN 978-3-319-70703-7 (eBook)
https://doi.org/10.1007/978-3-319-70703-7

Library of Congress Control Number: 2017957657

Printed on acid-free paper

This Springer imprint is published by Springer Nature
The registered company is Springer International Publishing AG
The registered company address is: Gewerbestrasse 11, 6330 Cham, Switzerland

Preface

Climate change is known to impact coastal areas in a variety of ways. According to the 5th Assessment Report produced by the Inter-Governmental Panel on Climate Change (IPCC), coastal zones are highly vulnerable to climate change and climate-driven impacts may be further exacerbated by other human-induced pressures.

Apart from sea level rise which poses a threat to both human well-being and property, extreme events such as cyclones and storm surges lead not only to significant damages to property and infrastructure, but also to saltwater intrusion, the salinisation of groundwater, and intensification of soil erosion, among many other problems. There are also many negative impacts on the natural environment and biodiversity, which include damages to important wetlands and habitats that safeguard the overall ecological balance, and consequently the provision of ecosystem services and goods on which the livelihoods of millions of people depend.

These impacts are particularly acute in the developing countries and island States in the Pacific, Caribbean, Latin America and Asian region, since they have limited access to the funding and technologies needed to allow them to be more resilient and recover from the damages caused by hurricanes, floods and other extreme events.

The above state of affairs illustrates the need for a better understanding of how climate change affects coastal areas and communities, and for the identification of processes, methods and tools which may help the countries and the communities in coastal areas to adapt and become more resilient. There is also a perceived need to showcase successful examples of how to cope with the social, economic, and political problems posed by climate change in coastal regions.

It is against this background that this book has been prepared. It is a truly interdisciplinary publication, with contributions from scholars, representatives from social movements, practitioners and members of governmental agencies, undertaking research and/or executing climate change projects in coastal areas and working with coastal communities.

The focus of the book is 'managing climate change in coastal regions', meaning that it serves the purpose of showcasing experiences from research, field projects and best practice to foster climate change adaptation in coastal areas and communities, which may be useful or implemented elsewhere.

The aims of this book are as follows:

i. to document the influences of, the damages and the threats posed by climate change to estuaries and coastal communities;
ii. to introduce approaches, methods, initiatives and projects which demonstrate how coastal communities can successfully meet the challenges climate change poses to them. Here, an emphasis is given to the latest research, but also on infrastructure projects, demonstrations on the use of technologies and natural and artificial means to reduce the impacts of extreme events and sea level rise to coastal communities;
iii. to introduce funding schemes and mechanisms which can finance climate change adaptation in coastal areas.

Last but not least, a further aim of the book is to document and disseminate the wealth of experiences available today.

The book is structured along two parts:

Part 1-Principles, Approaches and Projects on Change Adaptation in Coastal Areas
Part 2-Case Studies on Climate Change Adaptation in Coastal Areas

I thank the authors for their willingness to share their knowledge, know-how and experiences, as well as the many peer reviewers, who have helped us to ensure the quality of the manuscripts.

It is hoped that this publication will help to outline the need for integrated approaches towards handling the impacts of climate change to coastal areas, and hence contribute towards advancing this field of work even further.

Enjoy your reading!

Hamburg, Germany
Spring 2018

Prof. Walter Leal Filho
B.Sc., Ph.D., D.Sc., D.Phil., D.Ed., D.L.

Contents

Part I
Principles, Approaches and Projects on Change Adaptation in Coastal Areas

"We're not Refugees, We'll Stay Here Until We Die!"—Climate Change Adaptation and Migration Experiences Gathered from the Tulun and Nissan Atolls of Bougainville, Papua New Guinea

Johannes Luetz and Peni Hausia Havea

Abstract Atoll island communities are naturally vulnerable to flooding hazards such as king tides, storm surges and overtopping, among others. Climate change can be expected to catalyse the susceptibility to flooding through extreme weather events, sea level rise (SLR) and other climate related pressures. Further, population growth in coastal proximity can exacerbate vulnerabilities by degrading ecosystems such as mangroves and coral reefs that island communities tend to rely on for protection. The net effect: More and more people are congregating in a high-risk zone for floods and storms, but are less and less protected from them. Conventional adaptation wisdom suggests three responses: (1) plan a managed retreat (e.g., move inland); (2) accommodate the changes (e.g., build stilt houses); (3) resist the intrusion (e.g., build sea walls). On the Carteret Islands of Bougainville/Papua New Guinea (PNG), also known as the Tulun or Kilinailau Atoll, none of these adaptation measures have so far enabled the islanders to adapt in situ to mounting people and sea level pressures, resulting in both ad hoc and planned out-migration responses. Drawing on pilot research conducted on the Tulun and Nissan Atolls of Bougainville/PNG, this paper examines the linkages between climate change and human movement. It extends previous research by expressly inviting the grassroots perspectives of atoll communities of origin and mainland communities of destination in Tinputz and Buka. The research develops recommendations in areas of education, livelihood security, government planning and countering misinformation. Experiences and lessons gathered in this paper will be useful for both policy and practice serving the cause of climate change adaptation in Pacific island communities. Working towards a

J. Luetz (✉)
CHC Higher Education, Brisbane, Carindale 4152, QLD, Australia
e-mail: jluetz@chc.edu.au

P. H. Havea
Pacific Centre for Environment and Sustainable Development (PaCE-SD),
The University of the South Pacific, Laucala Campus, Suva, Fiji

J. Luetz
University of New South Wales (UNSW), Sydney 2052, NSW, Australia

W. Leal Filho (ed.), *Climate Change Impacts and Adaptation Strategies for Coastal Communities*, Climate Change Management,
https://doi.org/10.1007/978-3-319-70703-7_1

better understanding of climate change related vulnerabilities in coastal areas will also enable better adaptation responses.

Introduction: Geographic, Demographic and Climate Change Issues

Over recent years, remote atolls to the north and northeast of Bougainville, an autonomous region of Papua New Guinea (PNG), have been described numerously in public discourse as being in danger of inundation and permanent submersion (Roberts 2002; Vidal 2005; Gupta 2007; MacFarquhar 2009; Harman 2013; Beldi 2016). Regional atolls include the Nissan (Green), Nuguria (Fead) and Malum groups to the north of Buka Island, the largest of the Solomon Islands archipelago, and the Tulun (Carterets, Kilinailau), Takuu (Mortlock) and Nukumanu (Tasman) groups to the northeast (O'Collins 1989, pp. 248–250). To date several video documentaries have been produced which discuss both regional climate change vulnerabilities and present and anticipated resettlement from these low-lying islands in the Pacific (UNU 2009a; OTLP 2010; Metzger and Redfearn 2011; Light Studios 2016).

Anecdotal islander eyewitness reports typically point to extraordinary flooding events that can momentarily overcome human coping capacities: "[s]torm surges regularly overtop our islands—then the sea and low-lying land become 'level.' My home island—the Island of Huene—was sliced in two.[1] The time for adaptation and mitigation has run out. The time for migration and relocation has come. Resettlement is underway. It is so sad to leave." (U. Rakova, pers comm, 17 March and 10 April 2008, cited in Luetz 2008, p. 20). A map from 1964 provided by the Geography Department of the University of Papua New Guinea still depicts Huene Island as one unfractured island (O'Collins 1989, p. 252), thus revealing the splicing in two to be a more recent event.

The multiplicity of precursory problems to island abandonment typically comprise progressive pressures depicted in Fig. 1. As shown, islands have disproportionately large coastal areas compared to total land available (Nunn and Kumar 2006), and storm surges and island "overtopping" can cause erosion and freshwater lens contamination (Barnett and Campbell 2010, p. 172), while population growth and decreasing land area may exacerbate water stress, making rain catchment and freshwater storage systems increasingly indispensable to safeguard water security in rain water dependent localities (Fig. 2). Taken together, these figures align with research participant responses that commented on polluted wells on the Tulun Atoll no longer serving as a viable means of meeting the drinking water requirements of the islanders: "the well water is salty, we don't use [it] anymore" (P9).

[1]A photo of the two islets Huene One and Two is available online: http://planetprepare.blogspot.com.au/2010/11/phd-pilot-study-carteret-islands.html.

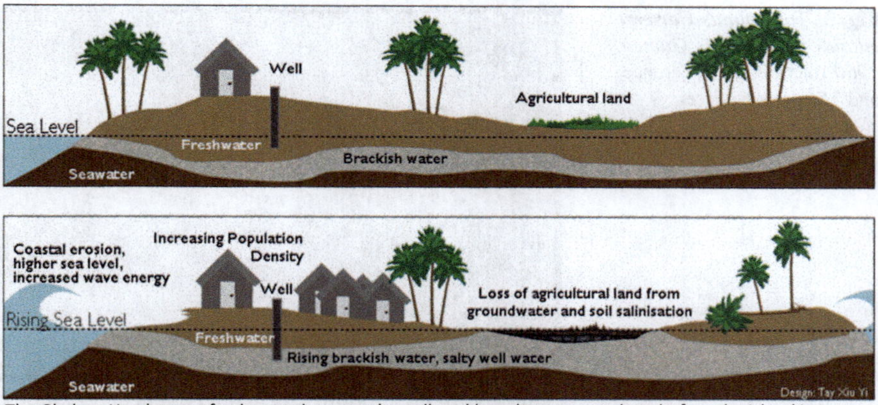

The Ghyben-Herzberg or freshwater lens may be polluted by salty seawater long before the island is ultimately submerged (Barnett & Campbell 2010, p. 172).

Fig. 1 Schematic representation of island subsistence (normal sea level) and progressive island submergence (rising sea level). (Illustration © World Vision, quoted from Luetz 2017, p. 5; adapted from Aung et al. 1998, p. 97)

Fig. 2 Rainwater catchment system on the Island of Huene One, Tulun Atoll

Being heavily reliant on rainwater harvesting, steady freshwater supplies present significant challenges for low-lying atoll islands and coastal communities. (Photo taken in 2010 by Luetz)

Figure 2 also illustrates environmental consequences on the Island of Han where "the ocean water is now seeping up through the ground, creating swampy breeding grounds for malaria-carrying mosquitoes … The results are rotting roots, degraded arable land and tiny taros." (Luetz 2008, p. 20; cf. UNU 2009b). Photographs taken on the Tulun Atoll in 1960 and available through the National Library of Australia (Fig. 3) show taro plants in apparently healthy size and condition (Spencer and Spencer 1953), thus suggesting that the erosion of food security could be an SLR related phenomenon experienced predominantly since the 1960s in conjunction with population growth (O'Collins 1989; see also Connell 1990).

A number of explanations for the progressive inundation of the affected islands have been offered, including above average rates of SLR affecting the western

Fig. 3 *Taro plants Carteret Islands, Papua New Guinea, 1960* [picture slide]/Terence and Margaret Spencer

Source with attribution: (Spencer & Spencer 1953, held in National Library of Australia, NLA)

Pacific (Amos 2012); the possibility that the "old volcano [is] … sinking back into the sea" (Gupta 2007, para. 4); dynamite fishing and consequent reduction of reef protective capabilities (Roberts 2002, para. 4, 14); and seismic activity on account of shifting plate tectonics (Traufetter 2012; Connell 2015, pp. 5–6), among others.

Notwithstanding, none of these factors alone seems to adequately explain the high rate of locally experienced SLR. Chalapan Kaluwin, Professor of Environmental Science at the University of Papua New Guinea (pers comm, 20 March 2008), has pointed to some of the complexities surrounding sea level science: "The causes of sea level rises are manifold, intertwined and acting in concert. Research requires a prolonged, multidisciplinary investigation with scientific contributions from different fields of study—glaciologists, oceanographers, geodesists, geologists, meteorologists, and climate scientists." (cited in Luetz 2008, p. 21). A comprehensive discussion of the causes of SLR lies beyond the scope of this research paper. However, given the high incidence of SLR experienced in the region of investigation, relative to other locations in the world, the following discussion is useful.

According to Bamber and Riva (2010), Pacific Islands have been shown to be subject to some of the highest rates of "regional sea-level rise due to land-ice melt only" (cited in World Bank 2012, p. 33), a point both corroborated and explained by Professor John Schellnhuber CBE, founding director of the Potsdam Institute for Climate Impact Research (PIK) and chair of the German Advisory Council on Global Change (WBGU), who discussed the "injustice" of SLR at the 2011 Melbourne conference "Four degrees, or more?"[2]:

> This [is] the injustice of sea level rise … on average you have about a metre of sea level rise by 2100, … all over the globe. But the … very vicious thing is, that this sea level rise will be distributed in a highly inhomogeneous way across the planet. […] Elementary physics – if Greenland is losing mass, that means its gravitational pull for seawater will be diminished

[2]Transcribed from audio file of Session 1 @ 51:00 min previously posted at http://www.fourdegrees2011.com.au.

– that means, around Greenland, sea level may even drop, in particular for the north-eastern part of the American continent, while … the Pacific Islands … that haven't done anything to contribute to global warming, will again get the brunt of it, will get all the water which is released from Greenland. […] And those who are most responsible for that, northern Europe, northern America, will be spared sea level rise, at least for a while. So you see nature can be extremely unfair, if humanity is sort of provoking that injustice (cited in Luetz 2013, p. 231).

According to the Intergovernmental Panel on Climate Change (IPCC), sea levels have been relatively stable for 2000–3000 years and "did not change significantly … until the late 19th century … Estimates for the 20th century show that global average sea level rose at a rate of about 1.7 mm $year^{-1}$. Satellite observations available since the early 1990s provide more accurate [global] sea level data [which show] that since 1993, sea level has been rising at a rate of around 3 mm $year^{-1}$, significantly higher than the average during the previous half century." (IPCC 2007, p. 409). Satellite altimetry also provides "unambiguous evidence of non-uniform sea level change in open oceans, with some regions exhibiting rates of sea level change about five times the global mean" (ibid., p. 411).

Research in Papua New Guinea has shown SLR on the order of 7.0 mm $year^{-1}$ (AusAID 2010, pp. 2, 12), and monitoring in the Pacific has shown enormous variability, including rises in sea level of 7–10 mm $year^{-1}$ (Albert et al. 2016) or even 25 mm $year^{-1}$ (Kaluwin 2008, p. 8), far above IPCC global SLR averages discussed above. To summarise, "[r]egional variations, significant discrepancies, and some uncertainty remain in measures of SLR, according to sources and analytical techniques, but *a distinct overall upward trend exists throughout the western Pacific*." (Connell 2015, p. 5 emphasis added; attributed to Merrifield 2011; Zhang and Church 2012). Further, "it is reasonable to assume that the rate of SLR is increasing, [and] that it will increase further in the future" (ibid., attributed to Church and White 2006; Horton et al. 2014).

With SLR rates in the region "amongst the highest globally" (Albert et al. 2016, p. 2), it may perhaps not be surprising that most media accounts present migration responses from the Tulun Atoll to locations in Bougainville as a new, recent, and predominantly climate change related phenomenon (Roberts 2002; Vidal 2005; Gupta 2007; MacFarquhar 2009; Harman 2013; Beldi 2016; cf. Barnett and Campbell 2010, p. 173). However, complicating the picture, there is evidence from a UNEP compendium that resettlement as a "response to land loss and population growth" was considered even before the 1990s (O'Collins 1989, p. 247), with the research reporting that collective relocation on account of food shortages and people pressures was contemplated as far back as 1968 (ibid, p. 250), at which point in time the Carteret Islanders "were reported to have a unanimous desire for resettlement" (ibid.). Hence in summary, it is not unreasonable to anticipate that in light of climate change related pressures now progressively bearing on these and other affected regional atolls, which Campbell and Warrick (2014, p. 29) already described as "urgent cases", historical human resettlement intentions flagged almost three decades ago (Connell 1990) could be increasingly accentuated further. Expressed in simple language, this research seeks to better understand climate change related vulnerabilities in coastal areas, since a better understanding in this area will likely lead to better coping and adaptation responses.

Rationale for and Intended Contribution of This Research

Island abandonment is not a new phenomenon (Arenstam Gibbons and Nicholls 2006; Albert et al. 2016; cf. Leon 2016), and macro-managed human migrations within and across atolls have been documented before, for example in the Maldives (Luetz 2017). While challenging for the islanders involved, impelled migration experiences can also be seen as useful in that they can inform policy makers to assume a more forward thinking posture of migration policy development, where lessons are gleaned retrospectively, to help benefit policy research and development prospectively, thereby enabling the preparation of appropriate anticipatory migration management mechanisms, which can then be implemented before they are needed as instruments to assuage future island abandonment situations (UNISDR-UNDP 2012). Given the diverse vulnerabilities of atoll environments discussed above, climate change impacts seem to be closely related to the erosion of food, water and human security (Barnett and Adger 2003), wherefore incidences of island depopulation and abandonment could conceivably increase, especially if recent island submersions in the Solomons Islands are to be taken as a guide to the future (Leon 2016, cf. Albert et al. 2016). Arenstam Gibbons and Nicholls (2006) have cautioned: "[l]ooking to the future, … many small low-lying islands could be abandoned due to sea-level rise long before they become physically uninhabitable" (p. 40). Moreover, Barnett and Campbell (2010) have advocated that "communities that are likely to be exposed to climate change effects might be encouraged to consider relocation *as an adaptive process*. If it is necessary to relocate communities, *it should be proactive and planned*" (p. 173, emphasis added). In short, the benefits of foresight and preparedness in adaptation research seem to be straightforward and have been advocated both in the literature (e.g., Blanco et al. 2009; Foresight 2011), and by UN agencies tasked with human development: "[h]oping—and working—for the best while preparing for the worst, serves as a useful first principle for adaptation planning" (UNDP 2007, p. 198); "early preparedness could also help avert a humanitarian catastrophe by promoting orderly movements of affected populations and increasing the viability of the move" (UNHCR 2009, p. 3; see also Leighton 2012, pp. 703, 718). Given the sheer size of the Pacific Island region comprising 22 countries and territories, a total land area of 551,385 km^2, and a total population of 9,498,829 (Barnett and Campbell 2010, p. 5; attributed to SPC 2008), it seems prudent to learn lessons from past and present migration experiences to prepare vulnerable coastal communities for future scenarios.

This research represents an inquiry into what lessons may be learned from the experiences of forced human movements from, to and between atolls in Bougainville, PNG. This study extends previous research by expressly inviting the participation of both migrants and hosts in communities of origin and destination. Moreover, given a certain tendency in some news media reporting toward sensationalist representations with headlines such as "Pacific Atlantis: First climate change refugees" (Vidal 2005), and recurrent characterisations of the Carteret (Tulun) Islanders as "the world's first environmental refugees" (Gupta 2007, para. 4) or "[t]he world's first climate change refugees" (Tweedie 2009; cf MacFarquhar 2009; Harman 2013; Beldi 2016), this research was also interested to learn more about the preferred self-description/s of affected islanders in communities of origin and destination. This offers potential benefits with respect to appropriately informing the terminologies of future policy and research discourses. In short, "[a]n alternative language ... may be more likely to lead to constructive outcomes." (Barnett and Campbell 2010, p. 174).

Research Design, Methodology, and Data Collection

This pilot study engaged participants aged 25–67 in six (6) locations: Port Moresby, Buka Island, Torotsian Island, Tulun Atoll, Tinputz, and Nissan Atoll in Bougainville, Papua New Guinea.[3] There was a total of 28 semi-structured face-to-face on-site interviews conducted, each having an average duration of approximately 1 h. Of those recruited to participate in the research there were more men (67.9%) than women (32.1%), with most participants aged between 40 and 50 years (Fig. 4). Given that interviewees were recurrently encountered in the company of by-standing conversation partners, a small number of impromptu focus group contexts were spontaneously accommodated, thus raising the total estimated number of respondents broadly informing this pilot study to between 55 and 60. Quantitative data collection and verbatim qualitative quotes were limited to interview participants.

Research participants were recruited for the study using a non-probability sampling framework,[4] snowballing technique and respondent driven sampling (Heckathorn 2011; Babbie 2010). Data collection occurred during October and November 2010, and the availability of locally sensitive guides and interpreters assisted in moderating discussions. Information was collected using key informant interviews (KII) and a survey interviewing questionnaire (SIQ). Research participants were recruited based on their personal or proximate migration experience/s, and all respondents in Tinputz were migrants from the Tulun Atoll. As a mixed method approach this research used an 'exploratory design' paradigm (Creswell 2013, 2014; Creswell and Plano Clark 2011), in the sense that the overall approach was weighted heavily on the qualitative study rather than the quantitative study.

Field research was also aided by a preceding research visit to Bougainville in March 2008, during which options for heightened community level disaster preparedness were explored, and data were gathered for the "World Vision Asia Pacific Annual Disaster Report 2008 Planet Prepare" (Luetz 2008). This earlier research visit broadly informed this pilot study, enhancing familiarity with cultural,

[3]Selected interview locations are available online: http://goo.gl/maps/byN0F.

[4]According to Earl Babbie, non-probability sampling represents "[a]ny technique in which samples are selected in some way not suggested by probability theory. Examples include reliance on available subjects as well as purposive (judgmental), quota, and snowball sampling" (Babbie 2010, p. 192).

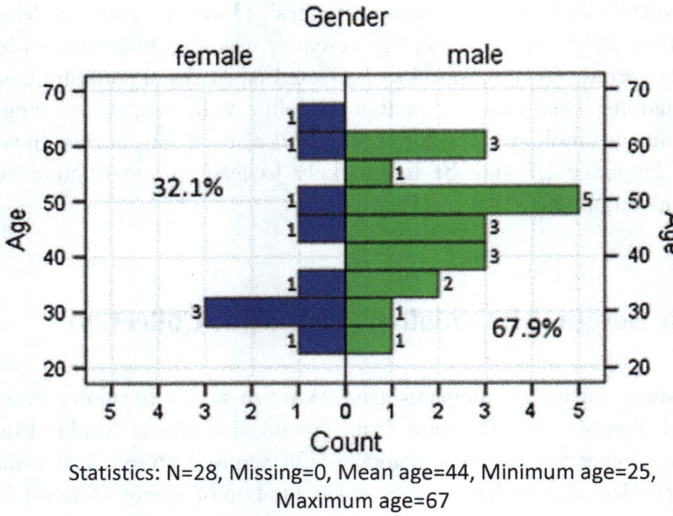

Fig. 4 Study population pyramid, 2010: total number of participants by age and gender

environmental, and socioeconomic issues, and laying a foundation for collaboration in areas of translation/interpretation, logistics and overall research support.

There were six overarching research questions: (1) What are the environment related push factors (ERPFs) and their respective contributions to migration? (2) What are the non-environment related push factors (NRPFs) and their respective contributions to migration? (3) What is the preferred terminology to describe/characterise forced migrants? (4) What is/are the problem situation/s? (5) What is/are the proposed solution/s? (6) What is/are the preferred migration destination/s?

Data Analytical Strategy

The data analysis was carried out using exploratory and thematic analytical strategy. The first phase of the analysis was done by transcribing and entering the data into the computer and then exploring it by focussing on the participants' responses regarding factors that cause or contribute to migration, including preferred migration destinations. The data were analysed qualitatively with Nvivo 11 (QSR International 2016) and a word clouding, tree mapping and clustered analysis was conducted. Results were then compared with the transcripts and themes iteratively. In the second phase of the analysis, the quantitative study was then explored using frequency analysis in SPSS 24 (George and Mallery 2016) and Likert plotting using r (RStudio Team 2016). The qualitative and the quantitative findings from the two phases of the analysis were then discussed with reference to prior research. Data analyses broadly incorporated approaches informed by grounded theory (Charmaz 2006) and phenomenography (Marton 1981).

Results and Key Findings

From the research, several key findings have been synthesised and consolidated below. These are presented chronologically in congruence with the six (6) research questions detailed above.

(1) *Environment related push factors (ERPFs) are contributing to migration*

As highlighted by quantitative data (Fig. 5), environment related push factors (ERPFs) appear to be strongly correlated with migration, with 71–96% of respondents considering ERPFs as either “important” OR “very important” factors in migration.

The contributing role of ERPFs in migration was also expressed qualitatively through interview data that highlighted both environmental primary causes and secondary environmental knock-on effects. During interviews two primary causes were recurrently mentioned, namely SLR and sudden onset disasters:

> “The sea is rising really fast. Every day the sea is eroding away land.” (P4) “Sea level rise is eating up the island, washing away the island. Towards the back of the island coconut trees are now in the sea, it happened during the last two years.” (P6) “Rises in sea level [are] the cause of everything – this is the main one.” (P14) “[The island] is only 80 m wide, so erosion matters.” (P22) “Rises in sea level [are] very important … because of currents, they are stronger than in the past.” (P27) “Very strong storms, no place to hide.” (P8) “Recurrent king tides… destroy all of the island” (P5).

These primary environmental causes then seemed to enhance secondary environmental problems:

> “Sudden natural disasters, cyclones, destroy buildings and mangroves, breadfruit … food trees.” (P21) “Higher tides bring sea water into swamps where mosquitos breed. Wind storms knock down coconut trees, this then results in higher waves… shortage of food and lack of land is what drives people to migrate away.” (P6) “Han Island is the worst place for

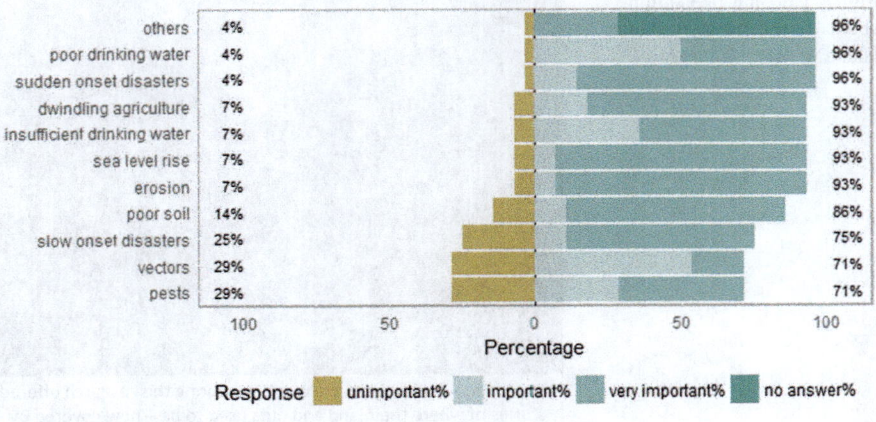

Fig. 5 Environment related push factors (ERPFs)

> mosquitos [I have] ever seen, and February to April 2009 was the worst mosquito season. High tide once a month … mosquitos! People complain and talk about moving on account of the mosquitos." (P12) "Mosquito problem has been getting worse, less fruit, organic matter increasingly thin, salinisation." (P2) "sea water, infertile soil, thin layer of soil" (P12) "Saltwater content in soil very high: 80% – borehole wells [are not possible], World Vision did a feasibility study." (P25) "Diseases [are] enhanced by poor nutrition, people don't eat greens, malaria [is] quite common." (P12) "Soil fertility can't be improved" (P14).

Research participants also pointed out what they interpreted to be signs of erosion (Fig. 6) and evidence of sea level rise and decreasing island size (Fig. 7).

(2) *Non-environment related push factors (NRPFs) are contributing to migration*

As highlighted by quantitative data (Fig. 8), non-environment related push factors (NRPFs) are also strongly correlated with migration, although views were more

Fig. 6 Coconut trees on the Island of Han, Tulun Atoll

Research participants pointed to coconut trees being gradually eroded away as evidence of the vulnerability of their coastal atoll environment. (Photo taken in 2010 by Luetz)

Fig. 7 Coconut tree stump about 10 m from shore on the Island of Han, Tulun Atoll

Numerous older islanders encountered during this research offered stories of where their land and huts used to be – now covered by water like this coconut tree stump. (Photo taken in 2010 by Luetz)

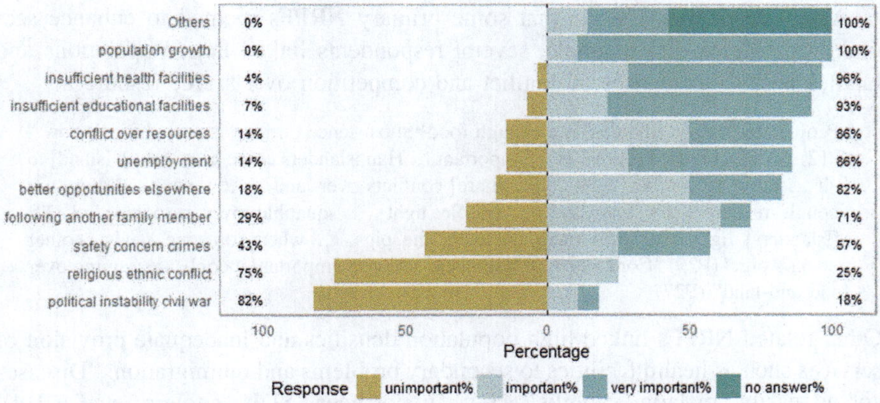

Fig. 8 Non-environment related push factors (NRPFs)

nuanced and category dependent, with between 18 and 100% of respondents considering NRPFs as either "important" OR "very important" factors in migration.

A number of NRPFs were considered paramount, first and foremost "population growth", which all respondents unanimously considered either "important" or "very important" for migration. At times population growth and overcrowding were linked in discussions to lack of available land: "Overcrowding [is] very important, the average number of children [is] seven… so many kids coming up and little available land… big problems coming up." (P25). "Population growth, overcrowding very important… average family size per couple: seven to ten children" (P28).

The average number of children per participant in this research is 4.65 (Fig. 12), and the result shows that 39.3% (11) of the participants have six children or more (Fig. 9).

There was also the sense that population growth appears to work in concert with ERPFs presented above, exemplified by the following respondent commentary: "[There is] less land, less food, more people" (P14).

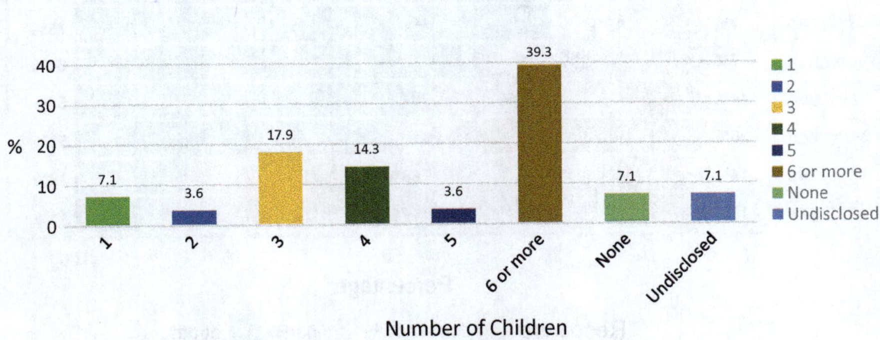

Fig. 9 Number of children per participant

Again, there was a sense that some primary NRPFs seemed to enhance secondary problems. For example, several respondents linked large populations and landlessness to incidences of conflict and competition over scarce resources:

> "People are hungry, [there is] not enough food! Subsistence farming is very difficult now." (P12) "Conflict over resources is … important … Han islanders come here [Iolasa Island] to fish … building conflict." (P9) "[There are] conflicts over land … too many children, not enough resources for everyone … people fight … squabble over resources." (P20) "[Islanders] fight over who owns the land, the pigs, … when someone steals another person's pig." (P25) "Conflicts over resources [are] very important, people are arguing over food and land" (P27).

Other related NRPFs linked high population densities and inadequate provision of services such as health facilities to secondary problems and outmigration: "Diseases spread easily, malaria, syphilis, AIDS, gonorrhoea, SDIs, cholera, etc." (P14). "Malaria [is] quite common here [on the atolls], people then go to Buka." (P12).

(3) *Respondents resist the label "refugees" and seem to prefer local terminology*

As highlighted by quantitative data (Fig. 10), respondents overwhelmingly rejected the representations "refugees" (79%) and "exiles" (68%) to refer to migration from atolls off the coast of Bougainville, and instead preferred local terms (100%) that were suggested during the semi-structured interviewing process.

The following local *Tok Pisin* terms were offered as counter-proposals:

(1) ***tripman** (male)/**tripmeri** (female)* "someone who wanders from place to place" (Verhaar 1990, p. 355). The term conveys "flexibility" and is "not a sealed identity", being reminiscent of a "passenger", "drifter", "traveller" (P4). If people return home, they "would no longer be called *tripman* or *tripmeri*" (ibid.).
(2) ***Turangu*** "victim, something happened to you, you need help." (P18). It "conveys empathy. Popular term in PNG, meaning, 'I am sorry for what happened to you.' The word reaches out from one person to another. Not a

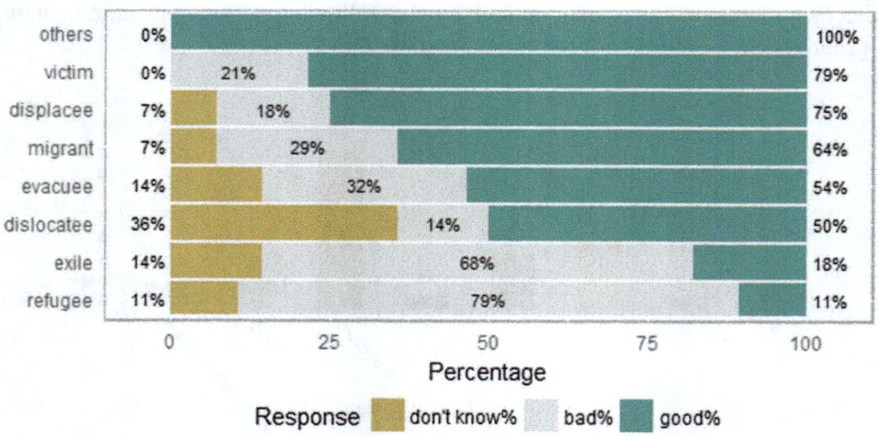

Fig. 10 Preferred nomenclature: characterisations and emotional responses

permanent label, but only a situational term. People like being a 'Turangu', but wouldn't want to stay one." (P23). "Oh, poor one!—Covers almost every individual person or place, all *'manmeri'*. Sweeping word that captures many contexts, droughts, displacement … and strikes people's hearts in Pidgin language, used throughout PNG. For example: 'your friend, John, died yesterday.' Response: 'oh, poor one, Turangu!'" (P25).

(3) ***Mekim wokabout*** "contextualised term that is easier to understand in Pidgin than 'migrant'" (P26).

Some respondents also suggested temporal fluidity and flexibility as important requirements: "there is no one-size-fits-all name … You cannot give a single permanent term to a person. Time and situation influence the usage" (P1).

Finally, more than three quarters of respondents commented positively on the term "victim" (79%), as the following commentary exemplifies: "Victim is the best word because the disaster has forced the people to leave the island, for example, climate change" (P6).

(4) *The problem compound is complex and comprises dissimilar contributing factors*

The following issues were mentioned by participants as constituent parts of an overall complex problem commixture: Environment; climate change; geography (remoteness, inaccessibility); poor governance; disputes over customary land ownership; unemployment (limited cash economy); subsistence development context; limited access to reliable news, information and education; among others.

> "The problem is we have more mouths to feed than what we are producing. More people go hungry … because of the issue of climate change there is sea level rise, even just a few centimetres … is a contributing factor, limiting the ability to produce enough food supply for the people. High deposit of salt on the surface of the soil [means] we cannot grow food crops that will sustain the escalating population. People are starving. This is inevitable, the only solution is a forced relocation to … Bougainville." (P26) "Soil structure in Carteret Islands not conducive to intense farming" (P1); "severe erosion on the island" (P2); "on balance more land [is] eroded away than [is] deposited elsewhere" (P20); "no VHF radio, no telephone, no internet" (P9); "not enough tanks for rain catchment" (P20); "land problems are delaying [the] move" (P4); "there are two relocation sites" (P7) "Autonomous Bougainville Government [ABG] and Tulele Peisa [TP] are coordinating two different locations: Tinputz [TP] and Carola Plantation [ABG], big problem relocating everyone to one site… land issue!" (P26); "government wants young couples to leave - their first priority" (P9); "too much rubbish, no waste disposal" (P20); "sea walls are washed down. Higher tides bring sea water into swamps where mosquitos breed" (P6); "population is growing, island is shrinking" (P13).

Traversing Han Island on foot, and circling it by boat, also revealed limits to coastal in situ adaptation[5] as illustrated both by Fig. 11 and the following research

[5]Limits of sea walls and coastal in situ adaptation on the Tulun Atoll are also shown in this United Nations University (UNU) (2009b) edited documentary: *Sinking Paradise, Carteret Islands, PNG*, https://youtu.be/Hgw4HTtokgk?t=1m15s.

Fig. 11 Flooded sea walls made of "gabion baskets"

"Gabion baskets", wire-enforced structures filled with rocks on the Island of Han, Tulun Atoll. (Photos taken in 2010 by Luetz)

participant commentary: "ultimately, sea walls, gabion baskets and mangroves failed" (P2).

(5) *The proposed solutions may be subdivided into "soft" and "hard" solutions*

"Soft" solutions are presented first and include attitudes, social research, consultation, governance, and education:

> "Elderly are the hardest group of people to convince; they never move out, even after disasters – cultural bonds are strong, and prior experience – [they] have experienced bad things and disasters before and are the highest risk takers, therefore hard to convince – you can preach but they're not listening. A solution is to integrate climate change adaptation into the curriculum … focus on the young people in advocacy and education. Older people can hold back young people as they have a shorter remaining life expectancy. Psychological issue – people won't be listening until information comes from many directions." (P1) "Protect their cultural identity" (P3); "[should resolve] the land issues so

> the land is ready for resettlement" (P4); "more research is needed to hear from the Carteret Islanders what they want to do before, during and after the move: research at origin and destination communities" (P3); "limit their families to two kids because of overcrowding" (P21); "industrialised countries that are responsible for the cause of this should help" (P13); "proper consultation [is] essential among all stakeholders: host communities, land owners, Government (ABG and National Government), churches, re-settlers. Everything must be documented to make [the process] transparent, ensure longevity, leave evidence" (P17); "[more] consultation with the community" (P11); "create micro finance for seafood – bêche-de-mer is big business, but people don't know how to commercialise" (P5); "the government should do more speaking and more listening. The communication from the government is not reaching the village" (P22); "technical expertise, education, build up local pool of experts" (P24).

In contrast, "hard" solutions include land reclamation and engineering approaches:

> "Waste from copper mine should be brought and dumped in the islands. The mine [is] not now operating, but waste is still there. They've done that in Bougainville in Kobeinan, central Bougainville, – sea walls from the mined waste – big rocks!" (P5)."Building sea walls works with limited temporary success" (P2); "concreting and land reclamation … Using mining waste could really work … Dependent on reopening Panguna Mine. Very promising adaptation potential. Right now barges wait at dump sites, trucks load up the barge and dump the waste in a sea trench" (P17).

(6) *Bougainville Island is the most preferred migration destination*

There are two aspects in which islanders do not wish to leave: (1) they do not wish to leave their atoll islands, however, *if* forced to do so, then (2) they do not wish to leave their Bougainville region, language, culture, clan and family attachments. This may be visualised as concentric circles, with islanders preferring familiar and proximate over unfamiliar and non-proximate destinations:

> "I would like to go someplace where the environment is similar to where I'm living and the way of living is the same so that I will feel that I did not lose my home but that my home is still here … want to stay in Bougainville, where it's culturally similar, but maybe, if there is no more space left, then moving from close and known to wider and different: 1st choice: Bougainville; 2nd choice: PNG; 3rd choice: Australia; 4th choice: Germany" (P24).

Elderly islanders especially emphasised that they would "stay here [Tulun Atoll] until we die" (P5), a finding aligned with previous research that "[t]here has been reluctance to leave, especially among older islanders" (Luetz 2008, p. 20).

Some respondents explicitly stated that they did not wish to move to Australia (P12), as "in Australia it's all about money" (P9), or that they would be "afraid of the tall buildings there" (P13). "Australia? Not unless you got nowhere else to go." (P18). One research participant elaborated that "educated people can go further afield to pursue opportunities. Less educated people tend to stay much closer to where they've relocated from to safe areas known to them. The vast majority will choose Bougainville. Very few people will go overseas [to] Vanuatu, Solomon Islands, Fiji" (P17).

Discussion of Core Issues: "We're not Refugees"

Although environmental factors (Fig. 5) seem to have a stronger contributing effect on migration than non-environmental factors (Fig. 8), research made it quite clear that ERPFs and NRPFs are interrelated and cannot be meaningfully discussed in isolation of each other, thus raising the need for holistic development and migration approaches. An example of this is the following problem description, which links ERPFs and NRPFs:

> "[There is] too much rubbish, no waste disposal; [atoll islanders] just pile up rubbish at the shore, cover with sand … this also provides breeding grounds for mosquitos … too many breeding places, when the sea comes up, it doesn't drain out" (P20) "Malaria [is] quite common here [on the atolls], people then go to Buka" (P12).

The example highlights a complex interrelationship between waste management practices, environmental effects, related health outcomes, and corresponding migration consequences. As this pilot study has discovered, reasons for migration are numerous, interdependent and complex, making it impossible to disaggregate the intertwined mix of reasons why people decide to move. Hence a monocausal attribution of reasons why people migrate appears methodologically unsound, given that reasons for migration are inclusive, not exclusive (Barnett and Campbell 2010, pp. 170–174; Betts 2010, p. 378; Brown 2008, p. 9; CCEMA 2010; Hamilton et al. 2016).

Extreme population pressures on atolls have been documented long before climate change ever registered as a problem in mainstream public awareness (Bayliss-Smith 1974, 1975), and even the Nukumanu (Tasman) Atoll to the northeast of Bougainville "experienced a considerable population reduction in the 1870s following overpopulation." (Connell 2015, p. 17; attributed to Bayliss-Smith 1975, pp. 312–323). Hence a single cause of migration arising uniquely and exclusively from the effects of climate change, as sometimes simplistically suggested in the mainstream media through headlines such as "Pacific Atlantis: First climate change refugees" (Vidal 2005), appears to fall well short of encapsulating the complex interplay of environmental, sociodemographic and historical realities that have long shaped the region through a combination of both "physical factors (tectonic movements, SLR, ENSO events, cyclones) and human factors (particularly intense where populations are growing)" (Connell 2015, p. 20).

However, this should not be misconceived as implying that climate change is not a major causal factor in migration, or that its contribution to the depopulation of islands in the region is not growing (Albert et al. 2016). While conceiving of climate change and migration as a chain of cause and effect is too simplistic, de-linking the two issues as causally unrelated seems similarly untenable, seeing that "absence of evidence about a problem does not imply evidence of absence of a problem" (Myers and Kent 1995, p. 29). On the contrary, it seems that physical factors and human factors are inseparably intertwined, and that the contribution of climate change to migration, relative to other factors, could be growing:

> "[W]hile the evidence for a distinctively anthropogenic 'climate change signal' in forced migration so far is circumstantial, *it is mounting*. And with all available scenarios predicting accelerating climate change impacting growing populations and more people living on marginal land, *forced climate migration is certain to increase*" (Brown 2007, p. 18; emphasis added).

This could be particularly pertinent for coastal communities on atolls, seeing that "there is already a strong awareness of the risk of accelerated sea-level rise (e.g., Church and Gregory 2001; Nurse and Sem 2001), [which] could accelerate the process of abandonment … as people lose faith in the future of low-lying islands based on the thought of the impacts of sea-level rise." (Arenstam Gibbons and Nicholls 2006, p. 46). This seems to be especially urgent in view of recent scientific evidence that in the nearby Solomon Islands "[s]ea-level rise has claimed five whole islands" (Leon 2016; cf. Albert et al. 2016).

In this context continued population increase remains a significant human pressure point, and this has had little or nothing to do with climate change (Fig. 12), and more to do with the levels of development and educational attainment that are realisable in remote atoll environments where even the constancy of electricity can present significant challenges. Further, fertility has a significant positive correlation with age and household size but not education (Fig. 13), thus suggesting that higher levels of education are conducive to better family planning. Expressed differently, the results presented in Figs. 12 and 13 seem to suggest that higher levels of

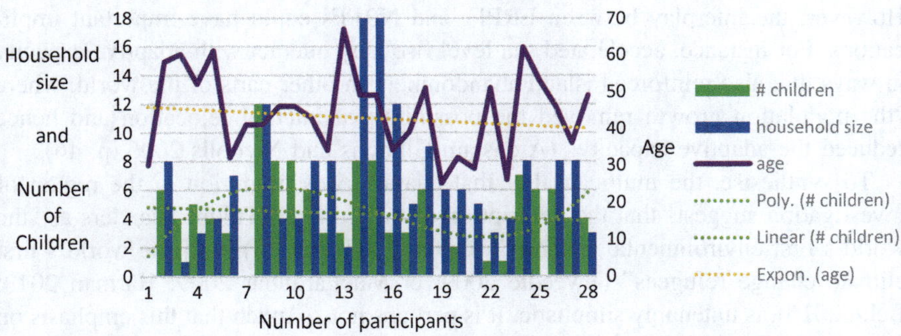

Fig. 12 Study participants and number of children: on average, each participant had 4.65 children

	Children	Household	Age	Education
Children	----	.521**	.453*	-.438*
Household	.521**	----	-.063	-.341
Age	.453*	-.063	----	-.114
Education	-.438*	-.341	-.114	----

N=28, *P<0.05, **P<0.01.

Fig. 13 Correlations: Participant fertility and educational attainment. The average level of education is secondary education (grade 10)

education lead to fewer children, and that lower levels of education lead to more children. Hence the situation on the Tulun Atoll seems to be influenced not only by climate change and ERPFs but also by developmental matters observed elsewhere in the world where higher levels of education are commonly shown to be inversely related to childbearing (UN-DESA 1997; Basu 2002). This makes education a promising adaptation response.

Relatedly, this field research encountered one atoll islander who already had 12 children, and his wife was pregnant at the time with his 13th child. While this number of children may seem unusually high in any environmental context, both the smallness of the island where the face-to-face encounter took place (measuring little over 100 m in diameter) and its limited subsistence potential, visually illustrate the geospatially confined limits to population growth. Further, it is worth remembering that contemporary population growth dovetails historical fertility trends flagged as locally unsustainable long before climate change entered mainstream global consciousness:

> In 1964, [a] survey team had recorded a population of 721, noting that this represented an average annual growth rate of 2.4% since the 1954 census figure of 581. They warned that by 1984 the population would be well over 1,000 … this prediction proved correct but increasingly out-migration has been a factor reducing the rate of resident population growth. (O'Collins 1989, p. 253; see also Figure "Population projection for the Carteret Islands" depicted on p. 254).

In short, high fertility rates are not newly experienced in the region of investigation. However, the interplay between ERPFs and NRPFs could have important implications. For instance, accelerated sea level rise may interact with adaptive capacity in ways that also reinforced island abandonment in other parts of the world, where "the population growth removed the possibility of internal relocation and hence reduced the adaptive capacity" (Arenstam Gibbons and Nicholls 2006, p. 46).

To synthesise, the multicausality that characterises migration in the region of investigation suggests that the description of the Carteret (Tulun) Islanders as "the world's first environmental refugees" (Gupta 2007, para. 4) or "[t]he world's first climate change refugees" (Tweedie 2009; cf MacFarquhar 2009; Harman 2013; Beldi 2016), is untenably simplistic. It is perhaps not so much that this emphasis on climate/environment in migration is causally incorrect but rather that it is causally incomplete, as Adichie (2009) cautions in her reflections about "the danger of a single story" (para. 42): "The single story creates stereotypes, and the problem with stereotypes is not that they are untrue, but that they are incomplete. They make one story become the only story".

Additionally and importantly, a further compelling reason to reconsider the continued use of the term "refugee" in referring to migrating atoll islanders is that local respondents overwhelmingly resist this characterisation. While this understanding is not newly discovered (McNamara and Gibson 2009), it accompanies the well-established recognition that in a strictly legal sense affected migrating atoll islanders are, in fact, not "refugees" at all, being denied the very status of protection that this term was originally legislated to bestow (McAdam 2010).

Although islanders overwhelmingly meet the "well-founded fear" requirement stipulated by the Refugee Convention in Article 1 A.(2), as shown above in the results section, they are evidently not "persecuted for reasons of race, religion, nationality, membership of a particular social group or political opinion", and also do not find themselves "outside the country of [their] nationality" (UNHCR 2010, p. 14), to mention only a few of the major impediments to the legal correctness and applicability of the term "refugee" in this socioenvironmental context. While alternative constructs involving "climigrants" and "climigration" (Hamilton et al. 2016) may manage to steer clear of the legal difficulties surrounding the term "refugees", such terms nevertheless also fall short of conveying the multicausality inherent in migration where human movement is typically triggered or supported by a complex combination of human and physical, push and pull factors, which are impossible to disentangle or quantify (Barnett and Campbell 2010, pp. 170–174).

Against this background it is conjectured that the "refugee" label, whether linked to "environmental refugees" (Rakova et al. 2009), "climate change refugees" (Tweedie 2009), "sea-level refugees" (WBGU 2006, p. 61), or other alternative types of "refugees" that have been suggested in the literature,[6] owes its enduring popularity to the relative ease of comprehension among non-expert audiences that are typically addressed by the media or NGOs in the hopes of enlisting their advocacy for social justice (Luetz 2008, p. 121).

Clearly, terms such as "climate refugees" are far more easily comprehended in the public space than more precise albeit more cumbersome academic constructs such as "climate change-related migrants" (Campbell and Warrick 2014, p. 2), or people "forced to leave their homes due to sudden-onset climate-related natural disasters" (UN-OCHA 2009, p. 15).

This ease of use may also explain the continued utilisation of this term by both scholars (e.g., McNamara and Gibson 2009, p. 475; Hartmann 2010, p. 233; Bettini 2013), UN development groups (e.g., Campbell and Warrick 2014, p. 24; Brown 2007, pp. 3, 8), and nongovernmental organisations (e.g., Bauer 2010; Luetz 2008, p. 121[7]), who perpetuate variations of the term "climate refugees" in inverted commas, apparently recognising the widespread comprehension that this construct enjoys among readerships, and yet without conceding personal assent.

As suggested by Barnett and Campbell (2010), "[a]n alternative language, based on notions of risk, and focusing on adaptation and adaptive capacity, may be more likely to lead to constructive outcomes." (p. 174). This research suggests that local

[6]See "2.2 Nomenclature: Definitional difficulties" in Luetz (2013, pp. 29–30) for a list of terms that have been conceptually proposed in the literature to describe climate change related human movement.

[7]Endnote (a) in Luetz (2008, p. 121) states: "The term 'climate refugee' is used in this report on the understanding that World Vision does not endorse this term but rather recognises its usage by various commentators. World Vision acknowledges the legal definition of refugees as rendered by the 1951 Convention on Refugees".

contexts, dialects and expressions (e.g., “Turangu”) have much to offer with respect to judiciously informing the nomenclature of future policy and research discourses that are contextually precise and popularly accepted.

Research Limitations and Opportunities for Future Research

The apparent comparative popularity of the term “victim”, relative to all other non-local alternatives, should be researched further as it appears to challenge positions presented in other research that “[i]slanders have no wish to be seen as … tragic victims, even if that elicits empathy and support elsewhere” (Connell 2015, p. 19; attributed to McNamara and Gibson 2009). More research is needed to better understand the apparent acceptance among participants of the term “victim”. The small participant sample remains as a limitation to the robustness of the inferred possibility that “victimhood” may perhaps not be quite as unpopular among some Pacific islanders as previously described (Campbell and Warrick 2014, p. 10).

Moreover, more research is needed to better understand migration between rural and urban atoll island contexts. While this research found “population growth/ overcrowding” to be a very important migration push factor in the context of the densely-settled atoll environment of autonomous Bougainville, research in the Maldives discovered “low/dwindling populations” to be a very important migration push factor in the context of sparsely settled atoll environments characterised by declining populations (Luetz 2017). Hence policy research stands to benefit from a better understanding of some of the differences respecting rural-urban atoll migration, an opportunity for future research separately highlighted in Campbell and Warrick (2014, p. 19), who surmise that “[i]t is not unreasonable to anticipate that rates of rural-urban migration may be increased by climate change” (attributed to Locke 2009).

Conclusion

This research has found that environmental and non-environmental issues are working in concert with other drivers to stimulate migration from atolls to the north and northeast of Bougainville. The mix of individual and interrelated factors seems to be inseparably amalgamated within a complex compound of physical and human, historical and contemporary, climate change and non-climate change related, and push and pull factors, which are difficult to disaggregate, and whose individual contributions to migration seem to be impossible to quantify. However, given that environmental factors (Fig. 5) seem to have a stronger contributing effect on migration than non-environmental factors (Fig. 8) also suggests that climate change remains an undeniable concern of growing significance.

The overall complex situation makes holistic and proactive migration management and sustainable development approaches preferable to narrow piecemeal portrayals of "climate refugees" that seem to be incomplete, isolated, extraordinary, foreign or even frightening (Bettini 2013). Relatedly, climate change-related migration and adaptation strategies should pursue education and countering misinformation as pathways to propagating non-extraordinary portrayals of migration, where the depopulation of islands is realigned within "crowd psychology" (Diamond 2005, pp. 435–436) as more natural, normal, attainable and potentially positive, and importantly, in line with historical precedent (Kirch and Rallu 2007; Arenstam Gibbons and Nicholls 2006). However, historical precedent also remains as a warning that, contradictory to some media portrayals discussed above, the Tulun Atoll is *not* the first location where human habitation on account of environmental factors, resource depletion and population pressures is seemingly coming to an end (Diamond 2005, p. 426; Green and Green 2007). This understanding makes forward-thinking policy preparedness indispensable and should reinvigorate decade-old relocation endeavours going as far back as the late 1960s (O'Collins 1989, p. 253).

With respect to preferred migration destinations, migrating islanders prefer familiar and proximate destinations over unfamiliar and non-proximate ones, and overwhelmingly prefer to remain in the greater Bougainville region. Given that forced community relocation costs also rise with increasing social, political and actual distance from the point of origin (Campbell and Warrick 2014, p. 22), makes proximate locations the destinations of first choice. This research also suggests that local contexts, dialects and expressions (e.g., "Turangu") have much to contribute terminologically with respect to more appropriately informing the definitional and conceptual constructs of policy and research discourses. Finally, expansive education is advised as an adaptation strategy as this offers the dual promise of both lowering levels of population growth and concurrently raising in situ and ex situ adaptive capacity as vulnerable coastal communities prepare for migration.

Acknowledgements The authors wish to thank Kirsty Andersen for her copy-editorial support, Karen du Plessis for assistance with data analysis, John Connell for constructive comments, and Boniface Wadari for his research assistance in Bougainville. Grateful acknowledgment for relevant Ph.D. research support is also made to John Merson, Eileen Pittaway, Curt von Boguslawski, Richard Rumsey, Geoff Shepherd, and to the international development organisation World Vision. Finally, the authors wish to thank the people of Bougainville Region for generously sharing their stories, struggles, experiences and perspectives.

References

Adichie, C. N. (2009). The danger of a single story. *TEDGlobal 2009*. https://www.ted.com/talks/chimamanda_adichie_the_danger_of_a_single_story/transcript?language=en. Accessed May 15, 2017.

Albert, S., Leon, J. X., Grinham, A. R., Church, J. A., Gibbes, B. R., & Woodroffe, C. D. (2016). Interactions between sea-level rise and wave exposure on reef island dynamics in the Solomon

Islands. *Environmental Research Letters, 11*(5), 1–9. http://iopscience.iop.org/article/10.1088/1748-9326/11/5/054011/pdf.

Amos, J. (2012, September 24). Satellites trace sea level change: A major reassessment of 18 years of satellite observations has provided a new, more detailed view of sea-level change around the world. *BBC News*. http://www.bbc.com/news/science-environment-19702450. Accessed April 19, 2017.

Arenstam Gibbons, S. J., & Nicholls, R. J. (2006). Island abandonment and sea-level rise: An historical analog from the Chesapeake Bay, USA. *Global Environmental Change, 16*(1), 40–47.

Aung, T. H., Kaluwin, C., & Lennon, G. W. (1998). *Climate change and sea level: Curriculum modules for the Pacific schools. Part one: Physical Science.* Bedford Park: National Tidal Facility, Flinders University of South Australia.

AusAID—Australian Agency for International Development. (2010). *Pacific country report on sea level & climate: Their present state. Papua New Guinea.* Canberra: Australian Agency for International Development. http://www.bom.gov.au/ntc/IDO60029/IDO60029.2010.pdf. Accessed May 20, 2017.

Babbie, E. (2010). *The practice of social research* (12th Ed.). Belmont: Wadsworth Cengage Learning.

Bamber, J., & Riva, R. (2010). The sea level fingerprint of recent ice mass fluxes. *The Cryosphere, 4*(4), 621–627.

Barnett, J., & Adger, N. W. (2003). Climate dangers and Atoll nations. *Climatic Change, 61*(3), 321–337.

Barnett, J., & Campbell, J. (2010). *Climate change and small island states: Power, knowledge and the South Pacific.* London: Earthscan.

Basu, A. M. (2002). Why Does Education Lead to Lower Fertility? *A Critical Review of Some of the Possibilities, World Development, 30*(10), 1779–1790.

Bauer, S. (2010). "Climate Refugees" Beyond copenhagen: Legal concept, political, implications, normative considerations. *Analysis, 12.* Stuttgart: Diakonisches Werk der EKD. https://www.brot-fuer-die-welt.de/fileadmin/mediapool/2_Downloads/Fachinformationen/Analyse/analyse_12_englisch_climate_refugees.pdf. Accessed April 28, 2017.

Bayliss-Smith, T. P. (1974). Constraints on population growth: The case of the Polynesian outlier Atolls in the precontact period. *Human Ecology, 2*(4), 259–295.

Bayliss-Smith, T. P. (1975). The central Polynesian outlier populations since European contact. In Carroll, V. (Ed.), *Pacific Atoll populations* (pp. 286–343). Honolulu: University Press of Hawaii.

Beldi, L. (2016, August 7). Carteret climate refugees seek home: A Grassroots Group in Bougainville is scrambling to relocate the Carteret islanders before rising sea levels swallow their land forever, Pacific Beat. *ABC News.* http://www.abc.net.au/news/2016-08-07/carteret-climate-refugees-new-home/7693950. Accessed May 15, 2017.

Bettini, G. (2013). Climate Barbarians at the gate? A critique of apocalyptic narratives on 'climate refugees'. *Geoforum, 45,* 63–72.

Betts, A. (2010). Survival migration: A new protection framework. *Global Governance: A Review of Multilateralism and International Organizations, 16*(3), 361–382.

Blanco, H., Alberti, M., Olshansky, R., Chang, S., Wheeler, S. M., Randolph, J., et al. (Eds) (2009). Shaken, shrinking, hot, impoverished and informal: Emerging research agendas in planning. *Progress in Planning*, *72*(4), 195–250.

Brown, O. (2007). *Climate change and forced migration: Observations, projections and implications.* A background paper for the Human Development Report 2007/2008. Geneva: Fighting Climate Change: Human Solidarity in a Divided World. http://www.iisd.org/sites/default/files/publications/climate_forced_migration.pdf. Accessed May 16, 2017.

Brown, O. (2008). Migration and climate change. *IOM Migration Research Series*, No. 31. Geneva: International Organization for Migration, IOM. http://publications.iom.int/bookstore/free/MRS-31_EN.pdf. Accessed April 30, 2012.

Campbell, J., & Warrick, O. (2014). *Climate change and migration issues in the Pacific*. Fiji: United Nations Economic and Social Commission for Asia and the Pacific (UN-ESCAP). http://www.unescap.org/sites/default/files/Climate-Change-and-Migration-Issues-in-the-Pacific.pdf. Accessed April 26, 2017.

CCEMA—Climate Change, Environment and Migration Alliance. (2010). *Climate change, environment and migration: Frequently asked questions.* http://www.iom.int/jahia/webdav/shared/shared/mainsite/activities/env_degradation/CCEMA_top_10FAQs.pdf. Accessed April 21, 2017.

Charmaz, K. (2006). *Constructing grounded theory: A practical guide through qualitative analysis*. London: Sage.

Church, J. A., & Gregory, J. M. (2001). Changes in sea level. In J.T. Houghton, Y. Ding, D. J. Griggs, M. Noguer, P.J. van der Linden, X. Dai, K. Maskell, & C.A. Johnson (Eds.), *Climate change 2001. The scientific basis: Contribution of Working Group I to the third assessment report of the intergovernmental panel on climate change* (pp. 639–694). Cambridge: Cambridge University Press.

Church, J. A., & White, N. J. (2006). A 20th century acceleration in global sea-level rise. *Geophysical Research Letters, 33*(1). Article No. L01602. http://onlinelibrary.wiley.com/doi/10.1029/2005GL024826/epdf. Accessed April 19, 2017.

Connell, J. (1990). The Carteret islands: Precedents of the greenhouse effect. *Geography,* 75(2), 152–154. https://www.jstor.org/stable/pdf/40571812.pdf. Accessed April 19, 2017.

Connell, J. (2015). Vulnerable islands: Climate change, tectonic change, and changing Livelihoods in the Western Pacific. *The Contemporary Pacific, 27*(1), 1–36.

Creswell, J. W. (2013). *Qualitative Inquiry and research design: Choosing among five approaches* (3rd ed.). Thousand Oaks: Sage.

Creswell, J. W. (2014). *Research design: Qualitative, quantitative, and mixed methods approaches* (4th ed.). CA, USA: Thousand Oaks.

Creswell, J. W., & Plano Clark, V. L. (2011). *Designing and conducting mixed methods research* (2nd ed.). CA: Thousand Oaks.

Diamond, J. (2005). *Collapse: How societies choose to fail or succeed.* New York: Penguin.

Foresight. (2011). *Migration and global environmental change: Final project report.* London: The Government Office for Science. https://www.gov.uk/government/publications/migration-and-global-environmental-change-future-challenges-and-opportunities. Accessed October 21, 2011.

George, D., & Mallery, P. (2016). *IBM SPSS statistics 23 step by step—A simple guide and reference* (14th ed.). New York, Abingdon, Oxon: Routledge.

Green, V. J., & Green, R. C. (2007). An accent on Atolls in approaches to population histories of remote Oceania. In P.V. Kirch, & J.-L. Rallu (Eds.), *The growth and collapse of pacific island societies: Archaeological and demographic perspectives* (pp. 232–256). Hawaii: University of Hawaii Press.

Gupta, S. (2007, July 31). Pacific swallowing remote island Chain. *CNN.* http://edition.cnn.com/CNN/Programs/anderson.cooper.360/blog/2007/07/pacific-swallowing-remote-island-chain.html. Accessed May 15, 2017.

Hamilton, L. C., Saito, K., Loring, P. A., Lammers, R. B., & Huntington, H. P. (2016). Climigration? Population and climate change in Arctic Alaska. *Population and Environment, 38*(2), 115–133.

Harman, G. (2013, December 6). Can Carteret Atoll climate-change refugees bank on chocolate? As the islanders Abandon their homes to rising sea levels, many look to the volatile but dearly-in-demand Cacao Crop. *The Guardian.* https://www.theguardian.com/sustainable-business/carteret-atoll-island-climate-refugees-cacao-chocolate. Accessed May 15, 2017.

Hartmann, B. (2010). Rethinking climate refugees and climate conflict: Rhetoric, reality and the politics of policy discourse. *Journal of International Development, 22*(2), 233–246.

Heckathorn, D. D. (2011). Snowball versus respondent-driven sampling. *Sociological Methodology, 41*(1), 355–366.

Horton, B. P., Rahmstorf, S., Engelhart, S. E., & Kemp, A. C. (2014). Expert assessment of sea-level rise by ad 2100 and ad 2300. *Quaternary Science Reviews, 84,* 1–6.

IPCC—Intergovernmental Panel on Climate Change (2007). In S. Solomon, D. Qin, M. Manning, Z. Chen, M. Marquis, K. Averyt, M. Tignor, & H.L. Miller (Eds.), *Climate change 2007: The physical science basis*. New York: Cambridge University Press. https://www.ipcc.ch/pdf/assessment-report/ar4/wg1/ar4_wg1_full_report.pdf. Accessed April 17, 2017.

Kaluwin, C. (2008). *Security for development: Confronting threats to survival and safety.* The Ninth Annual Global Development Conference of the Global Development Network. 27 Jan–2 Feb 2008. Research Paper on file with Author Presented at https://www.jica.go.jp/jica-ri/GDN/english/gdn/conference/pdf/9th_brisbane_program.pdf.

Kirch, P. V., & Rallu, J.-L. (Eds.). (2007). *The growth and collapse of Pacific island societies: Archaeological and demographic perspectives*. Hawaii: University of Hawaii Press.

Leighton, M. (2012). Population displacement, relocation, and migration. In M. B. Gerrard & K. Fischer Kuh (Eds.), *The law of adaptation to climate change: United States and international aspects.* Chicago: American Bar Association. https://auca.kg/uploads/Migration_Database/Leighton%20Population%20Displacement%20Relocation%20and%20Migration%20Ch%20 20%20in%20The%20Law%20of%20Climate%20Adaptation%202012%20(2).pdf. Accessed May 15, 2017.

Leon, J. (2016, May 7). Sea-level rise has claimed five whole islands in the Pacific: First scientific evidence. *The Conversation*. http://theconversation.com/sea-level-rise-has-claimed-five-whole-islands-in-the-pacific-first-scientific-evidence-58511. Accessed May 20, 2016.

Light Studios. (2016). *Fading sands*, a trailer for the film, produced by Light Studios in partnership with World Vision. https://www.fadingsands.com/; https://vimeo.com/100607503. Accessed October 20, 2017.

Locke, J. T. (2009). Climate change-induced migration in the Pacific region: Sudden crisis and long-term developments. *The Geographical Journal, 175*(3), 171–180.

Luetz, J. M. (2008). *Planet prepare: Preparing coastal communities in Asia for future catastrophes, Asia Pacific Disaster Report*. Bangkok, Thailand: World Vision International. http://luetz.com/docs/planet-prepare.pdf. Accessed April 16, 2016.

Luetz, J. M. (2013). *Climate migration: Preparedness informed policy opportunities identified. During field research in Bolivia, Bangladesh and Maldives*. Ph.D. Dissertation, University of New South Wales, NSW, Australia. http://handle.unsw.edu.au/1959.4/52944. Accessed May 31, 2016.

Luetz, J. M. (2017). Climate change and migration in the Maldives: Some lessons for policy makers. In W. Leal Filho (Ed.), *Climate change adaptation in Pacific countries: Fostering resilience and improving the quality of life*. Berlin, Germany: Springer.

MacFarquhar, N. (2009, May 28). Refugees join list of climate-change issues. *The New York Times*. http://www.nytimes.com/2009/05/29/world/29refugees.html?_r=0. Accessed May 15, 2017.

Marton, F. (1981). Phenomenography: Describing conceptions of the world around us. *Instructional Science, 10*(2), 177–200.

McAdam, J. (Ed) (2010). *Climate change and displacement: Multidisciplinary perspectives*. Oxford, UK: Hart Publishing.

McNamara, K. E., & Gibson, C. (2009). We do not want to leave our land: Pacific Ambassadors at the United Nations resist the category of climate refugees. *Geoforum, 40*(3), 475–483.

Merrifield, M. A. (2011). A shift in Western Tropical Pacific Sea level trends during the 1990s. *Journal of Climate, 24*(15), 4126–4138.

Metzger, T., & Redfearn, J. (2011). *Sun come up: The story of the world's first climate refugees.* A trailer for the film, produced by Red Antelope Films, Brooklyn, NY, USA. http://redantelopefilms.com/project/sun-come-up/. Accessed May 15, 2017.

Myers, N., & Kent, J. (1995). *Environmental exodus: An emergent crisis in the global arena.* Washington, D.C, USA: Climate Institute.

Nunn, P. D., & Kumar, R. (2006). Coastal evolution in the Asia-Pacific region. In Harvey, N. (Ed.), *Global change and integrated coastal management: The Asia-Pacific Region* (pp. 93–116). Dordrecht: Springer.

Nurse, L., & Sem, G. (2001). Small island states. In J.J. McCarthy, O.F. Canziani, N.A. Leary, D. J. Dokken, & K.S. White (Eds.), *Climate change 2001: Impacts, adaptation and vulnerability: Contribution of Working Group II to the third assessment report of the intergovernmental panel on climate change* (pp. 843–875). Cambridge: Cambridge University Press.

O'Collins, M. (1989). Carteret islanders at the atolls resettlement scheme: A response to land loss and population growth. In J.C. Pernetta, P.J. Hughes (Eds.), *Implications of expected climate changes in the South Pacific Region: An overview. UNEP Regional Seas Reports and Studies,* No. 128 (pp. 247–269). Nairobi, Kenya: UNEP. https://wedocs.unep.org/rest/bitstreams/13341/retrieve. Accessed April 18, 2017.

OTLP—On The Level Productions. (2010). *There once was an island: Te Henua e Nnoho.* A trailer for the film, produced by On the Level Productions. https://vimeo.com/11017386. Accessed May 15, 2017.

QSR International. (2016). *NVivo 11 Pro for Windows: Getting started guide.* Australia: QSR International. http://download.qsrinternational.com/Document/NVivo11/11.3.0/en-US/NVivo11-Getting-Started-Guide-Pro-edition.pdf. Accessed May 15, 2017.

Rakova, U., Patron, L., & Williams, C. (2009, June 16). How-to guide for environmental refugees. *Our World.* Tokyo: United Nations University. https://ourworld.unu.edu/en/how-to-guide-for-environmental-refugees. Accessed April 21, 2017.

Roberts, G. (2002, March). Islanders face rising seas with nowhere to go. *Sydney Morning Herald.* http://www.smh.com.au/articles/2002/03/29/1017206152551.html. Accessed May 15, 2017.

RStudio Team. (2016). *RStudio: Integrated development for R.* Boston: RStudio Inc. http://www.rstudio.com/. Accessed May 15, 2017.

SPC—Secretariat of the Pacific Community. (2008). *Pacific Island populations 2008: 2008 populations and demographics indicators.* Noumea: SPC.

Spencer, T. E. T., & Spencer, M. (1953). *Taro plants Carteret islands, Papua New Guinea, 1960,* picture slide, Call No. PIC/3803/777 LOC Cold Store Pic SPE, Spencer collection of slides of Papua New Guinea, 1953–1978, National Library of Australia. http://nla.gov.au/nla.obj-145611866. Accessed April 19, 2017.

Traufetter, G. (2012, June 15). Climate change or tectonic shifts? The mystery of the sinking South Pacific Islands. *Spiegel Online.* http://www.spiegel.de/international/world/south-pacific-islands-threatened-by-more-than-just-rising-sea-levels-a-838675.html. Accessed April 19, 2017.

Tweedie, N. (2009, December 7). The world's first climate change refugees: The inhabitants of the Carteret islands do not relish their New-found Notoriety. *The Telegraph.* http://www.telegraph.co.uk/news/earth/carteret-islands/6752962/The-worlds-first-climate-change-refugees.html. Accessed April 19, 2017.

UN-DESA—United Nations Department of Economic and Social Affairs. (1997). *Linkages between population and education: A technical support services report.* New York: Population Division, Department of Economic and Social Affairs (DESA). http://www.un.org/esa/population/pubsarchive/tsspop/tss976/gbc976.htm#tab1. Accessed May 18, 2017.

UNDP—United Nations Development Programme. (2007). *Human development report 2007/2008. Fighting climate change: Human solidarity in a divided world.* New York, NY:

United Nations Development Programme. http://hdr.undp.org/sites/default/files/reports/268/hdr_20072008_en_complete.pdf. Accessed April 16, 2016.

UNHCR—United Nations High Commissioner for Refugees. (2009). *Climate change and statelessness: An overview*. Submitted by UNHCR, supported by the International Organization for Migration (IOM) and the Norwegian Refugee Council (NRC) to the 6th session of the Ad Hoc Working Group on Long-Term Cooperative Action (AWG-LCA 6) under UNFCCC, 1–12 June 2009, Bonn, Germany. http://www.unhcr.org/4a1e50082.html. Accessed May 15, 2017.

UNHCR—United Nations High Commissioner for Refugees. (2010). *Convention and protocol relating to the status of refugees*. Geneva: UNHCR Communications and Public Information Service. http://www.unhcr.org/protect/PROTECTION/3b66c2aa10.pdf. Accessed April 26, 2017.

UNISDR-UNDP—United Nations International Strategy for Disaster Risk Reduction, United Nations Development Programme. (2012). *Disaster risk reduction and climate change adaptation in the Pacific: An institutional and policy analysis*. Suva: United Nations International Strategy for Disaster Risk Reduction, United Nations Development Programme. http://www.unisdr.org/files/26725_26725drrandccainthepacificaninstitu.pdf. Accessed April 16, 2016.

UN-OCHA—United Nations Office for the Coordination of Humanitarian Affairs. (2009). *Monitoring disaster displacement in the context of climate change: Findings of a study by the United Nations Office for the Coordination of Humanitarian Affairs and the Internal Displacement Monitoring Centre*, The Internal Displacement Monitoring Centre, The United Nations Office for Coordination of Humanitarian Affairs, Geneva. http://www.internal-displacement.org/assets/publications/2009/200909-monitoring-disaster-displacement-thematic-en.pdf. Accessed April 26, 2017.

UNU—United Nations University. (2009a). *Local solutions on a sinking paradise, Carterets islands, Papua New Guinea*. Documentary film produced for United Nations University (UNU) by UNU Media Studio & Tulele Peisa. https://vimeo.com/4177527. Accessed May 15, 2017.

UNU—United Nations University. (2009b). *Sinking paradise, Carteret Islands, PNG*. Documentary film produced by United Nations University (UNU) in collaboration with Hakata, N., Peisa, T. & Rakmova, U. https://youtu.be/Hgw4HTtokgk?t=1m15s. Accessed May 19, 2017.

Verhaar, J. W. M. (Ed). (1990). *Melanesian Pidgin and Tok Pisin: Proceedings of the First International Conference of Pidgins and Creoles in Melanesia*. Amsterdam: J. Benjamins Publishing Co.

Vidal, J. (2005, November 26). Pacific atlantis: First climate change refugees. *The Guardian*. https://www.theguardian.com/environment/2005/nov/25/science.climatechange.

WBGU—German Advisory Council on Global Change. (2006). *The future oceans—warming up, rising high, turning sour*. Berlin: Special Report, German Advisory Council on Global Change (WBGU). http://www.wbgu.de/fileadmin/user_upload/wbgu.de/templates/dateien/veroeffentlichungen/sondergutachten/sn2006/wbgu_sn2006_en.pdf. Accessed April 21, 2017.

World Bank. (2012). *Turn down the heat: why a 4 °C warmer world must be avoided*. International Bank for Reconstruction and Development/The World Bank, Washington, D.C, USA. http://documents.worldbank.org/curated/en/865571468149107611/pdf/NonAsciiFileName0.pdf. Accessed April 18, 2017.

Zhang, X., & Church, J. A. (2012). Sea level trends, interannual and decadal variability in the Pacific Ocean. *Geophysical Research Letters*, *39*(21). Article No. L21701. http://onlinelibrary.wiley.com/doi/10.1029/2012GL053240/epdf. Accessed April 19, 2017.

Author Biographies

Dr. Johannes M. Luetz is Senior Lecturer, Postgraduate Coordinator and Research Chair at CHC Higher Education in Brisbane, Australia, and Adjunct Lecturer at the University of New South Wales (UNSW) in Sydney in the School of Social Sciences, where he also earned his Ph.D. in Environmental Policy and Management with a thesis on forced human migration. Dr. Luetz has worked extensively with World Vision International on research projects raising awareness of the growing effects of climate change on poor and vulnerable communities in Asia, Africa and Latin America. His previous international development work experience includes the role of Lecturer in Development Studies at UNSW in Sydney.

Mr. Peni Hausia Havea is a Ph.D. candidate in climate change who holds a Master degree in Public Health majoring in International Health from the University of Melbourne, Australia. He is a finishing Ph.D. candidate from the University of the South Pacific.

Sustainable Small-scale Mariculture Ventures as a Comparative Climate Friendly Livelihood Alternative in Pohnpei, Federated States of Micronesia

Simon Ellis, Maria Haws, Jasmine Mendiola and Mikelson Hemil

Abstract Like most Pacific island nations, the Federated States of Micronesia (FSM) faces threats to livelihoods, food security, and health from the impacts of climate change. The island of Pohnpei faces specific climate related threats linked to over fishing, sedimentation from deforestation, and a poorly developed economy, which restricts rural residents access to livelihood alternatives. With assistance from the Pacific American Climate Fund (PACAM); and funding from United States Agency of International Development (USAID), the Marine and Environmental Research Institute of Pohnpei (MERIP) is more than half way through a three-year program to reduce climate impacts in Pohnpei. The project goal is to reduce vulnerabilities associated with climate change faced by Pohnpei's coastal communities, inshore reefs and Marine Protected Areas (MPA). One of the primary strategies employed is reducing dependency on fishing and unsustainable farming practices by increasing number and sizes of aquaculture farms growing sponges and marine ornamentals. Engaging rural fishers and farmers in alternative livelihoods will reduce pressure on natural resources and make communities less vulnerable to the effects of climate change on these resources over time. This paper examines the benefits and constraints of the introduced livelihood alternatives—farming of sponges and marine ornamental invertebrates and their impact on traditional fishing activities. In addition, comparisons are made with the growing of *Piper methysticum,* or sakau, a narcotic root crop. Growing of sakau is Pohnpei's most widespread, but comparatively environmentally destructive, form of rural income generation and is widely recognized as the main cause of the islands upland deforestation and resulting lagoon sedimentation. Results show that growing the new adaptive commodities provide similar or greater incomes than the less climate friendly traditional activities of fishing of growing sakau. However, expansion of these activities or additional livelihood alternatives need to be developed, to further reduce environmental degradation.

S. Ellis (✉) · M. Haws · J. Mendiola · M. Hemil
Marine and Environmental Research Institute of Pohnpei, P.O. Box 2031,
Pohnpei, FM 96941, Federated States of Micronesia
e-mail: microellis@gmail.com

W. Leal Filho (ed.), *Climate Change Impacts and Adaptation Strategies for Coastal Communities*, Climate Change Management,
https://doi.org/10.1007/978-3-319-70703-7_2

Introduction

The geographic focus of this project is Pohnpei Island in the Federated States of Micronesia (FSM). The FSM has a vast ocean EEZ (2.9 million km^2) but only 700 km^2 of landmass. Pohnpei is the capitol island of the FSM with 35,000 residents and Pohnpei Island is the third largest island in Micronesia. Per capita GDP is a low USD3000 (192nd in the world) while unemployment is 16.2% (CIA World Factbook 2016). The FSM is an extremely difficult place to do business, and this has dire effects on the less educated and fortunate members of its society. The World Bank ranks the FSM as fourth from last out of 26 East Asian and Pacific Island nations in ease of doing business and 152nd in the world, out of 190 nations (World Bank 2017). Poor governance and immense geographical separation are the primary contributors to this status, a problem shared by many small-island developing states, or SIDS. The FSM ranks particularly badly in the following categories: registering property; protecting investors; enforcing contracts and resolving insolvency. Economic development lags far behind the latent potential of their aquaculture, agriculture and agro-forestry resources and skills. Manifold factors affect commercial business development in these islands, including: non-Western values; geographic separation; poor infrastructure and transportation; and poor business and technical capacity. Climate change is likely to drive per capita GDP lower when compounded with these development constraints.

The lack of income earning opportunities for coastal communities relegates many members of society to a semi-subsistence lifestyle with almost no access to the cash economy other than fishing and farming. The result has been increasing pressure on inshore fisheries leading to declines in fish stocks (Rhodes and Tupper 2007; Rhodes et al. 2008; Hopkins and Rhodes 2010; Rhodes et al. 2016) and increasing concerns about food security. In a household survey of fishing communities around Pohnpei (Hopkins and Rhodes 2010) the mean number of people per household were 8.3 while only 1.4 people per household were employed in any way. A preliminary follow up report, on their initial market survey in 2006, by Rhodes et al. (2016), showed sharp declines in diversity of species caught and a shift toward catching fish which feed in the lower trophic levels, both indicators of over fishing.

Climate change is an increasingly growing threat to livelihoods in Pohnpei and the FSM. According to predictions from the Pacific-Australia Climate Change Science and Adaptation Planning Program (PACCSAPP 2015) there is a very high statistical confidence of increases in the following climatic areas, by 2030, for the FSM: surface air temperatures; sea surface temperatures; annual and seasonal rainfall; intensity and frequency of extreme heat days; intensity and frequency of extreme rainfall days; ocean acidification; and mean sea level rise. Incidence of drought and tropical cyclones are not predicted to increase and may even decrease but may be more severe when they occur. The Secretariat of the Pacific Community (Bell et al. 2011) also predicts that these changes will lead to a 40% decrease in coral cover and 65% decrease in fish abundance and diversity by 2100. Given the already decreasing fish stocks in Pohnpei, from overfishing, it will be key for coastal communities to have viable options to fishing, and a robust and healthy marine environment, to build resiliency to the effects of climate change in the coming years.

One particular threat to Pohnpei's environment is farming *Piper methysticum*, a root crop with narcotic properties, known locally as sakau, but better known globally as "kava". This root based drink is enjoyed in many areas of the Pacific such as Fiji, Vanuatu and the US state of Hawaii. Kava or sakau is a significant plant for the people of Pohnpei (Scrimgeour and Gallen 2003). It is used in ceremonial feasts, settling of family disputes, and more recently as a recreational pastime where people gather to drink and exchange news. The latter activity has led to a large increase in sakau cultivation around the island. Traditionally, sakau is planted in upland forests, which are cleared for growing the crop. So intense is the demand for sakau that between 1976 and 1996 the upland forests of Pohnpei shrank by 42% to just 15% of the total land area of the island (Trustum 1996). Introduction of alternative methods for growing sakau in the lowlands and the establishment of a watershed line protecting the upland forest in some of Pohnpei's municipalities has slowed the rate of deforestation (Dahl and Raynor 1996). In municipalities where the watershed line was not implemented, deforestation continues and the demand for sakau continues to grow (Jasmine Mendiola, personal communication). Most farmers still prefer to plant in upland areas due to a lower incidence of insect pests and a higher potency end product (Merlin and Raynor 2005). This deforestation often results in landslides and soil erosion (Victor et al. 2006) with subsequent sedimentation of rivers, mangroves and inshore coral reefs. This has led to substantial coral mortality in the lagoon (Victor et al. 2006), impacting fisheries habitat. The combination of forest clearing and sedimentation makes the upland growing of sakau the most environmentally destructive activity in Pohnpei, and one of the most destructive activities for inshore fisheries. The cultural and recreational importance of sakau combined with a high local and foreign demand is a major hindrance to establishing more environmentally sustainable alterative livelihoods in Pohnpei.

Since 2005, the Marine and Environmental Research Institute of Pohnpei (MERIP) has been working with coastal communities in Pohnpei to develop alternative income generating forms of aquaculture and to protect their fish stocks through community-based marine protected areas (MPA's). MERIP works with rural fishing communities based around five Marine Protected Areas (MPA's) in Pohnpei, to promote marine conservation and climate change adaptation through coral, giant clam and sponge farming. The premise behind these activities is that engaging communities will reduce stress on over exploited inshore fisheries and upland forests and will also make communities less vulnerable to the effects of climate change in the long run. Corals and giant clams are grown for export to the marine ornamental trade while sponges are grown for export to the bath and beauty industry. All methods of rearing are highly sustainable and have very low impact on the environment. Communities learn about marine and coral conservation and make income from selling their products. Only community members that have engaged seriously with protected areas and other conservation activities are invited to become part of the farming program. In addition, the links between farming opportunities and MPA and marine conservation are also stressed during community meeting so that people are aware of the origin of the opportunity being presented. Income from sales of products stimulates the communities to fish less and to respect the MPA regulations. MERIP has been very successful in developing export markets for products being farmed and presently exports to more than 10 locations globally. Since 2015, MERIP has been receiving support for these activities from the Pacific American Climate Fund (PACAM) with funds originating from the United States Assistance for International Development (USAID). One of the main purposes of this grant is to increase the number of people on Pohnpei engaged in climate friendly aquaculture activities in the form of giant clam, coral and sponge farming. Using PACAM funds MERIP has expanded the number of farmers engaged in sustainable aquaculture practices to 63 individuals.

This paper examines the climate change adaptation benefits and constraints of the introduced livelihood alternatives—farming of sponges and marine ornamental invertebrates. In addition, socioeconomic comparisons are made with inshore fishing and the growing of sakau.

Methods

MERIP employs the nucleus estate model for working with local farmers for farming all the commodities grown. The farmers receive all materials and assistance necessary to start their farm at no cost. In return they receive a lower price for their harvested products. MERIP also maintains central farms for some of the commodities grown which guarantees throughput and provides a place to conduct applied research on new species and farming techniques. Post-harvest handling, marketing and export also take place through MERIP. Descriptions of farming methods are as follows.

Sponge Farming

Farmers currently grow two species of natural sponges for the bath and beauty trade—the Micronesian wool species (*Cosinoderma matthewsi*) which is used for body bathing and the smaller, softer *Spongia matamata* used primarily for facial cleaning. Wool sponge and facial sponge farming is based on sustainable fragmentation of sponge explants (Ellis et al. 2005). Farms are started using wild collected broodstock colonies, which are sustainably collected by cutting just the top half of the sponge and leaving the bottom half to regenerate. Farmers then set aside one third of their explants for grow-out as future cutting stock. Farm structure is simple, consisting of submerged mainlines 6 mm in diameter that are anchored to rocks some 2–3 m below the surface. Mainlines are generally about 30 m long and the farm may be made up of anywhere from 2 to 5 mainlines. Lighter grow-out lines are then strung perpendicular to the mainlines about 1.5 m apart. Onto these grow-out lines, explants are individually hung about 30 cm apart. The mainlines have sufficient slack to allow them sag about 8–10 m below the surface, also allowing them to also be raised close to the surface using floats so that the farmers can work without repeated free diving. Sponge farms need intermittent cleaning approximately every 4–6 weeks and grow-out time is about 2.5 years for the wool sponge and 9–12 months for the facial sponge. Processed sponges are sold to a variety of customers primarily in New Zealand, the United States, Australia, Singapore, Malaysia and Hong Kong. Sponges are sold without any chemical whitening and most customers are retailers of eco-friendly products.

Coral Farming

Farmers currently grow 31 species of hard and soft corals for export to the marine ornamental trade. All corals are grown using sustainable fragmentation techniques. Broodstock colonies are started from wild collected stock and explants are then sustainably cropped from these colonies. Grow-out time ranges from 2–6 months depending on the species being grown and the corals require very little maintenance during that time. Corals are grown on metal tables equipped with steel mesh trays or flat plastic mesh sheets, which house the explants. The tables are placed at a depth of 3–7 m depending on the site location and the species being grown. Harvested explants are exported live to two wholesalers in the region, one in Kosrae, FSM, and one in Majuro, Republic of the Marshall Islands. These wholesalers then combine these products with their own for further export, primarily to the United States and European Union. MERIP handles all export of products from Pohnpei.

Giant Clams

One species of giant clam is raised for export, the colorful *Tridacna maxima*. This species does not grow much larger than 40 cm in the wild and has a minimum market size for the marine ornamental trade of just 4–5 cm. *T. maxima* are spawned at the MERIP facility and larvae reared to a size of 2 cm using raceways receiving flow-through seawater. At a size of about 2 cm the clams are allowed to attach to concrete plates for lagoon-based grow-out. The concrete plates are placed into plastic cages designed primarily for grow-out of edible oysters in Southern Australia. These cages are resistant to almost all giant clam predators except for the Ranellid snail *Cymatium spp.*, which settle onto the clams from the plankton. Cages are placed on metal trestles in the lagoon in about 2–3 m of water so the clams receive sufficient sunlight. To reduce predation from *Cymatium* snails the cages and clams must be cleaned and inspected every 3 weeks. Grow-out time, in the lagoon, for the *T. maxima* ranges from 6–9 months. The giant clams are exported live to the same wholesalers in Kosrae and Majuro for onward shipping to markets in the United States and European Union.

Data regarding number of hours worked by farmers was collected from interviews with MERIP extension agents and farmers themselves. Income earned by farmers was collected from records of payments by MERIP to farmers over a two-year period from January 2015 through December 2016. Information on farmer's other livelihood activities and perceptions on conservation and the marine environment were gathered from socioeconomic surveys conducted in May 2016.

Limitations

Grow-out methods for sponges and corals are well understood, as these commodities have been adopted by farmers since, or before 2010. Research by MERIP on lagoon culture methods for the giant clam, *T. maxima,* has been ongoing for a number of years but only carried out by local farmers since 2015. Therefore, results for this commodity may not be as robust as for the sponges and corals, which have been established for much longer. Data for farmers engaged for less than 6 months was not included in the data analysis.

Results

Income and Time Commitment

Monthly earnings (USD), number of months engaged, and earnings per hour (USD) for the four main commodities being farmed (giant clams, hard corals, soft corals and sponges) are presented in Table 1.

Soft and hard coral farmers had the highest monthly income of all the commodities. However, mean monthly income was almost twice as high for soft coral farmers at $83.81 than hard coral farmers at $43.21. Sponge farmer and giant clam farmer incomes were approximately a quarter of soft coral farmers and half of hard coral farmers at $24.65 and $16.95 respectively. When hourly income was calculated hard and soft coral farmers had similar incomes at $7.23 and $8.38 respectively. Likewise sponge and giant clam farmers had similar hourly incomes of $3.08 and $2.83 respectively, but these returns were both less than half of the coral farmers.

Mean number of hours worked monthly by farmers was highest for soft coral (10 h) and sponge farmers (8 h). Hard coral and giant clam farmers worked lower hours at 6.2 and 6 h/month respectively. Mean engagement time of hard coral, soft coral and sponge farmers were similar ranging from 15.22 to 17.53 months. Giant clam farmer average engagement time was lower at a mean of 12.85 months.

Table 1 Income and time commitment data for farmers engaged in mariculture activities in Pohnpei, FSM (n = number of farmers)

	Commodity			
	Giant Clams (n = 10)	Hard Corals (n = 12)	Soft Corals (n = 16)	Sponges (n = 14)
Average Number of Hours Worked Monthly	6	6.2	10	8
Mean Monthly Income (± St. Dev.)	\$16.95 ± 10.34 Range \$6.35–\$38.18	\$43.21 ± 25.78 Range \$21.42–\$102.77	\$83.81 ± 36.82 Range \$11.11–\$134.78	\$24.65 ± 10.44 Range \$7.77–\$42.57
Mean Hourly Income (± St. Dev)	\$2.83 ± 1.72 Range \$1.06–\$6.36	\$7.23 ± 2.53 Range \$4.29–\$15.36	\$8.38 ± 3.68 Range \$1.11–\$13.48	\$3.08 ± 1.31 Range \$1.55–\$4.52
Mean Number of Months Engaged in Farming (± St. Dev)	12.85 ± 5.02 Range 7–20	15.22 ± 6.48 Range 7–24	17.53 ± 5.44 Range 7–24	16.93 ± 7.11 Range 6–24

Socioeconomic Data

A total of 38 aquaculture farmers were interviewed in May 2016 to gather data on their sources of income, livelihood activities and outlook on fisheries/coral reef conservation.

Following are the main outcomes of the surveys:

- There were 16 coral farmers, 13 sponge farmers and 9 giant clam farmers interviewed
- Mean number of people per farmer household was 7
- Primary occupations of aquaculture farmers are: farming sakau and other crops—42%; fishing—38%; paid employment—9.3%; home maker 10.7%
- Since starting aquaculture farming: 66% fish less; 17% fish about the same; and 17% fish more
- 60% of respondents claimed aquaculture farming had brought about the change in their fishing habits
- 75% of respondents felt it was harder to catch fish than in the recent past
- 70% of respondents felt fish are smaller than in the recent past
- 100% of respondents felt having a marine protected area was important and that fisheries management was essential to a healthy future
- 97% of respondents felt that coral reefs are important in protecting the land from storm activity.

Discussion

The main point of discussion of this paper is the efficacy of sponge, coral and giant clam farming as climate friendly adaptations and alternatives to fishing and sakau farming. In terms of income, corals present the best option for farmers with an annual income of between \$500–\$1000. Sponge and giant clam farming are less attractive with an annual income of \$200–300. However, it should be noted that none of these activities take up more 12% of a typical working week of 40 h. In Pohnpei, as in many areas of the Pacific and developing world, many rural residents do not have a full-time occupation but rather subsist or make income from a range of different activities. In this regard, climate friendly aquaculture farming is an ideal alternative to provide a supplemental income, or in some case a replacement income, for rural residents of Pohnpei.

Average earnings per hour for all the new commodities are above the Pohnpei State minimum wage of \$1.75 per h. For coral farmers, average earnings per hour are more than 4 times the minimum wage of \$1.75. Estimated annual income from fishing in Pohnpei was estimated in 2011 to range from \$535 to \$780 (Jeff Kinch personal communication) and fishers on average worked 1.8 days per week on this activity (Hopkins and Rhodes 2010). This equates to an hourly income of \$0.71 to \$1.04, much lower than even the giant clam farmers in this study, who earn \$2.83 per h.

It is harder to get an estimate of average income for sakau farmers as the industry is very informal and many people only grow sakau for personal consumption and cultural purposes. In 1996, it was estimated that more than 450,000 kg of sakau, worth around \$3 million, entered the commercial economy in Pohnpei (Merlin and Raynor 2005) and that around 4000–5000 people were engaged in the industry by the year 2000 (Merlin and Raynor 2005). Extrapolating these numbers indicates that average annual income per farmer would be around \$600–\$750. In a 2003 study of the use value of forests in Pohnpei, Scrimegour and Gallen estimated that households from two typical villages in Pohnpei spent an average of 18 h per month in the forest looking for land and planting and tending to sakau. This equates to an hourly income of between \$2.78 and \$3.47. This is much higher than fishers or the Pohnpei state minimum wage but only equivalent to giant clam and sponge farmers, but much lower than farmers who grow corals.

Average annual per capita income for Pohnpei was \$2093 in 2005; the last time a household survey was conducted in the FSM (FSM 2005). However, farming and fishing tend to be the occupation of members of society with lower levels of education and less opportunity so annual incomes of less than \$1000 are to be expected. Scrimegour and Gallen (2003) stated, “most of the (sakau) farmers are young without any formal education”. This is also indicated in the FSM Houshold Income Survey of 2005 where 33% of households in Pohnpei had per capita income of less than \$1300 (FSM 2005). The same survey also showed data that per capita income dropped as the number of people per household increased. The mean number of people per household of sponge, giant clam and coral farmers is seven.

It is clear that in terms of hourly income rates, sponge, giant clam and coral farming can provide an equivalent or greater income to people engaging in the more environmentally harmful activities of fishing and sakau farming. However, to date only 63 farmers have been engaged in farming of these climate friendly commodities. By comparison estimates of sakau farmers in Pohnpei ranges from 4000–5000 (Merlin and Raynor 2005) and fishers who sell at least a part of their catch from 2500–3000 (Hopkins and Rhodes 2010). Constraining the growth of these climate friendly sustainable commodities is the demand for corals and giant clams and the need to upscale production of sponges. Giant clam and coral demand from Pohnpei, based on MERIP exports, are estimated at approximately 1000 and 2000 pieces per month respectively. Corals exports from MERIP averaged 2150 pieces per month during the calendar years of 2015 and 2016. Giant clam farming demand has not yet been met and could be as high as 1000 pieces per month (Jacob Applebaum and Martin Selch personal communication). Because of this the number of giant clam and coral farmers is unlikely to increase by more than 10 individuals in the coming 2–3 years. Demand for sponges on the other hand, remains high and more farmers could be engaged. Farm production of sponges could expand greatly. In 2015 and 2016, 3867 sponges were exported by MERIP. Clearsight Consultants, MERIP's biggest customer for sponges, based in New Zealand, estimated just their demand for sponges to exceed 18,000 pieces per year (Carina Sim-Smith personal communication). It is estimated a further 20–30 sponge farmers could be engaged before market demand was met. In total a maximum number of around 100 farmers could be engaged for these 3 commodities.

Socioeconomic data gathered from interviews in 2016 show that there is an impact in reduced fishing and increased income among aquaculture farmers, although no data was collected on the impact on sakau farming. In addition, aquaculture farmers displayed a good understanding of environmental issues through training they had received from MERIP staff during semi-annual meetings held during 2015 and 2016.

The value or gaining qualifications, either formally or informally, as a means of adapting to, and reducing vulnerability to, climate change is clearly demonstrated by the work undertaken by MERIP. Community members who engaged as aquaculture farmers were not only well informed about how the effects of climate change will affect their lives in the future, but are also making valuable income from their activities. Even the lower paying activities of giant clam and sponge farming make approximately the same per hour as sakau farmers and more than fishers in Pohnpei, while coral farmers can make 3–4 times more per h.

Best Practices

Key to the development of new climate friendly economic and livelihood adaptations in Pohnpei and the rest of the Pacific region is adherence to sustainable best practices for development. In the context of this paper it is useful to look at the lessons learned from a review of government led livelihood diversification projects in the Pacific by Gillett et al. (2008). In this report 22 livelihoods projects, were analyzed for valuable lessons. Many of these findings have been adopted by MERIP and can equally apply to just about any sustainable development project in the Pacific, especially those relating to climate change, as follows.

Strong Private Sector Involvement. MERIP has sought to make strong linkages with private sector wholesalers and buyers. This has helped the development and long-term sustainability of climate friendly aquaculture farming because prices obtained for products are realistic and based on world market prices. This gives the farming operations a much better chance of long-term viability.

Long Term Support by a Local NGO. There is strong evidence that livelihood diversification initiatives that are successful in the Pacific require a long time to achieve profitability and eventual profits are characteristically modest rather than spectacular (Gillett et al. 2008). MERIP has sought not to treat the development of climate friendly aquaculture farming as a "project" but rather a long-term investment toward a more sustainable future. The future of MERIP's support for the activity is not linked to a finite funding cycle but is ongoing. Additionally, MERIP plays the role of an "honest broker" between private sector interests and the communities. Businesses are generally better than fisheries or other government departments or NGO's at identifying and/or developing opportunities, but often have difficulties in spreading benefits. They are also better at community relations, hence the need for somebody to smooth the interface between business and community (Gillett et al. 2008).

Community Training, Awareness and Education. Awareness raising on climate change, environmental education and technical training for community members is essential to the long-term success of any climate change adaptive development. Without this, participants will not be able to understand the context of their involvement versus less sustainable activities. In addition, it helps participants to understand their business better and make the link between their own activities and conservation and climate change. Constant communication with the involved communities also helps to dispel mistrust of the intentions of outside assistance providers.

Sustainability and Continuity. Heavily subsidized livelihood diversification activities run the risk of failing once the subsidy is reduced or removed (Gillett et al. 2008). In addition, promotion of climate friendly livelihoods that do not have a plan for transitioning of activities into an entirely local business structure from the outset run this same risk of failure. Therefore, interventions should be planned to slowly transition out subsidies and to ensure local businesses are involved from the outset. Because MERIP is a locally based corporation it can provide both long-term extension and export support to ensure farmers transition smoothly to self-sufficiency over an appropriate time frame.

Conclusions and Future Prospects

The development of climate friendly aquaculture farming has provided a comparative income for 63 rural residents of Pohnpei and a viable alternative to the less environmentally friendly activities of fishing and sakau farming. Coral farming provides an hourly income of 3–4 times that of sakau farming or fishing, while sponge and giant clam farming are provided comparable, or greater, hourly incomes. All the aquaculture commodities provide an hourly income greater than the Pohnpei state minimum wage while allowing farmers to pursue other cultural and subsistence activities.

There is potential for these types of aquaculture to accommodate another 30–40 individuals before market constraints are met. However, the sheer scale of people involved in these less sustainable activities accentuates the need for development of more climate friendly alternatives. In addition, the traditional and recreational demand locally for fish and sakau continue to make them attractive commodities to disadvantaged and less educated members of Pohnpeian society. However, these resource-based activities are increasingly under threat from the combined effects of over exploitation and climate change. Equipping community members with formal or informal qualifications in alternative livelihoods and knowledge of how climate change will affect them, will greatly increase their resilience and reduce vulnerability in the long run.

Lessons learned from this work have broad scale application across the Pacific. In general, there is environmental degradation of some sort and the effects of climate change are reasonably uniform among these tropical and subtropical islands nations. In addition, the development and cultural constraints are also fairly constant across the Pacific. Applying some of the best practices learned from this work may assist other organizations working on climate change adaptation and reducing vulnerability in other Pacific Island nations.

Given the predicted changes to Pohnpei's inshore fisheries and forests because of climate change, the need for sustainable alternative livelihoods is greater than ever. Future climate change adaptation projects should therefore focus on providing alternatives to fishing and sakau farming such as niche agriculture products and development of finished, easily exportable high value commodities. In addition, value adding for existing fishery and forest products can help in increasing per unit revenues. The presence of community and environmentally minded businesses or non-state actors with business and marketing capacity can greatly enhance the success of small-scale climate friendly sustainable ventures in the future.

Acknowledgements The authors would like to thank the people of Pohnpei for their trust, support and assistance in the development of this project. Financial and technical support was provided by: USAID—Pacific American Climate Fund (Award #: PACAM-14-0001); National Oceanic and Atmospheric Administration award #NA15NMF4270339 through the University of Hawaii at Hilo Pacific Aquaculture and Coastal Resources Center (UHHPACRC); the New Zealand North Pacific Development Fund; Misereor; and University of Hawaii Sea Grant.

References

Bell, J., Johnson, J., & Hobday, A. (2011). *Vulnerability of tropical Pacific fisheries and aquaculture to Climate Change*. Noumea, New Caledonia: Secretariat of the Pacific Community.

Dahl, C., & Raynor, B. (1996). Watershed planning and management: Pohnpei, Federated States of Micronesia. *Asia-Pacific Viewpoint, 37,* 235–253.

Ellis, S., Ellis, E., Lohn, W., Haws, M., & Fong, Q. (2005). *Farming the Micronesian Wool sponge*. Hawaii, United States: University of Hawaii.

Federated States of Micronesia. (2016). The CIA World Factbook. https://www.cia.gov/library/publications/the-world-factbook/. Accessed March 14, 2017.

FSM. (2005). Household income and expenditure survey analysis report. Division of Statistics, Palikir, Pohnpei, FM 96941, Federated States of Micronesia.

Gillett, R., Preston, G., Nash, W., Govan, H., Adams, T., & Lam, M. (2008). Livelihood diversification as a marine resource management tool in the Pacific Islands: Lessons learned. SPC Fisheries Newsletter #125—April/June 2008. Noumea, New Caledonia.

Hopkins, K., & Rhodes, K. (2010). A Field and household assessment of non-commercial fishing, per capita consumption and trade patterns for coral reef fishery management improvement in Pohnpei, Micronesia. Final Report. NOAA Coral Reef Conservation Grant NA08NM F4630458. United States.

Merlin, M., & Raynor, W. (2005). Kava cultivation, native species conservation, and integrated watershed resource management on Pohnpei. *Pacific Science, 59*(2), 241–260.

PACCSAPP—Pacific-Australia Climate Change Science and Adaptation Planning Program. (2015). Current and future climate of the Federated States of Micronesia. Australian Bureau of Meteorology, Australia.

Rhodes, K., & Tupper, M. (2007). A preliminary market-based analysis of the Pohnpei, Micronesia, grouper (Serranidae; Epinephelinae) fishery reveals unsustainable fishing practices. *Coral Reefs, 26,* 335–344.

Rhodes, K., Tupper, M., & Wichilmel, C. (2008). Characterization and management of the commercial sector of the Pohnpei, Micronesia, coral reef fishery. *Coral Reefs*. doi:10.1007/s00338-007-0331.

Rhodes, K., Hernandez-Ortiz, D., Ioanis, M., Washington, W., & Ladore, R. (2016). Preliminary market findings: Overview, status and initial comparison (2006–2016) of the Pohnpei inshore fishery. A preliminary report prepared for the Pohnpei State Government December 20, 2016. Pohnpei, Federated States of Micronesia.

Scrimgeour, F., Gallen, J. (2003). *Use values of forests in Pohnpei.* Contributed paper presented to 47th annual conference of the Australian Agriculture and Resource Economics Society at Fremantle Australian February 12–14, 2003.

Trustum. (1996). *Pohnpei's watershed spatial planning and management guidelines*. Contributed paper in The Nature Conservancy Consultants Report: T.A. No. FSM-1925, Watershed Management and Environment. The Asian Development Bank, Manila, Philippines.

Victor, S., Leinson, N., Golbuu, Y., Wolanski, E., & Richmond, R. (2006). Sedimentation in mangroves and coral reefs in a wet tropical island, Pohnpei, Micronesia. *Estuarine, Coastal and Shelf Science, 66,* 409–416.

World Bank. (2017). *Doing business 2017: Equal opportunity for all.* Washington, DC: World Bank. doi:10.1596/978-1-4648-0948-4. License: Creative Commons Attribution CC BY 3.0 IGO.

Planning and Urban Informality—Addressing Inclusiveness for Climate Resilience in the Pacific

Colleen Butcher-Gollach

Abstract The Pacific Island Countries (PICs) are well documented as being amongst the most vulnerable countries in the world to both the long term effects of climate change and to short-term natural disasters. The proportion of land area located in low elevation coastal zones (LECZs) is four times larger than that of Asia, and the urban population living in LECZs is almost double that of the rural populations. In countries where there has been little to no forward planning or investment in service infrastructure to open up new and safe land for many decades, and where registering land in the formal administrative systems is cumbersome, costly and time consuming, urban low income earners have few options but to rent in the existing housing stock (leading to chronic overcrowding) or to pay for the use of customary land and self-build their houses in unplanned, extra-legal settlements. The trend of unplanned and unauthorized settlements continuously expanding onto hazardous and unsafe land has been documented in many of the Pacific's urban and peri-urban areas. The damage and losses incurred by extreme weather events in the past three years alone have pointed to far heavier damage to the housing stock of the urban poor living in urban informal settlements than in adjacent formal city neighborhoods. The paper scopes the extent and nature of urbanization in the Pacific region and documents the rapid expansion of unplanned settlements—increasingly located in low elevation coastal zones at risk of sea level rise and coastal erosion, that are a consequence of a number of factors including inefficient and costly land registration systems; inappropriate normative and static spatial planning instruments; and lack of serviced subdivisions, both safe and affordable by low income earners and close to employment centers. It goes on to identify that the dearth of location specific climate risk data, the lack of policies at national or local levels, and a commonly-held antipathy towards urban settlements have contributed to planning systems and approaches that are not only failing to meet the housing needs of the burgeoning urban populations - and in particular low income groups with limited options, but inadvertently are contributing towards the rapid expansion

C. Butcher-Gollach (✉)
Faculty of Architecture, Building and Planning, University of Melbourne, Melbourne, Australia
e-mail: butcher@unimelb.edu.au

W. Leal Filho (ed.), *Climate Change Impacts and Adaptation Strategies for Coastal Communities*, Climate Change Management,
https://doi.org/10.1007/978-3-319-70703-7_3

of informal, unplanned urban settlements into hazard prone areas. The paper identifies a number of possible entry-points for planners and urban managers to break this cycle of urban deterioration, founded as it is on exclusion and multi-dimensional vulnerability. It concludes with a suggested set of key approaches as guidance for practitioners to better understand and plan with rather than against informality in the urban Pacific.

Problem

It is now well established that the Small Island Developing State (SIDS) are amongst the most vulnerable in the world to the effects of climate change (Antarctic Climate & Ecosystems Cooperative Research Centre 2008; Szoldra 2014) and yet, paradoxically, 'face an unfair burden of responding to a problem – climate change – that they did not create' (Bassett and Scruggs 2013, p. 32). A 2007 global study of the population living in low elevation coastal zones, calculated that 16% of the total land area of the SIDS fell within low elevation, coastal zones as compared to, for example, only 3% in the case of countries in Asia (McGranahan et al. 2007). A summary of the impacts of climate change in the SIDS prepared by the UN Framework Convention on Climate Change (UNFCC) stressed that as most settlements and infrastructure located near the coasts would be affected by sea level rise and coastal erosion, human habitation 'and thus sovereignty of some states [would be] threatened due to reduction in island size or complete inundation' (UNFCCC 2007 cited in Baker and Week 2011, p. 16). It is therefore both timely and an opportunity for further pressing the SIDS' case on the international stage, that the Republic of Fiji recently has become the first of the SIDS to take on responsibilities of Presidency for the 23rd session of the Conference of Parties (COP23, 2017) to the UNFCCC.

In addition to this long-term existential threat, a number of SIDS, in particular, the Pacific Island Countries (PICs) are located in high natural hazard risk areas and regularly experience multiple, more immediate-term natural disasters. According to work by Birkmann and Welle (2015) evaluating exposure and susceptibility to natural hazards, and vulnerability based on lack of coping and adaptive capacities for 171 countries, Vanuatu (index value: 36.72) and Tonga (index value: 28.45), are ranked respectively as first and second on the World Risk Index. There are seven PICs in the top 20 most exposed and vulnerable countries in the world.

However, less well documented is that it is oftentimes the urban population, and in particular urban low-income earners, who suffer the brunt of the seemingly ever increasing intensity of natural disasters in the small island countries. The Damage and Loss Assessment carried out in the post-Mataniko River flooding event in the Solomon Islands in 2014, and the Post Disaster Needs Assessments in the aftermath of Tropical Cyclone Pam in Vanuatu (2015) and Tropical Cyclone Winston in Fiji

(2016) found the damage to housing stock in the unplanned informal settlements in urban and peri-urban areas was between double to three times the extent of damage in adjacent 'formal city' neighborhoods (Government of Solomon Islands 2014; Government of Vanuatu 2015; Government of Fiji 2016a). Such findings are not entirely unexpected. Whilst traditional house and meeting place construction techniques are known to incorporate risk reduction measures (Milbank 2015; Government of Fiji 2016b), such knowledge is becoming weakened over time and dissipated when rural or outer island populations move to cities. Similarly, the social bonds that have helped communities in far flung and remote outer islands to foretell, withstand, and recover from extreme weather events for centuries[1] tend to break down when populations migrate to new and urban islands. Furthermore, the residents of informal and squatter settlements are generalized as being poor and unable to afford permanent or durable materials for house construction that might meet any Building Code construction standards.

As the scale and rate of urban growth spreads across the Pacific's coastal towns and cities, there is a growing awareness that the Island Countries' land use planning and urban management systems are failing to keep pace with either the rate of growth and levels of exposure and vulnerability to natural hazard risks. This paper argues that furthermore, the failure to acknowledge and grapple with urban informality by Island governments and physical planners in fact may be contributing towards maladaptation and exacerbating the vulnerability of the urban poor.

Purpose

The purpose and structure of the paper is fourfold. Firstly, the paper briefly describes the extent and nature of urbanization in the Pacific Island region. Secondly, it seeks to better understand the location priorities and housing needs of low income earners, documents examples of existing formal planning instruments, and discusses how the two parallel systems together are giving rise to the rapid growth of informal, unplanned settlements and on hazardous land. Thirdly, the paper deconstructs sub-sets within the 'urban problems' being faced. Finally, it identifies entry-points and approaches that might be constructively employed by urban policy makers and practitioners as means to plan for, rather than against, urban informality and in so doing, to strengthen the resilience of urban populations, and in particular that of low income earners.

[1]Personal communication with Mr. Leveni 'Aho, Director, National Disaster Management Office, Tonga, April 3, 2014.

Method

The paper draws on a desk review of the literature and studies of the impacts of climate change and natural hazard risks facing the urban areas in the Pacific Island Countries. It relates these perspectives to data available from government population censuses and national health surveys, and the urban planning policies, laws and practices in the Island countries. The research draws on data and insights from semi-structured interviews and discussions by the author with local experts and focus groups in eight of the Pacific Island countries, post-disaster assessments in housing, infrastructure and land use planning sectors, and designing urban development and housing reconstruction programs in the region. The recommendations and approaches in the final part of the paper are based on actual case studies and development projects which the author has either led or been associated with and which might usefully be adapted by other urban and climate practitioners in the field.

Extent and Nature of Urbanization in the Pacific Island Region

Urbanization is not a new phenomenon in the Pacific. Rapid rates of urban in-migration were well underway in the 1960s and 1970s and, for the past three decades, urban growth rates have been higher than rural growth rates in almost all of the island countries (Butcher-Gollach 2012). The Federated States of Micronesia, French Polynesia and Samoa are the exceptions (The Pacific Community (SPC) 2016) where urban centers do not hold populations but are seen to act as springboards for international migration (Rallu 2009). By 2017, more than half of the countries in the region have more than 50% of their population now living in urban areas including adjacent peri-urban settlements (UNESCAP Statistical Division 2015). Whilst the rate of urban growth is equally apace in all of the Island Countries (see Fig. 1), it should be noted that there are differences between countries regarding the total proportion of population that is urban, varying for example from over 50% in Fiji, the Marshall Islands and Kiribati to only 25% in the Melanesian countries of Vanuatu and Papua New Guinea (Rallu 2009; Pacific Institute of Public Policy 2011). However, a number of challenges underlie consistency of the official data, not the least being that the designated urban boundaries are infrequently updated and as such, most national population censuses do not enumerate large, adjacent peri-urban settlements as being 'urban', for example, 'greater Nuku'alofa' 'greater Apia' or 'greater Port Vila'.[2]

[2]For example, based on the Vanuatu 2009 Census (Government of Vanuatu 2009), the population of Port Vila Municipality is estimated as 65,300 people in 2017. However, taking into account the

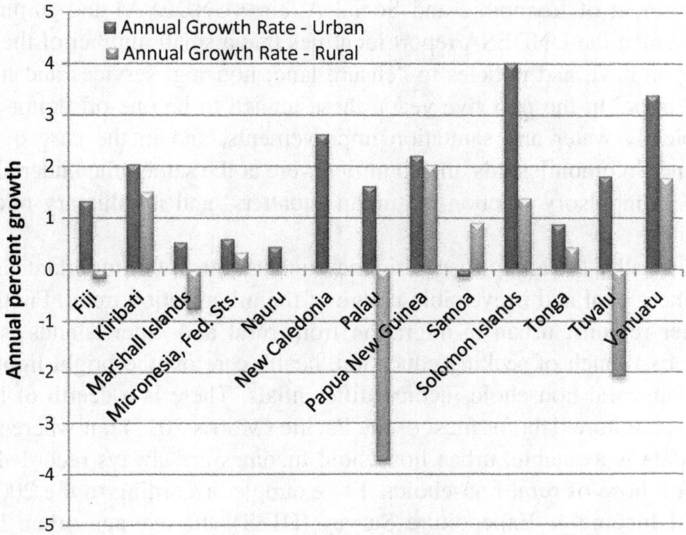

Fig. 1 Pacific Island Countries' Urban and Rural Population Growth Rates, 2015. *Source* Based on data query generated by Author from World Development Indicators Database, The World Bank Group (2017b)

The underlying drivers of this urbanization are not well understood or much accepted by the Island governments and their practitioners. The recent *Preliminary Report on the 2015 Population and Housing Census* for Kiribati, for example, enumerated that 57% of the population now lives in the urban areas of South Tarawa and Kiritimati Island with "people continually leaving their home islands to find paid jobs on S Tarawa or to be closer to schools or hospitals". The Report then goes on to conclude that "The challenge therefore is how to stop or even reverse this trend" (Government of Kiribati 2016, p. 3).[3] Similar sentiments—seeking to not only influence and maintain rural settlements through strong investments in infrastructure and facilities but even to reverse in-migration, are common in the literature and domestic policies, as highlighted by a recent report compiled by the

adjacent peri-urban wards, the 2017 population of 'greater Port Vila' would be around 109,000 people (pers. comm. PhD Candidate Alexei Trundle).

[3]The anti-urban political bias in the Pacific is rooted in political-economy. The political power in countries such as Kiribati, Solomon Islands, Tonga and Vanuatu is in all cases weighted in favor of the outer islands. Within Kiribati, for example, the 50% of the population living on the outer islands is represented by 39 of the 45 Members of Parliament. The remaining MPs represent the 50% of the population living in South Tarawa and Kiritimati Island urban areas (Member of Parliament for South Tarawa, *pers.com*, January 2007). Similarly in Vanuatu, the urban constituencies of Port Vila and Luganville account for 24% of the national population (excluding the de facto much larger urban population living in the adjacent peri-urban areas of 'greater' Port Vila) but only have eight Members of Parliament compared to 44 MPs for the remainder of the country (Marango 2016).

UN Department of Economic and Social Affairs (UNDESA) and summarized in Table 1. Whilst the UNDESA report identifies that a small number of the countries in the region have had policies to "ensure land, housing, services and livelihoods for urban poor" in the past five years, these tended to be one-off donor-led initiatives, typically water and sanitation improvements, and in the case of Vanuatu, Kiribati and Solomon Islands, the countries were at the same time undertaking slum clearance, compulsory eviction of 'urban squatters' and involuntary relocation to other islands.

This antipathy towards urban residency under-states two important factors that underlie the global and irreversible nature of the urbanization trend. Firstly that, as in all other regions, urban in-migration from rural and outer islands is a consequence not so much of seeking education, health care or 'the bright lights' as it is due to urban-rural household income differentials. There is a dearth of household income-expenditure data for most of the Pacific (Morris 2011) but where individual earnings data is available, urban household incomes are always recorded as being higher than those of rural households. For example, according to the 2008–09 Fiji Household Income & Expenditure Survey (HIES), the average urban household income (FJD23,036) was double that of the average rural household income (FJD11,608)[4] According to the subsequent Fiji HIES 2013/14, only 20% of urban households were below the basic needs poverty line compared to 37% of rural households (Government of Fiji, HIES 2013/14). Similarly large urban-rural household income differentials have been recorded in Solomon Islands (Government of Solomon Islands 2006), Kiribati (Government of Kiribati 2007), Samoa (Government of Samoa 2008) and Tonga (Government of Tonga 2009). Under circumstances of static or declining incomes in the rural and outer islands, there is therefore a strong and economically rational incentive for individuals and families to move to urban centers in search of improved economic opportunities.

Economists have long argued the positive effects of agglomeration, namely: deeper networks of buyer/suppliers and specialized inputs (sharing); a greater chance of finding the right worker, supplier or investor (matching); and more opportunities for knowledge spillovers (learning) (Marshall 1890 cited by Deichmann 2016) resulting in higher productivity and translating to higher incomes. Notwithstanding their relatively small economies and distance to markets, the concentration of people in the urban areas and towns in the Pacific, as elsewhere, lends itself to wealth creation through the typical processes of sharing inputs and ideas, proximity to customers, matching supplies to products and markets, and learning—factors that are not possible at a rural dispersed scale (Glaeser and Xiong 2017). Global studies have unequivocally demonstrated that 'There is a near-perfect correlation between urbanization and prosperity across nations. On average, as the share of a country's population that is urban rises by 10%, the country's per capita output increases by 30%. Per capita incomes are almost four times higher in those countries where a majority of people live in cities than in those countries where a

[4]Fiji Bureau of Statistics—Key Statistics June 2012, Table 13.3.

Table 1 Human settlement policy, strategy or measure adopted (2011–2015). *Source* (UNDESA 2016, p. 8 and p. 18)

Country	Policy, strategy or measure adopted (2011–2015)			
	Reduce migration from rural areas to urban areas	Provide incentives to establish or retain industries in rural areas	Develop rural infrastructure and facilities	Ensure water and sanitation services for urban poor (2007–2011)
Melanesia				
Fiji	√	√	√	√
Papua New Guinea	√	√	√	√
Solomon Islands	√	√	√	√
Vanuatu	√	√	√	√
Micronesia				
Kiribati	√	X	√	√
Marshall Islands	√	√	√	X
Micronesia F.S.	√	√	√	X
Nauru	–	–	–	X
Palau	X	X	√	X
Polynesia				
Cook Islands	√	X	√	X
Niue	–	–	–	–
Samoa	√	√	√	X
Tonga	√	X	√	–
Tuvalu	√	√	√	√

majority of people live in rural areas' (Glaeser 2011, p. 7). Additional research taking into account variables such as level of education, infrastructure investment, political openness would need to be undertaken to better understand the trend within the Pacific Island Countries. Notwithstanding the constraints of using GDP as a metric for prosperity or economic well-being (White et al. 2012; Jones and Klenow 2016), data for many of the Pacific Island countries shows a similarly strong correlation between increasing urbanization and growth in GDP per capita over the past 25 years, as illustrated in a range of the Island countries in Fig. 2.

The second factor that is misconstrued is with regards to the source of urban growth. National population census data in almost all of the PICs confirms that urban growth is not only the consequence of in-migration from outer islands but increasingly is the result of natural population growth of an already urbanized and youthful population base. A high proportion of the current urban population was

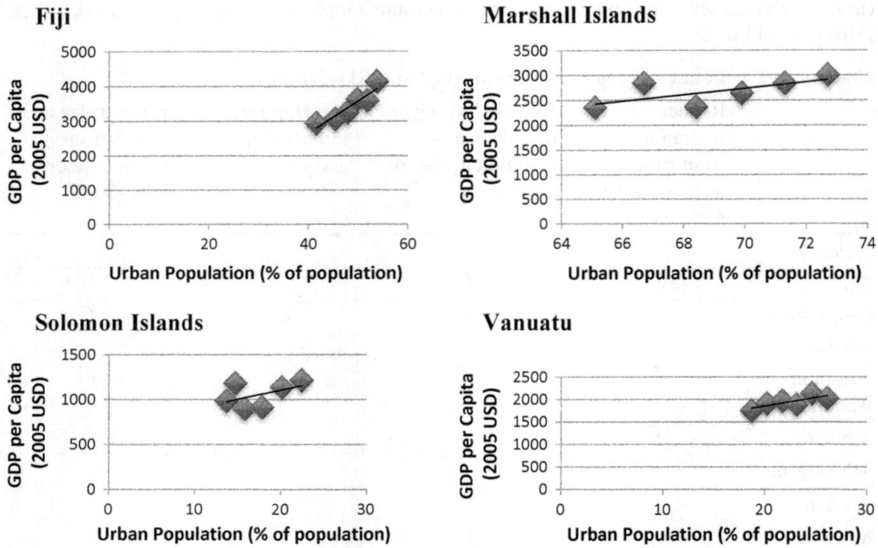

Fig. 2 Changes in GDP Per Capita and Urban Population (1990–2015). *Source* Based on data query generated by Author from UNESCAP Online Statistical Database based on data from the World Urbanization Prospects–2014 revision, 5 August 2014. Available from http://data.unescap.org/escap_stat/. Accessed 28 May 2017

born in the towns and cities or has lived there for many decades[5] and once again, this indicates a trend that is unlikely to be reversed.

Increasing Vulnerability, Location Priorities and Planning Responses

If the data concerning the size, growth and underlying trends towards urbanization in the Pacific are not well understood, neither too are the location priorities of urban residents nor the factors that influence urban settlement form. The towns and cities of the Pacific and associated living conditions are variously described in terms of urban squalor, overcrowding, environmental degradation and problems. A further worrying aspect is the extent to which the settlements are littoralized, given that the local economies predominantly are based on fishing, tourism, industry, and trade (ports) that are all clustered along coastlines. Andrew et al (2017), making use of LandScan 2015 Global Datasets supplemented by national population censuses, have calculated that between 75–100% of the population live less than five

[5]For example, "One striking feature of South Tarawa's population is that … over three quarters of South Tarawa residents (76%) were born here" (Government of Kiribati 2012, p. 6).

kilometres from a coastline in all 19 Pacific Island Countries that they studied. More than 50% of the population in all countries (other than Niue and Guam) live within one kilometre of the coast. However, to better understand the level of risk facing the Islands' populations, accurate elevation data is also needed. Densely settled populations located at low elevations are at high risk of sea level rise and coastal erosion, increased height and frequency of storm surge inundation, changes in tropical storm and hurricane/cyclone frequency and intensity, saline intrusion into groundwater, and degradation of the coastal and marine ecosystems that provide natural protection to the coastline.

The study by McGranahan et al. (2007) was one of the first to comprehensively calculate the number of people living in low elevation **and** coastal zones in 224 countries around the world. By overlaying population data for 1990 and 2000 (and estimates for 2100) with the land area in low elevation coastal zones (LECZ[6]) by country, region and economic grouping, the study calculated that globally, 10% of the world's population (based on 2000 estimates) lives within the LECZs, equating to 634 million people, of whom 360 million are urban. The study data was expressed at one kilometer resolution and as such, the small islands states displayed inconsistencies due to inaccuracies associated with geo-coding of their small land areas and finally were omitted from a number of the analyses. Notwithstanding these limitations, the study identified that out of the 21 countries where 50% or more of their population lived in the LECZs, 16 were SIDs and, in almost all cases where data was available, the urban population was more at risk in terms of location than the rural population, as shown in the PICs covered by the study in Table 2.

For the 13 Pacific Island countries for which there was reliable elevation and population data, the study calculated a total 248,695 urban residents who were living at an elevation of ≤ 10 m or lower (2010 population), as shown in Table 3.

According to available modelling, sea-level rise is not expected to reach 10 m above the current mid-tide elevations in the foreseeable future and the 10 m elevation provides a large safety margin even with regard to storm surge (McGranahan et al. 2007). As a gross rule-of-thumb, the population most at risk would be those close to the coastline and within reach of the high astronomical tides + 0.4 m (the typical reach of storm surge waves), roughly equating to elevations below the 2 m–2.5 m above mean sea level contour line.[7] It should be noted that sea level rise and storm surge would still cause damage to people living above this elevation, for example as a result of saline intrusion into groundwater reserves. Figure 3 graphically illustrates the number of urban residents living at extremely low elevations (≤ 1 m, ≤ 3 m and ≤ 5 m elevations) within coastal zones by country for which data was available (2010 population estimates).

[6]Low elevation coastal zone was defined in the study as being the land area contiguous with the coastline 10 m or less in elevation and closer than 100 km from the sea.

[7]Personal communication with Mr. Lawrie Carlson, Project Director, Tonga Cyclone Ian Reconstruction and Climate Resilience Project, August 2014.

Table 2 Proportion of urban and rural population living in LECZs in four Pacific Island countries. *Source* McGranahan et al. (2007)

Country (Total popn. in LECZ)	% Urban Popn. in LECZ	% Rural Popn. in LECZ
New Caledonia (62,000)	19	10
Samoa (37,000)	18	7
Solomon Islands (87,000)	10	12
Fiji (143,000)	12	6

Table 3 Urban population living within 10 m or less (2010 population est.). *Source* Center for International Earth Science Information Network (CIESIN)/Columbia University. 2013. Urban-Rural Population and Land Area Estimates Version 2. Palisades, NY: NASA Socioeconomic Data and Applications Center (SEDAC). http://sedac.ciesin.columbia.edu/data/set/lecz-urban-rural-population-land-area-estimates-v2. Accessed May 22, 2017

	Urban population at elevations (no. people, 2010)					
Country	≤ 1 m	≤ 3 m	≤ 5 m	≤ 7 m	≤ 9 m	≤ 10 m
Timor-Leste	21	466	1436	2252	3290	3754
Cook Islands	50	99	242	720	1294	1643
Fiji	2146	4567	25,840	45,383	57,475	63,398
Kiribati	540	600	1761	3741	5904	6824
Micronesia F.S.	744	1305	3231	5478	7498	8440
Nauru	4	28	100	249	486	571
New Caledonia	4451	10,265	20,973	32,199	41,270	45,050
Palau	74	129	223	347	607	942
Papua New Guinea	1377	3323	10,086	23,926	44,472	59,821
Samoa	1781	1783	2037	7128	17,940	22,891
Solomon Islands	1329	3576	5733	11,900	16,025	18,575
Tonga	224	709	3059	6466	10,803	13,022
Vanuatu	608	675	920	1225	2952	3764
Total Affected People	13,349	27,525	75,641	141,014	210,016	248,695

The findings of the 2007 McGranahan et al. global study calculated for the first time, the quantum of people (urban and rural) living in different countries in the Pacific in high risk coastal locations, not only as a function of proximity to coastlines but also at low elevations. The study concluded with a set of somewhat generic recommendations regarding the need for mitigation, inland migration and international support for 'Vulnerable settlements in low-income countries … to adapt to climate change' (McGranahan et al. 2007, p. 37) but without any detail of how such measures might be designed or implemented.

Where more detailed country or city specific hazard risk assessments have been undertaken (and they are few), the enormity of the challenge for the Pacific's small but poorly managed and high risk towns becomes even more challenging. By way of example, recent probabilistic hazard and risk mapping prepared for urban areas in Vanuatu for multiple natural geophysical and hydro-meteorological hazards

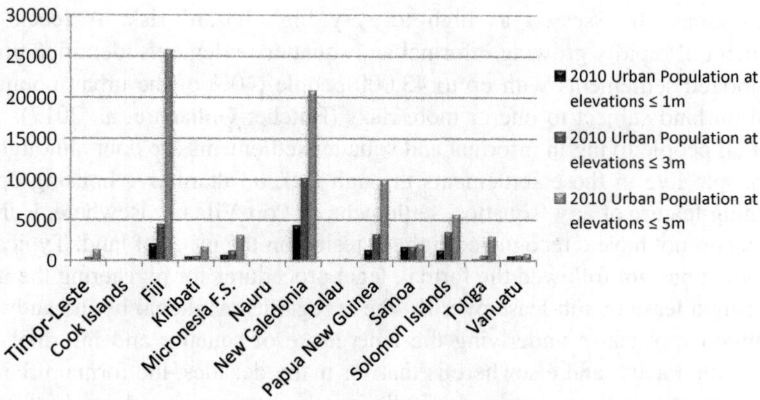

Fig. 3 Urban population living in high risk (≤ 5 m elevation) coastal zones by country (2010 population estimates). *Source* Center for International Earth Science Information Network (CIESIN)/Columbia University. 2013. Urban-Rural Population and Land Area Estimates Version 2. Palisades, NY: NASA Socioeconomic Data and Applications Center (SEDAC). http://sedac.ciesin.columbia.edu/data/set/lecz-urban-rural-population-land-area-estimates-v2. Accessed May 22, 2017

(volcanic eruptions, earthquake, cyclone, coastal and riverine flooding, landslide and tsunami) assessed the following over a 100-year mean return period:

- Greater Port Vila: approximately 85% of the buildings and population are exposed to moderate levels of risk from earthquakes and cyclones (wind) and 15% are exposed to high levels of risk. Around 6% of the buildings and population are exposed to moderate to very high risk from coastal inundation. 20% of the Port Vila flood study area (flood modeling was limited to areas where LiDAR-based elevation data were available[8]) is at moderate to very high risk.
- Luganville: Approximately 98% of the buildings and population are exposed to moderate levels of earthquake risk and 99% are exposed to a high level of risk for cyclones (wind). Around 4% of the buildings and population are exposed to moderate to high risk from coastal inundation and 26% at moderate to high risk of riverine flooding.

(Beca International Consultants Ltd; GNS Science; National Institute of Water and Atmospheric Research (NIWA) 2015).

Intersecting with this high risk setting are the very high urban growth rates in Port Vila: 4.1% growth rate in the inter-censual period between 1999–2009, and **74% of this growth is taking place in the peri-urban, *kastom*-owned, unplanned and informal settlements** outside of the municipal boundary to the north and east of the city for example, Blacksands-Manples-Tagabe Bridge, Etas, and Teouma

[8]Flood hazard datasets were prepared for the Mele and Sarakata river catchments only and excluded Teouma.

Valley—areas all assessed as high to very high hazard risk. A recent rapid assessment of rapidly growing informal and squatter settlements identified up to 21 unauthorized settlements with up to 43,000 people (40% of the urban population) and all on land subject to one or more risks (Butcher-Gollach et al. 2017).

Not all people living in informal and squatter settlements are poor although most poor people live in these settlements through lack of alternative housing options. A defining feature of any 'squatter' settlement, in Port Vila or elsewhere, is that the residents do not have a recognized right to reside on the piece of land. Typically, if the resident has not followed the formal, legal procedures for registering the land or registering a lease or sub-lease over it, s/he is regarded as illegal by the authorities. A common root cause underlying the emergence of squatter and informal settlements in the Pacific and elsewhere is that for many decades, the formal urban land and housing markets for low and middle-income earners have been heavily constrained by a number of supply-side factors. These include inefficient and costly land lease registration systems, outdated land use planning, zoning and building regulations, unrealistic planning and building standards (regarding minimum plot size, house size, road hierarchies, quality of materials), lack of publically serviced and affordable subdivisions close to employment centers, no housing finance for low income and informal sector-employed households, and inefficient and costly transport systems.

The burgeoning urban population therefore has little option but to crowd into the existing housing stock (overcrowding) or to self-build through informal arrangements with private landowners, either as informal tenants or through traditional rights secured through kinship and sealed with 'key-money' (e.g. *vakavanua* in Fiji). Both unplanned informal (extra-legal) and squatter (illegal) settlements have been a feature of rapidly growing urban and peri-urban areas in the Pacific for many decades. Such informal occupancy or tenure arrangements are precarious for the majority of urban low and middle income population. As a result, households have no long term incentive to invest in constructing durable buildings even though vulnerable to riverine or coastal flooding and erosion, and exposed to cyclonic wind speeds and other natural hazards. In addition, as the settlements are not officially recognized by the authorities, they are not provided with basic infrastructure services such as storm water and flood drainage, piped water, sanitation, or solid waste collection that serve to reduce or offset exposure to natural hazards.

In towns where there is little to no guided land development or use of land use zoning by the authorities, households with limited options regarding access to land are at risk of constructing in high risk areas with no official caution or sanction. Figure 4 shows the vulnerability of villages in South Tarawa to the effects of climate change (sea level rise) based on a Coastal Zone Management study by Elrick and Kay (2009) for the Kiribati Adaptation Program Phase II (KAP II).

According to the 2015 Kiribati Population and Housing Census, South Tarawa grew at a relatively rapid rate of 2.9% p.a. between 2010–2015. The 2015 Census recorded very high growth rates in a number of villages in Fig. 4 that have been identified as being at 'high-extreme' risk in the absence of adaptation measures and

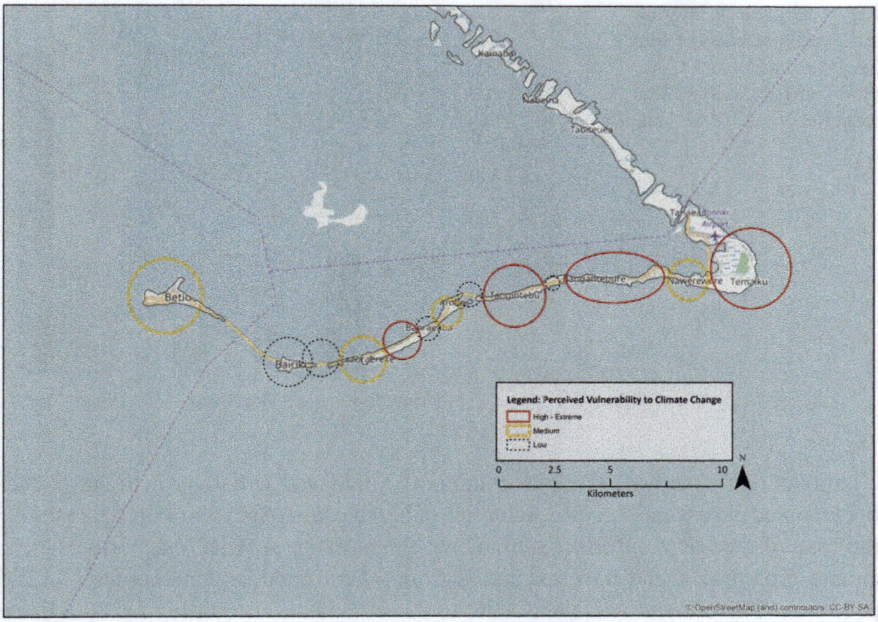

Fig. 4 Vulnerability to climate change, South Tarawa coastal communities. *Source* ArcMap Layout View generated by Author 5 June 2017, adapted from Elrick and Kay (2009) (Based on a subjective assessment or perceived vulnerability to climate change of different coastal communities (villages) by local experts, taking into account climate variables (for example, temperature, wind and sea level), possible changes to climate variables (in particular, sea level rise), potential impacts and resultant risks including community upheavals as a result of impacts on livelihoods, increased maintenance costs of public assets, and increased coastal instability and erosion); Base map data from © OpenStreetMap contributors and available from https://www.openstreetmap.org

coastal protection structures, namely: Antebuka (10.4% p.a.), Eita (2.6% p.a.), Abarao (slow growth of 1.4% p.a.), Bikenibeu (3.6% p.a.) and Temaiku (6.8% p.a.) (Government of Kiribati 2016, p. Table 2).

Unplanned and unauthorized settlement typologies and increasing expansion into hazardous locations similarly have been documented in greater Honiara where more than 35% of the urban population lives in 30 or more informal squatter settlements that are growing at a rapid 6% annually (UN-Habitat 2012). The informal settlements (Temporary Occupation License areas) have been recorded by Reuben (2013) as having expanded tenfold from an area of 71 ha in 1984 to 721 ha in 2010. He goes on to point out that approximately 20% of these settlements are located in low-lying flood plain of the Lungga River Delta or along the coastline (such as Lord Howe/Mamana Water and Fishing Village) and that over the past three decades, the settlements have continuously expanded unguided and onto unsafe lands susceptible to flooding (in the east of the city) and landslides (central and western parts of the city), as tracked in Fig. 5.

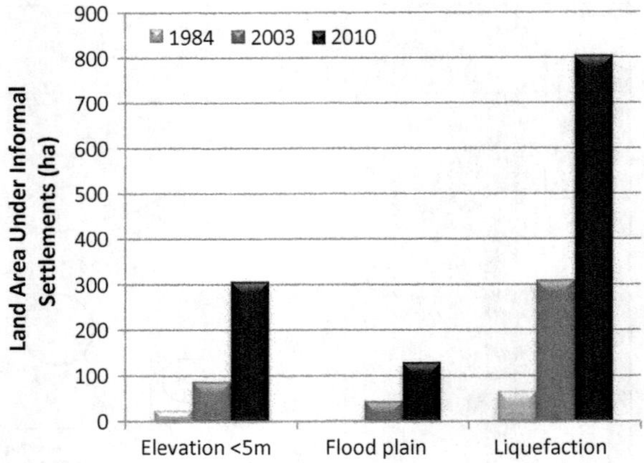

Fig. 5 Growth of informal settlements in areas of high hazard risk, Honiara City (1984–2010). *Source* Data from Reuben 2013, p. Table 2

Similar case studies of rapid and poorly managed urban growth have been documented across the Pacific, including in Nuku'alofa, Apia and Port Moresby. In the case of the latter, informal settlements are growing at twice the 'formal' city's annual growth rate and have expanded from 34 informal settlements with 11,270 people in 1980 (Chand and Yala 2006) to 79 settlements accounting for 40% of the city's population by 2013 (UN-Habitat 2013).

In the face of ongoing urban growth and visibly uncontrolled expansion of the cities into high risk areas, a small number of the Pacific Island countries are preparing and adopting national urban policies. The *National Urbanization Policy for Papua New Guinea 2010–2030* for example, is aimed at better managing the country's urban growth and sets out plans to upgrade unplanned settlements, develop new greenfield serviced subdivisions, enhance the economic base of the country's secondary towns, and to strengthen urban management capabilities (Government of Papua New Guinea 2010). The Policy is brief on hazard risk analysis or the impacts of climate change and it's effectiveness has been questioned for lack of action to date in actual implementation (Keen and Barbara 2015), although the PNG Government's Office of Urbanization recently has announced the launch of two pilot serviced subdivisions targeting middle income public sector employees. Samoa too has adopted a *National Urban Policy*—an urban design-type guide aimed at controlling development within the City of Apia to be 'more workable, livable, inclusive, competitive and sustainable' (Government of Samoa 2013, p. 3).

Fiji stands out in the Pacific for its adoption of a well-articulated *National Housing Policy* with the vision of providing 'Affordable and Decent Housing for All' (Government of Fiji 2011). The Policy encompasses an ambitious and broad range of measures to address Fiji's housing challenges, including housing assistance for rural villagers, providing security of tenure in squatter/informal settlements, coordinated provision of trunk infrastructure to open up new land for affordable housing, provision of housing finance, and regulatory reforms. Under the active stewardship of the Department of Housing and a dedicated Squatter Resettlement (sic, in practice Upgrading) Unit in the Ministry of Local Government, Housing &

Environment, the Fiji Policy is arguably the only pro-poor and 'living document' being implemented in any of the Island countries to date. Notwithstanding these initiatives, the challenge is enormous, as an estimated 20% of Fiji's urban population live in about 212 informal settlements, largely in and around the rapidly growing Lami-Suva-Nausori and Nadi-Lautoka-Ba urban corridors and in Labasa town. The settlements on state land (including settlements established 40 or more years ago) tend to be located within town boundaries; settlements on *i-Tuakei* land are found both within urban areas and in the less regulated peri-urban areas; and there are a number of settlements on private freehold land, each requiring different approaches towards negotiations (different types of landownership) and ultimate regularization (Shaw et al. 2016). Solomon Islands and Vanuatu too, have recently established Task Groups to investigate and recommend options for addressing the need for safe and affordable urban land and housing.

Entry Points for Breaking the Cycle of Deterioration

There are a number of reasons why countries in the Pacific might consider undertaking regularization of the unplanned and extra-legal settlements that are located in areas of low or 'acceptable' hazard risk. Firstly, a 'human rights based approach' has gained new currency with the adoption in 2016 of the *New Urban Agenda* at the UN Conference on Housing and Sustainable Urban Development (Habitat III) in Quito, Ecuador. The Agenda promotes the '**Right to the City**', described as the "equal use and enjoyment of cities and human settlements, seeking to promote inclusivity and [ensuring] that all inhabitants, of present and future generations, without discrimination of any kind, are able to inhabit and produce just, safe, healthy, accessible, affordable, resilient and sustainable cities and human settlements to foster prosperity and quality of life for all" (UN General Assembly 2016).

A second reason is that by not recognizing the settlements and withholding the provision of basic infrastructure services such as potable water, sanitation, drainage, and solid waste collection to large parts of the cities and towns, **public health** is put at risk. Port Moresby, for example, has the highest incidence of tuberculosis in the country as a result of urban overcrowding (John 2015 cited by Keen and Barbara 2015) and according to one public health officer working in Tarawa, under present circumstances "the people on South Tarawa will die of preventable diseases long before sea level rise".[9]

A third compelling reason is that relating to the skewed impacts in **economic damage and losses** as a result of disasters. The damage incurred by the April 2014 Mataniko River flooding event in Solomon Islands, the 2015 Tropical Cyclone Pam in Vanuatu and the 2016 Tropical Cyclone Winston in Fiji all resulted in far heavier damage to the housing stock in the urban informal settlements than in adjacent

[9]Personal communication with World Health Organization liaison officer, May 2007 cited in (Butcher-Gollach C., 2012).

neighborhoods of the 'formal' cities' (Government of Solomon Islands 2014; Government of Vanuatu 2015; Government of Fiji 2016). This trend is not only a consequence of poverty in the settlements but also as a result of the well-documented fact that households are unwilling to invest personal savings and resources into houses if they lack security of tenure. International experience in post disaster reconstruction calls on governments to invest in pre-disaster risk reduction measures as both effective and cost-efficient—for every one dollar spent on prevention and building resilience before an event, four dollars is saved in recovery (IFRC 2016).

Approaches for Addressing Urban Informality

In the face of rapid urban growth rates, high hazard risk settings both now and increasing into the future, and the multi-dimensional vulnerability of the urban poor, what might be appropriate planning and urban management responses to address the deteriorating living conditions in the coastal towns and cities of the Pacific Islands, to promote social and economic inclusion, and to make room for urban population growth in safer locations?

Better Understanding of Location Specific Climate Risk

In its guidance to countries on the selection of adaptation measures, IPCC-13 actively encouraged the comparison of options at critical stages and cautioned against measures that might inadvertently result in maladaptation. However, the spatial plans used in the PICs (master plans, local plans, town planning schemes, etc.) are generally not well risk informed. Given the large number of variables involved, long-term climate simulations are calculated at relatively coarse scales of around 100–200 km grid size and downscaling the low resolution, global climate models to the higher resolutions needed for urban ward or settlement-level planning requires high quality data sets that are not readily available for the small island countries (Butcher-Gollach 2015). As such, the plans, where they do exist and are used, rely on limited and subjective assessments of risk and fall back on coarse 'rules of thumb' such as establishing 50 foot (15.24 m) buffer zones and setbacks from river courses and coastal foreshores. Typical climate change adaptation and disaster risk reduction strategies that are promoted by the planning agencies relie on the continuum of Do Nothing—Protect—Elevate—Retreat. Yet, for the majority of low income earners with limited location options, precarious tenure rights and lacking in personal resources, the option to retreat from a roughly proscribed and prohibited zone invariably results in loss of livelihoods, severe personal hardship, and is not easily enforced by the local authorities.

Under such circumstances and he high hazard risk faced by most ofthe PICs, the local planning agencies require dedicated and specialist scientific and engineering advice to (i) understand and quantify the climate change and natural disaster risk of specific locations; (ii) map and assess the risks and climate adaptation or risk reduction options; (iii) screen the level of acceptable risk given the probability of different hazards over different return periods; and (iv) identify a range of technically feasible low-regrets options with associated economic and social costs.[10] The technical judgement on acceptable risk is based on the likelihood of the hazard occurring and the resulting consequences in light of different community's vulnerability circumstances. The nature and size of the risk will be different in different towns and within different parts of each town and would call for different design solutions or interventions. For example, some risks, such as river flooding and coastal in inundation, might be localized and adaptation measures could include larger plot sizes so that buildings can be set back at a safe distance or, where this is not possible, for land in alternative lower risk locations to be made more easily available. Other risks, such as strong winds or cyclones can be managed to a degree by households and local builders being trained in low cost and affordable construction techniques to strengthen existing houses (Beca International Consultants Ltd; GNS Science; National Institute of Water and Atmospheric Research (NIWA) 2015).

Smarter Use of Spatial Planning Instruments to Guide Land Development

Statutory 'master plans' are the most commonly relied on planning instruments in the Pacific Island town planning systems and legislation, ranging from the (draft) Master Plans for Port Vila and Luganville to the Detailed Land Use Plans for the larger urban villages in South Tarawa. A Master Plan usually covers more than one neighborhood (it can be prepared to cover an entire city or a part of a city) and is meant to identify the relationships between various land uses, road and transportation routes, residential areas, social and community facilities, and open space corridors, including land reservations and easements for drainage and main infrastructure services. A Master Plan is a static, normative planning instrument - it usually consists of a map showing the desired, end-state (20 years) land use norms or zones of permitted, prohibited and permitted with special consent land use classes and an accompanying written statement. The use of master or comprehensive plans in rapidly growing cities in developing countries has long been questioned on the grounds that such plans do not adequately address implementation issues or take into account the complexity of land markets, are silent on the

[10]"Low-regrets" options include measures by households, communities, and local/national agencies that can be justified from economic, social, and environmental perspectives even if extreme weather events or climate change impacts take place and even if there is uncertainty about future conditions (OECD 2009).

role of public versus private sector actions, and fail to make linkages between spatial and financial planning (Farvacque and McAuslan 1992).

In order to better address the pressing questions regarding what are the future residential/housing land needs for low, middle and high income earners and where and how these needs can be met, i.e. be both safe from natural hazards and affordable, more appropriate use could be made of broader level, strategic (or concept) spatial plans that can be prepared at lower cost. Examples would include the use of Urban Reference Plans and Urban Grids. These are simplified, non-statutory planning documents that set out broad development options and outline the major urban road networks for future investment in areas most susceptible to the unplanned urban sprawl, so as to anticipate and channel the direction of future growth over a 5–10 year period. Ideally, in the vulnerable coastal settings, the reference or guide plans should identify not only 'no-build' zones but also assess and identify land that would be or could be made (engineered) to be safe for future expansion. For example, the Government of Vanuatu is currently preparing Guided Land Development Options for Greater Port Vila showing: administrative boundaries and urban services boundaries, main transportation and drainage networks, main employment nodes, low hazard lands, high hazard areas, land ownership characteristics, and indicative residential densities for future expansion, and low serviced existing settlements for regularization and in situ upgrading.

Address Administrative and Regulatory Bottlenecks in Accessing Land

The length of time taken and costs to access land through formal administrative procedures play as strong a role in shaping the direction of urban expansion as the spatial or land use zoning plans. Table 4 shows the number of processing steps and average length of time it takes to transfer an already surveyed and registered and correctly zoned property in a number of the Island cities.[11] It can be seen that the time taken can range from an exceptionally rapid three weeks in Apia to up to five months in Nuku'alofa and up to two years in Tarawa. Clearly there is room to improve and streamline the administrative steps involved to legally access land in many of the towns.

These times and costs are optimistic in that they assume that the property already has been incorporated into the formal land registration system of the country. Other than Apia, Honiara and Suva, there have been little to no formal new serviced subdivisions other than or high income earners, for many decades. The process for acquiring formal registration for a new lease/new subdivided property are not

[11]The steps are similar in most countries and include: payment of stamp duties; checking for encumbrances; obtaining Ministerial or Cabinet consent to transfer the lease; and applying for and obtaining registration of a deed of transfer or other form at a Department of Lands.

Table 4 Ease of formally registering a property (by city). *Source* The World Bank Group (2017a)

City	No. Procedures	Time (days)	Cost (% of property value)	Quality of land admin index (0–30)
Apia	5	15	3.8	12.5
Port Vila	4	58	7	18.5
Suva	4	69	3	19.5
Port Moresby	4	72	5.2	4.5
Majuro	5.4	74.4	4.5	13.4
Pohnpei	5.4	74.4	4.5	13.4
Honiara	10	86.5	4.7	11
Nuku'alofa	4	112	15.1	17
Tarawa	5	513	n.a.	9

routinely monitored by the countries but, by way of example, according to a recent study in Port Vila, the process would take around 632 days (28 months).[12]

In the case of a new and low income housing market entrant, s/he would be unlikely to be able to afford the lease fees/rentals of an already registered and serviced plot and could not afford to wait many months to follow the formal registration procedures. Other than renting one or more rooms in the existing housing stock, the low income earner's only alternative would be to obtain permission to construct a dwelling on customary owned, largely unregistered land. Access to land in the informal land market, based on family or kinship ties and nowadays sealed by the payment of 'key-money' and thereafter regular lease payments to the landowner(s) is rapid in comparison to the formal systems—for example, around 42 days in Tarawa (Butcher-Gollach et al. 2008) and a few months in Port Vila (Beca International Consultants Ltd; GNS Science; National Institute of Water and Atmospheric Research (NIWA) 2015). It is therefore not unsurprising that most urban growth in all of the towns is occurring in the unplanned, extra-legal settlements, outside of the influence of the local planning authorities, and with little knowledge or regard by households or customary landowners for the hazard risks of that location.

Whilst 'customary land' rights and responsibilities are put forward as opaque and an intractable barrier to better functioning land markets in the Pacific, it is clear that much more could be done to at least ensure that the formal land registration systems operate more efficiently, that registered urban lands are not underutilized or allowed to be held for speculative purposes, and that serious attempts be made to encourage and incentivize willing customary owners, particularly in safer locations, to enter

[12]Personal communication with former Director of Lands, 28 April 2017.

into some form of recorded/secure tenure for their informal occupants, successful examples of which are being piloted by the authorities in peri-urban Port Vila.[13]

Promote the Use of Prescriptive not Proscriptive Development Consent Approaches

According to Janoff-Bulman et al. (2009), "Prescriptive morality is sensitive to positive outcomes, activation-based, and focused on what we should do. … Proscriptive morality is sensitive to negative outcomes, inhibition-based, and focused on what we should not do".

The main elements of an appropriate, prescriptive urban planning approach would include the use of:

(a) *Passive planning in areas of high hazard risk*—The aim would be to allow on-going existing uses but minimize any infrastructure investments to basic lifeline services (such as clean water supply) and to not encourage or facilitate expansion through improved connectivity and transport links or encourage secure tenure arrangements. This passive planning approach is being carefully implemented by the Public Utilities Board in pilot water improvement zones in South Tarawa and is under consideration by the Department of Housing for informal settlements in high hazard risk locations in Fiji.

(b) *Promote voluntary, incremental retreat away from areas of high hazard risk*—The aim would be to proactively assist affected communities to identify vacant land in locations close to employment or livelihood opportunities and to facilitate access to that land in a timely way through negotiated security of tenure with landowners as, for example, the plans for the April Ridge serviced subdivision in Honiara.

Promote 'Making-Room on Safe Lands' Urban Planning Rather Than 'Containment' Planning

In rapidly growing urban areas, it is urgently necessary to plan ahead of growth rates and to open up new greenfield sites for planned urban expansion (Angel et al. 2012), particularly in safer locations that are on well serviced public transport routes. At the city and donor levels, this can be supported by prioritizing city-wide primary/trunk infrastructure investments (roads/transport links, power and water)

[13]For example through public private sector partnerships (including in some cases, negotiated land readjustment) where landowners provide land as equity and the public sector provide bridging or subsidy finance for infrastructure services.

to open up new land for urban growth for the lower-middle income groups who are able to afford associated transport costs. Recent positive examples of this include the planning of April Ridge in Honiara (Government of Solomon Islands 2015) and the upgrading and surfacing of the Temaiku ring road to the east of South Tarawa which will allow for large-scale future urban expansion onto the Temaiku state land, estimated at close to 25,000 people (Butcher-Gollach et al. 2008).

Involuntary Retreat and Relocation Should only be Used as a Last Resort

In cases where voluntary relocation or retreat is unavoidable due to insecure tenure or unacceptably high risk exposure, finding alternative safe **and** well located land (near employment nodes) is a primary consideration. Poor and extremely poor urban residents must live as close as possible to their source of livelihoods as, unlike wealthier groups, the poor cannot afford high transportation costs. Location —if possible within walking or public transport distance of employment and livelihood zones, drives the housing location decisions of poorer groups more than any other element of housing, such as the physical quality of the house or the risk of building on hazardous land, and explains why low income households rebuild after extreme weather events in the same, unsafe locations—because they have no other alternatives.

Conclusion

Regularizing and upgrading the existing poorly serviced urban villages, strengthening the resilience of extra-legal peri-urban informal settlements, and planning for safe urban expansion of the rapidly growing, small coastal cities and towns of the Pacific Island Countries are seemingly complex undertakings. In this paper, it has been emphasized that the task is made even more difficult in the small Island countries, that overall find themselves flying-blind into the face of an uncertain future on almost every variable. Although rapid urbanization has been apace in the PICs for the past three decades or more, basic demographic, household income/expenditure, and land/housing needs and priorities data for the urban and peri-urban populations, and in particular for low income earners, is outdated or not available. The economic roles played by the urban centers are poorly understood and there is a lingering and widespread antipathy towards recognizing the positive effects of agglomeration and the rationale that brings poor people to the cities and towns in search of employment and opportunity.

The PICs are well documented as being amongst the most vulnerable countries in the world to both the long term effects of climate change and to short-term natural disasters. In the Pacific region, the proportion of land area located in low elevation coastal zones is four times larger than that of Asia, and where, when data is available, the urban population living in LECZs is found to be almost double that of the rural populations. In countries where there has been little to no forward planning or investment in service infrastructure to open up new land for many decades, and where registering land in the formal administrative systems is cumbersome, costly and time consuming, the urban low income earners have little option but to rent in the existing housing stock (leading to chronic overcrowding) or to pay for the use of customary land and self-build their houses in unplanned, extra-legal settlements. The trend for unplanned and unauthorized settlements to continuously expand onto hazardous and unsafe land has been documented in many of the Pacific's urban and peri-urban areas. The damage and losses incurred by extreme weather events in the recent past all point to far heavier damage to the housing stock of the urban poor living in urban informal settlements than in adjacent formal city neighborhoods.

There are a number of promising approaches that could be implemented to upgrade existing settlements that are located in areas of low or 'acceptable' hazard risk and to plan ahead for urban expansion in safe and affordable locations. However, to do so, we will require new 'ways of seeing' and a lexicon that extends beyond the conventional mantra of 'Prohibit and Retreat'. The paper has set out are a number of entry-points that justify the need for intervention and that may vary depending on the particular country, ranging from the 'rights to the city' as promoted by UN Habitat (for example, Fiji and Samoa), to addressing public health risks (for example, Tuvalu and Kiribati), to reducing post-disaster reconstruction costs through ex-ante planning and risk reduction investments (for example, Vanuatu and Solomon Islands).

The approaches that then could be implemented to address the deteriorating living conditions and high hazard risk settings of the urban poor in the Pacific's rapidly growing coastal cities and towns need not be complex or costly. The paper has set out six key approaches. Firstly, invest time and effort in better modeling and understanding the climate risks of specific locations and to ensure that options are prepared and weighed at critical stages. Secondly, making smarter use of spatial instruments to guide land development. This may require discarding the static, normative master planning approach and making better use of plans that inform decision-making, in particular investments in new infrastructure to open up safer lands in well located areas of the city and close to employment nodes for low income earners. Thirdly, a suggestion was made for the Lands and Planning agencies to critically analyze the length of time taken and costs involved for a low income earner to access and obtain formal registration/leasehold for land on which to build a house, and to simplify and streamline administrative procedures where feasible. Fourthly, recognizing the limited resources available for proscribing and enforcing land use zoning plans and building construction codes, available resources could be concentrated on identifying vacant lands (either large blocks or infill) with secure tenure in safer locations and facilitating voluntary retreat by one or more households away from high hazard areas whilst at the same time, minimize investments to basic lifeline services only to settlements located on high risk lands. Fifthly, proactively promote the planning, protection and servicing of new land, affordable by low to middle income households for self-build housing. Finally, to resort to involuntary relocation only as a last resort and after other options have been discussed and rejected with the affected community.

Overall, an inclusive and risk-informed approach would encompass spatial planning, administrative reforms, and financing options to meet the affordability levels and location priorities of the urban poor.

Reference

Angel, S., Parent, J., Civco, D., & Blei, A. (2012). *Making room for a planet of cities.* Cambridge, MA: Lincoln Institute of Land Policy.

Antarctic Climate & Ecosystems Cooperative Research Centre. (2008). *Position analysis: Climate change, sea-level rise and extreme events: Impacts and Adaptation Issues.* Hobart: Antarctic Climate & Ecosystems Cooperative Research Centre. Retrieved January 25, 2017 from http://canute2.sealevelrise.info/access/repository/resource/a47f6934-6202-102f-a3d0-40404adc5e91/PA01_SLR_FIN_090616.pdf.

Baker, A. J., & Week, D. (2011). *Infrastructure and climate change in the Pacific.* Department of Climate Change and Energy Efficiency; Australian AID. Canberra: Government of Australia. Retrieved May 19, 2017 from https://www.environment.gov.au/system/files/resources/1f81b90c-1581-4460-865a-7b7306f7f528/files/infrastructure-report.pdf.

Bassett, T., & Scruggs, G. (2013). *Water, water everywhere: Sea level Rise and land use planning in Barbados, Trinidad and Tobago, Guyana, and Pará.* Lincoln Institute of Land Policy: Washington D.C.

Beca International Consultants Ltd; GNS Science; National Institute of Water and Atmospheric Research (NIWA). (2015). *Urban growth trends report—Port Vila and Luganville: Risk mapping and planning for urban preparedness.* Port Vila: Vanuatu Meteorological and Geohazard Department; The World Bank Group—Japan PHRD TF012184.

Birkmann, J., & Welle, T. (2015). The World Risk Index—An approach to assess risk and vulnerability on a global scale. *Journal of Extreme Events, 2*(1).

Butcher-Gollach, C. (2012). Health and the city: Consequences of the 'King Tides' of Urbanization in Kiribati, Central Pacific. *Peer reviewed paper presented at the 6th Urban Research and Knowledge Symposium October 8–10, 2012.* Barcelona, Spain: The World Bank. Retrieved June 22, 2017 from http://siteresources.worldbank.org/INTURBANDEVELOPMENT/Resources/336387-1369969101352/Butcher-Gollach.pdf.

Butcher-Gollach, C. (2015). Planning, the urban poor and climate change in the Small Island Developing States (SID): Unmitigated disaster or inclusive adaptation? *International Development Planning Review, 37*(2), 225–248.

Butcher-Gollach, C., Banes, C., Jones, P., Liousse, C., & Onorio, K. (2008). *Kiribati urban renewal scoping study. Vol. 2—Recommendations.* Wellington: The New Zealand Aid Programme.

Butcher-Gollach, C., Nari, R., & Harrison, M. (2017). *Preliminary screening of informal settlements in Greater Port-Vila: Options for guided land development for affordable and resilient urban expansion.* Port Vila: unpubd.

Chand, S., & Yala, C. (2006). *Informal land systems within urban settlements in Honiara and Port Moresby.* Making Land Work—Volume Two Case Studies #5.

Deichmann, U. (2016, December 19). Revisiting Economic Geogrpahy. (DEC Policy Research Talk, Interview).

Elrick, C., & Kay, R. (2009). *Mainstreaming of an integrated climate change adaptation based risk and response process into Government of Kiribati: Final report.* Report prepared for the KAP Project, Phase II. Perth: Coastal Zone Management Pty Ltd. Retrieved May 19, 2017 from http://www.climate.gov.ki/wp-content/uploads/2013/05/Mainstreaming-of-an-Integrated-CCA-Based-Risk-Diagnosis-and-Response-Process.pdf.

Farvacque, C., & McAuslan, P. (1992). *Reforming urban land policies and institutions in developing countries*. Washington, DC: World Bank Publications.

Glaeser, E. (2011). *Triumph of the city*. New York: Penguin.

Glaeser, E. L., & Xiong, W. (2017). Urban productivity in the developing world. *National Bureau of Economic Research*. Retrieved October 25, 2017 from http://papers.nber.org/tmp/7739-w23279.pdf.

Government of Fiji. (2011). *The National housing policy: Affordable and decent housing for all.* Suva: Ministry of Local Government, Urban Development, Housing and Environment.

Government of Fiji. (2016a). *Post disaster needs assessment: Tropical Cyclone Winston, February 20, 2016*. Suva: Government of Fiji, Strategic Planning Office.

Government of Fiji. (2016b). Revitalization of bure building skills in the district of Nanisu in Ovalau, Dec 2008. *Workshop Report on Developing Safeguarding Plans for Intangible Cultural Heritage (ICH) in Fiji.* Suva: Ministry of Education, Heritage and Arts; Ministry of iTaukei Affairs; UNESCO.

Government of Kiribati. (2007). *Analytical report on the 2006 Kiribati household income and expenditure survey*. Bairiki, Tarawa: Ministry of Finance National Statistics Office.

Government of Kiribati. (2012). *Republic of Kiribati Island Report Series: 6. South Tarawa.* Bairiki, Tarawa: Office of Te Beretitenti.

Government of Kiribati. (2016). *2015 population and housing census—Preliminary report.* Bairiki, Tarawa: National Statistics Office, MFED.

Government of Papua New Guinea. (2010). *National Urbanization Policy for Papua New Guinea.* Port Moresby: Government of Papua New Guinea.

Government of Samoa. (2008). *Household income and expenditure survey tabulation report*. Apia: Government of Samoa Bureau of Statistics.

Government of Samoa. (2013). *Samoa National Urban Policy*. Apia: Planning and Urban Management Agency.

Government of Solomon Islands. (2006). *Household income and expenditure survey 2005/6.* Honiara: Ministry of Finance & Treasury, Solomon Islands Statistics Office.

Government of Solomon Islands. (2014). *Solomon Islands: Rapid assessment of the macro and sectoral impacts of flash floods in the Solomon Islands, April 2014*. Honiara: Ministry of Development Planning and Aid Coordination; Ministry of Finance and Treasury; The World Bank.

Government of Solomon Islands. (2015). *Honiara's future: An investigation to identify land for urban expansion, main report.* Honiara: Ministry of Lands, Housing & Surveys.

Government of Tonga. (2009). *Househomd income and expenditure survey: report.* Nuku'alofa: Government of Tonga Statistics Department.

Government of Vanuatu. (2009). *2009 National Population and Housing Census*. Port Vila: Vanuatu National Statistics Office.

Government of Vanuatu. (2015). *Post disaster needs assessment: Tropical Cyclone Pam, 2015.* Port Vila: Vanuatu Prime Minister's Office.

IFRC. (2016). *World Disasters Report. Resilience: saving lives today, investing for tomorrow.* International Federation of Red Cross and Red Crescent Societies PS Centre publication. Available online: http://ifrc.org/.

Janoff-Bulman, R., Sheik, S., & Hepp, S. (2009, March). Proscriptive versus prescriptive morality: Two faces of moral regulation. *Journal of Personality and Social Psychology, 96*(3), 521–537.

John, L. (2015). *Challenges of TB Control in PNG.* 13th National Rural Health Conference, Darwin, 24–27 May.

Jones, C. I., & Klenow, P. J. (2016). Beyond GDP? Welfare across countries and time. *American Economic Review, 106*(9), 2426–2457.

Keen, M., & Barbara, J. (2015, February 25). *Pacific urbanisation: Changing times*. Australian National University. Retrieved May 22, 2017 from DEVPOLICYBLOG: http://ssgm.bellschool.anu.edu.au/sites/default/files/publications/attachments/2016-07/ib2015.64_keen_and_barbara.pdf.

Marango, T. (2016, January 12). *2016 Snap General Election Candidates*. Retrieved from Vanuatu Daily Post: http://dailypost.vu/news/snap-general-election-candidates/article_ab461322-7a95-55d2-9178-decc9bb7bc20.html.

Marshall, A. (1890). *Principles of Economics*. London: Macmillan; reprinted by Prometheus Books.

McGranahan, G., Balk, D., & Anderson, B. (2007). The rising tide: assessing the risks of climate change and human settlements in low elevation coastal zones. *Environment and Urbanization, 19*(1), 17–37.

Milbank, A. (2015). The Samoan fale and disaster risk reduction in the contemporary built environment. In *Proceedings of the Workshop on Revitalization of Indigenous Architecture and Traditional Building Skills* (pp. 64–78). Apia: Government of Samoa; UNESCO; CRIHAP.

Morris, M. (2011). *Measuring poverty in the Pacific*. Canberra: Crawford School of Public Policy, The Australian National University. Retrieved from https://devpolicy.anu.edu.au/pdf/papers/DP_9_-_Measuring_Poverty_in_the_Pacific.pdf.

Narsey, W. (2008). *The quantitative analysis of poverty in Fiji*. Suva: Vanuavou Publications.

OECD. (2009). *Integrating climate change adaptation into development co-operation: Policy guidance*. Paris: Secretary General of the OECD.

Pacific Institute of Public Policy. (2011). *Urban hymns: Managing urban growth*. Discussion Paper #18, Port Vila.

Rallu, J. (2009). Urban drift, urban growth, urban youth. In M. Rao, *A new generation of youth lifestyle. Influence and impact* (pp. 68–96). Hyderabad, India: Icfai University Press. Retrieved April 22, 2017 from http://www.unescap.org/EPOC/documents/R3.13_Study_3.pdf.

Reuben, R. (2013). *Spatial analysis of informal settlement growth and disaster management preparedness: A case study in Honiara City, Solomon Islands*. Pacific Center for Environment and Sustainable Development. The University of the South Pacific.

Shaw, V., Butcher-Gollach, C., Saito, K., Southcombe, D., & Basalusalu, T. (2016). Housing Sector. In *Fiji Post-Disaster Needs Assessment—Tropical Cyclone Winston, February 20, 2016* (pp. 73–77). Suva: Government of Fiji, Strategic Planning Office.

Szoldra, P. (2014, February 26). *Stunning map shows what a climate change nightmare scenario might actually look like*. Retrieved February 28, 2014 from www.businessinsider.com.au/climate-change-map-2014-2.

The Pacific Community (SPC). (2016, June). *PACIFIC ISLAND POPULATIONS—Estimates and projections of demographic indicators for selected years*.

The World Bank Group. (2017a). *Ease of doing business in ... registering property*. Retrieved May 22, 2017 from http://www.doingbusiness.org/data/exploreeconomies/kiribati#registering-property.

The World Bank Group. (2017b). *World development indicators*. Washington, DC: The World Bank. Retrieved May 1, 2017 from http://data.worldbank.org/products/wdi.

UN General Assembly. (2016). The New Urban Agenda. *Adopted at the United Nations Conference on Housing and Sustainable Urban Development (Habitat III)*. Quito. Retrieved May 15, 2017 from https://habitat3.org/the-new-urban-agenda/.

UNDESA. (2016). *Policies on spatial distribution and urbanization: Data booklet*. UN Department of Economic and Social Affairs, Population Division. Retrieved from http://www.un.org/en/development/desa/population/publications/pdf/policy/Data%20Booklet%20Urbanization%20Policies.pdf.

UNESCAP Statistical Division. (2015). Statistical Yearbook for Asia and the Pacific. Bangkok: United Nations ESCAP. Retrieved April 19, 2017 from www.unescap.org/stat/data/.

UNFCCC. (2007). *Climate change: Impacts, vulnerability and adaptation in developing countries*. Bonn: UNFCCC.
UN-Habitat. (2012). *Honiara City climate change vulnerability and adaptation assessment.* Suva.
White, S. C., Gaines, S. O., Jr., & Jha, S. (2012). Beyond subjective well-being: A critical review of the Stiglitz report approach to subjective perspectives on quality of life. *Journal of International Development, 24*(6), 763–776.

Author Biography

A/Prof. Colleen Butcher-Gollach holds an Honorary Appointment as Principal Fellow in the Faculty of Architecture, Building and Planning, University of Melbourne, Australia. As Associate (Adjunct) Professor since 2010, she has ongoing responsiblities for post-graduate teaching on 'Cities Without Slums' and related research. Colleen also works as a Staff Consultant in the World Bank's Social, Urban, Rural and Resilience Global Practice and holds appointments on the Programme Design Support Services Panel on International Development for The New Zealand Aid Programme, Ministry of Foreign Affairs and Trade and as a Senior Urban Development Expert on the Pacific Region Infrastructure Facility (PRIF) Advisory Panel of the Asian Development Bank. For the past 12 years in particular, she has worked on numerous infrastructure, land and housing delivery, and post-disaster needs assessments and reconstruction projects in Asia and eight of the Pacific Island Countries. The views expressed and any errors in fact or judgment in the paper are the responsibility of the author alone.

At the Frontline of Climate Change: Adaptation, Limitations and Way Forward for the South Pacific Island States

Dhrishna Charan, Kushaal Raj, Ravneel Chand, Lionel Joseph and Priyatma Singh

Abstract The indicators of climate change pose multiple challenges for the inhabitants of the Pacific Island Nations. The remoteness and the limited resource base of the South Pacific Island states makes them particularly vulnerable to the adverse effects of climate change. The response to sea level rise and extreme weather events due to climate change in the South Pacific region targets on adaptation in preference to mitigation. Over the past decade, community-based climate change adaptation has gained momentum and several successful projects have been implemented and evaluated. However, several publications still exist as grey literature. This paper attempts to congregate current best-practice in climate change adaptation, primarily focusing on community and ecosystem based approaches. Based on the available literature and case studies, widespread adaptation methods to overcome the vulnerability and conceivable impacts of climate change in the South Pacific region is determined. Conversely, with the evidence available, there exists some barriers that impede the reinforcements and implementations of climate change based adaptation strategies. Limits to adaptation lie within the South Pacific society, hence, are restrained by lack of knowledge, policies, funding, human resource constraints and ethics. The paper concludes by providing recommendations and the way forward for community-based climate change adaptation initiatives in the South Pacific region.

Introduction

South Pacific region is composed of fifteen countries categorized by the United Nations as Small Island Developing States (SIDS). All the fifteen countries are moderately constrained and highly vulnerable to the impacts of climate change (Barnet 2008). Climate change projections for the South Pacific Islands indicate

D. Charan · K. Raj · R. Chand · L. Joseph · P. Singh (✉)
Department of Science, The University of Fiji, Saweni, Lautoka, Fiji
e-mail: PriyatmaS@unifiji.ac.fj

W. Leal Filho (ed.), *Climate Change Impacts and Adaptation Strategies for Coastal Communities*, Climate Change Management,
https://doi.org/10.1007/978-3-319-70703-7_4

high ecological, economic and social implications. According to satellite data captured by NASA (Lynch 2016), trends in both the average global temperature, and the loss of the Arctic ice caps has been broken numerous times in 2016. Although, it is evident that climate change is prevalent globally, Pacific Island Countries (PICs) are considered to be at the "global front line"(Commission on Climate Change and Development 2009) of climate change. The South Pacific Island nations are the most vulnerable groups in the entire Pacific region (Secretariat of the Pacific Regional Environment Programme (SPREP 2016). The South Pacific Islands are susceptible to sea level rise affecting the low lying coral islands, however, the impacts are not limited to sea level rise, but include other associated effects such as storm surges, coastal erosion, inundation, changes in seasonal precipitation trends leading to drought, threats to water resources and agriculture and heightened incidents of food security issues due to extreme events (Gero et al. 2011).

The indicators of climate change pose multiple challenges for the inhabitants of the PICs. The remoteness and the limited resource base of the South Pacific Island countries make them particularly vulnerable to the adverse effects of climate change. Subsequently, the frequency of natural disasters challenge the investment used for development in PICs and the route to generating sustainable growth and adaptation become demanding. Presently, adaptation to climate change has gained increasing attention globally. The United Nations Framework Convention on Climate Change (UNFCCC) and other developed nations continue to provide aid to the developing regions. Past and current research has presented sufficient evidence that anthropogenic greenhouse gas (GHG) emissions are changing the climatic structure of earth, and yet, there remains inadequate information, international agreement and technology to participate in direct reduction of GHG emissions (IPCC 2007). Hence, adaptation is essential in light of international response to climate change (Van Aalst et al. 2008).

More so, community-based climate change adaptation approach is considered to be an effective means of coping with the adverse impacts of climate change. Community based approach constitutes trust and enhances participation; therefore, probability of success is high. Furthermore, transect walk is an effective manner to gain confidence of the local people, where researchers walk through the community with the locals, recording significant hazards induced by climate change(Van Aalst et al. 2008). In contrary, reaching out to a colossal number of communities is a human resource challenge; further impeding researchers who will be required to spend considerable time within the community. Another complication lies in the question; do the community members possess the knowledge, skills and technology to adapt to climate change? This does not only affect the outcomes of the multi-million dollar projects but also questions the sustainability aspects thereafter. However, despite the complications and the challenges, the local communities have the right to be informed about the consequences of climate change, as well as appropriate adaption strategies in relation to the impacts of climate change (Rojas Blanco 2006).

In addition, numerous projects are carried out in the Pacific to aid the recovery and revival of the most vulnerable sectors impacted by climate change. The initiative taken by development partners such as the United Nations, European Union, United States Agency for International Development (USAID), Department of Foreign Affairs and Trade (AusAID), Global Environmental Facility, Asian Development Bank and the World Bank (Pacific Climate Change Portal 2016), has been widely acknowledged for their contributions in improving the livelihood of the Pacific Islanders through various community and ecosystem based approaches. Thus, it is vital to examine the outcomes of the previous projects and to learn from its shortcomings. This paper will provide an overview of the major community based projects that have been implemented in the South Pacific region. This review also examines the common drawbacks of work that has been carried out in the five broad areas of: Climate Change Capacity-building, Awareness and Training, Water Resource Management, Disaster Risk Reduction (DRR), Coastal Zone Management & Climate Induced Relocation, and Food Security and Agriculture. The paper concludes by explicating the lessons learnt and establishing avenues for the way forward.

Methodology

This paper draws its information primarily from published reports, journal articles and other climate related documents. The search for literature was limited to publications in English. There was vast number of projects that were carried out in the region over the past few decades. Our research focused on major regional projects carried out and completed in the South Pacific region in the past fifteen years. These documents were mostly sourced from papers and reports produced by various national and international institutions and the websites of regional organisations having carried out these various adaption projects. Government websites and donor website were also explored as sources of information.

Results/Discussion

Climate Change Capacity-Building, Awareness and Training

Capacity-building, awareness, education and training are the stepping stones towards climate change adaptation programmes for almost all the South Pacific Nations. Many climate change adaptation projects demanded adequate prior knowledge of the impacts of climate change and it became a prerequisite to include capacity-building exercises and awareness into the adaption projects for appropriate action on climate change (Nunn 2009).

Capacity-Building for the Development of Adaptation Measures in Pacific Island Countries (CBDAMPIC) Project was implemented across 16 communities in Cook Islands, Fiji, Samoa and Vanuatu. The Final Report prepared for the Canadian International Development Assistance stated that the prime objective of the project was to carry out capacity-building activity that would enhance the Pacific Island countries institutions and people's capacity to reduce the impacts of climate change. Deliberations at Conference of Parties 1 (COP 1) specified 3 phases of adaptation. Phase 1 established planning and identification of vulnerable areas and policy implications for adaptive strategies and capacity-building; Phase 2 further emphasized capacity-building for adaptation and Phase 3 demonstrated adequate actions and measures for appropriate adaptation strategies (Cosbey et al. 2005). CBDAMPIC was the first project to proceed to Phase 3. This project successfully improved the receptivity of climate change adaptation issues for government administrators, policy makers and the local communities. One of the key outcomes of the project was the enhancement of the awareness levels of climate change adaptation at community level.

Cook Islands utilized television programs and climate change associated school competitions, while Samoa opted for innovative strategies of emphasizing the need for climate change adaption which included songs, dances and dramas. Capacity-building exercise such as workshops and seminars proved useful for government administrators and policy makers. In fact, Fiji has developed a climate change awareness tool kit which is being used by policy makers. Vanuatu conducted systematic awareness programmes that targeted specific audience. The contents of the awareness materials were designed specifically to meet the needs of the target groups such as Politicians, school children and local community members. Vanuatu also created awareness programmes through television, drama, pamphlets/booklets and newsletters (Nakalevu 2006). The project increased the capacity of the government administrators and policy makers to integrate climate change adaptation into national policies and planning.

Another similar project namely; Pacific Islands Framework for Action on Climate Change (PIFACC) is a regional framework established for Pacific Island Countries to enhance the capacity of people to the manage the impacts of climate change. As stated in the Report by the Secretariat of the Pacific Regional Environment Programme, the PIFACC project facilitated the production of an online tool called the Pacific Climate Change Portal (PCCP) and was successful in promoting climate change knowledge across PICs. Other key outcomes of the framework included increased capacity of key technical officers to properly advice decision makers, effective dissemination of relevant climate related information and more training and scholarship opportunities in climate science (Secretariat of the Pacific Regional Environment Programme 2015). Additionally, a project titled "Supporting the Regional Management of Climate Change Information in the Pacific (Pacific iCLIM)" was conducted in the three pilot countries of Fiji, Tonga and Vanuatu and focused on achieving milestones in the area of technical and capacity-building to efficiently manage climate change information and developed tools to interpret complex climate related data for policy makers (Pacific iCLIM 2014).

Coastal Community Adaptation Project (C-CAP) accomplished national-level climate change activities in nine Pacific Island countries including Fiji, Kiribati, Nauru, Papua New Guinea (PNG), Samoa, Solomon Islands, Tonga, Tuvalu, and Vanuatu. The key objective was to enable coastal communities to build their resilience towards climate change. The programme enriched capacity-building through several community-based trainings in the nine regional countries. Kiribati Adaptation Program Phase III (KAP III) was implemented following the successful implementation of Phases I and II. Part of KAP III achievements encompassed increased capacity of the communities to cope with the impacts of climate change which was made possible through a series of educational programs (Pacific Climate Change Portal 2016).

It is evident that climate change awareness and education is a critical enabling factor in almost all the projects implemented across the South Pacific Island countries. It has remarkably upgraded climate change adaptation knowledge at local community level as well as institutional level. Improved institutional capacity and community participatory engagements trigger appropriate behavioural change required for the effective implementation of climate change action for sustainable development (Vila et al. 2015).

Water Resource Management

Several of the projected impacts of climate change has been linked to water resource availability which in turn will impact sectors such as agriculture, fisheries, public health, infrastructure, energy production and biodiversity. Water conservation and the concept of Integrated Water Resource Management (IWRM) have attracted a lot of attention from Pacific Islanders. Also, water security has become a top priority for some of the Island Nations and is the major focus of adaptation.

The Pacific Integrated Water Resource Management is a leading regional water based Project (SPREP 2014). The Project involved 14 PICs and established standard sanitation guidelines in the region. The Project aims at improving waste water and water resource management by creating a balance in overused and scarce water reserves. To this effect, a national water apex body was set-up in several countries by their national IWRM to facilitate conservation of water reserves and enable stakeholder participation. Furthermore, it aimed to improve the assessment and monitoring of water resources, reduce water pollution, improve access to technologies and data, strengthen institutional agreements and legislations, and leverage additional financial resources (GEF 2012).

In addition, the Pacific Hydrological Cycle Observing System (Pacific HYCOS), another regional Project under the World Hydrological Cycle Observing System (WHYCOS) was designed to enhance water monitoring in 14 PICs. The outcomes of the Project are referenced from WHYCOS (2006) and SOPAC. Major initiative of this Project was to strengthen and provide a water monitoring and collection network to better protect and manage freshwater sources in the Pacific. This Project also established a regional database using an integrated catchment management system which can be utilized as a timely and critical decision making tool. Apart from this, maintenance work on water pumps, set-up of hydrological units, and the installation of borehole systems using hydrographs and electromagnetic survey was also accomplished. This assisted in targeting the best drill sites for communities. Additional data on sewerage, water monitoring, drafting or measuring flow and total rainfall data for monitoring and forecasting purpose was accomplished through simulations at National level. This work enhanced information sharing in the decision-making process at National level.

Apart from coastal management and food security, the Pacific Adaptation to Climate Change (PACC) Project which was implemented in 14 regional countries. The PACC Project enhanced the catchment basin reservoir system in The Republic of the Marshall Islands and installed solar water purifiers in hospitals in the outer islands. The project successfully promoted rainwater harvesting and storage (Cain 2014).

In addition, further works were conducted in water resource management by the Pacific Islands Climate Prediction Project (PICPP) and the National Adaptation Programme of Action (NAPA) which has country targets as part of The United Nations Framework Convention on Climate Change (UNFCCC) framework (PACE 2011). Ground water, surface water and rainwater management programmes including early warning flood systems, drought forecasting and hydropower management were conducted under this project (PACE 2011). NAPA (PACE 2011) implemented water management in Kiribati with a full time functioning Water Engineering Unit set-up to monitor the weather patterns, further devising plans in case of a drought. About 500 ground water wells were protected. Similarly, in Tuvalu, major works were executed in terms of community awareness of water management during drought, sustainable use of water and reducing groundwater contamination from human waste as part of country targets.

Likewise, the Water Demand Management Project for the PICs facilitated the set-up of the District Metered Area in the Solomon Islands with automated software and bulk flows to record discharge data (APAN 2012). Leak detection equipment was provided to reduce up to 40% loss of water in some areas (Kleppen 2009). Workshops and in-house training in Australia has been conducted for capacity-building (Butler & Memon 2006).

Disaster Risk Reduction

Disaster Risk Reduction (DDR) and Climate Change Adaptation share the same goal of reducing the vulnerability of communities to achieve sustainable development (United Nations International Strategy for Disaster Reduction 2004). Pacific Islands - Global Climate Observation System (PI-GCOS) Project was conducted in

Federated States of Micronesia, Kiribati, Nauru, Solomon Islands, Papua New Guinea, Vanuatu, Fiji, Western Samoa, Tokelau, Cook Islands, Niue, Republic of Marshall Islands, and French Polynesia (SPREP 2006; USP; PACE-SD 2011). The prime objective was to enact a robust PI climate observing system to meet long-term climate observation needs in the region to cope with climate change and variability. The target oriented in the implementation of technical support to systems such as: climate prediction, upper air observations, climate monthly newsletters, and weather forecasting trainings. Furthermore, by 2010, relevant regional activities included the completion of the review of Regional Meteorological Services, and the development of the Pacific Desk concept in participating Island countries (Secretariat of the Pacific Regional Environment Programme 2011). Robust GCOS provided necessary data on climate variability, capacity for the preparation of advanced warnings to minimize impacts on the economic, environmental, and social sectors throughout the region (USP; PACE-SD 2011).

As reported in (USP; PACE-SD 2011) Capacity-Building for the Development of Adaptation Measures in PICs Project administered capacity-building programmes to increase the competency of PICs in order to minimize climate-related risks. This was achieved by including DRR in national planning and budgeting by PICs involved. The policy and decision makers became cognizant of the threats posed by disasters to their economic sector as well as their citizens' livelihoods. Consequently, some countries included DRR and climate change risk management into national planning.

Climate Change Adaptation Program for the Pacific Project was formed to aid Cook Islands, Federated States of Micronesia (FSM), and other developing Pacific Member Countries to enhance their adaptive capacities and resilience to climate change and climate variability, including extreme events (USP; PACE-SD 2011). The success of this Project is commendable as the Cabinet for Cook Islands permitted the adaptation mainstreaming guidelines and this enhanced build-up of knowledge, raised awareness in the country, and enhancement of skills in relation to disaster risk reduction. As for FSM, the analysis indicated that the communities were at risk from high waves, sea surges and increased storm intensities, so a sea wall was constructed. FSM and Cook Islands offer an example to show mainstreaming risk-based approach at three categories namely: project activities, national development plans and sector programs.

Similarly, another Project namely; Finnish-Pacific Project to Reduce Vulnerability of the Pacific Island Countries' livelihoods to the effects of Climate Change (FINPAC) is a regionally coordinated initiative targeting the adaptation needs of 14 PICs to the adverse effects of climate change. The two prime objectives of FINPAC are to provide the National Meteorological Services with the proper capacity and tools to convey efficient climate and weather news to the community and to work with communities to assist them in utilizing the meteorological data to establish relevant plans to address disasters induced by climate change (Secretariat of the Pacific Regional Environment Programme 2016). This initiative reinforced the community's early warning system and aided in building community resilience. The Project empowered Pacific Island countries involved to generate precise weather information and early warnings allowing communities to take heed of severe climate and weather events (SPREP 2015). In addition, another regional Project namely; Building Disaster Response and Preparedness in the Pacific implemented in Fiji, Samoa, Kiribati, and Vanuatu was aimed at raising awareness through campaigns. The priority was to educate the Church Leaders on disaster risk reduction and enable them to deliver the information to the wider community (Caritas Australia 2008; Gero et al. 2011). Henceforth, the awareness campaigns offered by this Project were geared in the view of behavioural reforms of locals, further enhancing their capacity to cope with natural hazards.

Moreover, Pacific Community-Focused Integrated Disaster Risk Reduction (PCIDRR) is another DRR awareness based Project that was administered in Fiji, Solomon Islands, Tonga, and Vanuatu and funded by National Council of Churches Australia (NCCA) and AusAID (AusAID 2009; Gero et al. 2011). The key objective of this Project was to enhance awareness and knowledge of disaster risks at community level and increase community resilience to climate related hazards. The Project effectively increased community participation in disaster management trainings. Awareness campaigns were identified as an effective tool to educate the locals about climate and weather systems. However, in some regions it was evident that more time was required to properly disseminate climate change information. Another limitation was language and interpretation of information in some cases where the locals faced challenges in understanding climate change and DRR information.

Coastal Zone Management and Climate Change Induced Relocation

The coastal zones of the Pacific region are the most vulnerable, as they are often heavily populated. The increased urbanization of the coast has led to immense pressure on this fragile ecosystem with impacts such as coastal erosion, flooding, salt water intrusion and infrastructure damage, posing an increasingly critical challenge to many South Pacific Island communities to adapt (United Nations Development Programme 2015). Many coastal communities experience regular coastal inundation due to the climate induced sea-level rise. For some of these communities, relocating to another area might be the only feasible option remaining. The London School of Economics has estimated that across the Pacific Islands, which is home to 10 million people, 1.7 million could be displaced out of their homes due to climate change by the year 2050 (United Nations Framework Convention on Climate Change 2016).

One of the earlier projects focusing on coastal zone management was the Capacity-Building for the Development of Adaptation Measures in the Pacific Island Countries (CBDAMPIC). Coastal rehabilitation activities such as building a seawall, constructing coastal springs and mangrove conservation programme and other coastal protection infrastructure were carried out near the Samoan Coast for three communities. However, coastal management was not an option for two communities in Vanuatu due to extensive and irreversible damage to their coastline which resulted in them (Lateu and Panita) being relocated through funding from this project (Nakalevu 2006).

Using the lessons learned from CBDAMPIC, PACC programme took a comprehensive multilayered framework spanning from the community to the national level. Demonstration measures to reduce vulnerability in coastal areas have been carried out successfully in Cook Islands, Federated States of Micronesia, Samoa and Vanuatu under this project. The programme which spans from community level to national level responded to demands for an increasing need to build capacity on climate-related risks from the destructive effect of climate change on the Pacific coastal areas. The activities carried out under this project included building protective coastal structures and coastal vegetation in Fiji, Samoa and Vanuatu. Existing coastal infrastructure was reinforced through climate proofing of roads and harbours in Cook Islands, Federated States of Micronesia and Vanuatu. The project was also successful in changing coastal resource use patterns in some Samoan communities by reducing activities such as sand-mining which was being carried out (Secretariat of the Pacific Regional Environment Programme 2013).

The former Australian Department of Climate Change and Energy Efficiency implemented the Pacific Adaptation Strategy Assistance Program (PASAP) in 15 partner Pacific countries on the same principles as PACC. According to the Australian Department of the Environment and Energy (2017), there were varied outcomes of the programme ranging from improved governance at national level to several successful community projects in many different areas.

Communities in the Roviana region, which is in the Western Province of Solomon Islands, worked together with PASAP to map marine and coastal habitats, assess the health of coral reefs, sea grasses and mangroves, and survey coral bleaching and other diseases. This together with other activities were carried out to assess the vulnerability of their coastal environment to identify tangible ways to increase their resilience. In a similar attempt to improve the communities' coastal resilience, high coastal elevation data were captured using LiDAR technology for coastal modelling in Tonga, Papua New Guinea and Vanuatu [Australian Department of the Environment and Energy (no date)].

Also, the Global Climate Change Alliance: Pacific Small Islands States (GCCA: PSIS) was funded by the European Union and aimed at supporting climate change plans for governments of nine Pacific small Islands states (Cook Islands, Federated States of Micronesia, Kiribati, Marshall Islands, Nauru, Niue, Palau, Tonga and Tuvalu) and assisting them fulfil the adaptation needs of remote island communities. The four-and-a-half-year Project executed by SPC focused on a national and regional level. Working with communities near the Marshall Islands and Tongan coastline, the Project successfully implemented coastal protection structures including planting of coastal trees and shrubs, constructing causeway and groyne in several communities which directly benefited over 5000 community members (Pacific Community 2016).

An important dimension of the projects identified and all community adaptation projects in general is the level of understanding among the community members. The SPC/GIZ German government funded "Coping with climate change in the Pacific Island Region (CCCPIR) program" developed a "Coastal change toolkit" in 2014 to support community understanding of issues such as flooding and coastal erosion and assist them to identify the potential impacts and hazards that threaten their coastal areas (Pacific Community 2015).

Adaptation by remaining at the same site may not be an option for some communities. As the last resort, these coastal communities must relocate. The Australian Aid funded regional Pacific-Australia Climate Change Science and Adaptation Planning Program (PACCSAP) supported 14 Pacific Island counties by developing in-country adaptation responses through improved science and data through the years 2011 to 2015. In a quest to produce climate resilient settlements, a hazard analysis of coastal erosion, flood and groundwater risk was undertaken in Lifuka, Tonga by engaging the local community. A major outcome of this programme saw the development of a master plan for relocation of the community of Choiseul Bay Township in Solomon Islands (Pacific Climate Change Portal 2015). The plan for relocation was designed after extensive community engagements. Attachment to customary land, cultural ethics, communal unity, and traditional obligations to ancestors play a huge role in negotiations relating to relocation in the South Pacific Islands and it can be a strong barrier to relocation and climate change adaptation.

One of the ongoing projects which will have extensive impact on increasing the resilience of coastal communities in the Pacific is the Coastal Community Adaptation Project (C-CAP), which is in its fourth year of implementation and focuses on community level implementation. The Project is currently running in Fiji, Kiribati, Nauru, Palau, Papua New Guinea, Republic of Marshall Islands, Samoa, Solomon Islands, Tonga, Tuvalu, Vanuatu and the Federated States of Micronesia and has already engaged 77 communities across these 9 countries in completing asset mapping and infrastructure planning in the many different sectors. Some of these include coastal management through community coastal protection and erosion control projects.

These regional scale projects have brought about numerous benefits to our costal island nation communities with thousands benefiting in multiple ways that increase their resilience to climate change. Lessons can be learnt from the experiences of these regional scale projects that may be applicable to future projects.

Food Security and Agriculture

The consequences of climate change impact combined with increasing population and income growth gives urgency for addressing agricultural and food security adaptation more reasonably (Howden et al. 2007). Pacific Adaptation to Climate Change (PACC) Programme addresses the issue of directly improvising the response to climate change in the Pacific Islands. PACC is one of the first Projects initiated by United Nations Development Programme (UNDP) in the Pacific Islands to enhance the systemic and institutional capacity towards adaptation across the region. According to Dr. Netatua Pelesikoti, the Director of Climate change of Secretariat of the Pacific Regional Environment Programme, PACC is one of the largest regional Climate Change Adaptation Programmes that enabled implementation of practical and tangible results on the ground (Secretariat of the Pacific Regional Environment Programme 2014). The Project focuses on three key components: strengthening of climate change effect policies, carrying out demonstration Projects and developing tools and resources (Secretariat of the Pacific Regional Environment Programme 2014). PACC enhanced the resilience to the effects of climate change in agriculture (food production and security) in Island nations such as Papua New Guinea, Palau, Fiji, and the Solomon's.

In Fiji, the PACC Project assisted in the development of better drainage systems and guidelines caused by climate change effects. It also emphasized in testing cash crops for tolerance characteristics. In Palau, the Project Team worked closely with South Pacific Community (SPC) and the farmers of Ngatpang State to test and identify varieties of Taro in saline environments and create a dike to minimise salt water intrusion. In Papua New Guinea, the focus was on drought as a major cause of food shortage. The activities included seed supply system of drought-tolerant varieties and promoting conservation practices. Similarly, in the Solomon's, the PACC Project focused on improving the food production in low-lying areas through different farming systems. It successfully strengthened the institutional framework, policies and plans of the national government and community for sustainable resource development programmes. It also demonstrated innovative decision making system, practical and technological approaches to the adverse effect of climate change in selected regional countries. The scheme ensured that the lessons and results derived assists regional as well as global, citizens with the creation of new knowledge through the Project.

The Pacific Islands Climate Prediction Project (PICPP) was carried out in Fiji, Solomon Islands, Samoa, Cook Islands, Niue, Tuvalu, Kiribati, Tonga, Papua New Guinea and Vanuatu. The project was initiated to emphasize the importance of meteorological services in the Pacific Islands Countries. The emphasis of the project was on the development of a framework for incorporating climate change information into decision-making process for agencies and industries including farmers, tourism, health, and water resource management authorities affected by climate change variability. The Project assisted the meteorological services of Pacific Island Countries to move away from interpreting information to producing climate related information. Climate related information assisted the agricultural, fishing and tourism industries in planning and preparing the Pacific Island Countries to adapt to the potential impacts of climate change. Hence, at the completion of the programme, the selected countries (involving agencies and government) are believed to have tailored the software for predicting the climate change services to support climate-sensitive decision making. The climate sensitive industries will benefit from climate prediction risk management context. (Australian Bureau of Meteorology 2012).

The Pacific Climate Change Science Program (PCCSP) with the National Meteorological Services aimed to enhance the understanding of the nature and magnitude of climate change for better adaptation. Primarily, it worked to build an understanding on the South Pacific Convergence Zone, the Inter Tropical Convergence Zone and the Southwest Pacific Monsoon. The regional countries engaged in this Project were Cook Islands, FSM, Fiji, Kiribati and Nauru. The Research Programme made significant progress in identifying the gaps in management of climate data records, understanding the recent and current climate trends, tropical cyclones and understanding ocean processes and changes under global warming (Cambers 2012). The results achieved by PCCSP were important in terms of assisting the Pacific Island Countries to progress towards resilient and sustainable development in terms of generating information about climate change in the Pacific (Hunnam 2013).

Coping with Climate Change in the Pacific Island Region (CCCPIR) was supported by the Government of the Federal Republic of Germany to extend its commitment to support Pacific Island Countries in relation to climate change effects. The need for this Programme was to urge the need to progress towards adaptation and mitigation measures in the Pacific Island region. The focus on enhancing renewable energy and energy efficiency is crucial in increasing the resilience of their economies. The Programme was an extension of the regional Programme 'Adaptation to Climate Change in the Pacific Island Region'. The countries engaged in the Project were FSM, Fiji, Kiribati, Nauru and Palau. The Project focused on the agriculture, forestry, land use planning, fisheries, tourism, energy and education. More specifically, it strengthened the Pacific Island regional advisory and management capacity (Killmann 2012).

Pacific Islands Nations are most susceptible to climate change effects with considerable differences in the level of vulnerability in different countries. Sparse information was available to review the management processes in different projects. For instance, the PCCSP Project was succinct, with clear and precise information. However, it did not contain all the relevant information that might be helpful to organise similar program better or for reviewers to evaluate it efficiently (Hunnam 2013).

Limitations

Irrefutably, climate change adaptation projects have increased in the Pacific Island Nations. While, considerable amount of effort is put into increasing the capacity of communities to adapt to a changing climate, a number of factors impede effective implementation of climate change adaptation projects. A common limiting factor that emerged from different Projects was the lack of institutional policies and the skill set to effectively implement specific adaptation actions. This is mostly due to the lack of human resources and expertise to carry out technical activities. Several Projects have emphasized the need for a multi-sectoral approach engaging formal networks and partnerships to facilitate institutional leadership. Despite some progress observed in cross sectoral networking and collaboration, strong governance and direction at various levels are still needed to administer policies both at national and regional levels. In regards to the IWRM projects, a lack of stakeholder collaboration was a major drawback in some of the countries such as the FSM which could have attained better outcomes since stakeholders make a vital component of a system and can provide the much-needed resource in capacity-building.

Other projects such as the Pacific HYCOS lacked national water quality data and in some cases absence of monitoring and evaluation plan was identified as a major limitation towards the implementation of these projects. Lack of local workforce with essential technical expertise to implement scientific adaptation procedures is a matter of concern that was demonstrated across different projects. Funding constraints also fail to attract qualified personnel at management positions. Several projects encountered challenges in recruiting local people with technical skills to participate in the projects. Training programmes directed at increasing practical knowledge can provide a significant avenue for staff to develop their skills. It is imperative for the regional governments to increase their local capacity by investing in training programmes that provide formal qualification and certification.

One of the most significant aspects of various climate change adaptation projects has been the inability to efficiently document and disseminate findings across various platforms. One of the limitations faced was in extracting primary data due to poor response from Project Coordinators and implementing agencies. Despite these constraints, the importance of climate data management is being recognised as an integral factor in effective adaptation and resilience planning. Also, it is explicitly evident that the Pacific Island states have a progressive outlook towards tackling these issues through devising of relevant and applicable policies.

Recommendations and Way Forward

Community outreach and awareness programmes proved to be an effective mechanism to increase climate change knowledge amongst the communities. Consequently, it is highly recommended that community awareness and trainings are included in all the climate change adaptation projects that are implemented in the region. It is also important for the local communities to take ownership of their resources and establish guidelines for natural resource management. As demonstrated in some of the projects, water security is a high priority area in many PICs and national governments are encouraged to upgrade their water monitoring and evaluation plans to enhance the IWRM projects and policies.

The modus operandi for coastal programmes and projects should embrace all dimensions including an on-ground physical activity coupled with simultaneous capacity building of community members. This can be factored in after relevant assessments of the site conditions have been carried out. This is to be closely tied with governance by collating lessons learnt and using them for recommendations on mainstreaming into national level sector wise or standalone policies and planning. The relocation guidelines that are being prepared by some Pacific Island countries must incorporate socio-cultural aspects to ensure smooth relocation process.

One of the best ways to enhance the resilience and adaptive capacity of a coastal zone management project is to apply a combination of adaptation measures in communities, rather than a single adaptation response with the measures appropriately aligned to the anticipated outcomes. Many coastal based Projects now focus on multiple soft adaptation activities as opposed to a single large scale hard adaptation activity.

There are numerous communities lining the Pacific Coast and many of these communities are facing climate induced challenges. To study the coastal ecosystem, community structure and organisation for each of these communities separately and specifically would take the bulk of the projects resources. A comparative approach can be undertaken by Projects planning on spanning across many communities. The major coastal ecosystem processes and community demographics of one or two communities can be generalized to apply to other similar communities to put the project resources to the best use and cover many similar communities. This overview should be accompanied by a smaller and limited study of each community to ensure that the uniqueness of its coastal environment is not overlooked using this comparative technique. Using this comparative approach, the adaptation measures for one community can be replicated to other communities facing similar challenges.

A scientific foundation and access to relevant data is imperative to building the resilience of a community. Projects such as PACCSAP and PASAP have already engaged in developing adaptation responses through improved science and data with mapping, climate modelling and hazard analysis of coastal habitats. Some Projects recommended the use of meteorological data and climate modelling in effectively establishing relevant plans to address disasters induced by climate change. A great deal of emphasis was placed on the early warning system that would generate more accurate weather information enabling communities to take heed of severe climate and weather events.

Experience through diverse Projects depicts that capacity-building in the communities should be a mixture of top-down and bottom up approach where capacities of government officials and communities are simultaneously improved. However, some of the projects discussed noted that substantial efforts are needed to form this multidisciplinary team. This limitation needs to be surmounted to maximize the resources brought about by the project, be it in terms of improved human resources such as trained personnel or an improvement in infrastructure.

In synopsis, for climate change adaptation project of any nature to be sustained beyond the lifeline of the Project, local communities and stakeholders must be appropriately trained to be able to continue with the monitoring and evaluation of the project deliverables. Thus, it is evident from the review that local capacity-building is essential for climate change adaptation Projects. Without proper investment in local communities and its human resources, conducting future projects can be a huge challenge.

Conclusion

It is evident that the South Pacific Island states are vulnerable to the detrimental effects of climate change due to its remoteness and limitation of the available resources. Strenuous effort is required to combat the response to sea level rise and extreme weather events induced by climate change. However, this paper focused on the main practices carried out by some of the regional Projects on climate change capacity-building, awareness and training, water resource management, disaster risk reduction, coastal zone management, climate change induced relocation and food security and agriculture. The best practices are to be taken as important considerations that could be utilized as a reference. However, there existed limitations within each Project. Major limitations to adaptation stem from the lack of human resources to carry out the projects efficiently, ethical dilemma as in the case of relocation, delay in project timings due to extensive administrative processes, lack of knowledge and understanding, funding constraints and policy implications. Despite these limitations, several projects have been successfully implemented and recommendations for effective practices to be used with respect to community-based climate change adaptation initiatives in the South Pacific region are accentuated.

References

Australian Bureau of Meteorology. (2012). Pacific Islands Climate Prediction Project. Retrieved March 25, 2017 from http://www.asiapacificadapt.net/adaptation-practices/pacific-islands-climate-prediction-project-picpp.

Australian Department of the Environment and Energy. (2017). Pacific Adaptation Strategy Assistance Program. Retrieved March 20, 2017 from http://www.environment.gov.au/climate-change/adaptation/international-climate-change-adaptation-initiative/pasap.

Barnet, J. (2008). Food security and climate change in the South Pacific. *Pacific Ecologist, 14*, 32–36.

Butler, D., & Memon, F. A. (2006). *Water demand management*. London: Iwa Publishing.

Cambers, G. (2012). Pacific Climate Change Science Programme. Retrieved March 25, 2017 from http://www.asiapacificadapt.net/adaptation-practices/pacific-climate-change-science-programme-pccsp.

Commission on Climate Change and Development. (2009). Closing the Gaps: Disaster risk reduction and adaptation to climate change in developing countries.

Cosbey, A., Bell, W., Murphy, D., Parry, J. E., Drexhage, J., Hammill, A. & Van Ham, J., (2005). Which way forward? *Issues in developing an effective climate eegime.*

Gero, A., Méheux, K., & Dominey-Howes, D. (2011). Integrating community based disaster risk reduction and climate change adaptation: Examples from the Pacific. *Natural Hazards and Earth System Sciences, 11*(1), 101–113.

Howden, S. M., Soussana, J.-F., Tubiello, F. N., Chhetri, N., Dunlop, M., & Meinke, H. (2007). Adapting agriculture to climate change. *Proceedings of the National Academy of Sciences, 104*(50), 19691–19696. doi:10.1073/pnas.0701890104.

Hunnam, P. (2013). Review of the International Climate Change Adaptation Initiative Pacific Programs in Climate Science and Adaptation Planning. *Pacific Programs in Climate Change Science and Adaptation Planning*, 1–65.

IPCC. (2007). Climate Change 2007: The Physical Science Basis. Contribution of Working Group I to the Fourth Assessment Report of the Intergovernmental Panel on Climate Change. Retrieved March 11, 2017 from http://ipcc-wg1.ucar.edu/wg1/wg1-report.htmlS.

Killmann, W. (2012). Coping with Climate Change in the Pacific Island Region. Retrieved March 25, 2017 from http://www.asiapacificadapt.net/adaptation-practices/coping-climate-change-pacific-island-region-cccpir.

Lynch, P. (2016). 2016 Climate Trends Continue to Break Records: NASA's Goddard Space Flight Center in Greenbelt, Md. Retrieved March 17, 2017 from https://www.nasa.gov/feature/goddard/2016/climate-trends-continue-to-break-records.

Nakalevu, T. (2006). Capacity Building for the Development of Adaptation Measures in Pacific Island Countries (CBDAMPIC) project. *Final report prepared for the Canadian International Development Assistance*.

Nunn, P. D. (2009). Responding to the challenges of climate change in the Pacific Islands: management and technological imperatives. *Climate Research, 40*(2–3), 211–231.

Pacific Climate Change Portal. (2015). The Pacific-Australia Climate Change Science and Adaptation Planning Program. Retrieved March 17, 2017 from https://www.pacificclimatechange.net/project/pacific-australia-climate-change-science-and-adaptation-planning-program-paccsap.

Pacific Climate Change Portal. (2016). USAID—Pacific Islands Coastal Community Adaptation Project. Retrieved March 15, 2017 from https://www.pacificclimatechange.net.

Pacific Community. (2015). German Government co-funds new coastal change toolkit. [Press release]. Retrieved from http://www2.pazifik-infostelle.org/uploads/cccpir_press_release_coastal_change_toolkit_2.pdf.

Pacific Community. (2016). Global climate change alliance: Pacific small island states -Volume 1: Final Report. Retrieved March 16, 2017 from http://ccprojects.gsd.spc.int/wp-content/uploads/2016/11/web-Volume_1_Global_Climate_Change_Alliance_PICTs.pdf.

Pacific iCLIM. (2014). Supporting the Regional Management of Climate Change Information in the Pacific. Retrieved March 15, 2017 from https://www.griffith.edu.au/__data/assets/pdf_file/0006/667464/Pacific-iClim-Report.pdf.

Rojas Blanco, A. V. (2006). Local initiatives and adaptation to climate change. *Disasters, 30*(1), 140–147.

Secretariat of the Pacific Regional Environment Programme. (2011). Annual Report : 2010 (pp. 1–52). Apia, Samoa.

Secretariat of the Pacific Regional Environment Programme. (2013). Adapting to climate change in the Pacific: The PACC programme. Retrieved March 16, 2017 from http://www.sprep.org/attachments/Publications/CC/PACC_Programme.pdf.

Secretariat of the Pacific Regional Environment Programme. (2014). Regional five year climate change adaptation project reviewed in Fiji this week. Retrieved March 25, 2017 from http://www.sprep.org/climate-change/regional-five-year-climate-change-adaptation-project-reviewed-in-fiji-this-week.

Secretariat of the Pacific Regional Environment Programme. (2015). Pacific islands framework for action on climate change 2006–2015. Retrieved February 23, 2017 from http://www.sprep.org/climate_change/pycc/documents/PIFACC.pdf .

Secretariat of the Pacific Regional Environment Programme. (2016). Annual Report: 2015 (pp. 1–64).

United Nations Development Programme. (2015). Reducing vulnerability of island coasts. Retrieved March 19, 2017 from http://stories.undp.org/climate-change-in-the-pacific.

United Nations Framework Convention on Climate Change. (2016). How Fiji is impacted by climate change. Retrieved March 21, 2017 from http://newsroom.unfccc.int/cop-23-bonn/how-fiji-is-impacted-by-climate-change/.

Van Aalst, M. K., Cannon, T., & Burton, I. (2008). Community level adaptation to climate change: The potential role of participatory community risk assessment. *Global Environmental Change, 18*(1), 165–179.

Vila, P., Brown, R., Mackay, S., McGregor, K., Mackey, B., Gonelevu, M., et al. (2015). *Barriers to effective adaptation and resilience planning in the Pacific: An information management perspective: Griffith University*. Samoa: Queensland and SPREP.

Enhancing Adaptive Capacity and Climate Change Resilience of Coastal Communities in Yap

Murukesan Krishnapillai

Abstract The biophysical and socioeconomic conditions in the coastal zone of the Yap Islands is highly vulnerable to climate-related changes in precipitation, sea level, storm surges, coastal erosion, and salinity. The climatic changes are affecting every aspect in the lives of coastal communities due to the small size of the islands and atolls, their low elevation, and extensive coastal areas. Climate risks are further amplified by the regional El Niño Southern Oscillation phenomena that threaten the food and water security of island communities. Further changes are projected to manifest in the coming decades because of increased temperature, decreased rainfall, rising sea levels, and ocean acidification. Recurrences of disasters and crises threaten food security through impacts on traditional agriculture, causing the forced migration of coastal communities to highlands in search of better living conditions. As many of the projected impacts are now unavoidable, implementing some degree of adaptation is essential to enhance food security, strengthen livelihoods, and increase the resilience of coastal communities to future climate risks. This paper highlights the outcomes of an ongoing project on 'Climate Adaptive Agriculture and Resilience,' and presents a three-pronged adaptation model to enhance the adaptive capacity and climate resilience of coastal communities in Yap. The potential of sustainable soil management practices, water conservation and management, and mosaic restoration activities in enhancing the livelihood opportunities of displaced coastal communities is highlighted. Through this adaptation model, coastal communities can moderate the harm of current and future climate risks and take advantage of new opportunities.

M. Krishnapillai (✉)
Cooperative Research and Extension, College of Micronesia-FSM, Yap Campus, Colonia, Yap 96943, Federated States of Micronesia
e-mail: vazhaveli@hotmail.com

W. Leal Filho (ed.), *Climate Change Impacts and Adaptation Strategies for Coastal Communities*, Climate Change Management,
https://doi.org/10.1007/978-3-319-70703-7_5

Introduction

Coastal ecosystems that exist at the interface between terrestrial and marine environments are some of the most diverse and dynamic environments on Earth. Coastal zones' socioeconomic and ecological importance is virtually unparalleled. However, global climate change has significant consequences for the world's coasts,[1] coastal ecosystems, and coastal inhabitants. The impacts of climate change on resources have far-reaching effects on the livelihood of coastal communities.[2] It threatens food security and economic growth in complex ways (USAID 2009; Wong et al. 2014).

Global climate change already impacts and will continue to impact coastal communities, ecosystems, and many facets of people's lives in the coastal zone where approximately 2.7 billion people—over 40% of the world" population—live (Monirul and Mizra 2003; Dolan and Walker 2017). Among the changes, the accelerated sea-level rise has received much attention as it leads to coastal erosion, loss of land and property, increased flood frequency, increased saltwater intrusion, and a myriad of ecological changes. The coastal systems most vulnerable to climate change are low-lying islands, coastal areas, countries subjected to typhoons and hurricanes, and less-developed countries. Relative to other coastal areas, low-lying islands, including many Small Island Developing States (SIDS), are more vulnerable to the impacts of climate change because they have relatively scarce natural resources, and they have limited and high transportation costs. In addition, climate change interacts differently with a variety of human activities and other drivers of change along the coastlines of SIDS. Weather and climate extremes affect a wide range of economic activities supporting coastal communities and pose additional risk to many low-lying small islands and atolls (Wong et al. 2014).

Even without climate change, coastal areas face many problems associated with population growth, habitat change, resource over-exploitation and degradation,

[1]Coastal areas are commonly defined as the interface or transition areas between land and sea, including large inland lakes. Coastal areas are diverse in function and form, dynamic, and do not lend themselves well to definition by strict spatial boundaries. Unlike watersheds, there are no exact natural boundaries that unambiguously delineate coastal areas. Geologically, continental margins are of two types: (i) active margins where the edge of a continent happens to be at the edge of an oceanic plate, and (ii) inactive margins where the transition from continental lithosphere to oceanic lithosphere is within a plate rather than at a plate edge. Coastal areas are, therefore, characterized by the vertical accretion of near-shore land (Scialabba 1998). In relation to exposure to potential sea-level rise, the Low Elevation Coastal Zone (LECZ) is the contiguous and hydrologically connected zone of land along the coast that is less than 10 m above mean sea level (McGranahan et al. 2007; Neumann et al. 2015).

[2]In this article, the terms coastal communities and atoll communities are used interchangeably.

water pollution, and changes in freshwater flows (USAID 2009; Wong et al. 2014). Climate change is expected to amplify many of these and other stresses on coastal areas. This, in turn, increases the need and urgency to include coastal adaptation as part of effective coastal management options. Because of these realities, climate change is considered by many to be one of the most important challenges of the 21st century and a priority for immediate action in coastal areas (Hay 2013; IPCC 2014; Santha, 2015; Hay and Mimura 2013).

The biophysical and socioeconomic conditions in the coastal zone of the Yap Islands are vulnerable to climate-related changes in precipitation, sea level, storm surges, coastal erosion, and salinity. Local climate variability is influenced by the El Nino Southern Oscillation (ENSO) phenomena, which affects the position of the Western Pacific Warm Pool, the positions of the South Pacific Convergence Zone (SPCZ) and Intertropical Convergence Zone (ITCZ), as well as the penetration of the western Pacific monsoon (WPM). This results in severe localized droughts in some years and chronic floods and the frequency, strength, and location of tropical cyclones and sea-level height (Lough et al. 2016). Climatic changes affect every aspect of the lives of coastal communities in Yap due to the small size of the islands and atolls, their lower elevations, and their extensive coastal areas. Further changes are projected long into the future (Australia Bureau of Meteorology and CSIRO 2014; Lough et al. 2016) because of increased temperature, decreased rainfall, rising sea level, and ocean acidification. The recurrence of disasters and crises threaten food security through impacts on traditional agriculture. Many of the projected impacts are now unavoidable, making the implementation of some degree of adaptation essential to enhance food security, strengthen livelihoods and increase the resilience of coastal communities to future climate risks.

Realizing the potential of restoration agroforestry and based on previous success with a group of climate change-induced and climate change forced-migrants on Yap Proper (Krishnapillai 2017), a project was initiated in Yap under the umbrella of USAID's Pacific American Climate Fund[3] (PACAM) to scale up the multiple benefits of climate-smart adaptation strategies among the four displaced atoll community settlements on Yap Proper. This article highlights the outcomes of an ongoing project on *Climate Adaptive Agriculture and Resilience* (CAAR). The purpose is twofold: First, it considers the geographical features and vulnerabilities of coastal communities in Yap in different natural settings, including atolls, that lead to their displacement. From this, a three-pronged adaptation framework is

[3]The Pacific-American Climate Fund (PACAM) is a five-year USAID grant facility that provides support to assist Pacific island communities to adapt to the negative impacts of climate change and provide co-benefits or solutions to other development challenges, such as livelihood enhancement, improved health, food security, water availability, ecosystem conservation and improved governance.

presented that takes a more locally relevant approach to enhance the adaptive capacity and climate resilience of displaced coastal communities in a new terrain. Although various strategies of the adaptation framework are implemented on Yap Proper, this pragmatic and "multiple benefit" approach is equally applicable in other vulnerable coastal settings.

Yap's Geography

Yap is the westernmost State in the Federated States of Micronesia (FSM). It consists of a heterogeneous collection of high-elevation islands and atolls. Of all the FSM states, Yap State has the largest ocean area and spans the greatest distance from one end to the other (Fig. 1). The main islands of Yap, known as Yap Proper, consists of a group of four major conjoined islands (Fig. 2). In addition to the main islands, there are ten atolls, four reef islands, and one high limestone island distributed over the tropical western Pacific Ocean. These islands have been referred to as tectonically stable during the late Quaternary (Dickinson 2001). Yap Proper occupies 38 square miles whereas the outer islands and atolls collectively occupy about 7.32 square miles. The 2010 FSM Population Census recorded 4006 persons living in the outer islands of Yap State. The outer islands account for 35% of the state's population and about 16% of the state's land area (FSM Census Bureau 2010).

Richmond et al. (1997) identify three distinct island types within Yap State: (i) high islands of volcanic and metamorphic rocks, (ii) atolls and low-lying reef islands, and (iii) a high limestone island (Table 1). Owing to the heterogeneity in

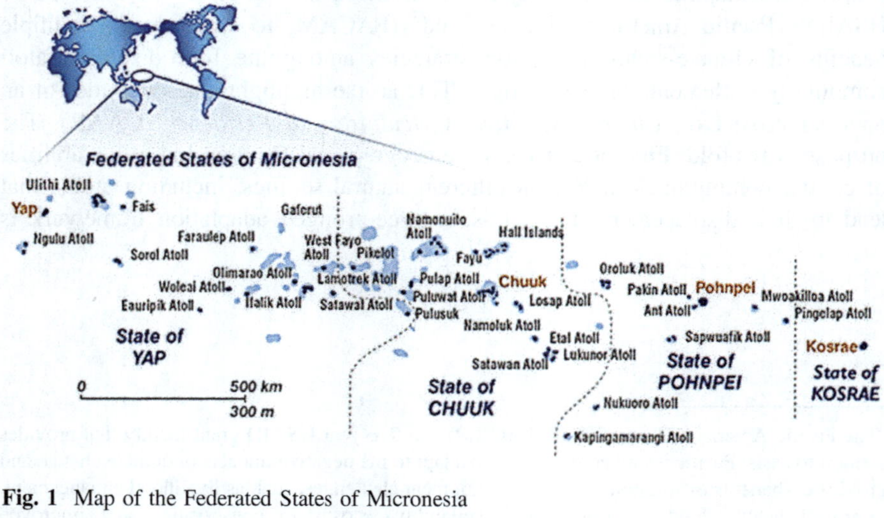

Fig. 1 Map of the Federated States of Micronesia

Fig. 2 Satellite imagery of Yap Proper

island geomorphology, culture, ecosystems, populations, and distinct coastal environments, each of these island types respond uniquely to the impacts of climate variability.

Table 1 Population and geographical features of Yap Islands (based on data from Richmond and Reiss 1994; Taborosi and Neth 2014)

Island	Island type	Population	Maximum elevation (m)	Land area (m^2)
Ngulu	Atoll	6	2	0.11
Yap Proper	High volcanic	7371	170	38
Sorol	Atoll	0	2	0.36
Ulithi	Atoll	847	5	1.93
Fais	High limestone	294	19	0.99
Eauripik	Atoll	114	2	0.09
Ifalik	Atoll	578	2	0.59
Faraleup	Atoll	193	2	0.23
Woleai	Atoll	1039	2	1.63
Lamotrek	Atoll	329	2	0.37
Elato	Atoll	105	2	0.26
Olimarao	Atoll	0	2	0.06
Gaferut	Reef island	0	2	0.05
Pikelot	Reef island	0	2	0.04
West Fayu	Atoll	0	2	0.02
Satawal	Reef island	501	2	0.53

High Islands of Weathered Volcanic Breccia and Metamorphic Rocks—Yap Proper

Yap Proper consists of four main high islands (*Marbaaq* or *Yap*, *Gagil-Tomil*, *Maap* and *Rumung*) situated closely together and encompassed by a fringing reef. These high islands are composed of accretional metamorphic rocks with smaller amounts of sedimentary (breccia) and volcanic rocks (Johnston et al. 1960; Nedachi et al. 2001). The shoreline area consists of three critical natural environments that have the potential to be substantially affected by the accelerated sea-level rise (Richmond et al. 1997). These are mangrove forests, sand beaches, and reefs.

Mangrove forests comprise about 10% of the total land area and are spread over 72 miles along the shoreline. The mangrove area has been expanding throughout the Holocene, and under a rising sea level, this will expand in a landward direction as low-lying coastal land becomes intertidal (Richmond and Reiss 1994). *Sandy beaches* and associated sand-rich coastal plains are limited in extent, occurring in exposed localities and comprising only about 0.25% of the total land area. However, sandy beaches form important settlement sites and are critical to modern and traditional Yapese culture. Sand beaches are extremely vulnerable to accelerated erosion, and a landward shift of the shoreline profile during a rising sea level is expected. *Coral reefs* surround Yap Proper and are a source of sediment for the beaches and adjacent fringing reefs. Reef response to accelerated sea-level rise is probably one of expanded vertical growth and increased carbonate production as shallow substrates are submerged (Richmond et al. 1997).

High Limestone Island

The high limestone island of Fais (Fig. 3) is an elliptical uplifted carbonate platform that has an area of approximately 0.75 m^2 with a maximum elevation of 19 m. Fringing reefs up to 150 m wide are adjacent to the beaches whereas shallow reefs are poorly developed and opposite to the cliffed shorelines. Archaeological

Fig. 3 Satellite imagery of Fais Island (post typhoon Maysak)

investigations have confirmed that Fais has been continuously occupied for nearly 2000 years (Intoh 1997). Regardless of its origin, the island shows evidence of exposure to different sea levels.

Atolls and Reef Islands

Atolls are islands composed of an annular reef rim surrounding a central lagoon, often with small, low-lying islets of reef-derived sediment occurring over a scattered upper-reef rim surface. Ten of the outer islands of Yap State are atolls, and four are single-reef islands. Reef islands are low-lying islands composed of reef-derived sediment but without a significant lagoon. The atoll islets and reef islands are Holocene in age and include a significant amount of storm deposits and possibly record deposition under a slightly higher than present (<2 m) mid-Holocene sea level (Richmond and Reiss 1994).

Climate Variability and Change in the Western Tropical Pacific

Climate change in the FSM is influenced by large-scale processes such as the El Niño Southern Oscillation (ENSO), South Pacific Convergence Zone (SPCZ), Intertropical Convergence Zone (ITCZ) and Western Pacific Monsoon (WPM). These large-scale climate controls and their seasonal variations result in the average climate typical of individual islands, with 73% of the islands influenced by the SPCZ, 36% by the ITCZ, 18% by the WPM, and 86% affected by tropical cyclones (Lough et al. 2011). Longer-term variability and climate change compound these impacts, particularly in terms of increased vulnerability to natural, climate-related disasters. The frequency of an extreme El Niño is expected to double due to climate change with the average frequency increasing from once every 20 years to once per decade (Cai et al. 2014). Various climate models indicate that El Niño and La Niña events will continue to occur and have a significant impact on inter-annual variability in the region. Some of the impacts of ENSO on rainfall (e.g., floods) may intensify in a warmer climate due to increased atmospheric moisture (Seager et al. 2012). Global warming is also expected to enhance average rainfall along the equator, and new research suggests it will also enhance El Niño-driven drying in the western tropical Pacific and El Niño-driven increases in rainfall over the central and eastern tropical Pacific (Power et al. 2013). Climate projections based on the analysis of about 26 new Global Climate Models (GCMs) in the Coupled Model Intercomparison Project Phase 5 (CMIP5) database show that for all emissions scenarios, temperatures will continue to rise in the FSM, as will sea level and ocean acidification (Australia Bureau of Meteorology and CSIRO 2014; IPCC 2014).

In addition, Yap's weather is influenced by the paired Hadley cells and Walker circulation (Chowdhury et al. 2010). Under El Niño conditions, the islands typically experience drought, and under La Niña conditions, the islands experience higher than normal rainfall. Studies indicate that this seasonal climatology of extreme events in the western Pacific is variable in spatial and temporal scales. Yap islands display considerably higher seasonal extremes than the other islands in the region for a 20-year return period due to typhoon-related storm surges. Decadal observations reveal the mean sea level in the western Pacific rose alarmingly (3.1 mm/year) from 1988 to 2008, which is in close correlation with the faster rate of rise as observed by the IPCC (2014).

Climate extremes are only some of the many challenges that climate variability and change present to local communities in Yap. The small size of the islands and atolls, their lower elevations and extensive coastal areas, their remoteness and limited financial resources, and poor economic and social decisions contribute to great ecosystem and human vulnerability to disasters (Shea et al. 2001).

Vulnerability of Yap's Coasts and Coastal Agriculture

Quantitative and site-specific data documenting climate-related impacts across Yap are meager. However, using profile data, environment maps, and assumptions based on sea-level rise and system response, Richmond et al. (1997) identified various coastal settings on the highlands and atolls of Yap State. This includes coastal plains, mangrove forests, and coral reefs.

Generally, there are three vulnerable areas in Yap that are significant under the accelerated sea-level rise scenarios.

Lowlands on Yap Proper: Yap Proper has a land area of 38 sq. mi., of which 13 sq. mi are low-lying coastal areas. These low-lying coastal plains, which are the sites of most of the coastal settlements, are particularly vulnerable to sea-level rise. Though much of the shoreline is bordered by engineering structures (e.g., in the Colonia area) or through traditional Yapese stonework platforms (*chabog*) (Nunn et al. 2017), many of those structures are barely above accelerated sea-level scenario still-water levels and, therefore, incur damage through direct wave attack.

Coastal plains: The loss of low-lying coastal sand plains, which are typically fronted by sandy beaches, has a dramatic effect on current settlement patterns on Yap Proper. Unprotected sand beaches are highly vulnerable to accelerated sea-level rise. For example, due to its lower elevations and the existence of lower-elevation gaps connecting the west and east coasts, extensive coastal erosion is evident in the southernmost part of Yap Proper (Fig. 4). The population throughout the FSM lives in the coastal zone, thus such coastal changes directly impact subsistence-crop production in coastal areas due to salinization of taro patches (Nunn et al. 2017).

Atoll islets and reef islands: Given their low elevations and open exposure, atolls and reef islands are particularly susceptible to accelerated sea-level rise. This is

Fig. 4 Coastal erosion and exposed shoreline at Gilmaan, South of Yap Proper. Persistent coastal erosion threatens agroforestry production

Fig. 5 Exposed shoreline in Falalop, Ulithi atoll showing wide carbonate sand and gravel beach. Islets of Ulithi atoll are severely affected by accelerated sea level rise owing to their open exposure and low elevations

Fig. 6 Aerial view (to the northeast) of a part of Fais Island showing shoreline alternating between carbonate sand beach and vertical hanging limestone cliffs. Settlements are located on beach ridges which form the narrow coastal plain and are subjected to potential land loss under accelerated sea level rise

recognized as a problem of increasing significance on many low-lying Micronesian islands (Keim 2010; Pam and Henry 2012). However, it is expected that small islands and atolls will show varying responses to sea-level rise due to island geomorphology, culture, ecosystems, and populations (Barnett and Campbell 2010; Nurse et al. 2014). Nevertheless, atolls characterized by extremely small size, coralline soils, and shallow freshwater lenses are highly exposed and sensitive to sea-level rise (Werner et al. 2017). Chronic coastal erosion combined with erosion associated with inundation events threatens the already narrow atolls and coastal plains (Fig. 5). Although Fais Island is an elevated carbonate platform, human settlements are located on a narrow coastal ridge and are, therefore, subject to land loss under accelerated sea-level rise (Fig. 6).

Climate models suggest that global warming will raise global sea levels significantly over the course of this century (Keener et al. 2012). Current projections of future sea level for the FSM suggest that sea level will rise at an average rate of between 8 mm/year between now and 2100 and ~9.5 mm/year between now and 2090, assuming in both cases the upper end of the RCP8.5 projections for 2090 relative to 1995 (Australian Bureau of Meteorology and CSIRO 2014), and almost certainly beyond that (Jevrejeva et al. 2012). Because of improved modelling, there is also greater confidence in future sea-level projections due to improved constraints on thermal expansion and ice sheet changes (Church et al. 2013). The effect of sea-level rise is most strongly felt in changes to extreme sea-level events. For example, a 50-cm increase in mean sea level (which corresponds to the low end of RCP8.5 projections for the coming century) can dramatically increase the frequency of the sea level exceeding a given threshold (Hunter 2010, 2012). Over the FSM, increases of two orders of magnitude or more have been suggested (i.e. with a

50-cm sea-level increase, a 1-in-100-year flooding event would typically occur every year). If these projections prove correct, it is likely that there will be profound changes to the coast of the islands in the FSM that will force changes to the ways that people currently interact with the coast (Nunn et al. 2017).

Despite the uncertainty that exists about the magnitude of climate change impacts, what is evident is that traditional agriculture in Yap is facing, and will continue to face, in the coming years from climate change (Taylor et al. 2016). Yap's sustainable food production system is dominated by an agroforestry system and taro patches (Falanruw 1993). Most subsistence crops are grown along the coastal plains. Food production in the outer atolls is also dominated by taro patch and mixed agroforestry culture. Taro patches along the coastal plains and atoll settings are especially vulnerable to sea-level rise, storm surges, and saltwater intrusion that are already occurring. Continued sea-level rise intensifies saltwater intrusion into coastal wetlands and groundwater systems (Fletcher and Richmond 2010). Taro patches near sea level, estuaries, low-lying agroforestry stands, and other ecosystems in coastal settings will experience increased wave energy and rising salinity over time. Observations and market data suggest that the traditional food production system has been reduced in extent and productivity over the last two decades.

The FSM State-Wide Assessments and Resource Strategies (SWARS) study reveals the impact of sea-level rise on areas more suitable for food production in Yap (FSM and USFS 2010). Figure 7 is a spatial analysis of the impact of a 1-m rise in sea level on areas more suitable for food production. It utilizes layers of vegetation, soil, rivers and riparian buffer zones, and slope. Areas of mangrove have been masked out. The remaining layers are ranked, and attributes are weighted to give a composite picture, indicating areas more suitable for food production in yellow and areas less suitable for food production in green. Under the impact of a 1-m rise in sea level, approximately 10% of coastal plains would become unsuitable for crop production. Given that most people in Yap occupy island coasts, and the taro patches are concentrated along the coastal plains, any coastal change under a 1-m rise in sea level would impact subsistence crop production significantly.

A 5-m storm surge based on the prediction of a 1–2 m rise in sea level and increased intensity of storms would impact up to 40% of Yap's coastal zone. Two major typhoons that hit Yap—Sudal in April 2004 and Maysak in March/April 2015—generated a storm surge of about 4 m in some parts of Yap Proper and Ulithi atoll (Figs. 8a, b). Except for Fais, Yap's atolls and reef islands lie within a 2–5 m zone of sea-level rise and storm surge, so increased salinization could result in accelerating decline in staple crop production in the short term (2030), with production potentially disappearing entirely by 2050 (Hoeke et al. 2013; Chui and Terry 2015; McGregor et al. 2016).

The 'big ocean, small islands' context of Micronesian islands contributes to the environmental and economic exposure and risks of the nation and its communities to natural disasters. It is probable that lower islands in the FSM will be rendered uninhabitable in 20–30 years' time, and their inhabitants will need to relocate, perhaps to higher islands. In extreme circumstances, sea-level rise and its associated

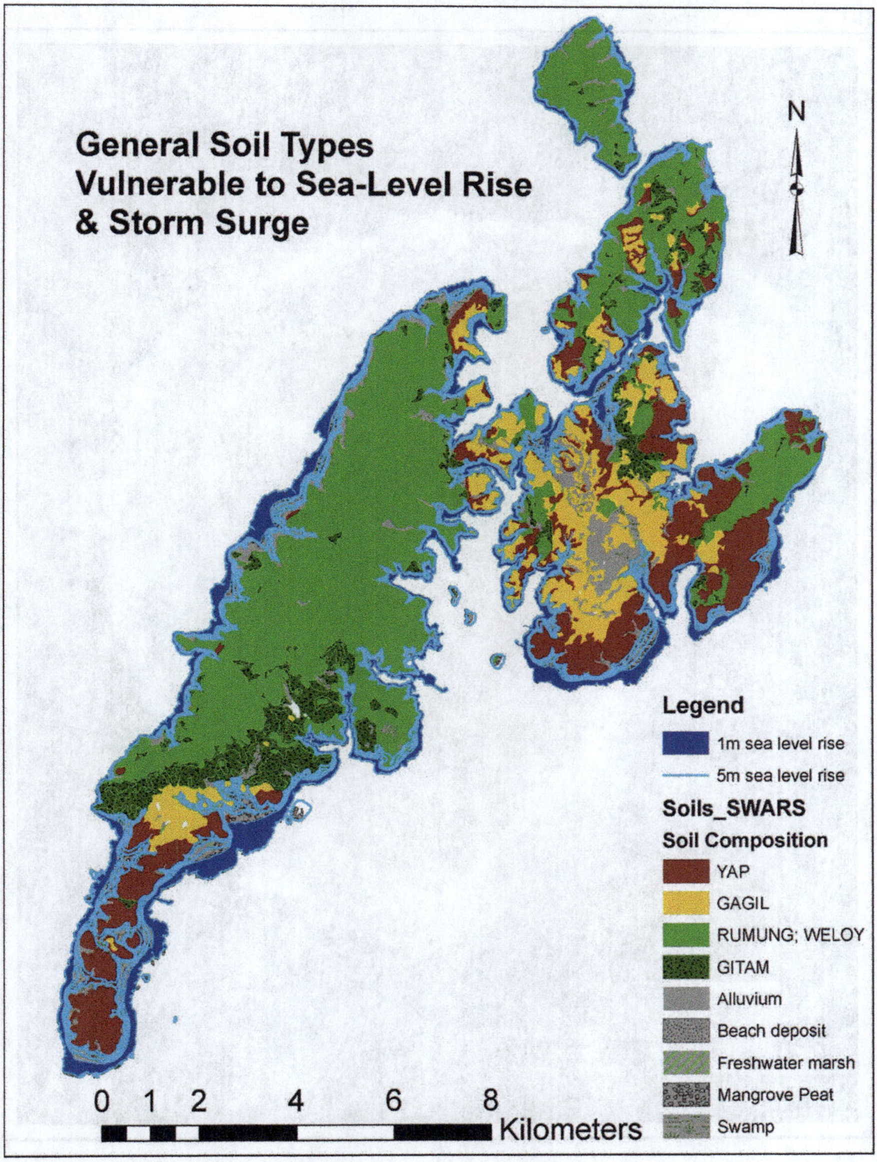

Fig. 7 Spatial analysis of the impact of a 1-meter and 5-meter rise in sea level on areas suitable for food production on Yap Proper. General soil types displayed with a blue overlay indicating the zone of 1-m sea level rise (in dark blue) and the areas vulnerable to a 5-m storm surge. Yap's most fertile soils are vulnerable to storm surge (Map by John Waayan)

consequences will trigger abandonment and significant off-island migration (Leatherman 1997; Nicholls and Mimura 1998; IOM 2008). Hoeke et al. (2013) showed that modern sea-level rise appreciably increased the severity of inundation

(a)

Fig. 8 **a** Satellite imagery of Falalop islet (post typhoon Maysak), Ulithi atoll, showing reef flats, reef crest (zone of breaking waves) and shallow submerged reef. **b** Typhoon Maysak passed directly over at peak intensity (175 mph) with storm surge of about 4 m causing extensive damage estimated over USD 8.5 million

events across the western Pacific and concluded that such events become more frequent. Adaptation along higher-island coasts, such as those of Yap Proper, is likely to involve relocation of the most vulnerable coastal settlements and their

(b)

Fig. 8 (continued)

lowland terrestrial food-producing systems to higher ground, perhaps in the immediate hinterland (Nunn et al. 2017). The cost of 1 m global mean sea-level rise in 2100 (considering land loss due to submergence and protection costs) is projected to be about 5.3% of national GDP for Micronesia, which is approximately 17 million USD (Anthoff et al. 2010).

Migration as an Adaptation Strategy

While the populations of several islands and island groups in the Pacific (e.g., Tuvalu, Carteret Islands, and Kiribati) have been perceived as the first probable victims of rising seas so that their inhabitants would become, and in some quarters already are seen to be, the first "environmental" or "climate change refugees" (Cochrane 2010; Luetz 2008), there are lesser known human migrations in the western Pacific owing to a myriad of climate change-induced impacts (Connell 2015; Krishnapillai 2017). Atoll communities in the FSM have been facing the implications of climate variability and change for the past two decades at least in four different ways: (i) the intensification of natural disasters, (ii) increased warming and drought and access to clean water, (iii) saltwater intrusion making subsistence crop production impossible, and (iv) accelerated coastal erosion and more frequent inundation making coastal areas uninhabitable. While sea-level rise alone is causing adverse impacts, combinations of extreme events have a profound impact on the livelihoods of communities on atolls. Although considerable variations occur in the physical manifestations of climate variability and change, climate stressors threaten the life-support systems of atoll communities (Campbell 2014). Failure of life-support systems—land security, food security, and habitat security—resulting from climate shocks and disasters propel vulnerable atoll communities into poverty traps. Low adaptive capacity will eventually force these communities to migrate to the highlands on Yap Proper in search of better living opportunities. Elsewhere in the region, certain thresholds have already been reached in Sorol atoll, where there is little choice left but to abandon and move. As the impacts of climate change are becoming more consequential and life is changing quickly for the atoll communities, islands are being lost below the surface of the ocean, and with them, unique and irretrievable cultures and languages are disappearing as well (Hoot et al. 2012).

The past two decades have seen a tremendous influx—induced and through forced mobility—of atoll communities to Yap Proper. Although such migrations were classified as "sojourning" and recorded in the past (Nelson 1976), today such movements are largely permanent and are synonymous with environmental migration. For atoll communities living in harsh environments, migration is the preferred option to maintain their livelihoods. It is important, however, not to view migration as simply the failure of atoll communities to adapt to climate change but instead as one of the ways in which atoll communities have chosen to adapt to changing environments (IOM 2009).

Displaced atoll communities on Yap Proper reside principally on four settlements: Gargey, Daboch, Ruu, and Makiy. These settlements are located on the Gagil-Tomil plateau on Yap Proper (Fig. 9). The areas were barren until 2004 when displaced atoll communities began colonizing gradually after the devastation from typhoon Sudal. In the aftermath of this natural disaster, rehabilitation programs expanded significantly across the State, with hundreds of households established across four major settlements, however, without effective livelihood support initiatives. The vulnerability of displaced atoll communities was exacerbated by relocation[4] to a degraded landscape. The sudden movement of atoll communities into a degraded landscape on Yap Proper, which de jure belongs to local government but under the de facto control of atoll communities, placed additional strain on livelihood support systems, often exacerbated by atoll communities being unfamiliar with crop production practices on degraded landscapes. Following the events of natural disaster and subsequent displacement, there was an urgency to prioritize the creation of an action plan for these atoll communities and implement it to enhance their adaptive capacity and climate resilience.

[4]Community relocation refers to "the permanent or long-term movement of a community or a significant part of it, from one location to another" (Campbell et al. (2007).

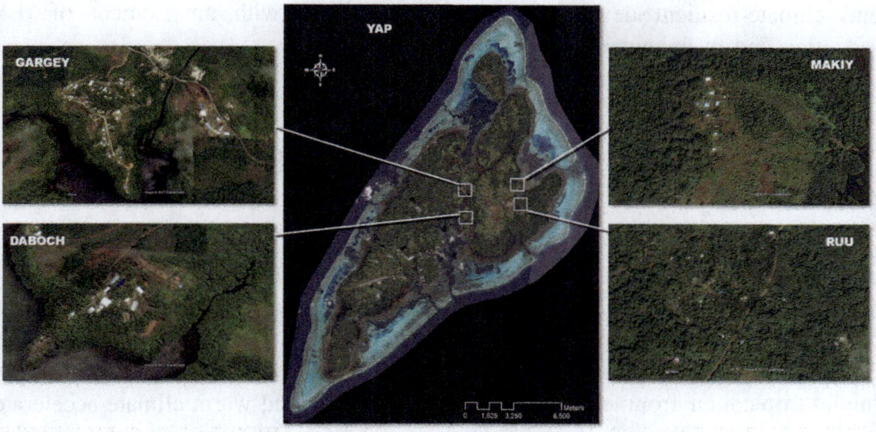

Fig. 9 CAAR Project's locations on Yap Proper

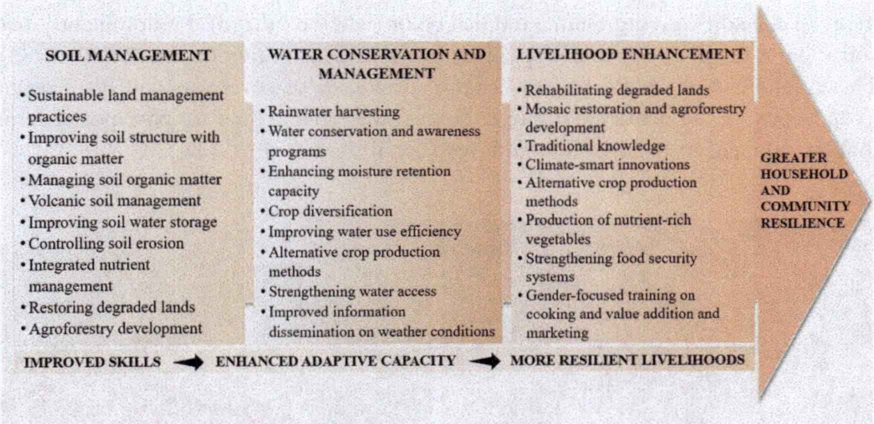

Fig. 10 CAAR Project adaptation framework. Atoll communities follow three strategies concurrently over three years and gain multiple benefits

The next section describes the work being applied to specific communities as part of the PACAM Project on *Climate Adaptive Agriculture and Resilience* toward enhancing the adaptive capacity and community resilience in the face of changing climate and variability. Based on a previously successful adaptation model implemented locally (Krishnapillai 2017), the project adopts a three-pronged approach for addressing the multiple challenges of food insecurity, water security, and livelihood enhancement of displaced atoll communities in degraded landscapes (Fig. 10). These adaptation strategies are designed to ensure short and long-term benefits, even under difficult conditions, given the current knowledge and anticipated futures. Adaptation strategies integrate activities that will generate sustainable and climate-resilient development benefits together with an element of risk reduction and management.

The Climate Adaptive Agriculture and Resilience (CAAR) Project

Atoll community settlements on Yap Proper are located on the Gagil-Tomil volcanic plateau. Soils in this region have developed in rocks that probably formed during the Miocene, but the soils may have formed in the Pleistocene (Johnston et al. 1960). That timeframe allows ample time for deep rock weathering in Yap's humid tropical environment. However, heavy rainfall and warm climate accelerate nutrient depletion through leaching. Dodson and Intoh (1999) suggested that major forest destruction, probably anthropogenic in origin, began about 3000 years ago, and the now extensive savanna may be an artifact of human impact on vegetation and soil. Leaching of nutrients causes a residual buildup of iron and aluminum in many of the soils, giving them a reddish color (when oxidized). Predominantly two soil types—Gagil Series and Yap Series—are present in this region (Smith 1983). These soils either sustain forests because of a lack of severe topsoil disturbance (Yap Series) or are degraded through topsoil removal and then support mostly ferns that are adapted to harsh soil conditions (Gagil Series) (Fig. 11).

Fig. 11 Soil profile of Yap Series (left) and Gagil Series (right) soils. These are highly weathered soils with some fertility under forest cover (Yap Series) or completely degraded (Gagil Series) (Yap Series photo by Dr. Bob Gavenda)

Soil Management

Climate-change-induced displacement threatens the food security of atoll migrants by affecting four dimensions: availability, accessibility, stability, and utilization. Hence, the core of the CAAR Project's approach is to enable atoll communities to meet food security challenges while utilizing degraded landscapes for sustainable crop production. Knowing the condition of soil and its properties is fundamental for making decisions about sustainable soil management practices that contribute to climate-smart land use. Therefore, the initial focus was to educate the communities in sustainable land-management practices. Since the Gagil series and the Yap series volcanic soils are relatively low in soil organic carbon, primary efforts were to increase the soil organic carbon (SOC) content by the addition of soil organic matter (SOM). All dimensions of soil health are linked to SOM. Communities adopted a locally improved, quick, and efficient hot composting method using 'brown and green alternative' for making compost and its use in household gardens. The addition of compost on a regular basis helped to improve the soil quality by reducing the bulk density and increasing porosity and water retention capacity. Eventually, this helped the communities to establish traditional crops such as cassava, taro, and banana around the settlement early into colonization.

The significance of traditional agroforestry systems in re-greening a degraded landscape as a sustainable food production system has been reported (Krishnapillai and Gavenda 2014; Krishnapillai 2017). Along with an emphasis on volcanic soil management, mosaic restoration[5] methods coupled with soil conservation measures, mulching, and nutrient management were also introduced to restore marginal lands. Atoll communities' customs, traditions, and local practices complement mosaic restoration at the local settings. Such efforts are aimed at strengthening resilience and keeping future management options open.

Coping with chronically-variable yields of food crops is critical for survival of vulnerable populations in marginal environments. Realizing the importance of nutrition and food security for enhanced action on adaptation, household sustainable food production systems are supported using climate-smart, alternative-crop production practices in complement with soil-restoration strategies. Sustainable intensification of vegetable production using methods such as raised-bed gardening, container home gardening, micro gardens, and small-plot intensive (SPIN) farming ensures the nutrition and food security of displaced communities.

[5]Mosaic restoration integrates trees into mixed-use landscapes, such as agricultural lands and settlements. Trees support people by improving water quality, increasing soil fertility, and boosting other ecosystem services. This type of restoration works best in deforested or degraded landscapes with moderate population density.

Water Conservation and Management

Water is the prime channel through which the impacts of climate change are felt by the coastal communities in Yap. Most of these impacts are from changes in the water cycle—through increased variability of rainfall, higher temperatures, and extreme weather events linked to ENSO phenomena. These climate events affect coastal communities' water supplies with little warning. Although water management as an adaptation strategy generally involves a range of options related to policies, investments, resource management, and institutional and technical factors, in smallholder systems, primary options demand implementation locally on fields and farms. Actions that increase the capacity to store and access water when needed increase the communities' resilience to climate variability. Therefore, the CAAR Project primarily focused on domestic rainwater harvesting[6] and storage.

Yap experienced below normal rainfall and severe drought associated with the 2015/16 El Niño. The climate throughout Yap during 2015 caused a sharp decline in rainfall. The 2015 fourth-quarter rainfall of 344 mm at the Yap Weather Office was the driest such three-month total in the post-WWII historical climate record at that observation site. While the annual rainfall of 2015 was 90.2% of normal (120.93″), 2016 data was 80.1% of normal recorded for Yap between 1981 and 2010. The effects of low rainfall included a complete drawdown of the municipal reservoir and wildfires that scorched roughly four percent of the land area of Yap Proper. While the atoll communities in Gargey, Daboch, and Ruu are connected to a public water system, the Makiy communities are exclusively dependent on rainwater. The El Niño-induced drought caused atoll communities to move afar to fetch water from distant locations.

The water-management strategy that the CAAR Project adopted involved a simple yet innovative rainwater-collection system for every household using a Bob Rainwater Bag. It is a trademark product developed by Relief International (Fig. 12). Bob offers 350 gallons of water storage and provides water right at the doorstep of every household. In terms of impact, this method proved to be simple, viable, replicable, and capable of bringing clean water to every household of the atoll communities. It also contributes to the growing body of knowledge about domestic rainwater harvesting, and it directly contributes to poverty reduction by improving access to safe water in every household and relieving women and children from the drudgery of fetching water while improving health and quality of life of household members.

Further, training and education on water conservation methods strengthened the communities to implement various field-water conservation methods to enhance crop growth using traditional practices. Water use efficiency on the field is

[6]Domestic rainwater harvesting refers to the collection, storage, and use of rainwater at the household level. Given the variable but ubiquitous rainfall on the island and the dispersed nature of rural settlements, domestic rainwater harvesting is an ideal way of providing access to water directly from community households.

Fig. 12 Bob rainwater collection bags, a trademark product developed by Relief International, help communities in collecting enough rainwater for doemstic use as well as for gardening

improved by minimizing losses of water from evaporation, runoff, or drainage and increasing water infiltration with methods such as maintaining a protective cover of mulch or the use of ridges and furrows (e.g., for taro or sweet potato cultivation), residue management, and promotion of agroforestry practices. Water conservation in raised-bed gardening and container gardening is achieved through plant choice, selection of an appropriate organic growth medium such as compost or coco pith, and efficient watering methods.

Livelihood Enhancement

Given the reliance of vulnerable atoll communities on environmental services for their livelihoods, another element of the adaptation approach implemented through the CAAR Project was mosaic restoration activities. For atoll communities living on fragile and degraded lands, deteriorating environmental conditions undermine their livelihoods and capacity to cope with disasters. Protecting and enhancing the natural services that buffer communities from climate impacts and providing them with a range of assets for coping with shocks not only addresses immediate development priorities but also improves local capacities to build resilient livelihoods. The mosaic restoration approach has the advantage of meeting immediate development needs while contributing to longer-term capacity development that creates a basis for reducing future vulnerabilities.

Being a pro-poor, pro-woman, pro-nature and locally adapted method, the mosaic restoration method increases community livelihood options. The agroforestry adoption at the local setting is improved by complementing it alongside local practices, atoll communities' cultural norms, and traditions. Integration of trees with traditional crops, household gardens, and other cultural land uses fulfills the 'double filter' requirements as recommended in forest landscape restoration (FLR)[7] efforts.

Less hunger, more cash from leafy vegetables was one of the concepts adopted to address food insecurity faced by most households that presented them an opportunity to tackle the dilemma of malnutrition—undernutrition, micronutrient deficiencies, obesity and diet-related non-communicable diseases—in the communities. The objective was to produce nutritious vegetables as part of a large crop portfolio using container gardening, raised bed gardening, SPIN farming, or micro gardens for reducing hunger and providing a source of alternate income. Focusing efforts on increasing the sustainable production of staple food crops confers significant nutritional benefits. In the process, communities learn new skills and how to grow more diverse crops along with traditional crops with a long-term goal of increasing livelihood resilience. Under the *Expanded Food and Nutrition Education Program's* (EFNEP) extension activity component, communities are also introduced to new cooking recipes for consumption of fresh, nutritious vegetables. Leafy vegetables like Chinese cabbage grow well in containers and mature in 4–5 weeks, and high demand in local markets increases communities' livelihood options. For many households, the sale of surplus vegetables raises household incomes. Atoll communities, who are mostly women, engaged with field staff during training and demonstration activities to gain new knowledge and methods of growing, thus ensuring potential sustainability beyond the project period.

As part of agroforestry-mediated restoration activities, intercropping of high-value crops such as pineapple with breadfruit, chestnuts, mango, or coconut has been encouraged to diversify incomes and increase resilience. Additionally, hands-on training with food preservation and the value addition of agriculture produce was provided. Food preservation enables communities to prepare food for times when the harvest is destroyed by natural calamities. Similarly, through

[7]The term forest landscape restoration (FLR) has been widely recognized as an important means of not only restoring ecological integrity at scale but also generating additional local-to-global benefits by boosting livelihoods, economies, food and fuel production, water security, and climate-change adaptation and mitigation. It combines adaptive management, participatory techniques, and new and not-so-new technologies to create a flexible and creative approach to the use of trees in degraded landscapes. The FLR initiative implies that any restoration efforts should improve the ecological functioning of a landscape and the well-being of the human communities that reside in that landscape. It is a forward-looking and dynamic approach, focusing on strengthening the resilience of landscapes and creating future options to adjust and further optimize ecosystem goods and services as societal needs change or new challenges arise (IUCN and WRI 2014).

hands-on training, community members have learned methods of making dry chips and flour from seasonal crops like breadfruit for food security during times of calamity.

Discussion

Rising sea levels and the associated impacts on coastal settings put low-lying islands and atolls in the Pacific region at the center of international attention as they are among the most vulnerable environments where climate change has imminent and catastrophic impacts (Weiss 2015). Episodic extreme water level events pose another serious threat to these tiny ecosystems (Fletcher and Richmond 2010; Wannier 2011; Hoeke et al. 2013; Connell 2015; Weatherdon et al. 2016; Werner et al. 2017). Higher than normal sea levels allow more wave energy to pass over reef systems (Keener et al. 2012). Combined with high waves, this increases inundation of coastal areas of low-lying islands and atolls that contribute to coastal erosion. This, in turn, leads to saltwater intrusion, which damages freshwater sources and crops (Werner et al. 2017) and affects atoll communities' life-support systems in many ways (Campbell 2014). Managing the effects of climate variability and change is thus a significant challenge for atoll communities. Developing an adaptation plan is crucial to enhance the resilience of these vulnerable communities in the face of potential climate impacts. However, a plan is not the final solution to climate change on coastal communities. How to take the plan into action and implement it in the local communities and find opportunities for enhanced resilience and development of coastal communities is essential. The PACAM Climate Fund-sponsored CAAR Project is a welcome move in this direction.

The diversity of Micronesian islands and atolls in both physical and human attributes, and their response to climate-related drivers means that climate change impacts, vulnerability, and adaptation are variable from one island to the other and among islands in each state. On small islands and remote atolls where resources are often limited, recognizing the starting point for action is critical to maximizing benefits from adaptation. They do not have uniform climate risk profiles (*high confidence*) and not all adaptations are equally appropriate in all contexts (Nurse et al. 2014). However, climate-induced internal displacement of atoll communities and their relocation to degraded landscapes in Yap provides an entry point for implementation of appropriate adaptation strategies. Atoll communities' displacement here is considered as part of an integrated adaptation strategy (Barnett and Webber 2010). With atoll environments already under pressure to sustainably provide livelihoods for their populations, it is possible that climate change effects act as triggers to intensify the processes of degradation and reduce access to livelihoods (Campbell 2014).

The IPCC (2014) describes adaptation as "the process of adjustment to actual or expected climate and its effects." Through adaptation, societies and communities can seek to moderate the harm of current and future climate risks or to take advantage of new opportunities. For atoll communities, adaptation options include steps that could be implemented locally (in situ) to facilitate continued occupancy in atoll islands or movement to locations that provide greater security as part of an integrated adaptation strategy (Campbell 2014; Barnett and Webber 2010).

Throughout history, Pacific islanders have adjusted to and coped with climate, climate variability, and extremes with varying degrees of success (Barnett and Campbell 2010). Micronesian cultural resilience is long-standing (Thomas 2015), and the stonework tradition and its intangible complement embody the long-standing interaction between island peoples and the changing coastline (Nunn et al. 2017). For example, Nunn et al. (2017) argues that "both 'defensive' and 'offensive' forms of coastal stonework (*chabog*)—the former conceived when sea level was rising, the latter celebrating subsequent sea-level fall—explains the current ambiguous response to coastal change in islands like Yap where the forms are mixed. By recognizing that the contemporary situation is one that now requires more defensive than offensive stonework may help convince coastal dwellers on high Micronesian islands like Yap to subscribe to long-term adaptation strategies." Although stonework practice such as *chabog* is embedded in Yapese culture for coastal protection in higher islands, upgrading or maintaining coastal defenses to protect human lives, properties, and infrastructure in low elevation atolls is often expensive in remote settings. This is partly because adaptation costs in small island developing states are relatively higher per capita due to the small size of their populations, as well as due to the vast geography of the region and dispersed populations (Betzold 2016). Protection of coastal settings necessitates versatile structures to resist the constant dynamic forces resulting from wave action. Often shoreline protection with hard structures in low-elevation atoll islands is ineffective or unsustainable (Betzold 2015). Moreover, when coastal communities are repeatedly affected by extreme events, a stoic in situ adaptation framework often becomes imprudent.

The CAAR Project's adaptation strategies are designed to ensure short and long-term benefits, even under harsh conditions on a degraded landscape. However, these methods are equally applicable in other vulnerable coastal settings since they are pragmatic and designed to protect both natural and human communities from the myriad of changes underway due to climate change. These strategies are both effective (appropriate to the environment where it is applied) and sustainable (atoll communities are involved in the activities, support them, and they are committed to sustaining them into the future). These strategies combine various activities that offer climate-resilient development benefits. For example, the mosaic restoration approach has the advantage of meeting immediate development needs while contributing to longer-term capacity development that creates a foundation for reducing future vulnerabilities. Agroforestry is "nature's green infrastructure" that provides many ecosystem services including shoreline protection, sustaining livelihoods besides sequestering carbon. To maintain their effectiveness over time, atoll communities are involved in all activities, support services, and information sharing. It is highly likely that atoll communities will face many intense climate-impacted issues and challenges in the future. Therefore, these adaption strategies are not a one-time event but an ongoing and iterative process that will be subjected to periodic evaluation of performance and fine tuning.

Outcomes and Challenges

Focused adaptation strategies—Integrated climate-smart adaptation strategies used in the CAAR Project have offered some broader insights and lessons to atoll communities for enhancing adaptive capacity and resilience on a degraded landscape. The coherent extension approach involved outreach, technical assistance, and education opportunities that were explicitly designed and involved all stakeholders. These activities are focused on strengthening the livelihoods of atoll communities by improving access to services, knowledge, and resources.

Soil management—Volcanic soil health management through composting, mulching, and incorporation of traditional crops and agroforestry systems gradually improves soil properties, making conditions more favorable for crop growth. Mulching protects soils from excessive heat, exposure to wind, and moisture loss.

Potential rates of carbon sequestration in response to improved soil management vary widely as a function of land use, climate, soil, and many other factors. With land (where displaced atoll community settlements are situated) being severely degraded, it is perhaps unreasonable to expect a higher rate in the beginning. The very low carbon levels in the Gagil and Yap series of soils (1–2% of organic matter) mean that a 1% increase in carbon content of soil mass is feasible. However, such an increase would take 50–60 years at a rate of 0.25 t C ha^{-1} $year^{-1}$ (a conservative rate, as per Lal and Bruce 1999). Maintaining an appropriate level of soil organic matter and biological cycling of nutrients is challenging in humid tropical conditions. In practice, however, the land is expected to be used for agriculture before that rate of C sequestration would apply.

Vegetable gardening—Incorporation of an alternative mode of crop production in conjunction with mosaic restoration activities increased production of nutritious vegetables for home consumption and provided an alternate source of income. Restoration agroforestry has great potential for re-greening degraded lands in a less expensive and participatory way, creating a basis for improved livelihoods, water provision, and sustainable food production.

Climate-smart agriculture—Integration of traditional and modern climate-smart agriculture practices maximizes the use of natural processes, enhances the diversity of production, and tailors production intensity to the capacity of the degraded landscape. Improving soil health and crop diversity and the important components of traditional farming systems can make a significant contribution to resilience and, at the same time, enhance adaptive capacity to better manage future challenges.

Rainwater harvesting—Installation of efficient rainwater harvesting Bob bags across the households reduced the communities' burden to fetch water from distant sources while improving their health and quality of life. Collection and conservation of rainwater and soil water management strategies helps in vegetable gardening and rehabilitating degraded land.

Agroforestry—Introduction of agroforestry systems and sustainable land management practices enhances crop diversity in the long run, and it increases food security and livelihood benefits. Agroforestry systems are locally based and incorporate indigenous species and techniques that have been used traditionally in the Pacific Islands for many generations. Fully established agroforestry systems will eventually reduce communities' dependence on purchased products and resources, as well as reducing vulnerability to changing market conditions. Exploring market opportunities for multipurpose tree products will have significant economic, social, and environmental benefits in the future.

Adaptation strategies such as protecting or restoring coastal ecosystems using 'soft' and 'green' infrastructure is always a better option than building seawalls, flood barriers, or other 'hard' structures that interrupt natural processes and degrade natural habitats. However, adaptation costs are relatively higher per capita due to the remoteness of atoll islands and isolated populations.

Food availability—Communities gain knowledge on food preservation and value addition trainings to meet food security challenges during off seasons or at times of adversity.

Continuous learning process—Bringing in new information to the atoll communities is indeed a challenge and a key barrier to adaptive capacity. However, local field staff serve as conduits for information and support to the dissemination of information in the local vernacular to non-literate members of the community. They ensure that outreach services are strengthened and responsive to community needs. The Project's gender plan provides focused formal and informal trainings to female members of the household and enhances their participation. Women's participation in adaptation strategies contributes to their empowerment as they earn their own income.

Multiple benefits—A climate-smart adaptation strategy is a "multiple benefit" approach as it typically builds climate resilience alongside other benefits. The instituted methods are fully scalable. However, context-specific priorities need to be determined. By promoting strategies, revitalizing degraded lands, and scaling up innovative adaptation methods, crop production can be enhanced to support the food security requirements of atoll communities living on marginal lands.

Replicable—The fact that these adaptation methods can be easily replicated in other geographic locations increases the importance of the present work. Furthermore, the activity and its results are measurable (effectiveness), it produces results with a reasonable level of resources and time (efficiency), the activity addresses priority climate risks (relevance), and it can be implemented over a long period of time (sustainable).

Although adaptation strategies under the CAAR Project are moving in the right direction, it is too early to assess whether the project has fully enabled atoll communities to effectively cope with the climate extremes they face.

Conclusion

The coastal systems of Yap Islands are experiencing climate change-induced damages to human and economic development with adverse effects already imposing a great threat to the small and dispersed communities living in highly vulnerable, low-lying atoll islands. Climate change-induced damages cause irreversible loss of habitable areas across this region. Yap Islands' natural resource base and livelihood opportunities of coastal communities are seriously undermined by the combined effects of sea-level rise, extreme weather events, rapidly progressing coastal erosion, increasing crop damage, and reductions in freshwater supply. Atoll communities are disproportionately affected with poverty and hardship, which serve as both the drivers and consequences of inadequate livelihoods that in turn lead to forced migration from coastal settings. An informed and timely response plan and its implementation has, therefore, becomes indispensable to enhance the resilience of vulnerable communities. The CAAR Project addresses vulnerability issues of displaced coastal communities in Yap and an implemented adaptation framework aims to strengthen their adaptive capacity and enhance resilience to future threats.

The project demonstrates three focused climate-smart adaptation technologies and practices targeting the displaced atoll communities residing at four settlements on Yap Proper. It is developed from the traditional knowledge and practices of the atoll communities coupled with climate-smart agriculture and built upon three vulnerable sectors—traditional agriculture, water, and community development. This program directly benefits over 100 households and about 1000 individuals indirectly (e.g., extended family members, clans). This approach ensures economic viability, improved biodiversity, and ecosystem services benefiting all island communities.

The motivation for this project supports displaced atoll communities to better manage and adapt to extreme climate pressures in the context of food security through community-based climate smart adaptation. Food is not only a basic need for these climate migrants, it is the single (and often fragile) support they have for maintaining their livelihood. This project, therefore, promotes the production of household-level food production systems utilizing traditional and climate-smart methods.

Adaptation benefits results from the catalytic and innovative nature of a previous successful attempt with a group of displaced atoll communities in Gargey settlement and the valuable lessons learned and information generated. By its concurrent focus on enhancing food security through traditional crops coupled with nutrient-rich vegetables, promotion of rainwater-harvesting systems and water conservation, and promoting resilient household livelihood opportunities, the program brings together crucial elements needed to reduce vulnerabilities and cope with disasters and climate extremes while embracing the traditional culture. Community resilience is a pre-condition for achieving Sustainable Development Goals.

Empowering atoll communities by providing appropriate outreach assistance and extension services allows them to improve their livelihoods by harnessing new methods and knowledge. The CAAR Project's climate-smart interventions are highly location-specific and knowledge-intensive. Considerable efforts are underway to develop knowledge and capacities to make climate-smart adaptation approach a sustainable venture. The approach offers an opportunity for atoll communities to revitalize and overcome barriers while also adjusting to the new realities of climate extremes.

Future Prospects

Adaptation is a vital part of the climate-change framework. Actions to enable adaptation to climate change present opportunities to promote sustainable development. Since the CAAR Project's adaptation interventions encompass the local customs and traditions of atoll communities, it is likely that these approaches will be successful in the long run. By providing innovative adaptation options such as climate-smart, alternative crop-production methods, communities are empowered to increase and sustain crop productivity that is affordable and low-risk. When empowered, atoll communities are better positioned to adapt. This project is creating a secure environment for atoll communities to learn and apply methods and share results with those who are still enduring risk in atoll settings.

A fundamental insight emerging from the climate change adaptation work in Yap is that of the multiplying effect of the program's strategies and the communities' acceptance and strengthening adaptive capacity. Communities see connections and value of the program's impacts to be more forward looking after initial success with the Gargey Project (Krishnapillai 2017). This has paved the way to expand the current program (CAAR Project) to four displaced atoll community settlements on Yap Proper. The project is taking another big leap and expanding the adaptation work to five municipalities covering over 300 households under the umbrella of the Adaptive Community Transformation (ACT) on Yap project funded by USAID—Office of U.S. Foreign Disaster Assistance—Catholic Relief Services. These projects represent an ongoing learning process on climate change adaptation, and the knowledge will contribute to the growing number of successful adaptation programs in the Pacific and beyond.

Acknowledgements The CAAR Project is funded by the United States Agency for International Development (USAID) under Pacific American (PACAM) Climate Fund award #: PACAM-14-0005. The author wishes to thank the Chief of Party, PACAM for encouragement and support. Many of the project concepts presented here have emerged from previous projects funded by the USDA-NIFA under the College of Micronesia Land Grant Program and USDA-OAO under the OASDVFR Program. The leadership and assistance of Executive Director Dr. Singeru Singeo is greatly acknowledged. Dr. Bob Gavenda (formerly with NRCS-PIA) is thanked for assistance with soil information and photographs and Ms. Maria Kottermair (University of Guam) and John Waayan (Division of Land Resources, Yap State Government) for digital maps. Thanks are also due to the atoll communities for their active participation, enthusiasm, and enduring support.

References

Anthoff, D., Nicholls, R. J., & Tol, R. S. J. (2010). The economic impact of substantial sea-level rise. *Mitigation and Adaptation Strategies for Global Change, 15,* 321–335.

Australian Bureau of Meteorology & CSIRO. (2014). *Climate variability, extremes and change in the Western Tropical Pacific: New Science and updated country reports* (358 pp). Melbourne: Pacific-Australia Climate Change Science and Adaptation Planning Program Technical Report, Australian Bureau of Meteorology and Commonwealth Scientific and Industrial Research Organization.

Barnett, Jon, & Campbell, J. R. (2010). *Climate change and small island states: Power, knowledge, and the South Pacific*. London: Earthscan.

Barnett, J., & Webber, M. (2010, April). *Accommodating migration to promote adaptation to climate change.* Policy Research Working Paper 5270. Washington, D.C.: World Bank.

Betzold, C. (2015). Adapting to climate change in small island developing states. *Climatic Change, 133,* 481–489.

Betzold, C. (2016). *Aid and adaptation to climate change in Pacific island countries*. Development Policy Centre Discussion Paper #46, Crawford School of Public Policy. Canberra: The Australian National University.

Cai, W., Borlace, S., Lengaigne, M., van Rensch, P., Collins, M., Vecchi, G., et al. (2014). Increasing frequency of extreme El Niño events due to greenhouse warming. *Nature Climate Change, 4,* 111–116.

Campbell, J. R. (2014). Climate change migration in the Pacific. *The Contemporary Pacific, 26,* 1–28.

Campbell, J. R., Goldsmith, M., & Koshy, K. (2007). Community relocation as an option for adaptation to the effects of climate change and climate variability in Pacific Island Countries (PICs). Final report for APN project 2005-14-nsy-Campbell. Tokyo: Asia-Pacific Network for Global Change Research.

Chowdhury, M. R., Chu, P. S., Zhao, X., Schroeder, T. A., & Marra, J. J. (2010). Sea level extremes in the U.S.-Affiliated Pacific Islands—A coastal hazard scenario to aid in decision analyses. *Journal of Coastal Conservation, 14,* 53–62.

Chui, T. F. M., & Terry, J. P. (2015). Groundwater salinization on atoll islands after storm surge flooding: Modelling the influence of central topographic depressions. *Water and Environment Journal, 29,* 430–438.

Church, J. A., Monselesan, D., Gregory, J. M., & Marzeion, B. (2013). Evaluating the ability of process based models to project sea-level change. *Environmental Research Letters, 8,* 015051 (8 pp).

Cochrane, S. (2010). Floating land—Rising sea: arts and minds on climate change. *LiNQ, 37,* 93–103.

Connell, J. (2015). Vulnerable Islands: Climate change, tectonic change, and changing livelihoods in the Western Pacific. *The Contemporary Pacific, 27,* 1–36.

Dickinson, W. R. (2001). Paleoshoreline record of relative Holocene sea levels on Pacific islands. *Earth Science Reviews, 55,* 191–234.

Dodson, J., & Intoh, M. (1999). Prehistory and palaeoecology of Yap, Federated States of Micronesia. *Quaternary Journal, 59,* 17–26.

Dolan, A. H., & Walker, I. J. (2006). Understanding vulnerabilities of coastal communities to climate change related risks. *Journal of Coastal Research, 39,* 1317–1324.

Falanruw, M. V. C. (1993). Micronesia agroforestry: Evidence from the past, implications for the future. USDA Forest Service General Technical Report, PSW-GTR-140.

Federated States of Micronesia National Government (FSM) & U.S. Forest Service. (2010). Federated States of Micronesia state-wide assessment and resource strategy 2010–2015+.

Fletcher, C. H., & Richmond, B. M. (2010). *Climate change in the Federated States of Micronesia; food and water security, climate risk management, and adaptive strategies* (28 pp). Report, University of Hawaii, Honolulu.

FSM Census Bureau (2010). 2010 FSM-wide census of population and housing, Yap State. Office of statistics, budget & economic management, overseas development assistance and compact management (SBOC), FSM National Government.

Hay, J. E. (2013). Small island developing states: coastal systems, global change and sustainability. *Sustainability Science, 8,* 309–326.

Hay, J. E., & Mimura, N. (2013). Vulnerability, risk and adaptation assessment methods in the Pacific Islands region: Past approaches, and considerations for the future. *Sustainability Science, 8,* 391–405.

Hoeke, R. K., McInnes, K. L., Kruger, J. C., McNaught, R. J., Hunter, J. R., & Smithers, S. G. (2013). Widespread inundation of Pacific islands triggered by distant-source wind-waves. *Global Planetary Change, 108,* 128–138.

Hoot, W., Kázmér, M., & Taborosi, D. (2012, September 9). The abandonment of Sorol: The first depopulation of a Micronesian atoll in modern history. In H-Soz-Kult (Ed.), shrinking, sinking, resurfacing: Small islands and natural hazards in historical and current perspective, 22.11.2012–24.11.2012 Munich. www.hsozkult.de/event/id/termine-20506.

Hunter, J. (2010). Estimating sea-level extremes under conditions of uncertain sea level rise. *Climatic Change, 99,* 331–350.

Hunter, J. (2012). A simple technique for estimating an allowance for uncertain sea level rise. *Climatic Change, 113,* 239–252.

Intoh, M. (1997). Human dispersals into Micronesia. *Anthropological Science, 105,* 15–28.

IOM. (2008). *World migration report 2008: Managing labour mobility in the evolving global economy* (p. 562). Geneva: International Organization for Migration.

IOM. (2009). *Compendium of IOM's activities in migration, climate change and the environment* (p. 320). Geneva: International Organization for Migration.

IPCC. (2014). Summary for policymakers. In Field, C. B., V. R. Barros, D.J. Dokken, K. J. Mach, M. D. Mastrandrea, T. E. Bilir, et al. (Eds.), *Climate change 2014: Impacts, adaptation, and vulnerability. Part A: Global and sectoral aspects. Contribution of Working Group II to the Fifth Assessment Report of the Intergovernmental panel on climate change* (pp. 1–32). Cambridge and New York: Cambridge University Press.

IUCN & WRI. (2014). *A guide to the restoration opportunities assessment methodology (ROAM): Assessing forest landscape restoration opportunities at the national or sub-national level* (125 pp.). Working Paper (Road-test edition). Gland: IUCN.

Jevrejeva, S., Moore, J. C., & Grinsted, A. (2012). Sea level projections to AD 2500 with a new generation of climate change scenarios. *Global Planetary Change, 80–81,* 14–20.

Johnson, C. G., Alvis, R. J., & Hetzler, R. L. (1960). Military geology of Yap Islands, Caroline Islands. Prepared under the direction of the Chief of Engineers, U.S. Army by the Intelligence Division, Office of the Engineer, Headquarters United States Army Pacific with personnel of the United States Geological Survey.

Keener, V. W., Marra, J. J., Finucane, M. L., Spooner, D., & Smith, M. H. (Eds.). (2012). *Climate change and pacific islands: Indicators and Impacts. Report for the 2012 Pacific Islands Regional Climate Assessment*. Washington, D.C.: Island Press.

Keim, M. E. (2010). Sea-level-rise disaster in Micronesia: sentinel event for climate change? *Disaster Medicine and Public Health Preparedness, 4,* 81–87.

Krishnapillai, M., & Gavenda, R. (2014). From barren land to biodiverse home gardens. *Farming Matters, 30,* 26–28.

Krishnapillai, M. (2017). Climate-friendly adaptation strategies for the displaced Atoll population in Yap. In W. Leal Filho (Ed.), *Climate adaptation in Pacific countries: Fostering resilience and improving the quality of life* (pp. 101–117). Berlin: Springer.

Lal, R., & Bruce, J. P. (1999). The potential of world cropland soils sequester C and mitigate greenhouse effect. *Environmental Science & Policy, 2,* 177–185.

Leatherman, S. P. (1997). *Island states at risk: Global climate change, development and population* (p. 242). Florida: Coastal Education Research Foundation.

Lough, J. M., Meehl, G. A., & Salinger, M. J. (2011). Observed and projected changes in surface climate of the tropical Pacific. In J. D. Bell, J. E. Johnson, & A. J. Hobday (Eds.), *Vulnerability of tropical Pacific fisheries and aquaculture to climate change* (pp. 49–99). Secretariat of the Pacific Community: Noumea.

Lough, J., Gupta, A. S., Power, S. B., Grose, M. R., & McGree, S. (2016). Observed and projected changes in surface climate of tropical Pacific Islands. In M. Taylor, A. McGregor, & B. Dawson (Eds.), *Vulnerability of Pacific agriculture and forestry to climate change* (pp. 47–102). Secretariat of the Pacific Community: Noumea.

Luetz, J. (2008). *Planet prepare: Preparing coastal communities in Asia for future catastrophes.* Bangkok: Asia-Pacific Disaster Report, World Vision International.

McGranahan, G., Balk, D., & Anderson, B. (2007). The rising tide: Assessing the risks of climate change and human settlements in low elevation coastal zones. *Environment and Urbanization, 19,* 17–37.

McGregor, A., Taylor, M., Bourke, R. M., & Lebot, V. (2016). Vulnerability of staple food crops to climate change. In M. Taylor, A. McGregor, & B. Dawson (Eds.), *Vulnerability of Pacific agriculture and forestry to climate change*. Noumea: Pacific Community.

Monirul, M., & Mirza, Q. (2003). Climate change and extreme weather events: Can developing countries adapt? *Climate Policy, 3,* 233–248.

Nedachi, M., Maeda, T., Shigeyoshi, R., Inoue, A., Shimada, K., Higashi, M., et al. (2001). Geological environments of Yap Islands, Micronesia. In: *Kagoshima University Research Center for the Pacific Islands, Occasional Papers* No. 34, 69–76, 2001 Part 1, Section 2, Report 1. The progress report of the 1999 survey of the research project *social homeostasis of small islands in an island-zone*.

Nelson, J. M. (1976). Sojourners versus new urbanites: Causes and consequences of temporary versus permanent cityward migration in developing countries. *Economic Development and Cultural Change, 24,* 721–757.

Neumann, B., Vafeidis, A. T., Zimmermann, J., & Nicholls, R. J. (2015). Future coastal population growth and exposure to sea-level rise and coastal flooding—a global assessment. *PLoS ONE, 10*(3), e0118571. doi:10.1371/journal.pone.0118571.

Nicholls, R. J., & Mimura, N. (1998). Regional issues raised by sea level rise and their policy implications. *Climate Research, 11,* 5–18.

Nunn, P. D., Runman, J., Falanruw, M., & Kumar, R. (2017). Culturally grounded responses to coastal change on islands in the Federated States of Micronesia, northwest Pacific Ocean. *Regional Environmental Change, 17,* 959–971.

Nurse, L. A., McLean, R. F., Agard, J., Briguglio, L. P., Duvat-Magnan, V., Pelesikoti, N., et al. (2014). Small islands. In Barros, V.R., C.B. Field, D.J. Dokken, M.D. Mastrandrea, K.J. Mach, T.E. Bilir, et al. (Eds.), *Climate change 2014: Impacts, adaptation, and vulnerability. Part B: Regional aspects . Contribution of Working Group II to the Fifth Assessment Report of the intergovernmental panel on climate change* (pp. 1613–1654). Cambridge and New York: Cambridge University Press.

Pam, C., & Henry, R. (2012). Risky places: Climate change discourse and the transformation of place on Moch (Federated States of Micronesia). *Shima: The International Journal of Research into IslandCultures, 6,* 30–47.

Power, S., Delage, F., Chung, C., Kociuba, G., & Keay, K. (2013). Robust twenty-first-century projections of El Niño and related precipitation variability. *Nature, 502,* 541–545.

Richmond, B. M., & Reiss, T. E. (1994). Vulnerability of the natural coastal systems to accelerated sea level rise, Yap Islands, Federated States of Micronesia: Case study preliminary results. U.S. Geological survey open-file report 94–665. Geological Survey, U.S. Department of the Interior.

Richmond, B. M., Mieremet, B., & Reiss, T. E. (1997). Yap Islands natural coastal systems and vulnerability to potential accelerated sea level rise. *Journal of Coastal Research, 24,* 153–172.

Santha, S. D. (2015). Adaptation to coastal hazards: the livelihood struggles of a fishing community in Kerala, India. *Disasters, 39,* 69–85.

Scialabba, N. (1998). *Integrated coastal area management and agriculture, forestry and fisheries. FAO Guidelines* (256 pp.). Rome: Environment and Natural Resources Service, Food and Agriculture Organization of the United Nations.

Seager, R., Naik, N., & Vogel, L. (2012). Does global warming cause intensified interannual hydroclimate variability? *Journal of Climate, 25,* 3355–3372.

Shea, E .L., Dolcemascolo, G., Anderson, C. L., Barnston, A., Guard, C., Hamnett, M. P., et al. (2001). Preparing for a changing climate: The potential consequences of climate variability and change. A report of the Pacific Islands Regional Assessment Group for U.S. Global Change Research Program, 8 pp.

Smith, C. W. (1983). *Soil survey of islands of Yap, Federated States of Micronesia* (90 pp.). U.S. Department of Agriculture, Soil Conservation Service.

Taborosi, D., & Neth, Y. (2014). *Student atlas of the Federated States of Micronesia* (pp. 55). Palikir, Pohnpei: Island Research & Education Initiative.

Taylor, M., McGregor, A., & Dawson, Brian (Eds.). (2016). *Vulnerability of Pacific agriculture and forestry to climate change*. Noumea: Pacific Community.

Thomas, F. R. (2015). Marginal islands and sustainability: 2000 years of human settlement in eastern Micronesia. *Economic and Ecohistory, 11,* 64–74.

United States Agency for International Development. (2009). *Adapting to coastal climate change —A guidebook for development planners* (pp. 147). Washington D.C.: Global Climate Change Team, USAID.

Wannier, G.E. (2011). Threatened island nations: Summary of legal issues. *Climate Law Blog*. Columbia Law School, Center for Climate Change Law. http://blogs.law.columbia.edu/climatechange/2011/07/11/threatened-island-nations-summary-of-legal-issues/.

Weatherdon, L. V., Magnan, A. K., Rogers, A. D., Sumaila, U. R., & Cheung, W. W. L. (2016). Observed and projected impacts of climate change on marine fisheries, aquaculture, coastal tourism, and human health: an update. *Frontiers in Marine Science, 3,* 1–21.

Weiss, K. R. (2015). Before we drown we may die of thirst. *Nature, 526,* 624–627.

Werner, A. D., Sharp, H. K., Galvis, S. C., Post, V. E. A., & Sinclair, P. (2017). Hydrogeology and management of freshwater lenses on atoll islands: Review of current knowledge and research needs. *Journal of Hydrology*. doi:10.1016/j.jhydrol.2017.02.047.

Wong, P. P., Losada, I. J., Gattuso, J. P., Hinkel, J., Khattabi, A., McInnes, K. L., et al. (2014). Coastal systems and low-lying areas. In C. B. Field, V. R. Barros, D. J. Dokken, K. J. Mach, M. D. Mastrandrea, T. E. Bilir, et al. (Eds.), *Climate change 2014: Impacts, adaptation, and vulnerability. Part A: Global and sectoral aspects. Contribution of Working Group II to the Fifth Assessment Report of the Intergovernmental Panel on Climate Change* (pp. 361–409). Cambridge: Cambridge University Press.

Capacity Development and TVET: Accredited Qualifications for Improving Resilience of Coastal Communities—A Vanuatu Case Study

Tess Martin, Sarah Hemstock, Helene Jacot Des Combes and Charles Pierce

Abstract For countries like Vanuatu, climate change is the most significant single threat to sustainable development, in particular due to the large proportion of the population living in coastal communities. Additionally Vanuatu is the world's most at-risk country for natural hazards mainly affecting coastal communities (Birkmann and Welle in The world risk index, 2015). One of the key barriers to improving Pacific Island Countries' resilience to climate change impacts is the lack of local and regional capacity and expertise resulting from the absence of sustainable accredited and quality assured formal training programs in climate change adaptation (CCA) and disaster risk management (DRM) (Jordon et al. in Climate change policy in the European Union: confronting the dilemmas of adaptation and mitigation. Cambridge University Press, United Kingdom, 2010; Martin et al. in Training needs and gap analysis. Suva, Fiji, 2015). The European Union funded PacTVET project has partnered with The Pacific Community (SPC) and the German aid agency (GIZ) Coping with Climate Change in the Pacific Region (CCCPIR) programme to support the delivery of the first accredited TVET certificate in Climate Change and Disaster Risk Reduction (CCDRR) in the Pacific Islands region. This TVET qualification provides outcomes based learning specifically focused on coastal communities through practical activities and field work

T. Martin (✉) · H. J. D. Combes
PACE-SD, The University of the South Pacific, Suva, Fiji
e-mail: martin_t@usp.ac.fj; tesskmartin@gmail.com

H. J. D. Combes
e-mail: helene.descombe@usp.ac.fj

S. Hemstock
EDD, The Pacific Community, Nabua, Fiji
e-mail: sarahh@spc.int

C. Pierce
EU PacTVET, Albany, WA, Australia
e-mail: Charliepierce19@gmail.com

W. Leal Filho (ed.), *Climate Change Impacts and Adaptation Strategies for Coastal Communities*, Climate Change Management,
https://doi.org/10.1007/978-3-319-70703-7_6

involving vulnerable coastal areas throughout the provinces of Vanuatu. The delivery of the CCDRR course is being led by the Vanuatu government through the Vanuatu Institute of Technology and is leading regional and global developments in formal accredited TVET training for Climate Change and Disaster Risk Reduction.

Introduction

The small island states of the South Pacific have been described as a "global front line" in the struggle of developing countries to adapt to the adverse impacts of climate change (Ferris et al. 2011). Without addressing climate change, sustainable development cannot be achieved. At the international level, the Sendai Framework for Disaster Risk Reduction 2015–2030 (UNISDR 2015) is a 15-year non-binding agreement which recognizes that the State has the primary role to reduce disaster risk but that responsibility should be shared with other stakeholders including local government, the private sector and other stakeholders. The Sustainable Development Goals (SDGs) (United Nations 2016) and the Paris Agreement (United Nations 2015), are two other seminal landmark agreements that call for "capacity building in disaster risk reduction, sustainable development, and climate change adaptation and mitigation".

Education and capacity building have roles to play in achieving the alignment and delivering the outputs of these three global initiatives (the Sendai Framework, SDGs and the Paris Agreement). Findings from the needs and gap analyses of all 15 Pacific–African, Caribbean and Pacific (P-ACP) countries indicate that formal qualifications which account for local contexts are required to build national capacity to: accurately monitor and assess impacts of climate change and natural hazards; identify solutions to reduce these risks; and plan, manage and implement risk reduction projects to reduce damage and losses (Martin et al. 2015).

Responsive and accredited qualifications provide quality assurance and trust that in turn ensures interventions managed by those having qualifications in Resilience are really supporting sustainable development, thereby: limiting the impacts of climate change and natural hazards; empowering locals to become involved actors in their own development; and limiting maladaptation and generation of new risks.

The significance of capacity building for climate change adaptation to the sustainable development of the Pacific Island countries and territories was seen in the endorsement by the Forum Leaders of the Pacific Islands Framework for Resilient Development (UNIDSR 2015) which replaces the Pacific Islands Framework for Action on Disaster Risk Reduction and Disaster Management.

The Vanuatu government supports the global and regional direction in its national policy on Climate Change and Disaster Risk Reduction (2016–2030). A key priority for the Vanuatu government is stated as 'achieving sustainable and resilient development across all levels and sectors in the small island nation, by addressing the risks faced from climate change and disaster impacts' (Government of the Republic of Vanuatu 2015).

With support from the EU PacTVET project, The Pacific Community (SPC) and the German aid agency (GIZ) Coping with Climate Change in the Pacific Region (CCPIR) programme Vanuatu is currently delivering the first nationally accredited TVET qualification in Climate Change and Disaster Risk Reduction (CCDRR) in the region (and world). This paper presents this initiative in Vanuatu focused on coastal communities and also makes the case for using accredited regional and national TVET qualifications to support capacity development. Capacity development through TVET qualifications is a process of empowerment that comes with an understanding that practical skills can directly impact livelihoods, cultures and the environment. Capacity development is a foundational aspect of successful overseas development assistance and effectiveness in meeting long-term sustainable development goals. The current delivery of a Certificate 1 in CCDRR in Vanautu is leading developments addressing the development of effective climate change adaptation strategies for coastal communities.

TVET and Development

Increasingly practitioners and policy makers working across the globe are recognising the importance of bringing together disaster risk reduction and climate change adaptation. From studies across 15 Pacific in nations, a key barrier to improving national resilience to disaster risks and climate change impacts has been identified as a lack of capacity and expertise resulting from the absence of sustainable accredited and quality assured formal training programmes in the disaster risk reduction and climate change adaptation sectors. In the 2016 UNISDR Science and Technology Conference on the Implementation of the Sendai Framework for Disaster Risk Reduction 2015–2030, it was raised that most of the training material available are not reviewed either through a peer-to-peer mechanism or by the scientific community and are, thus, not following quality assurance standards. In response to these identified barriers, there has been a call for accredited formal qualifications for capacity development.

Education and training are at the heart of the post-2015 sustainable development agenda and are considered essential for the success of all sustainable development goals. A consensus exists around the world that qualification frameworks based on learning outcomes are appropriate tools for the reform and expansion of educational and training provision in ways that will raise skills levels, improve labour market productivity and contribute to sustainable development (UNESCO Institute for Lifelong Learning 2015). Qualifications are developed and delivered, assessed through formal structured quality assured mechanisms that enhance outcomes including opportunities for graduates.

Qualification Frameworks

The need for national and regional qualifications frameworks was raised at the regional level more than three decades ago and was highlighted by Bartram (2004). A Pacific Qualifications Framework (PQF) and a Pacific Register of Qualifications and Standards (PQRS) are now in place. The main purpose of the PRQS is to facilitate the benchmarking of nationally offered Pacific qualifications against international standards. Pacific Island Countries (PICs) are encouraged to develop national qualification frameworks and quality assurance systems and to link these to the Pacific regional qualification framework to enhance the quality standing of their national systems of education and training. To date four Pacific Island Countries[1] have developed and aligned national qualifications frameworks with the PQF. Subsequently the national qualifications accredited in these countries are listed on the PRQS (EQAP 2011).

Qualifications and Quality Assurance

In the Pacific region although there has been substantial ad hoc informal training in resilience over the past decade there is a recognized absence of formal vocational training in this field (Martin et al. 2015). This absence combined with the increasing demand for such expertise has created a skills shortage. TVET qualifications establish the all-important links between the world of work and the world of education; they create a common language which is understood by employers, learners and all stakeholders. Qualifications signal that a person possesses certain knowledge, skills and attributes and, thus enable employers to make use of approved, recognized and thus quality assured qualifications. Additionally regionally-acknowledged qualifications enhance labour mobility, although the intention of the EU PacTVET project is that the capacity built via quality assured training (in particular at the Certificate 1 and 2 levels) will primarily benefit communities within which the training has taken place. In order to do this it is recognised that employment opportunities will need to be available for graduates in order to make the most of their enhanced capacity. With this in mind engagement with the private sector and aligning with relevant industry standards has been an integral part of the design and delivery of this project.

[1]Tonga, Samoa, Fiji, Vanuatu.

The Pacific Regional Federation for Resilience Professionals (PRFRP)

The EU PacTVET project has established the first global and regional industry body for Resilience, the Pacific Regional Federation for Resilience Professionals (PRFRP). The PRFRP will raise the profile of climate change adaptation and disaster risk management and their linkages to development. Membership is open to organisations, training institutions, private sector green and sustainable environment focused businesses, industry associations, utilities, government departments, multi-lateral and bi-lateral development partners, international agencies; and individuals.

The PRFRP will support the work of the EU PacTVET project in monitoring and reviewing the initial qualifications and developing further quality assured regional qualifications and processes. It will also promote and facilitate an enabling environment for sustainable community climate change adaptation, disaster risk reduction and management, and the maintenance of ecosystem services in coastal communities.

EU PacTVET Project

One of the key barriers to improving Pacific Island Countries' resilience to climate change impacts is the lack of local and regional capacity and expertise (Jordon et al. 2010; Martin et al. 2015). Although there has been substantial ad hoc training led by development partners over the past two decades, the lack of expertise is the result of the absence of relevant curriculum, trained personnel and well-resourced and equipped training institutions to deliver on the required formal training programme. Limited availability of appropriate training related to climate change adaptation and disaster risk reduction has led to: lack of locally trained people to implement and monitor projects; use of donor funds to support foreign experts; unsuccessful projects causing maladaptation or vulnerability and risk.

Findings from a regional needs and gap analysis (Martin et al. 2015) that formal qualifications which account for local contexts are required to build national capacity to: accurately monitor and assess impacts of climate change and natural hazards; identify solutions to reduce these risks; and plan, manage and implement risk reduction projects to reduce damage and losses.

The EU PacTVET project (2014–2019) was designed specifically to address this gap. The project aims to enhance Pacific–African, Caribbean and Pacific's (P-ACPs) regional and national capacity and technical expertise to respond to climate change adaptation (CCA) and sustainable energy (SE). The EU PacTVET project partnered with the Fiji Higher Education Commission (FHEC) to facilitate the establishment of regional Certificates 1–4 in Resilience to ensure quality assurance and alignment with Pacific regional frameworks. The accreditation of regional qualifications is a process managed by the Education Quality Assessment Program (an arm of The Pacific Community) to confirm compliance with PQF levels and the Pacific Quality Assurance Framework (PQAF).

The EU PacTVET project is using TVET as a vehicle to support regional and national developments addressing the increasing vulnerability in the Pacific Islands region to climate change impacts and natural hazards. At the same time the project is enhancing national and regional developments for TVET within the education and workforce development sectors.

Context

Vanuatu's society, environment and economy are highly vulnerable to climate change and disaster risks. A 2012 United Nations report assessed Vanuatu as one of the most highly exposed countries in the world to disaster risks. The devastating consequences of category 5 tropical cyclone Pam in March 2015 and the recent strongest ever out-of-season cyclone to hit the Southern Hemisphere (May 2017) highlight the country's risk from natural hazards. Predicted increases in extreme weather from climate change means the country will face even greater impacts.

Though a tiny country by global standards, among South Pacific nations Vanuatu is relatively large, comprising a double chain of about 40 islands, and 40 islets and rocks of volcanic and coral origin (approximately 65 inhabited). The total land area of nearly 13,000 km^2 includes more than 2500 km of coastline (see Fig. 1). Over 65% of the nation's people rely on subsistence agriculture, while the remainder rely on a few key industries, of which the most valuable is tourism (ADB 2009). According to the World Bank in 2016, the total population of Vanuatu numbered more than 270,000, of which approximately 25% lived in urban areas in and around the capital Port Vila, and Luganville. The majority of rural Ni-Vanuatu are subsistence farmers and fishermen. Given the extensive coastal area in Vanuatu where the majority of the population live, coastal communities are the country's focus for economic activity. Best estimates of long term, systematic changes associated with climate change indicate that by 2050, sea level is likely to have increased by 20 cm thus increasing the challenges for the many people living in these vulnerable coastal communities of Vanuatu.

The EU PacTVET project has partnered with the SPC and GIZ[2] funded Coping with Climate Change in the Pacific Island Region (CCCPIR 2009–2018) programme which aims to ensure that the skills and capabilities of the local population, national governmental authorities and regional organisations are enhanced in order to cope with the effects of climate change and combat its causes. The CCCPIR also spans over the fifteen P-ACPs. These two projects (EU PacTVET and CCCPIR) have combined resources in Vanuatu and worked collaboratively with the

[2]Deutsche Gesellschaft fur Internationale Zusammenarbeit (German government aid agency).

Fig. 1 Map of Vanuatu. *Source* http://www.lonelyplanet.com/maps/pacific/vanuatu/map_of_vanuatu.jpg

government of Vanuatu to support the delivery of the first TVET accredited training in Climate Change and DRR in the Pacific Islands region. Delivery of accredited TVET training in this subject area is a global first and will provide a local graduate cohort skilled to contribute to effective climate change adaptation in coastal communities of Vanuatu. This remainder of this paper presents the progress to date with this initiative.

Certificate 1 in Climate Change and Disaster Risk Reduction—Overview

The development of the Climate Change and Disaster Risk Reduction (CCDRR) course commenced in April 2014 with a request from the Vanuatu Rural Development Training Centres Association (VRDCTA) to The Secretariat of the Pacific Community and GIZ to revise an existing regional training manual on Climate Change and Disaster Risk Management to suit the national and local context and further developed to suit specific industry needs (e.g. agriculture). Due to the large proportion of the population living in coastal communities in Vanuatu the field work component in the CCDRR course focuses on coastal regions. Funds to develop curriculum and resources were provided and managed by SPC/GIZ. A long-time teacher at the Vanuatu Institute of Teacher Education was recruited to develop the skills package in collaboration with the GIZ project team. Initial consultations were conducted over a four-week period with key industry, government, NGO, education and community personnel based in Port Vila. The course was planned and designed at certificate levels 1 and 2 on the Vanuatu Qualifications Framework (VQF) which is equivalent to the Pacific Qualifications Framework (PQF) (SPC, Pacific Qualifications Framework 2011; Vanuatu Qualifications Authority 2015). These levels provide the skills and knowledge a graduate would be expected to have on the successful completion of all learning. The resulting skills package comprised seven units of competency, with a Learner's Guide; Learner Workbook and Trainer/Facilitator Guide for each of the units. These resources were designed to also support the course as a 'Training of Trainers'. All resources were developed in both English and French. The units collectively provide a graduate profile for employment and/or community support. For example a graduate of the CCDRR course could be a key contact for overseas development partners seeking local input to project/programme design, an environment/conservation officer, lead community/provincial person advising government and NGOs on local climate change issues and/or a local advisor for research activities and government policy planning. An important aspect of the course was the recognition of traditional knowledge as a critical component in all areas for planning and implementing effective adaptation strategies. Another key feature was the focus on learner interaction and communication with, and support for local communities, in particular local coastal communities.

The delivery of this course first piloted in May 2015 at the Fisher Young Rural Training Centre (RTC) on Vanua Lava in the Banks Islands, Torba Province. This delivery targeted student learning at a certificate 1 level. 14 secondary school aged students attended (9 male and 5 female). These students were also enrolled in vocational courses at the RTC such as carpentry, business and tourism. The learning in the CCDRR course provided these students with skills and knowledge integrated with other industry sectors.

A second pilot training course was held in Luganville, Santo Island in September 2015 for trainers working at Rural Training Centres throughout Vanuatu. The focus of this ten-day training course was to 'Train the Trainers'. The 33 participants (23 male, 10 female) were predominantly RTC trainers from the 6 provinces in Vanuatu. 3 RTC managers and 3 VRDTCA staff attended along with development partner participants.

The learning focused on gaining an understanding through experimentation and practical activities. For example participants studied the impacts of increased concentrations of greenhouse gases on sea levels and ocean pH values by conducting experiments such as; the effect of heating on the water level in a sealed plastic bottle and the effect of immersing a shell in an acidic solution (white vinegar). Field trips were conducted in coastal communities as these are considered the most vulnerable sites to the impacts of climate change and natural hazards in Vanuatu. The delivery was presented by the same consultant who had designed the curriculum and who is fluent in the three main languages of Vanuatu: English, French and Bislama. The learning resources provided to the RTC trainers included soft and hard copies of learning guides for all seven units of the CCDRR course, the set of 16 pictures from “Learning about climate change the Pacific Way: A visual guide” (SPC-GIZ), the Teacher’s Guide for these pictures, and the video-clip “Klaod Nasara”. Students’ diagrams on traditional fishing and food preservation techniques developed through these pilot courses were subsequently included in the revised guides.

In late 2016 the consultant worked with the Vanuatu Qualifications Authority (VQA) to support a submission for national accreditation of the CCDRR course. The VQA process of accreditation determines the appropriate certificate level. The submission for the national accreditation was completed by the Vanuatu Institute of Technology in late 2016. At the same time the regional certificates 1 and 2 in Resilience (developed by the EU PacTVET project) were mapped to the outcomes in the CCDRR course to facilitate subsequent submissions for the national accreditation and recognition of these regional qualifications in Resilience (Martin et al. 2016). Both the CCDRR and the Certificates 1 and 2 in Resilience provide pathways to further study in Resilience (Certificates 3 and 4) which have eight elective strands including ‘Coastal Management’.[3]

CCDRR: A Mechanism to Address Climate Change Impacts for Coastal Communities in Vanuatu

The CCDRR course is currently being delivered to 29 students from the six provinces of Vanuatu. The course was advertised in December 2016 and a total of 95 applications were received. Selection of students was based on the criteria of island and location of residence, level of education, previous experience in the fields of disaster risk reduction and climate change, and opportunities to advance awareness of climate change and disaster issues in their local communities. There are 19 male and 10 female students participating in this course. Further details are provided in Table 1.

The delivery of the CCDRR course is being led by the same consultant who developed the curriculum and delivered the pilot training courses. He is being supported by a full-time VIT local trainer and administratively by VIT. There are periodic meetings with the VIT management team to report on all matters and facilitate capacity building and sustainable development for this course and subsequent ones of the same nature. The development partners (SPC-GIZ and EU PacTVET) agreed to a combined budget supporting a full scholarship for all students who would be based in Port Vila for a period of five months for the full-time

[3]Agriculture, Coastal Management, Energy & Infrastructure, Fisheries, Forestry, Health, Tourism, Water Resources.

Table 1 CCDRR student details

Age		Province		Highest level of education achieved	
18–20 years	7	TORBA	2	Year 10 secondary schooling	1
21–24 years	11	SANMA	6	Rural Training Centre (CCDRR trial course)	1
25–30 years	8	MALAMPA	3	Year 12 secondary schooling	2
31–40 years	3	PENAMA	2	Year 13/Foundation (university entrance)	11
40 years+	0	SHEFA	15	Certificate or Diploma in Agriculture	3
		TAFEA	1	Undergraduate university courses completed (1st and 2nd year)	9

course in the first half of 2017. Students from provinces and locations outside Port Vila are accommodated at the VIT student hostel boarding facilities with all costs met by the scholarship.

The students in this current course are gaining first-hand experience working with vulnerable coastal communities through extensive field work. Specific examples are visits to produce hazard risk maps for five peri-urban villages around Port Vila, visits to investigate how people in five villages along the coast of north-west Efate are contributing to greenhouse gas emissions through their daily activities, and visits to the offshore island of Pele to study various mitigation and adaptation measures in use. Further field studies are planned to investigate vulnerability to hazards at community and individual level, and assisting the coastal village communities to plan and implement adaptation strategies.

At the opening of the course on 27th February, 2017, the Director General of Education stated how proud the government was that the Vanuatu Institute of Technology will deliver the first-ever certificate course in Climate Change and Disaster Risk Reduction and additionally, the first ever such course in the Pacific region. The delivery of the CCDRR course in Vanuatu and the pending national accreditation at a Certificate 1 level is leading regional and global developments in formal accredited TVET training for Climate Change and Disaster Risk Reduction which will provide skilled local personnel to lead the development of effective coastal community adaptation strategies.

Vanuatu Quality Assurance Framework (VQAF)

Trust in training and qualifications plays a crucial role for all stakeholders. The accreditation and certification process becomes particularly important in this context, and quality-assurance mechanisms are essential to ensure that processes effectively generate credibility and trust at local, national and regional levels.

Vanuatu has a Quality Assurance Framework (VQAF). This was developed by the Vanuatu Qualifications Authority (VQA) established in 2014, formerly the Vanuatu National Training Council, to ensure credibility and transparency in the services of training providers throughout Vanuatu. VQA is legally responsible for the development and maintenance of the National Qualifications Framework (VQF) through standards and qualifications setting, quality assurance, accreditation, equivalency of qualifications and assessment including the recognition of prior learning.

The VQF underpins the VQA quality assurance processes by enabling consistent alignment of nationally recognized competency standards and courses to relevant qualification levels. It supports the identification of pathways that enable people to move between different education and training sectors and levels. The National QF enables alignment of Vanuatu qualifications to regional and international qualifications systems. The standards for provider registration and course accreditation embrace the assessment of training provider capabilities, and in particular, whether the providers effectively offer training and assessment services up to clients' expectations. The Vanuatu Institute of Technology as a registered national provider was successful in its application to VQA to deliver the CCDRR course in 2017.

The rigour and diligence applied by the VQA processes ensure quality assurance of national qualifications which in turn enhance mutual recognition and pathways with regional qualifications and other PICs nationally quality assured qualifications.

The EU PacTVET model for accreditation of regional qualifications requires potential training providers to submit applications to deliver complete or partial regional qualifications to the relevant National Qualification Authorities (NQA) in the Pacific Island country of delivery. The NQAs apply their own national approval processes. The NQAs subsequently take on a major role in monitoring the delivery, assessment and issuing of awards (certification) for the regional qualifications. Given that Vanuatu is one of only four countries with a national quality assurance framework aligned with the PQAF, and it is responsive to innovative developments to support capacity building and sustainable development, it is ideally situated to be the first of the Pacific Island countries to offer a dual qualification; nationally accredited Certificate 1 in CCDRR and regionally accredited Certificate 1 in Resilience. This work is currently in progress through a collaboration between the VQA, VIT and EU PacTVET.

Conclusion

The delivery of a nationally accredited Certificate 1 in Climate Change and Disaster Risk Reduction (CCDRR) in Vanuatu is providing participants with skills and knowledge to assist coastal communities to address the impacts of climate change and natural hazards through effective adaptation strategies. Trust in the capabilities of these trained personnel/graduates is enhanced through the quality assured certification of the training. The focus of the learning in this CCDRR course is on coastal communities (field trips and activities) which enables local personnel to contribute to real sustainable developments through the much needed capacity development for coastal communities. Thus this nationally accredited certificate is a tool which is promoting skills development that directly impacts livelihoods, cultures and the environment.

The regional Certificates 1–4 in Resilience developed by the EU PacTVET project also benefit communities within which the training has taken place enabling and empowering them to build resilience and proactively respond to the impacts of climate change and natural hazards. At the same time the EU PacTVET initiatives are supporting regional and national developments focused on developing and enhancing quality assured TVET qualifications.

TVET is being used as a vehicle to advance the aims of global, regional and national policies to enhance sustainable livelihoods and strengthen countries' capabilities to adapt to the adverse effects of climate change at national, provincial and local/community levels. Responsive and accredited regional qualifications should ensure that the interventions managed by those having accredited qualifications are really supporting sustainable development, thereby: limiting the impacts of climate change and natural hazards; empowering locals to become involved actors in their own development; and limiting maladaptation and generation of new risks. Vanuatu has provided the lead in using TVET qualifications to address the need to develop effective climate change adaptation for coastal communities.

References

ADB. (2009). *Vanuatu economic report 2009: Accelerating reform.* GFDRR (Global Facility for Disaster Reduction and Recovery) Vanuatu. Mandaluyong City, Philippines: In Disaster Risk Management Programmes for Priority Countries—East Asia and the Pacific.

Bartram, J. (2004, July). *A review of Technical and Vocational Education and Training (TVET) Policy and Planning in the Pacific Islands.* Retrieved from Pacific Association of Technical and Vocational Education and Training: http://unevoc.net/fileadmin/user_upload/pubs/TVETPolicyReviewPacific.pdf.

Birkmann, & Welle. (2015). *The world risk index* (U. o. Stuttgart, Producer). Retrieved April 2017, from http://www.uni-stuggart.de/ireus/Internationales/WorldRiskIndex.

EQAP. (2011). *The pacific qualifications framework.* Secretariat of the Pacific Board for Educational Assessment (now EQAP—Educational Quality Assessment Programme), Secretariat of the Pacific Community. Suva, Fiji: IBSN:978-982-00-0501-3.

Ferris, E., Cernea, M. M., & Petz, D. (2011). *On the front line of climate change and displacement learning from and with Pacific Island countries.* School of Economics Project on Internal Displacement, London: The Brookings Institution.

Government of the Republic of Vanuatu. (2015). *Vanuatu climate change and disaster risk reduction policy (2016–2030).* Suva: Secrtariat of the Pacific Community.

Jordon, A., Huitema, D., Van Asselt, H., Rayner, T., & Berkhout, F. (2010). *Climate change policy in the European Union: Confronting the dilemmas of adaptation and mitigation.* United Kingdom: Cambridge University Press.

Martin, T., Hemstock, S. H., Jacot Des Combes, H., Buliruarua, L.-A., Kua, N., & Satiki, V. (2015). *P-ACP Training needs and gap analysis. Synthesis report. European Union Pacific Technical and Vocatonal Education and Training in Sustainable Energy and Climate Change Adaptation project (EU PacTVET).* The Pacific Community and The University of the South Pacific. Suva, Fiji: Eu-PacTVET-FED/2014/347-438.

Martin, T., Sanerivi, L., Prasad, R., Jacot Des Combes, H., & Hemstock, S. L. (2016). Climate Change for Quality Assured Regional Qualifications in the Pacific: An innovative collaboration (EU-PacTVET & EQAP). *Asia Pacific Quality Network (APQN).* Nadi, Fiji.

SPC. (2011). *Pacific qualifications framework.* Suva, Fiji: Secretariat of the Pacific Community.

UNESCO Institute for Lifelong Learning. (2015). *Global inventory of regional and national frameworks.* European Centre for the Development of Vocational Training CEDEFOP. Hamburg, Germany: UNESCO Institute for Lifelong Learning.

UNIDSR. (2015). *Framework for resilient development in the pacific 2017–2030. An integrated approach to address climate change and disaster risk management.* Retrieved 2017, from www.pacificdisaster.net/docSRDP.pdf.

UNISDR. (2015, April). *Sendai framework for disaster risk reduction 2015–2030.* Geneva, Switzerland: UNISDR. (UNIDSR, Ed.) Retrieved 2017 from http://www.unisdr.org/files/43291_sendaiframeworkfordrren.pfd.

United Nations. (2015). *Report of the Conference of the Parties on its twenty-first session, held in Paris from 30 November to 13 December 2015. Addendum part two: Action taken by the Conference of the Parties at its twenty-first session.* Retrieved 2017 from http://unfcc.int/meetings/paris_nov_2015/meeting/8926/php/view/documents.php.

United Nations. (2016). *Sustainable development goals.* Retrieved 2017 from https://sustainabledevelopment.us.org/?menu=1300.

Vanuatu Qualifications Authority. (2015). *VQAF: Vanuatu Quality Assurance Framework.* Retrieved from www.vqa.edu.vu.

An Evaluation of Climate Change Effects on Fishermen and Adaption Strategies in Central Region, Ghana

Selorm Akaba and Samuel Akuamoah-Boateng

Abstract Government parastatals, nongovernmental organizations and civil societies in Ghana seem to have little access to detailed climate projections and seemingly little understanding of the uncertainties surrounding coastal climate change impacts and adaptations. The study examines climate change effects and adaption strategies of fishermen in the Central Region of Ghana. Through the application of mixed method qualitative and quantitative research, a sample of 116 fishermen were selected through multiphase sampling in three randomly selected administrative assemblies on the coast of the Central Region of Ghana. Respondents were sampled using accidental sampling technique from two randomly selected communities in each assembly. Face-to-face interviews were conducted using structured interview schedule. The data was analysed using descriptive statistics including frequency counts and percentages. The study revealed that erratic rainfall, strong wind/storm with high tides, hot sunshine and high temperatures, drought and flood and heavy precipitation were the main climate effects experienced by the respondents. The negative impacts of climate change experienced by respondents in the fishing business include increasing risk and uncertainty in fishing, fluctuation in fish and wild stock distributions, reduction in the duration of fishing seasons and high fish spoilage and mortality. These effects, the respondents opined, lead to low productivity, reduced income, food insecurity and labour emigration. The respondents have therefore resorted to changing fishing methods, seasonal migrations and livelihoods diversification as responses to the climate change effects. Under-capitalization, limited information, poor marketing systems, lack of research and extension services were the main constraints faced by the fishermen in responding to the climate change effects. More reliable, effective and accessible measures of adaptation are therefore recommended to improve infor-

S. Akaba (✉) · S. Akuamoah-Boateng
Department of Agricultural Economics and Extension, University of Cape Coast, Cape Coast, Ghana
e-mail: sakaba@ucc.edu.gh

S. Akuamoah-Boateng
e-mail: sakuamoa-boateng@ucc.edu.gh

W. Leal Filho (ed.), *Climate Change Impacts and Adaptation Strategies for Coastal Communities*, Climate Change Management,
https://doi.org/10.1007/978-3-319-70703-7_7

mation and knowledge through outreach on various platforms by stakeholders in ensuring sustainable fishing behavior among fishermen.

Introduction

From local to global levels, fisheries and aquaculture play important roles for food supply, food security and income generation. Some 43.5 million people work directly in the sector, with the majority in developing countries. Besides, those who work in associated processing, marketing, distribution and supply industries, the sector supports nearly 200 million livelihoods. Aquatic foods have high nutritional quality, contributing 20% or more of average per capita animal protein intake for more than 1.5 billion people, mostly from developing countries. They are also the most widely traded foodstuffs and are essential components of export earnings for many poorer countries (Cochrane et al. 2009).

Over the years, man has been able to predict the climate and its effects with some degree of precision because it has been relatively stable. Climate stability enables individuals and agencies to plan agricultural programmes for good yield. Global climate change is mostly felt among populations in developing countries. Their vulnerability to climate change comes both from being predominantly located in the tropics, and from various socioeconomic, demographic and policy trends limiting their capacity to adapt to change (Morton 2007). Sagoe (2006) indicated that climate change influences human health, agricultural yields, pest outbreaks, crop timing, flooding, and biological distribution and extinction. Understanding the impact of climate variability and adaptation strategies is therefore important for agricultural decision making at the farm, market or policy levels.

Climate change is a long-term shift in the statistics of the weather (National Oceanic and Atmospheric Administration 2007). Climate change is a multi-faceted challenge for today's societies through its impacts on human lives and the natural environment. Awareness and quality of knowledge on its existence and issues relating to climate change could reduce the impacts of the phenomenon (Acquah 2011).

In many parts of Africa, small-scale fisheries and related activities provide income to rural communities where alternative employment opportunities are scarce or even non-existent. In these situations, small-scale fisheries, fish processing and trade provide people with important, and sometimes crucial, form of safety-net that helps protect them against the effects of agricultural product price volatility, macro-economic crises, structural reforms, harvest failures, political turmoil and other factors that threaten rural stability and food security. In this way, small-scale fisheries substitute and/or complement other economic activities and help households sustain their standard of living and food purchasing power (Williams and Rota 2012).

Fish has always been and will continue to be an important source of protein in the diet of Ghanaians. The demand for fish has escalated because of rapid population growth (Lokko and Anson 2006). Therefore, anything that affects fishing and the fishing community should be a matter of concern to the nation. Despite the importance of fish in Africa, the Sub-Saharan region is the only region of the world where per capita consumption of food fish has fallen (FAO 2009).

In the fishing industry, one of the major effects of climate change is occurrence of unpredictable storms. The most serious impact of storms in the fishery sector is the risk to the lives of the fishermen and/or destruction of their capital such as boats and fishing gear. Storm-related damage to capital assets, including boats and fishing gear, means the loss of income and livelihood, especially for poor fishing households. These victims may not have adequate savings to replace their capital assets, and are likely to face food insecurity unless these are replaced immediately through public support. The disruption of fishing activities of such households could also affect the livelihood and food security of others, for instance, small traders who buy and sell fish in small local retail markets (FAO 2007).

Individual fish actively select and rapidly change living areas based on suitable temperatures, oxygen concentrations, and food availability. Cold-water fish will actively avoid temperatures that exceed their preferred temperature by 2–5 °C and seek out refuge areas of cooler water such as groundwater or seepage areas and headwater streams (Meisner 1990; Gunn 2002).

Williams and Rota (2012), noted that there are many adaptation strategies to climate change effects available, which benefit or provide an advantage to small-scale fishers and fish-farmers. These include direct adaptations to specific changes as well as actions that increase the resilience and adaptive capacity of communities and ecosystems, and environmental (over-fishing, habitat destruction, pollution) stresses that can significantly increase vulnerability of communities and ecosystems to the impacts of climate change (Cheung et al. 2009; IPCC 2007; Walther et al. 2002). Many fishing communities are dependent on stocks that exhibit regular fluctuations and so have already developed considerable coping capacity (Easterling et al. 2007).

Climate change is acidifying the ocean, which increases dissolved CO_2 and decreases ocean pH, carbonate ion concentration, and calcium carbonate mineral saturation in the ocean (Cooley and Doney 2009). Johnson and Marshall (2007) asserted that ecological effects of climate change on tropical marine systems are predicted to be diverse and long-lasting. Observations are already supporting projections of increasing sea and air temperatures, rising sea levels, acidifying oceans, intensifying storms, and changing rainfall patterns and ocean currents (Marshall et al. 2010). Similarly, Fabricius et al. (2007) reported that widespread degradation of coral reef ecosystems will result from mass coral bleaching and ocean.

The potential impacts of global climate change (such as unpredictable rainfall, increasing temperatures, and longer dry periods) add to the vulnerability of Ghanaian coastal fishing systems. Although the general consequences of climate change are becoming better known, great uncertainty remains about how climate change affects specific locations (Nutsukpo et al. 2013) and more especially the coastal areas.

There is considerable uncertainty surrounding climate change, with a wide range of climate projections for Ghana (Arndt et al. 2015). The uncertainty in climate change and projections, especially for future coping strategies, brings to fore the need to highlight it for useful decision-making. Government parastatals, non-govenmental organisations, and civil societies in Ghana seem to have little access to detailed climate projections and seemingly little understanding of the uncertainties surrounding coastal climate change impacts and adaptations. Assessing the impact of climate change are critical components of poverty reduction programmes (Stanturf et al. 2011). Yet a major constraint to mainstreaming climate change within development policies is the lack of empirical evidence to inform decision making. Coastal level activities are critical for determining specific and individual effects and vulnerabilities, which either reduce or aggravate the eventual outcomes. It is essential to consider the range of potential climate realizations to identify major risks and to avoid incurring large opportunity costs if certain projections are not realized.

The rate of social, economic, and technological change in the fishery sector will gradually transform the setting in which climate change is likely to interact with sensitive features of the food system. The current state of the sector and important trends that would transform it provides a baseline against which to examine the potential consequences of climate change (Sagoe 2006). Public awareness and knowledge on climate change is crucial to managing climate change and its related problems. Ghana is faced with growing challenges in managing coastal and marine resources, such as the dramatic decline of fish stocks and the degradation of coastal resources. The capacity to utilize coastal and marine assets, while sustainably protecting them from degradation is woefully inadequate. This research therefore seeks to identify the effects of climate change on fishing and fishermen in the Central Region of Ghana and evaluate measures adopted by the fishermen to this climate change effects.

Study Objectives

The general objective of the study is to assess the effects of climate change on fishing and fishermen in the Central Region of Ghana. Specifically, the study sought to:

1. Determine the level of understanding of climate change among fishermen in the region.
2. Evaluate the impact of climate change on fishing and fishermen in the study area.
3. Examine the measures the fishermen employ to adapt to climate change situations.
4. Examine the challenges of adapting to climate change situations in relation to fishing in the study area.

This paper tries to address three main questions related to climate change and adaption in coastal areas by fishermen. First, what are the level of awareness of climate change and its effects among fishermen in the coastal areas of the Central Region? Second, what are the impacts of climate change and what adaptation strategies do fishermen in the Central Region adopt during their fishing expeditions? Third, what challenges do fishermen face in employing measures to adapt to climate change? Quantified estimations of climate change effects are beyond the scope of this paper, since it serves as a formative research and intends to bring to fore the issues about climate change from the fishermen's perspective. The study was undertaken with the view to determine the effects of climate change on fishing and adaptive measures among fishermen in major coastal fishing community in the Central Region.

Method

The setting for this study was the coast of Central Region in Ghana which comprises both urban and rural communities. It comprises of 17 administrative assemblies out of which more than half of them are along the coast of the sea; and Cape Coast, which is the regional capital, was one time seat of government for the then Gold Coast from 1830 to 1877 before the capital was relocated to Accra. The Central Region is along the coast of Gulf of Guinea; the region is characterised by predominant fishing communities. The region is bounded on the South by the Atlantic Ocean (Gulf of Guinea), the East by the Greater Accra Region, the North by the Ashanti and the Eastern Regions, and the West by the Western Region in Ghana.

Cross-sectional survey research design was employed for the data collection with a mixed method approach of qualitative and quantitative data. Cross-sectional design is deemed appropriate as data was collected from a section of fisher folks to explain the research problem. Zakour and Gillespie (2013) attest that this approach enables the researcher to assess different variables at the same time. Furthermore, Creswell (2013) also asserts that quantitative data yields numeric value and lends itself to statistical analysis while the qualitative item yields text data and is often analysed in themes. An advantage of the mixed method is that the respondents have the chance to individually justify their selection in the quantitative data with open ended questions.

In search for better ways of understanding the research problem, pragmatism philosophy guided the study to enrich research. It is believed that individuals come to know the world through the practicality or usefulness of objects (or concepts). Pragmatists, as their name would suggest, adopt a practical approach, albeit with varying emphases (Hookway 2016).

Pragmatism is mixed methods research approach that is increasingly becoming popular in modern researches (see Johnson and Onwuegbuzie 2004; Molina-Azorin and Cameron 2010). Iaydjiev (2013) contends that issues often arise in the world of human practice with new problems; and these are more appropriately addressed through the pragmatic approach. From the pragmatic philosophical perspective, climate change and its implications turn out to be a complex and a sensitive issue to individuals (fishermen). Whereas some fishermen may wish to use whatever practices to get the maximum catch with no or little regards to economic, environmental and social consequences, some are of them have the conviction of practices that will enable them continue to utilize the available resources on sustainable basis. As occurrence of extreme climatic conditions rises steadily, many more people try to adopt different fishing systems that will enable them achieve better results amid the constraints associated with their livelihood activities. The prestige that is accorded best practices soon takes a different form when external factors such as precipitation and temperature leads to extreme drought, floods, or storms. This makes the farmers experience the effect of climate change differently. Essentially, this supports the pragmatic view calling for mixed methods for research on the climate change responses.

The theoretical population in this study comprised all the fisher folk resident in at the coastal communities in the Central Region of Ghana. Population has specific common characteristics; and in this study, the population is made up of adults who go for fishing on the Atlantic Ocean. It is this population that becomes vulnerable to climate change effects, more especially those related to the coastal zones. The assumption is that increased vulnerability through intrinsic factors in the phase of environmental hazards which form the extrinsic factors could result in livelihood threats in the fishermen.

Various methods of sampling have been recommended to access valid respondents in any research study. The choice however depends on the population and design of the study among other things. In order to get the nearest accurate population of the fishermen in the study area, it was necessary to use multiphase sampling in this study (Amoah and Eshun 2015). The demarcation of the administrative assemblies on the coast guided the stratification for the sampling. Stratified sampling was used to ensure that the respondents are obtained from representatives of the districts, municipal and metropolitan assemblies along the coast of the region. Simple random sampling was then used to select three administrative assemblies which forms a third of the number of administrative assemblies along the coast. Two coastal communities were then randomly selected form the sampled administrative assemblies. Due to the inability to access the list of the fishermen within the selected communities, accidental sampling was conducted to select 40 respondents from each selected community for the study. With the initial assistance of the Assembly Men/Women together with some opinion leading fishermen were identified in the communities. This exercise ensured that the right respondents were reached for the sample size. Data was then collected from 120 respondents in 6 communities. This number was not achieved although an appreciable number was obtained. The one-on-one interaction offered high response rate in the study. Reduction in the number was attributed to some of the research instruments which were rejected finally because of some inconsistencies detected during the data cleaning and editing. At the end of the validation, 116 interview schedules were used for analysis, giving a response rate of 96.7%. Fryrear (2015) recommends response rate of 80% and above for surveys of this kind. The raw data collected was organized and analysed using IBM SPSS version 21. Descriptive statistical tools, mainly frequency counts and percentages, were used to analyse the data. The findings are presented in tabular and graphical forms below.

Results and Discussion

Background Characteristics of Respondents

Out of the 116 fishermen interviewed in the selected communities, 37.1% of the respondents were below 30 years of age while 55.1% were within the ages of 30 and 59 years. About 8% of the respondents were above 60 years of age (Table 1). This implied that most of the fishermen were in their early adulthood, an indication that most of the fishermen were in their active working age. About 80% of the fishermen been in their active working age group may be due to the demanding nature of the fishing activities such as carrying heavy tools and equipment like outboard motors, fishing nets and pulling of boats which required more physical strength.

While the majority of the respondents never had any formal education (55.1%), only 44.9% of the respondents have had some form of formal education (primary, junior high and senior high schools). The low level of education among the respondents is most likely to have influence on their level of awareness and adaptation strategies to climate change effects. The majority of the fishermen had been fishing for 15 years or more (63.86%). Meanwhile 19 (16.4%) had been in the fishing business for 5 years and below. It is therefore expected of the respondents to have a lot of experience in the fishing business. This is collaborated by the fact that, fishing is the main economic activity of most of the respondents. Just a few (8.6%) were into non-agricultural works (e.g. small scale mining, boat carving and carpentry, trade and security works) as their main occupation. The nature of the fishing business demands that the individual spend more time and resources to survive in the business. It is therefore not surprising that a lot of the respondents asserted that it is their main economic activity.

Fishermen's Perspectives of Climate Change

Doss and Morris (2001) opined that the perspectives of the indigenous people, the way they think and behave in relation to climate change, as well as their values and aspirations have a significant role to play in addressing climate change (Table 2).

Almost all of the respondents showed some level of understanding of climate change and its effect on fishing. Some of the indicators that respondents used to describe change in climate and its effect include erratic rainfall, extreme temperature (very low or very high), drought, flood and heavy precipitation, and strong

Table 1 Background characteristics of the respondents

Background	Frequency	Percent	Cum. percent
Age distribution			
< 30	43	37.1	37.1
30–39	26	22.4	59.5
40–49	23	19.8	79.3
50–59	15	12.9	92.2
> 59	9	7.8	100.0
Total	116	100.0	
Education levels			
No formal education	64	55.1	55.1
Primary	21	18.1	73.3
Middle/Junior High School	17	14.7	88.0
Senior High School	11	9.5	97.4
Others	3	2.6	100.0
Total	116	100.0	
Main occupation			
Fishing	102	87.9	87.9
Crop farmer	4	3.5	91.4
Non-agriculture jobs	10	8.6	100.0
Total	116	100.0	
Number of years in fishing business			
< 5	19	16.4	16.4
5–9	12	10.3	26.7
10–14	11	9.5	36.2
15–19	17	14.7	50.9
> 19	57	49.1	100.0
Total	116	100.0	

Source Authors' Construct, 2016 n = 116

Table 2 Respondents understanding of climate change

Climate change understanding	Frequency	Percent
Hot sunshine and extreme temperature	115	99.1
Erratic rainfall	114	98.3
Strong wind with high tides	112	96.6
Drought	107	92.2
Flood and heavy precipitation	106	91.4

Source Authors' Construct, 2016 n = 116

wind with high tides. This finding supports a report by Taderera (2010) that most Africans are aware that weather patterns are changing. Lefale (2003) also examined the utility of traditional Samoan weather and climate forecasting and found out that the respondents were acutely aware of environmental signs leading to extreme events such as tropical cyclones.

Effect of Climate Change on Fishing and Fishermen

The study sought to find out form respondents the effects of climate change on fishing and fishermen. Table 3 presents the respondents' perception about the impact of climate change on the fishing and their livelihoods.

As shown on Table 3, the respondents affirmed that climate change has increased the risks and uncertainties associated with the fishing business in these communities. Some of them explained that due to changes in weather and climate, their prediction of when to go fishing, the kind of gears they must take along and the type of fish and amount can no longer be guaranteed. They mostly catch less fish than predicted. The uncertainties could partly be because there are fluctuations in fish and wild stock distribution. Thus, more than 8 out of every 10 respondents claimed they are experiencing fluctuations in fish and wild stock distribution. As corroborated by Cochrane et al. (2009), that climate change is affecting the seasonality of biological processes, altering marine and freshwater food webs, with unpredictable consequences for fish production.

On the issue of changing wild stock population due to the impact of climate change, Sumaila and Cheung (2010), reported that some fisheries have suffered declines in the number of fishers as the opportunities for fishing have diminished. Some specific fish species that have high market demand can be found in the waters only for a short period during the year. The time they would be expecting a particular type of fish, they cannot get it or only a few of such species will be caught. They stated that, the storm, the cold and the hot temperatures associated with the fishing business are because of the climate change and has a lot of implication on their livelihoods, food security and health. FAO (2007) reported

Table 3 Effect of climate change on fishing and fishermen

Effects	Frequencies	Percent
Increased food insecurity	111	95.7
Reduced revenue	110	94.8
Increasing risks and uncertainties in fishing	108	93.1
Reduction in productivity	107	94.0
Fluctuations in fish and wild stock distribution	95	81.9
Reduction in the duration of fishing season	93	80.2
Labour emigration	87	75.0
Decline in fish stock	84	72.4
High fish mortality	79	68.1

Source Authors' Construct, 2016 n = 116

similar result that rain storm damage capital assets, including boats and fishing gears. As many as 81.9% of the respondents said there were fluctuations in fish and wild stock distributions and that the duration of the fishing season has been shortened due to the effects of climate change (80.2%).

Food insecurity is the single greatest danger of climate change to vulnerable human populations and indeed to all humanity. That is, because there are multiple adverse impacts of global warming and climate disruption on agriculture—and these impacts will increase as the global temperature increases (Climate Change Emergency Medical Response 2013). As high as 95.7% of the respondents claimed that climate change effects increase their food insecurity levels. The results support the arguments by FAO (2007) and Kling et al. (2003) that the disruption of fishing activities by climate change effects could affect livelihoods and food security of the fishing households and others, such as the fishmongers and retailers, who are in the fish value chains. The world's poorest people are at risk of increased hunger, particularly those in the tropical and sub-tropical areas in the face of climatic change effect on agriculture (IPCC 2007).

Kling et al. (2003), posited that productivity of a fish population is related to the amount of water present and its thermal suitability and that the abundance and productivity of fishes increase with increased time spent at the optimal temperature. As high as 94.0% reported low productivity in their fishing business as a result of climate change effects. The results confirmed the findings of Sagoe (2006) that agricultural productivity has been found to be affected by climate change.

Changes in ocean temperatures, currents and productivity will affect the distribution and abundance of marine populations, with unpredictable consequences to marine ecosystems and fisheries. Increasing carbon dioxide levels could also trigger abrupt changes in thermohaline ocean circulation, circulation driven by differences in the density of sea water, controlled by the effects of temperature and salinity. This can result in massive and severe consequences for the oceans and for global climate (National Oceanic and Atmospheric Administration 2001).

Adaptation Strategies Adopted by the Fishermen

Cochrane et al. (2009), indicated that sea level rise, glacier melting, ocean acidification and changes in precipitation, groundwater and river flows will significantly affect coral reefs, wetlands, rivers, lakes and estuaries; requiring adaptive measures to exploit opportunities and minimize impacts on fisheries and aquaculture systems. To alleviate some of the negative impacts associated with climate change on the fishing business and the fishermen, the respondents have adopted a number of adaptations strategies. These are presented in Fig. 1.

As shown in Fig. 1, 86.2% of the fishermen interviewed in two communities seasonally travel to other locations to fish as an adaption strategy to poor catch in their locality. Most of these respondents change their methods of fishing (80.2%) due to the adverse effects the climate change is having on the normal fishing methods.

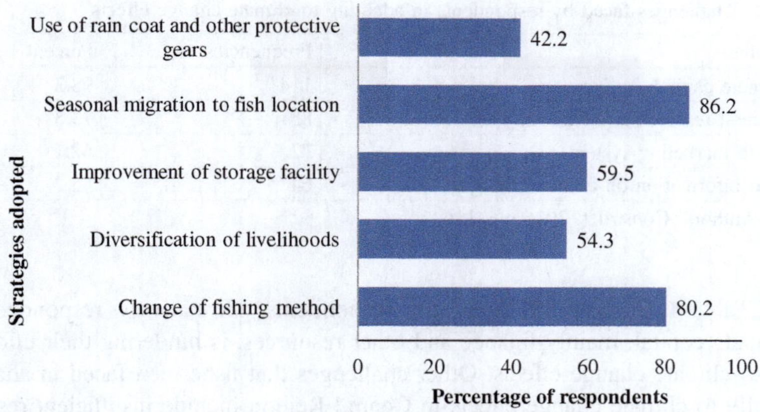

Fig. 1 Adaptation strategies adopted by the fishermen in the two communities *Source* Authors' Construct, 2016 n = 116

A publication by GNA (2013a) confirmed that some of the fishermen from Central Region are found in the Jomoro District of the Western Region due to bumper catch of fish in that region. Some of the methods employed as adaptation measures include changing of fishing nets, use of light (bulb) to attract the fish and use of machine to detect the presence of fish. Some of these methods are dangerous and illegal practices which can result in negative consequences in the near future (GNA 2013b).

As an adaptation strategy, the respondents claimed to have improved on their method of storing the fish so that the impact of the harsh temperature could be reduced. To this end, about 6 out of every 10 respondents used one or more methods of storing their catch to prevent fish spoilage before the fish get to their customers. This is achieved using deep freezers, ice blocks or covering the fishes with thick materials like cold jute sack to control the amount of direct sunshine and exposure to high temperature.

More than half of the respondents (54.3%) adapted to climate change by diversifying into other businesses as a livelihood strategy. Some of them do this by integrating fishing and aquaculture (40.5%) or leasing their fishing gears to others during periods when they anticipate low catch (37.9%). This finding is consistent with a revelation by Sumaila and Cheung (2010) that as fish stocks decline, some fishers in both developed and developing countries have attempted to diversify their income by engaging in other non-fishing livelihood activities like aquaculture.

Challenges in Adapting to Climate Change Effects

In their efforts to adapt to the climate change effects, a number of constraining factors prevent the respondents from attaining their optimum level of adaptation strategies. Challenges constraining the optimum adaption of appropriate adaptation strategies are presented in Table 4.

Table 4 Challenges faced by respondents in adapting to climate change effects

Challenges	Frequencies	Percent
Inadequate capital	114	98.3
Insufficient research and extension services	85	73.3
Poor fish marketing system	72	62.1
Limited information on climate change	64	55.2

Source Authors' Construct, 2016 n = 116

As Table 4 reveals, 98.3% of the respondents affirmatively responded that inadequate capital, mainly finance and other resources, is hindering their efforts to adapt to climate change effects. Other challenges that fishermen faced in adapting optimally to climate change effects in Central Region include insufficient research and extension services (73.3%), poor fish marketing system in such as price fluctuations, difficulty in transportation, and inadequate preservation and storage facilities (62.1%), and limited information on climate change and appropriate mitigation measures (55.2%).

Conclusion

The fishermen were relatively young but with substantial experiences in the fishing business. With the necessary extension and appropriate climate information system to provide localised and accurate climate intelligence, these fishermen will improve livelihoods, build resilience and use sustainable practices. Most of the respondents were highly aware of climate change and its effects on their fishing business and their livelihoods. These effects dispose the fishermen to low productivity, reduced revenue and food insecurity.

The difficulties facing the coastal fishermen are being aggravated by climate change and its effects. Extremes of weather are increasingly making fishing more and more risky business. To cope with these effects, the respondents resorted to seasonal migration of fishing locations, changing fishing methods, improvement in storage facilities, use of protective gears and diversification of livelihoods. However, a substantial number of respondents were unable to fully utilize the various adaptation strategies due to certain constraints. Some of these adaptations measures are detrimental to the sustainability of the fishing business in the area.

Government parastatals, non-governmental organisations and civil societies in Ghana should intensify non-formal education for the fishermen to enable them access current and relevant information on climate change, its effects and good adaptation strategies, and appropriate fishing equipment that will not have negative consequences on the sustainability of the fishing business in the area. Financial assistance, including saving, micro-credit and micro-insurance, should be extended to fishermen by both governmental and non-governmental organisations to enable them better adapt to the impact of unfavourable climate change on their fishing business. These will empower them and enhance their adaptive capacities against negative shocks due to climate change in the future.

References

Acquah, H. (2011). Public awareness and quality of knowledge regarding climate change in Ghana: A logistic regression approach. *Journal of Sustainable Development in Africa, 13*(3). Pennsylvania: Clarion University of Pennsylvania.

Amoah, S. A., & Eshun, P. (2015). *Research methods*. Winneba: Institute of Educational Development Extension, University of Education.

Arndt, C., Asante, F., & Thurlow, J. (2015). Implications of climate change for Ghana's economy. *Sustainability, 7*(6), 7214–7231.

Cheung, W. W. L., Lam, V. W. Y., Sarmiento, J. L., Kearney, K., Watson, R., Zeller, D. & et al. (2009). Large-scale redistribution of maximum fisheries catch potential in the global ocean under climate change. *Global Change Biology, 16*, 24–35. Retrieved from http://onlinelibrary.wiley.com/doi/10.1111/j.1365-2486.2009.01995.x/full.

Climate Change Emergency Medical Response. (2013). Food security and climate change. Retrieved from http://www.climate-change-emergency-medical-response.org/food-security.html.

Cochrane, K., De Young, C., Soto, D., & Bahri, T. (2009). Climate change implications for fisheries and aquaculture: Overview of current scientific knowledge. FAO Fisheries and Aquaculture Technical Paper No. 530. FAO, Rome.

Cooley, S. R., & Doney, S. C. (2009). Anticipating ocean acidifications economic consequences for commercial fisheries. *Environ. Res. Lett, 4*, 024007.

Creswell, J. W. (2013). *Steps in conducting a scholarly mixed methods study DBER speaker series*. Paper 48. Retrieved from http://digitalcommons.unl.edu/dberspeakers/48.

Doss, C., & Morris, M. (2001). How does gender affect the adoption of agricultural innovations? The case of improved maize technology in Ghana. *Agricultural Economics, 25*, 27–39.

Easterling, W. E., Aggarwal, P. K., & Batima, P. (2007). Food, fibre and forest products. In M. L. Parry, O. F. Canziani, J. P. Palutikof, P. J. van der Linden & C. E. Hanson (Eds.), *Climate Change 2007: Impacts, Adaptation and Vulnerability. Fourth assessment Report of the Intergovernmental Panel on Climate Change* (pp. 273–313). Cambridge, UK: Cambridge University Press. Retrieved from http://www.ipcc.ch/pdf/assessment-report/ar4/wg2/ar4-wg2-chapter5.pdf.

Fabricius, K. E., Hoegh-Guldberg, O., Johnson, J., McCook, L., & Loug, J. (2007). Vulnerability of coral reefs of the Great Barrier Reef to climate change. In *Climate change and the Great Barrier Reef: A Vulnerability Assessment* (pp. 516–554). Australia: Great Barrier Reef Marine Park Authority and Australian Greenhouse Office. Retrieved from http://www.cakex.org/virtual-library/652.

FAO. (2007). Building adaptive capacity to climate change: Policies to sustain livelihoods and fisheries. *New Directions in Fisheries—A Series of policy Briefs on Development,* (8), p. 16. Rome, Italy: FAO. Retrieved from ftp://ftp.fao.org/docrep/fao/010/a1115e/a1115e00.pdf.

FAO. (2009). Climate change implications for fisheries and aquaculture: Overview of current scientific knowledge. In K. Cochrane, De Young C, D. Soto, T. Bahri (Eds.), *FAO Fisheries and Aquaculture Technical Paper No. 530* (p. 212). Rome, Italy: FAO.

Fryrear, A. (2015). *Survey response rates*. Retrieved from https://www.surveygizmo.com/author/afryrear/.

GNA. (2013a). *Bumper catch of fish at Half Assini*. Retrieved from www.ghananewsagency.org/economics/bumper-catch-of-fish-at-half-assini-63597.

GNA. (2013b). *Fisherman arrested for suspected light-fishing*. Retrieved from ghananewsagency.org/human-interest/fisherman-arrested-for-suspected-light-fishing-62458.

Gunn, J. M. (2002). Impact of the 1998 El Nino event on lake char, Salvelinus namaycush, population recovering from acidification. *Environmental Biology of Fishes, 64,* 343–351.

Hookway, C. (2016). Pragmatism. In E. N. Zalta (Ed.), *The Stanford encyclopedia of philosophy* (Summer 2016 Edition). Retrieved from https://plato.stanford.edu/archives/sum2016/entries/pragmatism/.

Iaydjiev, I. (2013). *A pragmatic approach to social science*. Retrieved from http://www.e-ir.info/2013/03/01/a-pragmatic-approach-to-social-science/.

IPCC. (2007). Climate Change 2007: Climate change impacts, adaptation and vulnerability. In R. K. Pauchauri & A. Reisinger (Eds.), *Contribution of Working Groups I, II and III to the Fourth Assessment Report of the Intergovernmental Panel on Climate Change* (p. 104). Geneva, Switzerland.

Johnson, J. E., & Marshall, P. A. (2007). The Great Barrier Reef and climate change: Vulnerability and management implications. *Great Barrier Reef Marine Park Authority and Australian Greenhouse Office, Australia.*

Johnson, R. B., & Onwuegbuzie, A. J. (2004). Mixed methods research: A research paradigm whose time has come. *Educational researcher, 33*(7), 14–26.

Kling, G. W., Hayhoe, K., Johnson, L. B., Magnuson, J. J., Polasky, S., Robinson, S. K., Shuter, B. J. & et al. (2003). Confronting climate change in the great lakes region: Impacts on our communities and ecosystems. Cambridge, MA: Union of Concerned Scientists and Washington, DC: Ecological Society of America.

Lefale, P. F. (2003). A Samoan perspective on the role of traditional knowledge in adapting our understanding of climate change. In E. Helander & T. Mustonen (Eds.), *Snowscapes, dreamscapes: Snowchange book on community voices of change* (pp. 495–515). Finland: Fram Oy, Vaasa.

Lokko, P., & Anson, S. (2006). Baseline studies of some fishing communities in Ghana - A comparative assessment. *Ghana Journal of Science, 42,* 11–18.

Marshall, N. A., Stokes, C. J., Howden, S. M., & Nelson, R. N. (2010). Enhancing adaptive capacity. In C. Stokes & S. M. Howden (Eds.), *Adapting agriculture to climate change: preparing Australian agriculture, forestry and fisheries for the future* (pp. 245–256). Canberra, Australia: CSIRO Publishing.

Meisner, D. J. (1990). Potential loss of thermal habitat for brook trout, due to climatic warming in two southern Ontario streams. *Transactions of the American Fisheries Society, 119,* 282–291.

Molina-Azorin, J., & Cameron, R. (2010). The application of mixed Methods in organisational research: A literature review. *The electronic Journal of Business Research Methods., 8*(2), 95–105.

Morton, J. F. (2007). *The impact of climate change on smallholder and subsistence agriculture.* Natural Resources Institute: University of Greenwich, Kent, UK.

National Oceanic and Atmospheric Administration. (2001). Coral reefs, wetlands, shoreline communities and fisheries under threat due to potential impacts of climate change. Retrieved from http://www.publicaffairs.noaa.gov/releases2001/jan01/noaa01011.html.

National Oceanic and Atmospheric Administration. (2007). National weather service 2007. Retrieved from www.nws.noaa.gov/os/brochures/climate/climatechange.pdf.

Nutsukpo, D. K., Jalloh, A., Zougmore, R., Nelson, G. C., & Thomas, T. S. (2013). Ghana. In A. Jalloh, G. C. Nelson, T. S. Thomas, R. Zougmore, & H. Roy-Macauley (Eds.), *West African agriculture and climate change: A comprehensive analysys.* Washington DC: IFPRI.

Sagoe, R. (2006). *Climate change and root crop production in Ghana.* A report prepared for environmental protection agency (EPA), Accra, Ghana. Retrieved from http://www.nlcap.net/fileadmin/NCAP/Countries/Ghana/ROOT_TUBERS_DRAFT_FINAL_REPORT.pdf.

Stanturf, J. A., Melvin, L. W., Charnley, S., Polasky, S. C., Goodrick, S. L., Armah, F., et al. (2011). *Ghana climate change vulnerability and adaptation assessment*. Accra: USAID.

Sumaila, U. R., & Cheung, W. W. L. (2010). *Cost of adapting fisheries to climate change.* Development and climate change discussion paper, 5. World Bank,Washington, DC.

Taderera, D. (2010). *South African's awareness of climate change*. Briefing Paper 235. Southern African Catholic Bishops' Conference, Parliamentary Liaison Office, South Africa.

Walther, G. R., Post, E., & Convey, P. (2002). Ecological responses to recent climate change. *Nature, 416,* 389–395.

Williams, L., & Rota, A. (2012). *Impact of climate change on fisheries and aquaculture in the developing world and opportunities for adaptation*. IFAD Fisheries Thematic Paper: Tool for Project Design. Retrieved from http://www.ifad.org/lrkm/pub/fisheries.pdf.

Zakour, M. J., & Gillespie, D. F. (2013). Community disaster vulnerability: Theory, research, and practice. doi:10.1007/978-1-4614-5737-4_5. New York: Springer Science+Business Media.

Retreat or Rebuild: Exploring Geographic Retreat in Humanitarian Practices in Coastal Communities

Rosetta S. Elkin and Jesse M. Keenan

Abstract This chapter identifies an emerging discourse among humanitarian actors with regard to the planning and design of post-disaster geographic retreat of coastal communities. This chapter attempts to synthesize transdisciplinary knowledge in geography, landscape ecology and climate adaptation in order to offer insight into the unrecognized factors shaping the planning and design of the environments and landscapes of coastal retreat. This chapter evaluates the proposition that managing the cultural, environmental and economic byproducts of the former inhabited coastal geography are central to transformative adaptation. Particularly in light of sea level rise, autonomous landscape processes are central to these transformations. Beyond mere relocation, this chapter seeks to expand the definition of coastal retreat by examining the consequences and implications of existing frames defined by disaster resilience and housing, land and property issues. This chapter highlights two external case studies from Japan and the Philippines that provide qualitative evidence in an affirmation of the underlying theoretical proposition.

Introduction

This chapter identifies an emerging discourse among humanitarian actors with regard to the planning and design of post-disaster coastal retreat. Whether it is a greater recognition of natural hazards or an emerging understanding of climate change impacts, humanitarian actors are increasingly recognizing the broad array of social, economic and environmental complexities associated with the managed coastal retreat from one geographic area to another. What is less understood is the extent to which transdisciplinary knowledge from geography (French 2004;

R. S. Elkin · J. M. Keenan (✉)
Graduate School of Design, Harvard University, Gund Hall, 48 Quincy Street, Cambridge, MA 02138, USA
e-mail: jkeenan@gsd.harvard.edu

R. S. Elkin
e-mail: relkin@gsd.harvard.edu

W. Leal Filho (ed.), *Climate Change Impacts and Adaptation Strategies for Coastal Communities*, Climate Change Management,
https://doi.org/10.1007/978-3-319-70703-7_8

Agyeman et al. 2009; Jones and Clark 2014), landscape architecture and ecology (Ellis 2008; Nordstrom et al. 2015), and climate adaptation (Cooper and Pile 2014; Hino et al. 2017) offer insight into the unrecognized factors shaping the planning and design of the environments and landscapes of retreat. Beyond mere relocation, this chapter seeks to expand the definition of retreat by examining the consequences and implications of existing frames defined by disaster resilience and housing, land and property (HLP) issues.

Retreat is defined here as a process of relocating people, settlement infrastructure and materials to a new geographic condition, while at the same time giving recognition to the cultural, environmental and economic byproducts of the post-settlement landscape. This chapter evaluates the proposition that managing the aforereferenced cultural, environmental and economic byproducts of the former inhabited coastal geography are central to a successful geographic, economic, cultural, psychological and ecological transition (Niven and Bardsley 2013; Alexander et al. 2012). The caveat is that the definition of success is relative, subjective and invariably defined by trade-offs by and between the capture of various externalities and internalities associated with these post-landscape byproducts. In particular, this chapter highlights two case studies from coastal communities in Japan and the Philippines in order to explore the utility of the underlying proposition.

From shelter organization to architectural vernacular and from subsistence practices to environmental stewardship, the disruptions of managed retreat are increasingly understood as dynamic mechanisms of the expansion and contraction of human settlement (Dickson et al. 2007). Particularly in light of the impacts of sea level rise, societies are increasingly tasked with designing shelter and environments whose local stability is understood to be limited to well within the useful life of the contexts of conventional practices (Mengel et al. 2016). In this light, the challenge increases from managed relocation of existing settlements, to the design of recovery in a manner that acknowledges the transient nature of the defining social, political and environmental characteristics (Tadgell et al. 2017).

From a humanitarian actor's point of view, is it desirable or possible to imitate or capture aspects of memory, material and socioecological synchronicity that are attendant to design practices? Or, is the capture of these social constructions inherently subject to an inadequate understanding of collective preferences and an incomplete framing of dynamic socioecological and biophysical processes? While anthropogenic adaptation has always shaped the success and failure of human settlements, climate change is poised to accelerate change processes with little to no historical antecedent. Because global humanitarian actors are on the front lines of complex manifestations of social and environmental change—including coastal disasters—the relevance of this paper is grounded in a necessary critique of existing response strategies that are more functionally driven by a preservation bias rather than a critical discourse that acknowledges the inevitability of certain adaptations. As such, it reveals the potential to include ecologists, landscape architects and other professionals who spatialize temporality and long term adaptation, in order to work towards the design of post-disaster geographic retreat.

The relevance of this conceptual development is orientated toward a desire to provide a more comprehensive understanding of the human-environment interactions shaping coastal retreat. To date, much scholarship has focused on decision-making within sociological (Lo 2013), sociotechnical (Azevedo de Almeida and Mostafavi 2016), governance (Bardsley and Niven 2013; Buckley 2013), and management (Weinstein et al. 2007; Hino et al. 2017) framings. Much of this research has not accounted for the roles of humanitarians in post-disaster context, as the focus has been largely placed on vested political interests (Gibbs 2016). However, as global coastal communities becoming increasing reliant on external assistance, this assistance will likely bring transient power regimes—such as humanitarians—that are arguably tasked with an opportunity to balance human and environmental factors because of their lack of an immediate political constituency that may bias their actions to preference sociotechnical solutions in isolation.

Methodology

The research design supporting the development of the theoretical propositions evaluated in this chapters, as well as the identification of supporting illustrative empirical evidence, was based on a two-prong approach. First, over the course of a year and a half, semi-structured and informal interviews and focus groups were convened with humanitarian management officials from governments and non-profits (Harrell and Bradley 2009). This group of stakeholders specifically included humanitarian actors who were expert specialization in design, landscape architecture, civil engineering, social policy, community planning and public administration (Krueger and Casey 2014). In particular, government, non-profit and for-profit representatives from the Philippines with recent post-disaster experience from Typhoon Haiyan (2013) participated in an ongoing conversation concerning coastal retreat policies and post-disaster reconstruction in the Philippines. Second, these activities were developed in parallel various literature reviews and grey-literature reviews that help shape the emerging discourse of the consortium (Fink 2013). The collective outputs of these activities formed the basis for the development of this chapter.

The primary limitation to this research design is an over reliance on data collected in the Philippines. Given the country's unique geographic, political, economic, cultural and risk attributes, there is a limited generalizability to the data. However, the researchers sought to diversify this bias with a dedicated literature review of coastal communities throughout the Pacific, as well as other global coastal communities. In addition, the recent experiences in the Philippines may also be clouded by a lack of historical perspective given the short period of time between commencement of the research and the occurrence of Typhoon Haiyan. An additional limitation is the lack of independently derived empirical evidence that may have import on the validity of the propositions developed herein. This gap offers an opportunity for future research to examine the post-retreat coastal landscape for a more nuanced and detailed understanding of the processes supporting the transitional and transformational aspects of resettlement, restoration and adaptation.

Relocation as an Imperfect Substitution

The natural hazards that are driving population relocation are often subject to a chain of exacting hazard measurements that quantify risk and result in a series of procedures that represent a type of contradiction to the qualitative risks experienced pursuant to the associated human tragedies (Keller and DeVecchio 2016). In the aftermath of disasters, the primary response is to rebuild that which was lost (Daly and Feener 2016). Through short-term interventions, the institutions of humanitarianism have disproportionately dictated long-term recovery pathways despite the objective, empirical understanding that some land may actually be exposed to chronic biophysical and environmental stresses beyond the capacity of local populations to cope, mitigate and adapt (King et al. 2016). Natural hazards are not synonymous to natural disasters; rather hazards manifest as disasters when they encounter the built environment (Smith 2006). The relationship between the hazard itself and the extents of its damage is fundamentally an outcome of characteristics of the built environment, such as population density, urbanization and the economic operations that lay claim to land. Natural disasters are generally viewed as moments in time and space. Outside of psychology and psychiatry, they are not usually conceptualized as an ongoing chain of events linked by exposure, vulnerability and impact (Parkinson 2000; Clayton et al. 2015). Because natural disasters are episodically understood as a cause and effort to a known stimulus (i.e., typhoon), it can be argued that the slow and disconnected violence of climate change operates to cloud social leaning and multi-generational knowledge that has historically dictated the long-term limitations of the capacity of exposed land to lay the foundation for durable human settlements.

Revealed in the statistics of a rapidly changing climate, extreme events are escalating and intensifying such that each hazard serves to compound the resource constraints driving strategic adaptation in the name of recovery (Okuyama and Santos 2014). Therefore, it is increasingly critical that society understands and explores this range between the costs and benefits that underscores decision making. This can be achieved only to the extent that such actions narrow or expand the capacity of society to adapt to conditions beyond the threshold of the status quo between response and recovery. In cases where the risk is specifically predictable and chronic, there are opportunities for post-disaster response that are built upon the conventional operations of design, planning and humanitarian aid that are often based on practices that are both deterministic and probabilistic. However, where empiricism in a chaotic, if not random, natural environment and/or social system falls short, these operations struggle with the compounded risk of uncertainty and the potential for the waste of capital (Moser 2005; Heal and Millner 2014). Thus conceived, retreat is contextualized as a potentially viable post-disaster strategy vis-à-vis its capacity to guide the adaptation of response and recovery operations that currently are driven by cycles of development that tend to prioritize built-environment capitalism (Smith 2006). To this end, retreat, as defined herein, offers a broader array of human, social and environmental capital constructions that complete an economic picture whose opaqueness is blurred by the immediacy of financial capital. Perhaps, the operations and procedures of humanitarian actors will diversify in their orientation to capital, if retreat captures a more pluralistic understanding of capital accumulation and reorganization.

Within this context, landscape as a deterministic ecological manifestation of resource and design is substituted for an autonomy of geographic morphology that is free of the strictures of financial capital or of a cultural hegemony reinforced by such capital accumulation. As Sauer noted,

> Under the influence of a given culture, itself changing through time, the landscape undergoes development, passing through phases, and probably reaching ultimately the end of its cycle of development. With the introduction of a different—that is alien—culture, a rejuvenation of the cultural landscape sets in, or a new landscape is superimposed on remnants of an older one (1963, p. 343).

Superimposition operates at different scales for both the relocated landscape and the landscape left behind. As previously questioned, the extent to which landscapes as a cultural construction of human and environmental geography can be transplanted or superimposed vis-à-vis material and memory is yet to be fully theorized. From one perspective, it is a matter of substitution and the transaction costs of monuments and places that have no discrete substitution value or opportunity cost. From another perspective, culture is geographically specific and design can only imitate superficial elements, offering an incomplete substitution.

As such, the process of resettlement includes, but is not limited to, an instrumental shedding of identity and the cloaking of a nascent cultural identify. In this newly constructed or modified built environment, it can be argued that design would likely only imitate the environmental and place specific attributes that ultimately serve as inferior substitutes. This is revealed in the urge to re-store, re-build and other supposed return state projections. To this end, perfect substitution is impossible because the priorities, preferences and identity of the newly relocated culture are inherently unequal and misaligned with those defining the predicate state. This argument is reinforced by the proposition that relocation is designed and resourced almost exclusively by considerations of economy. The lack of substitution is reinforced by problems associated with the reduction of ecological systems to generalities and theories. Hilderbrand, Watts and Randle note that, "[a]lthough we cannot function without theory and conceptual models, their creation often ignores the variability that is so important to accurately describe, predict, and recreate current and future system attributes. In essence, restoration ecology strives to (re-)create complex systems from simplified guiding principles or myths." (2005, p. 1). It is these principles and cultural myths that are the contested elements of retreat.

Superimposing Humanitarian Institutions

Among humanitarian actors, the proliferation of acronyms, terms and policies framing long-term, post-disaster recovery is currently galvanize around two specific themes: resilience planning (Barrett and Constas 2014; Daviron et al. 2014; Sanderson et al. 2016) and housing, land, and property (HLP) issues (*see generally*, Leckie and Huggins 2011; da Costa and Pospieszna 2015). These themes represent two different scales of bottom-up and top-down actor orientation, respectively. Coded by expertise such as design, planning and humanitarian aid, these efforts tend to promote rebuild operations whereby reconstruction in situ and the reestablishment of land rights—even on hazardous land—is deemed preferable to the potential complications or disruptions of long-term withdrawal, relocation or retreat. As such, rebuilding procedures in the context of natural hazards *superimpose* building procedures on remnant conditions. In this sense, the techniques of rebuilding are often blinded by technological innovations and material economies relative to the nuances in location and landscape that have guided long-term local decision-making regarding land capacity and suitability. This disconnect is exacerbated by the relatively slow progress made by many humanitarian actors in adapting their response to the particularities of urban contexts (Pantuliano et al. 2012).

Interviews with senior officials at various humanitarian organizations have suggested that the prospect of relocation is considered only as a last resort. In such cases that warrant consideration, the option is considered 'extreme' and in all cases requires a more active intervention that can only be addressed through larger, more cumbersome policy frameworks.[1] In many cases, the immediacy of such a response does not address the plurality of the underlying hazards and socioeconomic risks. Thus, relocation strategies inherently place any alternative configuration outside the scope of recovery operations. More specifically, it arguably calls into question the viability of existing recovery and rebuilding strategies that may or may not consider the long-term livelihood of subject beneficiaries. Interviewees suggest that this has a destabilizing effect that runs the risk of undermining or second guessing existing recovery efforts that are focused on my immediate life-safety considerations. Currently, existing models of practice within HLP standards and resilience planning only make a de facto procedural response possible.

Often, the mere mention of retreat is associated with a sense of defeat, which suggests that the term normalizes failure. From this sense of perceived hopelessness, there emerges an intellectual conflict over how to explain the process of

[1]The Norwegian Refugee Council (NRC) is the globally designated Focal Point Agency for Housing, Land and Property (HLP) within the humanitarian coordination system. The HLP Area of Responsibility (AoR) was established in 2007. Urban resilience planning has largely been defined by the initiatives of the Rockefeller Foundation, as a means to help cities become more adaptable to physical, social and economic shocks. The term is seeing increased used across both sectors with little consistency.

recovery from natural hazards and the ensuing rebuild techniques that assume people (and places) can adapt to chronic risk. It can be argued that the current language of resilience is grafted to notions of rebuilding that perpetuate capital agendas and obscure varying local needs (Graham et al. 2016; Gillard 2016). As such, resilience's predominating frames (e.g., disaster resilience) prioritizes and preferences the stability functions of the status quo that may operate in contradiction to transformations necessary for long-term vulnerability reduction (Olsson et al. 2015; Keenan et al. 2016). In addition, fixed structural solutions—from strengthening a building code to reinforcing a levee—can deceive communities into a false sense of security (White 1945; Amundsen 2012).

Arguably, the potentially deceptive character of these measures—and the ways in which they contribute to an insistence on restoring infrastructure—operate to continue the cycles of capital 'rebuildism.' Financial capital investments that are tied to specific land, political and economic structures are prefaced to gain value from existing configurations, all of which serve to further lock-in rebuilding agendas and corresponding institutions focused on maintaining settlements in their present locations. But, as repeated failures of protective infrastructure around the world indicate, the built environment has a threshold and a certain brittleness for managing episodic events and/or chronic stresses.[2] Given the comparative ability for landscape to more readily adapt to perturbations than the entirety of the built environment, there lies the possibility that the autonomous processes of landscape have the capacity to harness biophysical forces to facilitate retreat (Elkin 2017). In this sense, the design of retreat seeks a different form of stability by balancing socioecological dynamics without regard to the investiture of special interests who utilize accumulations of human and financial capital with little regard for the distributive benefits of rebuilding or retreating. By removing some degree of human agency of retreat in favor of the autonomy of ongoing landscape and environmental adaptations, one arguably mitigates some degree of risk associated with exploitation from vested interests.

In this context, narrowing the definition of retreat can help to reconcile the theories and practices of recovery. Arguably, retreat is a form of recovery. A more nuanced definition must take into consideration a key tenant of ecological resilience theory—the ability to value and adapt to the prospect of an alternate configuration (Holling 1973). An ecological resilience frame that acknowledges a multi-equilibrium state (Walker et al. 2006; Davidson et al. 2016) runs in contradiction to the disaster resilience frames utilized by the humanitarian community that are based on a single equilibrium understanding of resilience processes (Matyas and Pelling 2015; Meerow et al. 2016). Within the built environment, when fixed structures, patterns and arrangements are disturbed repeatedly, one runs that risk that ultimately the chronic stress will exceed the capacity of single equilibrium

[2]For example, recent notable failures of protective infrastructure include power outages post-Sandy (New York City, 2012); levees post-Katrina (New Orleans, 2005); pump systems in monsoon season (Bangkok, 2011); and, evacuation highway collapse (Haiti, 2010).

resilience processes—at which point a system either adapts or fails (Keenan 2015). The divergent nature of these two theories of resilience between the stationarity of capital accumulations and the dynamic processes of socioecological systems reflects the fundamental divide shaping retreat decisions. In particular, the technocratic determinism of rebuilding often runs in contradiction to environmentally focused methods for assessing the suitability and capacity of land. To this end, a normative framing of disaster resilience by the humanitarian community is arguably on track to reinforce an inherently conservative 'rebuildism,' which operates to marginalize alternative pathways consistent with retreat (Collinson et al. 2010; MacKinnon and Derickson 2013; Brown 2014). In particular, emergent models for digital humanitarianism, though seemingly non-spatialized, have operated to perpetuate a neoliberal surveillance that further links benefits with geospatial stationarity (Duffield 2016).

Unlike displacement, retreat is defined by some measure of social agency, as well as some degree of autonomy of biophysical processes in the landscape, which operate in isolation from existing power structures and cultural regimes. Therefore, retreat is both the recognition of the technical limits of rebuilding procedures and a sensitivity to the forces of the living environment. Retreat emerges through a human capacity to acknowledge that the resettlement of hazard prone land is likely to cause an increase in social, economic and environmental vulnerability. From a geomorphological perspective, retreat pays specific attention to the terrestrial properties that manifest on the surface of the earth, the thin layer of human settlement. Implicit in the notion of retreat is a dynamic substitution between landscape conditions. Recognizing the value and viability of retreat challenges existing paradigms in disaster preparedness, response and planning, but it also opens the possibility of guiding future patterns of settlement within more nuanced forms of resilience and adaptation beyond disaster resilience (Smith 2006). The following case studies highlight the tensions inherent in balancing the social and environmental aspects of the substitutions of landscape and environments.

Case Studies: Japan and the Philippines

During the early morning hours of March 03, 1933, a historic tsunami in the Sanriku region of Japan generated enough institutional detestation to effectuate change in government policy, reflecting a list of measures and countermeasures that would set the course for global Tsunami research (Uchida et al. 2016). Significantly, the policies called for a unique combination of assisted relocation and cultural memorial (Shuto and Fujima 2009). In areas of significant damage and where loss of life was predicated on coastal proximity, a new configuration was presented, not only as a strategy, but also as a transaction between built and living form. Rather than simply reconstruct housing, the government established a series of control forests in the areas of devastation and offered this new landscape as a public amenity to the stricken community. As both a new site of mourning and a landscape that could dampen and attenuate wave action, the forests became a prominent feature in the effort to restructure a sense of community with the demands of daily life (Saravanan 2016). Each meter or kilometer planted imposes a meaningful setback for development, delineating a historical relationship between landscape dynamics, human settlement and a choice to follow the promise of change as opposed to a strategy of resistance. The forests offer an alternative configuration from settlement in that they anticipate the physical processes of erosion and renewal. Instead, each tree planted represents a significant form of cultural restoration, assisting in the reconstruction of the community. Rebuilding is no longer conceived of as the action of returning something to a former condition, and relocation is not a last resort. Rather, the disturbance is transformed into an opportunity to intensify the value of transformational change within cyclical patterns of settlement and resettlement. The forest as a cultural amenity serves as a living place and a monument to disaster. The permanence of reforestation provides a substitution of landscape that guides not only the physical aspects of retreat but also those aspects defined in cultural and environmental terms (Fig. 1).

Nearly 80 years later, the path of Typhoon Haiyan devastated the Philippines. In a country where urban growth and acute poverty fringe the archipelago, over four million people were immediately displaced (Yamada and Galat 2014). Following the storms and the surges, this tenuous ground between land and water became the site where levels of inequality manifest, as widespread destruction complicates HLP signatures that merely promote hazard resistance and sustainable building reconstruction (Fitzpatrick and Compton 2016). Within weeks, the government instigated a regulatory policy of no build zones (NBZ) through proposing a buffer system that prohibited the construction of dwellings and buildings (Thomas 2015) (Fig. 2).

The proposal detailed a 40-meter setback of mangrove plantations, as both as a means to justify relocation and to reduce future devastation (Republic of the Philippines, 2014). The ensuing estimations were rapid and tended to unfold across the vast scale of devastation, astonishing local governance, global media and non-governmental actors (Ong et al. 2016). While this abrupt combination of speed and scale is often blamed for the breakdown of NBZs by interviewees, it is conceivable to imagine that it was the simple disregard for the land being abandoned that contributed most significantly to its failure. Rather than offering landowners, squatters or regional authorities alike a positive phased outcome, the transmission prioritized codes, indexes, documents and classifications that revealed the cost of relocation and the linear homogeneity of a planted monoculture. Arguably, a mangrove plantation may only offer limited ecological services and little cultural value, while at the same time it may impede long-term livelihoods from fishing and trade. Further, it necessitates years of cultivation whereby access to humans is denied. While promoting a so-called 'restorative' practice, the policy did little to offer residents anything in return. In this sense, there was no articulation of a landscape substitution, creative reuse or restructuring whereby manipulation, design, recreation and livelihoods could punctuate coastal communities and evolve a public landscape in which people and plants could contribute in unique and mutually productive ways (Fig. 3).

Fig. 1 The Pine forest of Niji-no-Matsubara, Karatsu city, Fukuoka, Japan. Similar pine forests can be found in Kehi-no-Matsubara of the Fukui prefecture, and Miho-no-Matsubara, located in the Shumizu Ward of Shizuoka City. *Credit* Google Earth V. 7.1.8.3036. Japan, 33 27′ N 129 58′ E [May 25, 2017]

Fig. 2 The city of Tacloban, Eastern Visayas, Philippines. Tacloban is among the coastal cities that are imposing a no build zone (NBZ) following the devastation of Typhoon Haiyan in 2013. *Credit* Google Earth V. 7.1.8.3036. Philippines, 11 14′ N 125 00′ E [May 25, 2017]

Fig. 3 Mangrove (Rhizophora spp.) restoration projects are typically built in 5-meter wide cells. This standard exemplified by a 15-meter (3 cells) plantation. The current no build zone (NBZ) in Philippines proposes a 40-meter (8 cell) wide plantation, without consideration of access. Professor Niall Kirkwood, Bangkok, January 2014

The significant difference between the case of public control forests and restorative monocultures is revealed in the intentions of civic space. In the case of control forests, path systems, fishing docks, seating, memorial areas and leisure space contribute to the rebuilding of the social and cultural life without resorting to an optimized engineering methodology (Ong et al. 2016; de Leon and Pittock 2016). One argument in support of an explanation of the failure of the NBZ buffer system is the overt reliance on the proposition that a singular restorative practice can be a meaningful substitution for local livelihoods, culture and memory, which are embedded in both the site and its former inhabitants. Retreat—rather than merely relocation—considers the opportunity of disaster as means to generate novel outcomes that can increase stability and lower vulnerability (Kousky 2014). In the case of the Philippines, economic and social vulnerability may actually be increased by virtue of a relocation strategy that is not sensitive to the complexities of retreat.

Conclusions

This chapter propositions that future coastal recovery operations, designs and practices would be well informed to frame retreat beyond the rhetoric of defeatism. Specifically, this chapter has argues that managing the cultural, environmental and economic byproducts of the post-settlement landscapes are critical to the retreat process. This inevitable reformation would be based on an agenda that is starkly opposed to repeating the formulas that offer 'building back better' as a flawed framework for progress in coastal communities. This chapter has provided illustrative evidence from Japan that demonstrates scenarios where managing the post-retreat coastal landscape has helped ease the successful transition of coastal settlements. Conversely, the Philippines provide evidence where a strict conservation and restoration approach to post-retreat coastal landscapes—based in part on stationary principles—has manifested as a counter-productive severance of complex human-environment interactions.

The signature humanitarian practices associated with HLP and resilience planning have found an accomplice in the practices of restoration so that response is more reliant on known states and less reliant on changing, emergent conditions. Attention to the living environment is especially significant, and is even amplified as a type of biophilia, as the climate changes and human disasters increase in their frequency and intensity (Tidball 2012). To this end, conventional humanitarian practices have largely failed to internalize emergent practices and knowledge associated with climate adaptation in coastal areas. Through the single equilibrium lens of disaster resilience, the ensuing results emboss a fixed reading of the world, mobilized by the inability of recovery design and humanitarian aid to stray from fixed procedures and the stationarity of power regimes (Urwin and Jordan 2008). While these procedures are an outcome of optimization in the name of efficiency, they often fail to recognize that social and environmental outcomes evade optimization and are based on a set of complex criteria that are weighted and negotiated (Abel et al. 2011; Rulleau and Rey-Valette 2017). Given the conservative operations of disaster resilience that preference existing power regimes, notions of equity cloud both recovery and retreat in coastal communities. As such, this chapter has also argued for a secondary proposition that suggests that the autonomous biophysical processes of landscape offer a mechanism of adaptive transformation that is somewhat independent of social agency and the prospects of exploitation.

The humanitarian sector has become so focused on systematic collapse and disaster that it has lost sight of the broader range of prospects that may define a future condition. While there is arguably a local determinism for shaping these alternative futures, there is also an ethical obligation of external humanitarian actors to engage retreat as one of many options that are framed by the long-term stability, survivability and adaptability of any given settlement and society. The idea of resorting to a former state, fixing it, and elevating or defending territory only rehearses known itineraries, rather than working with the disturbance regimes we inhabit to inform the duration of the responses we cultivate. That challenge of future research is to expand the definition of retreat so as to introduce meaning and value to the procedures of response that bind practices—a prospect that offers the opportunity to strengthen society's capacity to adapt to extreme events and climate change. Ultimately, retreat in the humanitarian context is not only a matter of a calibration of practice, it will more fundamentally become a matter of absolute necessity.

References

Abel, N., Gorddard, R., Harman, B., Leitch, A., Langridge, J., Ryan, A., et al. (2011). Sea level rise, coastal development and planned retreat: Analytical framework, governance principles and an Australian case study. *Environmental Science & Policy, 14*(3), 279–288.

Agyeman, J., Devine-Wright, P., & Prange, J. (2009). Close to the edge, down by the river? Joining up managed retreat and place attachment in a climate changed world. *Environment and Planning A, 41*(3), 509–513.

Alexander, K. S., Ryan, A., & Measham, T. G. (2012). Managed retreat of coastal communities: Understanding responses to projected sea level rise. *Journal of Environmental Planning and Management, 55*(4), 409–433.

Amundsen, H. (2012). Illusions of resilience? An analysis of community responses to change in Northern Norway. *Ecology and Society, 17*(4), 46. Retrieved from https://www.ecologyandsociety.org/vol17/iss4/art46/.

Azevedo de Almeida, B., & Mostafavi, A. (2016). Resilience of infrastructure systems to sea-level rise in coastal areas: Impacts, adaptation measures, and implementation challenges. *Sustainability, 8*(11), 1115.

Bardsley, D. K., & Niven, R. J. (2013). Transforming society to govern planned retreat: Responding to "The contested nature of coastal climate change". *Regional Environmental Change, 13*(1), 215–217.

Barrett, C. B., & Constas, M. A. (2014). Toward a theory of resilience for international development applications. *Proceedings of the National Academy of Sciences, 111*(40), 14625–14630.

Brown, K. (2014). Global environmental change I: A social turn for resilience? *Progress in Human Geography, 38*(1), 107–117.

Buckley, R. (2013). The contested nature of coastal climate change—commentary to Niven and Bardsley. Planned retreat as a management response to coastal risk: A case study from the Fleurieu Peninsula, South Australia. *Regional Environmental Change, 13*(1), 211–214.

Bulkeley, H., Edwards, G. A. S., & Fuller, S. (2014). Contesting climate justice in the city: Examining politics and practice in urban climate change experiments. *Global Environmental Change, 25,* 31–40. doi:10.1016/j.gloenvcha.2014.01.009.

Clayton, S., Devine-Wright, P., Stern, P. C., Whitmarsh, L., Carrico, A., Steg, L., et al. (2015). Psychological research and global climate change. *Nature Climate Change, 5*(7), 640–646.

Collinson, S., Elhawary, S., & Muggah, R. (2010). States of fragility: Stabilisation and its implications for humanitarian action. *Disasters, 34*(s3), S275–S296.

Cooper, J. A. G., & Pile, J. (2014). The adaptation-resistance spectrum: A classification of contemporary adaptation approaches to climate-related coastal change. *Ocean and Coastal Management, 94,* 90–98.

Daly, P., & Feener, R. M. (Eds.). (2016). *Rebuilding Asia following natural disasters: Approaches to reconstruction in the Asia-Pacific Region*. New York, NY: Cambridge University Press.

Davidson, J., Jacobson, C., Lyth, A., Dedekorkut-Howes, A., Baldwin, C., Ellison, J., Holbrook, N., et al. (2016). Interrogating resilience: Toward a typology to improve its operationalization. *Ecology and Society, 21*(2): Art 27. Retrieved from http://www.ecologyandsociety.org/vol21/iss2/art27/.

Daviron, B., Duffield, M., Béné, C., Ancey, V., Pesche, D., & Bousquet, F. (2014). Resilience and development: Progress for human development or for humanitarian governance? In *Resilience and Development: Mobilising for Transformation*. Villeurbanne, FR: Centre pour la Communication Scientifique Directe, p. 535. Retrieved from http://agritrop.cirad.fr/574944/.

da Costa, K., & Pospieszna, P. (2015). The relationship between human rights and disaster risk reduction revisited: Bringing the legal perspective into the discussion. *Journal of International Humanitarian Legal Studies, 6*(1), 64–86.

de Leon, E. G., & Pittock, J. (2016). Integrating climate change adaptation and climate-related disaster risk-reduction policy in developing countries: A case study in the Philippines. *Climate and Development*, 1–8. doi:10.1080/17565529.2016.1174659.

Dickson, M. E., Walkden, M. J., & Hall, J. W. (2007). Systemic impacts of climate change on an eroding coastal region over the twenty-first century. *Climatic Change, 84*(2), 141–166.

Duffield, M. (2016). The resilience of the ruins: Towards a critique of digital humanitarianism. *Resilience, 4*(3), 147–165.

Elkin, R. S. (2017). Beyond restoration: The resilience of planted infrastructure. In W. Leal & J. M. Keenan (Eds.), *Climate change adaptation in North America: Experiences, case studies and best practices*. New York, NY: Springer Nature.

Ellis, M. H. C. (2008). Managed retreat: Coastal development in an era of climate change. *Landscape Architecture, 98*(3), 74–82.

Fink, A. (2013). *Conducting research literature reviews: From the internet to paper*. Thousand Oaks, CA: Sage Publications Inc.

Fitzpatrick, D., & Compton, C. (2016). Land tenure systems as a challenge for disaster recovery: Adapting to extreme weather events after Typhoon Haiyan. In K. Kheng-Lian, I. Kelman, R. Kibugi & R. Eisma Osorio (Eds.), *Adaptation to Climate Change: ASEAN and Comparative Experiences* (pp. 433–454). London, UK: World Scientific.

French, P. W. (2004). The changing nature of, and approaches to, UK coastal management at the start of the twenty-first century. *The Geographical Journal, 170*(2), 116–125.

Gibbs, M. T. (2016). Why is coastal retreat so hard to implement? Understanding the political risk of coastal adaptation pathways. *Ocean and Coastal Management, 130,* 107–114.

Gillard, R. (2016). Questioning the diffusion of resilience discourses in pursuit of transformational change. *Global Environmental Politics, 16*(1), 13–20.

Graham, L., Debucquoy, W., & Anguelovski, I. (2016). The influence of urban development dynamics on community resilience practice in New York City after superstorm sandy: Experiences from the lower east side and the rockaways. *Global Environmental Change, 40,* 112–124.

Harrell, M. C., & Bradley, M. A. (2009). *Data collection methods. Semi-structured interviews and focus groups*. Santa Monica, CA: Rand National Defense Research Institute.

Heal, G., & Millner, A. (2014). Reflections uncertainty and decision making in climate change economics. *Review of Environmental Economics and Policy, 8*(1), 120–137.

Hilderbrand, R., Watts, A., & Randle, A. (2005). The myths of restoration ecology. *Ecology and Society, 10*(1): Art. 19. Retrieved from http://www.ecologyandsociety.org/vol10/iss1/art19/.

Hino, M., Field, C. B., & Mach, K. J. (2017). Managed retreat as a response to natural hazard risk. *Nature Climate Change, 7,* 364–370. doi:10.1038/nclimate3252.

Holling, C. S. (1973). Resilience and stability of ecological systems. *Annual Review of Ecology and Systematics, 4,* 1–23.

Jones, N., & Clark, J. R. A. (2014). Social capital and the public acceptability of climate change adaptation policies: A case study in Romney Marsh. *UK Climatic Change, 123*(2), 133–145.

Keenan, J. M. (2015). From sustainability to adaptation: A case study of goldman Sach's corporate real estate strategy. *Building Research & Information, 43*(6), 1–15.

Keenan, J. M., King, D. A., & Willis, D. (2016). Understanding conceptual climate change meanings and preferences of multi-actor professional leadership in New York. *Journal of Environmental Policy & Planning, 18*(3), 261–285.

Keller, E. A., & DeVecchio, D. E. (2016). *Natural hazards: Earth's processes as hazards, disasters, and catastrophes*. New York, NY: Routledge.

King, D., King, D., Gurtner, Y., Gurtner, Y., Firdaus, A., Firdaus, A., et al. (2016). Land use planning for disaster risk reduction and climate change adaptation: Operationalizing policy and legislation at local levels. *International Journal of Disaster Resilience in the Built Environment, 7*(2), 158–172.

Kousky, C. (2014). Managing shoreline retreat: A US perspective. *Climatic Change, 124*(1–2), 9–20.

Krueger, R. A., & Casey, M. A. (2014). *Focus groups: A practical guide for applied research*. Singapore: Sage Publications Inc.

Leckie, S., & Huggins, C. (2011). *Conflict and housing, land and property rights: A handbook on issues, frameworks and solutions*. New York, NY: Cambridge University Press.

Lo, Alex Y. (2013). The role of social norms in climate adaptation: Mediating risk perception and flood insurance purchase. *Global Environmental Change, 23*(5), 1249–1257.

MacKinnon, D., & Derickson, K. D. (2013). From resilience to resourcefulness: A critique of resilience policy and activism. *Progress in Human Geography, 37*(2), 253–270.

Matyas, D., & Pelling, M. (2015). Positioning resilience for 2015: The role of resistance, incremental adjustment and transformation in disaster risk management policy. *Disasters, 39* (s1), s1–s18.

Meerow, S., Newell, J. P., & Stults, M. (2016). Defining urban resilience: A review. *Landscape and Urban Planning, 147,* 38–49.

Mengel, M., Levermann, A., Frieler, K., Robinson, A., Marzeion, B., & Winkelmann, R. (2016). Future sea level rise constrained by observations and long-term commitment. *Proceedings of the National Academy of Sciences, 113*(10), 2597–2602.

Moser, S. C. (2005). Impact assessments and policy responses to sea-level rise in three US states: An exploration of human-dimension uncertainties. *Global Environmental Change, 15*(4), 353–369.

Niven, R. J., & Bardsley, D. K. (2013). Planned retreat as a management response to coastal risk: A case study from the Fleurieu Peninsula, South Australia. *Regional Environmental Change, 13*(1), 193–209.

Nordstrom, K. F., Armaroli, C., Jackson, N. L., & Ciavola, P. (2015). Opportunities and constraints for managed retreat on exposed sandy shores: Examples from Emilia-Romagna, Italy. *Ocean & Coastal Management, 104,* 11–21.

Okuyama, Y., & Santos, J. R. (2014). Disaster impact and input–output analysis. *Economic Systems Research, 26*(1), 1–12.

Olsson, L., Jerneck, A., Thoren, H., Persson, J., & O'Byrne, D. (2015). Why resilience is unappealing to social science: Theoretical and empirical investigations of the scientific use of resilience. *Science Advances, 1*(4), e1400217. doi:10.1126/sciadv.1400217.

Ong, J. M., Jamero, M. L., Esteban, M., Honda, R., & Onuki, M. (2016). Challenges in build-back-better housing reconstruction programs for coastal disaster management: Case of Tacloban City. *Philippines. Coastal Engineering Journal, 58*(01), 1640010. doi:10.1142/S0578563416400106.

Pantuliano, S., Metcalfe, V., Haysom, S., & Davey, E. (2012). Urban vulnerability and displacement: A review of current issues. *Disasters, 36*(s1), s1–s22.

Parkinson, F. (2000). *Post-trauma stress: A personal guide to reduce the long-term effects and hidden emotional damage caused by violence and disaster*. Cambridge, MA: DaCapo Press.

Republic of the Philippines. (2014). *Joint DENR-DILG-DND-DPWH-DOST memorandum circular: Adoption of hazard zone classification in areas affected by Typhoon Yolanda and providing the guidelines for activities Therein*. Manilla: Republic of the Philippines. Retrieved from http://pcij.org/wp-content/uploads/2015/01/Joint-DENR-DILG-DND-DPWH-DOST-Adoption-of-Hazard-Zone-Classification.pdf.

Rulleau, B., & Rey-Valette, H. (2017). Forward planning to maintain the attractiveness of coastal areas: Choosing between seawalls and managed retreat. *Environmental Science & Policy, 72,* 12–19.

Sanderson, D., Kayden, J. S., & Leis, J. (Eds.). (2016). *Urban disaster resilience: New dimensions from international practice in the built environment*. New York, NY: Routledge.

Saravanan, D. (2016). Structural and nonstructural mitigation measures in coastal area threats. *International Journal of Oceans and Oceanography, 10*(2), 141–148.

Sauer, C. O. (1963). The morphology of landscape [1925]. In J. Leighley (Ed.), *Land and life selections from the writings of Carl Ortwin Sauer* (p. 343). Berkeley, CA: University of California, Berkeley Press.

Shuto, N., & Fujima, K. (2009). A short history of tsunami research and countermeasures in Japan. *Proceedings of the Japan Academy, Series B, 85*(8), 267–275.

Smith, N. (2006). There's no such thing as a natural disaster. In *Understanding Katrina: Perspectives From the Social Sciences*. Social Science Research Center. Retrieved from http://understandingkatrina.ssrc.org/Smith/.

Tadgell, A., Doberstein, B., & Mortsch, L. (2017). Principles for climate-related resettlement of informal settlements in less developed nations: A review of resettlement literature and institutional guidelines. *Climate and Development*, 1–14. doi:10.1080/17565529.2017.1291401.

Tidball, K. (2012). Urgent Biophilia: Human-nature interactions and biological attractions in disaster resilience. *Ecology and Society, 17*(2). doi:10.5751/ES-04596-170205.

Thomas, A. R. (2015). Post-disaster resettlement in the Philippines: A risky strategy. *Forced Migration Review, 49,* 52–54.

Uchida, N., Kirby, S. H., Umino, N., Hino, R., & Kazakami, T. (2016). The great 1933 Sanriku-oki earthquake: Reappraisal of the main shock and its aftershocks and implications for its tsunami using regional tsunami and seismic data. *Geophysical Journal International, 206* (3), 1619–1633.

Urwin, K., & Jordan, A. (2008). Does public policy support or undermine climate change adaptation? Exploring policy interplay across different scales of governance. *Global Environmental Change, 18*(1), 180–191.

Walker, B., Gunderson, L., Kinzig, A., Folke, C., Carpenter, S., & Schultz, L. (2006). A handful of heuristics and some propositions for understanding resilience in social-ecological systems. *Ecology and Society, 11*(1), 13–25. Retrieved from http://hdl.handle.net/10535/2570.

Weinstein, M. P., Baird, R. C., Conover, D. O., Gross, M., Keulartz, J., Loomis, D. K., et al. (2007). Managing coastal resources in the 21st century. *Frontiers in Ecology and the Environment, 5*(1), 43–48.

White, G. F. (1945). *Human adjustment to floods: A geographical approach to the flood problem in the United States*. Chicago, IL: University of Chicago.

Yamada, S., & Galat, A. (2014). Typhoon Yolanda/Haiyan and climate justice. *Disaster Medicine and Public Health Preparedness, 8*(05), 432–435.

[illegible] Preparing for [illegible] Earthquake [illegible]
Rubin, O., & [illegible] (20[illegible]). [illegible] linkage between [illegible] Journal of Contingencies and Crisis Management, [illegible]
[illegible] (20[illegible]). [illegible] New York, NY: Routledge.
[illegible] (20[illegible]). [illegible] International Journal of Disaster Risk Reduction, [illegible]
[illegible] Berkeley, CA: University of California, Berkeley Press.
[illegible] Proceedings of the [illegible]
Smith, N. (2006). [illegible] Understanding Katrina: Perspectives from the Social Sciences. Social Science Research Council. Retrieved from http://understandingkatrina.ssrc.org/Smith/
[illegible] (2017). [illegible] doi:10.1016/[illegible]
[illegible] (20[illegible]). [illegible] doi:10.1146/[illegible]
[illegible] Philippines [illegible] Review, 49, [illegible]
[illegible] (2015). [illegible] 1923 [illegible] Geomorphological [illegible] doi:10.1016/[illegible]
[illegible] support [illegible]
[illegible] Ecology and Society, [illegible]
[illegible] (2007). [illegible]
White, G. F. [illegible] University of Chicago.
[illegible] Medicine [illegible]

“God and Tonga Are My Inheritance!”—Climate Change Impact on Perceived Spritiuality, Adaptation and Lessons Learnt from Kanokupolu, ‘Ahau, Tukutonga, Popua and Manuka in Tongatapu, Tonga

Peni Hausia Havea, Sarah L. Hemstock, Helene Jacot Des Combes and Johannes Luetz

Abstract Climate change is a significant threat to health and well-being, in particular to people’s spiritual well-being, in coastal areas and communities. This paper describes the findings of a 2013 study on impacts of climate change on people’s spiritual well-being in 5 coastal communities in Tongatapu, Tonga: Kanokupolu, ‘Ahau, Tukutonga, Popua and Manuka, using a concurrent convergence parallel triangulation design. Information was collected from a group of 460 participants aged 15–75 via self-administered questionnaire, in-depth interview (IDI), focus group discussions (FGD) and key information interviews (KII). There was a statistically significant difference between people whose spirituality was affected and those unaffected ($p < 0.005$). People who were worried and whose physical well-being were affected, emerged as the most strongly affected ($\chi2(4) = 15.780$, $p < 0.005$). The same factor was explored qualitatively using thematic analytical strategy, and concluded that climate change had affected people’s spiritual well-being, negatively. Lessons learnt from this paper will be useful for people invested in policy practice, and pastorate serving the cause of holistic Climate

P. H. Havea (✉) · H. J. D. Combes
PaCE-SD, Lower Campus, Laucala, USP, Suva, Fiji
e-mail: ilaisiaimoana@yahoo.com

H. J. D. Combes
e-mail: descombe_h@usp.ac.fj

S. L. Hemstock
SPC, 2nd Floor Lotus Bldg, Nabua, Private Mail Bag, Suva, Fiji
e-mail: sarahh@spc.int

J. Luetz
School of Social Science, Christian Heritage College, 322 Wecker Rd, Carindale, QLD 4152, Australia
e-mail: jluetz@chc.edu.au

W. Leal Filho (ed.), *Climate Change Impacts and Adaptation Strategies for Coastal Communities*, Climate Change Management,
https://doi.org/10.1007/978-3-319-70703-7_9

Change Adaptation (CCA) in Tonga: preaching, fasting and praying to God to reverse climate change and its impacts, help State and Church to adapt more effectively and comprehensively.

Introduction

Consequences of global climate change impacts are already observed in coastal communities (Australian Bureau of Meteorology and Commonwealth Scientific and Industrial Research Organisation (CSIRO) 2014; IPCC 2014a, b, c; Michaelis 2011) and have also significant impact on people's spirituality (Ministry of Environment and Climate Change & National Emergency Management Office 2010; Nunn et al. 2016). Evidence suggests that research on perceived impact of climate change on spiritual well-being, however, is still very limited (Barnes et al. 2012; Havea et al. 2017; McIver et al. 2014; Woodward et al. 1998; World Health Organisation (WHO) 2011). To fill in this gap in knowledge in Tonga and the Pacific, a research was conducted in 2013 to better understand how climate change has affected people's well-being using a mixed method approach (Havea et al. 2016, 2017). This paper focuses on assessing the impacts of climate change on people's perceived spiritual well-being and the spiritual adaptation strategies needed for the five coastal communities in Tongatapu, Tonga: Kanokupolu, 'Ahau, Tukutonga, Popua and Manuka.

As Tonga is a Christian nation and with one of the highest rate of Christianity in the world (99.5%) (Statistic Department of Tonga 2016), when climate change affected people's physical environment and their church activities, it affects them spiritually, since they blame everything on their God. This system of belief is particularly significant for people's well-being in Tonga because the two major factors perceived to be important in terms of national well-being, are religion and education (Havea et al. 2016; Matangi Tonga Online 2001), which was mandated by King George Tupou I, who had dedicated Tonga to God in 1839 (Sekona 2014), through his motto, that "God and Tonga are my inheritance" (Lātūkefu 1975, 2014). So, when climate change affected church and/or their physical environment, (Government of Tonga 2015), it affected them spiritually (Bible Society of the South Pacific 1966), as opposed to Australia, USA or UK (Government of Tonga 2016; Leiserowitz et al. 2015; Morrison et al. 2015).

As this legacy is significant to Climate Change Adaptation (CCA) and Disaster Risk Management (DRM) (Government of Tonga 2015) in Tonga, by improving people's spiritual well-being through climate change adaptation in such a religious nation is therefore needed. God and Tonga are my inheritance (Koe 'Otua mo Tonga ko hoku Tofi'a), is the most significant element in people's spiritual well-being because of church, and linking it to climate change adaptation and disaster management is vital. This helps to refine and perfect the course of adaptation better, identify what forms of spiritual resilient healing and prayer is needed, and how these sustainable measures can be used to protect people's spirituality and inheritance. Most importantly, this may help fostering their fellowship with God for protection and peace, thus contributing to make Tonga a more resilient and a sustainable nation, by 2035 and beyond (Government of Tonga 2016).

Methods

Methodology

Participants were Tongans aged 15–75 from Kanokupolu, 'Ahau, Tukutonga, Popua and Manuka in Tongatapu, Tonga. The study assessed the perceived impacts of climate change on spirituality amongst 460 participants, who were selected to complete a self-administered questionnaire. Then, 24 participants were randomly chosen for in-depth interview (IDI) for more than 30 min, and 28 participants were randomly selected for focus group discussion (FGD) in the form of workshop for one day in Vakaloa beach resort in Kanokupolu, Tongatapu. In parallel with this collect of information, 12 key informant interviews (KII) were also conducted with people more than 75 years old in these five coastal communities, senior government officials in the Government of Tonga (GOT) and a representative from a church organisation in Tonga (Havea et al. 2017).

The mixed method approach used 'concurrent convergence parallel triangulation design' because the data was collected at the same time—concurrently—and then merged—combined or blended. Parallel corresponds to the fact that it was equally weighted (no emphasis) during the analysis, and triangulation—since the implementation of the data collections technique has used more than one method (Creswell 2013, 2014; Creswell and Plano Clark 2007, 2011). The study was conducted purely in Tongan to facilitate discussions and the data collection process in the communities and has been assisted by a local research assistant.

Data Analysis Strategy

Data was concurrently analysed using convergence analytical approach—that is a hallmark of mixed method research (Creswell 2013, 2014; Creswell and Plano Clark 2007, 2011). In the first phase, the authors determined those whose perceived spiritual well-being was affected and those who were unaffected, using binomial and chi-square goodness-of-fit tests for comparisons of proportions, respectively, and correlation analyses computed with Kendall's tau-b (T_b) (Peck and Devore 2012; Springer 2012; Sullivan 2013). Then, a binary logistic regression (Peck and Devore 2012; Springer 2012; Sullivan 2013) was used to describe the relationship between the 4 predictors (gender, worry, mental well-being (relating to the happiness of the mind), and physical well-being (relating to impact on physical environment) and the perceived spiritual well-being impacts. Altogether, there were 109 questions on well-being (including livelihoods and health) with nominal and Likert scale using 5-point scale type of response from strongly disagree to strongly agree. Data analysis was carried out with the use of SPSS 23 (IBM Corporation, New York, USA), Minitab 17 (Minitab Inc., Pennsylvania, USA), and JMP 11 (SAS, North Carolina, USA).

In the second phase of data analysis, a thematic analysis (Bazeley 2007; Miles et al. 2014; Ozkan 2004) was used to derive a theme related to the impact of climate change on perceived spirituality using QSR N10 (QSR International, Melbourne, Australia). Information was collected using semi-structured face-to-face interviews with the participants. All interviews were audio- and video-recorded and then translated and transcribed into the computer for analysis. The focused was on responses to interviews and focus group questions that related to the discussion of the perceived impact of climate change on spirituality. These themes were then related to the quantitative analysis.

Study Limitations

Although the research has reached its aim, there was some limitations. First, the research would have gain more weight if the president from churches in Tonga had been involved, because people in these communities were very religious. The reason is that church leaders are more powerful and trusted in delivering messages on climate change impacts and adaptation strategies to people in these coastal communities than scientists and local government (Nunn et al. 2014, 2016). Second, is the limited funding for this study does not enable the researcher to go back to the communities in Tonga to present the findings and collect feedbacks from the participants. Thirdly, because the study was specifically design for Tonga as a Christian state, it may not be applicable to secular countries. Finally, the research did not assess the impact of improving spirituality on health or how to use it to eliminate impact of diseases that are climate change-related on people's health.

Results

Quantitative Analysis and Results

Gender

Of the 460 participants recruited to participate in the study, there were 244 (53%) female and 216 (47%) male. This percentage are very close to the general population in Tonga (male = 49.4%, female = 50.6%) (Statistic Department of Tonga 2016; Tonga Department of Statistic 2011, 2013b). Meaning that the view of people whose perceived spirituality were affected by climate change in this study, the adaptation and disaster management strategies proposed by them can be used to represent of all women and men in Tonga.

Number of participants in the 5 studied coastal communities

The study recruited 40 participants from Kanokupolu, 52 from 'Ahau, 77 from Tukutonga, 246 from Popua and 45 from Manuka (Fig. 1). This distribution however is almost the same as the distribution of the general population in these five coastal areas and communities (Statistic Department of Tonga 2016; Tonga Department of Statistic 2011, 2013b).

Religious denomination—Christian (Including the Bahai Faith)

This study found participants belonged to 14 Christian churches, dominated by the Free Wesleyan Church (42%), (Fig. 2). Although the study indicated that Tonga is a Christian state, non-Christian religions are practiced as well. This is important because although the sample were 100% Christian, the general population is not (Statistic Department of Tonga 2016; Tonga Department of Statistic 2008, 2011, 2013a, b, 2016; Tonga Department of Statistic & Secretariat of the Pacific Community (SPC) 2008).

Comparing people whose spiritual well-being was affected with non-affected to inform adaptation strategies in coastal communities

Out of the 460 participants in the study, 98% (451) indicated that climate change had affected their spiritual well-being, while only just 2% (9) indicated that their perceived spiritual well-being was not affected.

Further to the frequency analysis, the collected response data were also analysed using binomial tests (Table 1), (Peck and Devore 2012; Springer 2012; Sullivan 2013) and a sample goodness-of-fit test (Table 2), (Peck and Devore 2012; Springer 2012; Sullivan 2013).

Based on this sample, the study found that the mean proportion of Tongans aged 15–75 in Kanokupolu, 'Ahau, Tukutonga, Popua and Manuka, whose perceived spiritual well-being were affected by climate change and those who were not, were statistically significant different, $\chi^2(1, N = 460) = 424.704$, $p < 0.005$ (Tables 1 and 2). People whose spiritual well-being were affected value church significantly, are very worried about climate change and have their physical well-being affected much more than those unaffected.

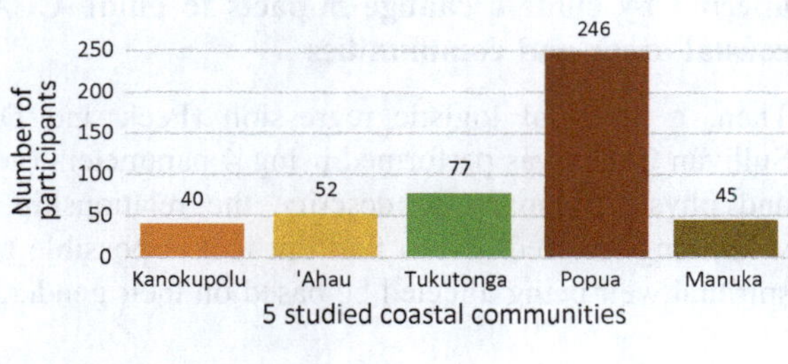

Fig. 1 Total number of participants in the 5 studied coastal communities

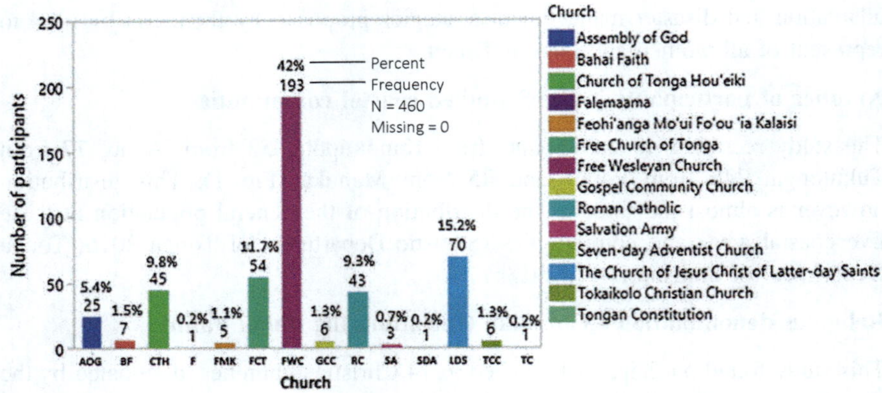

Fig. 2 Number of reported Christian religions adherents

Table 1 Perceived impact on spirituality

		Category	N	Observed prop.	Test prop.	Exact sig. (2-tailed)
Spiritual well-being affected	Group 1	Yes	451	0.98	0.50	0.000
	Group 2	No	9	0.02		
	Total		460	1.00		

Table 2 Chi-square goodness-of-fit—perceived spiritual impact

	Spiritual Impact		
Chi-square	424.704[a]		
df	1		
Asymp. sig.	0.000		
	Expected N	Observed N	Residual
No	9	230.0	−221.0
Yes	451	230.0	221.0
Total	460		

[a]0 cells (0.0%) have expected frequencies less than 5. The minimum expected cell frequency is 230.0

Using the predictors of people whose spiritual well-being was negatively affected by climate change impacts to guide CCA strategy needed for these coastal areas and communities

Then, a binomial logistic regression (Peck and Devore 2012; Springer 2012; Sullivan 2013) was performed using 4 parameters (gender, worried, mental impact and physical impact) to describe the relationship between them and spiritual well-being affected and to find out if it is possible to predict who will have their spiritual well-being affected be based on their gender, how worried they were, how their mental well-being were affected and/or how their physical well-being were affected. The first part of the analysis was the Omnibus Tests of Model Coefficients (George and Mallery 2016), which shows that this model was statistically significant, $\chi^2(4) = 15.780$, $p < 0.005$. The second part of the analysis presented the Model Summary (George and Mallery 2016) using the Cox and Snell R Square and Nagelkerke R Square values and it shows that 19.2% (Nagelkerke R^2) of variability in spiritual well-being negatively affected by climate change can be explained by the differences in people's gender, how worried they were, and whether physical or mental well-being were affected or not.

As these results show that 80.2% of the variability cannot be explained by these 4 parameters, other factors can be used to rule out spiritual well-being impact as well. These factors could be how emotional and how religious the participants were, have had a death experienced from an impact of natural disasters (e.g. earthquake or tsunamis) on a family member before, water supply deficits, health affected (e.g. asthma, shortness of breath, pneumonia, influenza), and participants denominations'. Although these factors are very important to predict spiritual well-being affected, but they may or may not be a significant predictor and therefore they might or might not added a significant contribution to predict people whose spirituality will be affected as people who were worried and have physical impact did (Table 3).

The last part of the analysis, is the Variables in Equation (George and Mallery 2016) presenting the 4 predictors that influences the perception of the people that

Table 3 Binary regression analysis summary for negative impact of climate change on perceived spirituality

Omnibus tests of model coefficients

		Chi-square	df	Sig.
Step 1	Step	15.780	4	0.003
	Block	15.780	4	0.003
	Model	15.780	4	0.003

Model summary

Step 1	−2 Log likelihood	Cox and snell R square	Nagelkerke R square
	72.855[a]	0.034	0.192

Classification Table[b]

	Observed		Predicted		
			Spiritual impact		Percentage correct
			No	Yes	
Step 1	Spiritual impact	No	1	8	11.1
		Yes	0	451	100.0
	Overall percentage				98.3

[a]Estimation terminated at iteration number 8 because parameter estimates changed by less than 0.001

[b]The cut value is 0.500

their spiritual well-being was affected. The study found that perceived worried ($p = 0.035$) and physical well-being affected ($p = 0.039$), are the most important predictors whereas gender ($p = 0.45$) and mental well-being affected ($p = 0.64$) did not. The odds of people who perceived their physical well-being was affected by climate change to be affected spiritually were 8.09 times greater than those whose perceived their physical well-being was unaffected by climate change. For those who were worried by climate change, the odds were 2.69 times more likely to be affected spiritually than those who were not worried. This means that based on this sample, the needs of the people who are worried by climate change and considered that their physical well-being is also affected by climate change need to be addressed in adaptation strategies (Table 4).

To support the above stated results, a contour plot (Ryan et al. 2013) was conducted to show how these 2 important parameters (worried, physical impact) were related to spiritual well-being impact. The emerald green region indicates higher spiritual well-being impact and the blue region with lower spiritual impact (Fig. 3). This relationship is significant and has yielded in a positive direction as indicated by the contour plot value to be positive and pointed to the right, with p-value table using Kendall's tau-b (Table 5).

Table 4 Results of binary regression analysis on impact on perceived spirituality

Variables		B	S.E.	Wald	df	Sig.	Exp(B)	95% C.I.for EXP(B)	
								Lower	Upper
Step 1[a]	Gender	0.555	0.739	0.563	1	0.453	1.742	0.409	7.417
	Worried	0.989	0.468	4.466	1	0.035	2.689	1.074	6.730
	Mental impact	−0.646	1.393	0.215	1	0.643	0.524	0.034	8.034
	Physical impact	2.091	1.012	4.274	1	0.039	8.096	1.115	58.800
	Constant	−2.241	1.876	1.427	1	0.232	0.106		

[a]Variable(s) entered on step 1: Gender, Worried, Mental impact, Physical impact

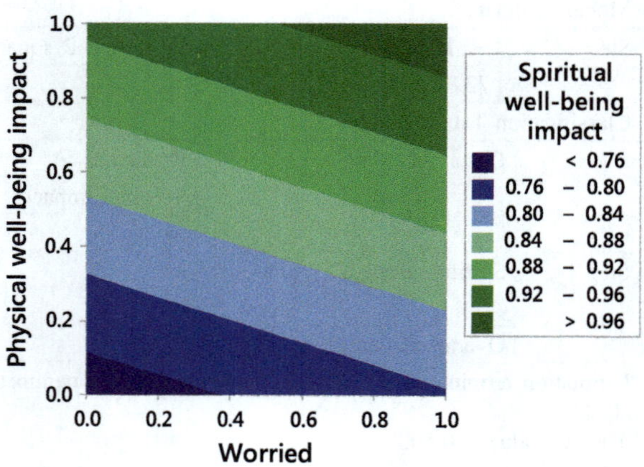

Fig. 3 Contour plot of spiritual well-being impact vs physical well-being affected, worried

Table 5 Correlations with gender, worried, mental, physical and spiritual well-being impact

	Gender	Worried	Mental impact	Physical impact	Spiritual impact
Gender	–	−0.001	0.010	0.024	0.024
Worried	−0.001	–	0.276**	0.233**	0.136**
Mental impact	0.010	0.276**	–	0.466**	0.174**
Physical impact	0.024	0.233**	0.466**	–	0.260**
Spiritual impact	0.024	0.136**	0.174**	0.260**	–

*$p < 0.05$, **$p < 0.01$

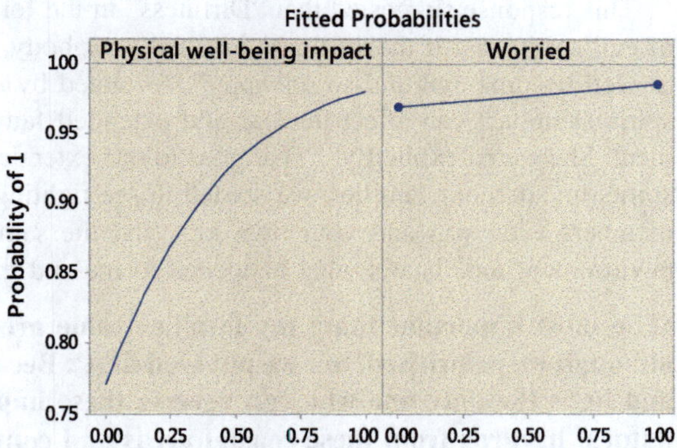

Fig. 4 Main effects plot for spiritual well-being impact

Then, a factorial plot (Ryan et al. 2013) was conducted to assess which of these 2 significant parameters (worried, physical impact) would have the most significant interactions with perceived spiritual well-being impact. The results show that the magnitude of physical well-being impact appears to be larger than the magnitude for worried (Fig. 4), with an exponentiation of the B coefficient (odds ratio) of 8.096 higher than 2.689. This has indicated that in terms of adaptation strategy, actions to reduce the perception of physical well-being being affected should be given priority compared to actions to reduce worry about climate change.

Qualitative Analysis and Results

Themes that Emerged in Interviews and FGD

Several themes emerged from a careful review of the transcripts in reference to the perception of climate change impact on spirituality and adaptation strategies. Here, six of these themes are described, which helps guide the spiritual adaptation strategies for these coastal areas and communities.

'Dirtiness is a representation of evil spirits and if that evilness/evil spirits invaded your body, your spiritual well-being has as well become evil and so unhappy': A guide to spiritual adaptation for these coastal areas and communities and Tonga

In several of the transcripts, participants expressed a belief that "heavy rainfall" and "flooding" affect spiritual well-being. For example, a female participant from Popua stated: that she "don't know what to do" especially "when the water started to be coming under the main door inside our house." The problem here is that because when the "heavy rainfall finished, it brings a lot of rubbish and different kind of debris surge. If we can't try our best to clean it as quickly as possible, it will cause diseases and may threaten the health of our children as a consequence of leaving those rubbishes and dirt at home."

This response suggested that "Dirtiness" in the form of filthiness "is a represent of evil spirits and if that evilness invaded your body, your spiritual well-being has as well become evil and so unhappy". As stated by the participant, this perceived spiritual impact can affect nuclear and extended families, as well as the environment. She stated explicitly: "That goes to our extended families as well. When evil spirits invaded our families, we started to see fighting and conflict between family members even you and your own kids and the same thing happen to your own environment too. That's what happened to me and my life in my own families."

'The most important thing my families value are the church and education, although we prioritised, but we put God first': Because God made this universe and he is the only one who can reverse these impacts of climate change and natural hazards from these coastal areas and communities and Tonga

This theme emerged when participants discussed how their spiritual well-being were affected by climate change and/or natural hazards. This has often come up especially in response to the interviewer's question: "What about the impact of climate change on the sources of well-being of you or your families in your household have? In this context participants always referred to God and the Bible. As one female participant from Manuka has put it: "The most important thing my family values are the church and education, although we prioritised, and put God first. Another male participant from Manuka, added on to this discussion by stated implicitly: 'Our family priority, is going to church. God controls everything in this universe."

"For example, if you want wisdom and understandings, and you asked God, then God will give it to you. So, the best overall well-being is to set your priority right, and put God first. I mean, you have to keep your relationship and fellowship with God, by worshipping and reading God scriptures. You see, for me, I am a Minister here in the Assembly of God, and now I am the Assistant Minister in our church. For me, it is such an important thing to read my Bible and praying to God in secret, before going to work in the morning. When I drive in the road to go to work, I know that God is protecting me. By doing that, I know that God is giving me, what I was asking for, so to improve my overall well-being."

From a spirituality point of view, this may mean that participants who perceived God send these impacts, also believed that he is the only one who can reverse them from affecting their coastal communities as well as the general population. Therefore, by keeping their faith and praying to God, the Christian God can reverse these impacts from these coastal communities and save them, their families and to all the people of Tonga. This Tongan worldview and interpretation of the Bible, however, are different from the perception of most of the western world in Europe, USA and even Australia although they have been reading the same book (Ipsos MORI 2014; Leiserowitz et al. 2015; Morrison et al. 2015; Nunn et al. 2016).

'The churches were destroyed, and most people commend that in this day, they believed that it is more like the judgement day for them and this is a part of God's plan': A sign of the second coming of Christ, therefore repent and turn to God for adaptation and safety as Christian in these coastal areas and communities perceived as what is impossible for men are possible with God

Several participants mentioned that they perceived climate change to be an indication of the judgement day, which affect their spirituality indirectly. A female participant from Kanokupolu was saying: "I still remember in the tropical cyclone Isaac in the 1980s, in that year in 1982, the tropical cyclone Isaac threatened and devastated almost every household in this community, livestock's, as well as individuals. The sea water pushed them to the other side of the village with everyone floating in the water. The churches were destroyed too, as far as I remembered. And most people commend that in this day, they believed that it is more like the judgement day for them especially the older people in their 60s and 70s. We were trying to safeguard them to a safer place." When she was asked to reflect on climate change as Christian, she said: "I know that I am a Christian, because of the church that I belong to. I do believe these climate change and impacts are part of the God's plan. Because there is nothing that God cannot do—"ko e 'Otua 'oku ne mafeia 'a e me'a kotoa pe". What we need to do though as Christian is to follow exactly what the scriptures dictated us to do. Nature is a true representation of God himself. If we try to solve nature (e.g. climate change) by our human intelligence, we can't do it. Like Tonga, we should be proud that our great King George Tupou I have devoted that my "God and Tonga are my inheritance" (Ko e 'Otua mo Tonga ko hoku Tofi'a). So, worship and prayer is always part of our culture so we need God to help us understand nature and solve this problem."

'A form of punishment to teach us a lesson to change our lives because we know that Tonga is so different now'—So to turn to God and ask him to reverse the impacts of climate change and natural hazards from these coastal areas and communities, thus help making a Resilient Tonga to 2035 and beyond as we are God's children

This theme emerged when participants perceived that climate change is a punishment from God. A female participant from Popua was saying: "To me, this impact of climate change is more like a form of education to educate us or to punish us so to open up our mind and know what to do. God does not want us to suffer.

It is more like God is educating us to look after our families better. It is also like a form of punishment to teach us a lesson to change our lives because we know that Tonga is so different now. We all know. Our young people now is way more different from what we've seen in the past. God doesn't want us to struggle. But it is like God is teaching us a lesson that we know what to do and how to behave in such a manner as we are his children." From the adaptation point of view, this means that people in Tonga should turn to God and pray to him to reverse these impacts from affecting them because he is the only one who can do it, and most importantly contributed to make Tonga a more resilience nation by 2035 (Government of Tonga 2016) and beyond. This interpretation of the Bible by the Tongan Christians in terms of climate change impacts and adaptation seems to be different from other nations, although they have read the same book (Ipsos MORI 2014; Leiserowitz et al. 2015; Morrison et al. 2015; Nunn et al. 2016).

'The churches were destroyed' and 'If I missed going to church', 'I don't feel spiritually occupied on that day'—So using Church (and State) to care for the most vulnerables in these coastal areas and communities and for Tonga is significant

This theme connotes that when climate change affected people's going to church to praise God it affects them spiritually. A female participant from Manuka was saying: "I've always prioritised church to come first. I am a preacher, and since, I was young, I've always put God first. I am the principal of the Sunday school in our church, and we are the maintenance steward. Every day, I read my Bible and always do my prayer in secret, as in daily bread devotions. I have attended all the church services, although, there will be climate change like rainfall and so on. Sometimes, however, I missed it due to some extreme natural events, but when I stayed home, I never go to sleep, rather, I keep on reading the scriptures. I don't feel completely spiritually, fully occupied on that day, because I wanted to go to church and praise God by singing." As indicated in this transcript because church play a significant role in satisfying the participant's spiritual well-being, by using it as a tool to tackle impacts of climate change on the ground would help people spiritually, and most importantly find peace, happiness and/or solutions to their problems effectively.

'Sometimes warming climate usually got me sick so easily like flu viral illness, asthma, and shortness of breath'—Similarly people spiritual health and/or well-being will be affected, concurrently and this need spiritual support and adaptation

This theme denotes impact of climate change on physical well-being in the form of perceived impact of climate change on health which simply affected a person's spiritual well-being. A female participant from Kanokupolu stated: "The only time that I usually get sick is when there is climate change like a lit bit hotter. Sometimes warming climate usually got me sick so easily like flu viral illness, asthma, and shortness of breath. And it usually happened somewhere around November, December up to February during our wet seasons. During these times of the year, I usually get those illnesses. And even nowadays, I felt that climate change is getting worse, because it feels even hotter, during days and at night times, so there is no much of a difference at all. And especially flu, common cold, coughing and rhinorrhoea, it usually affects me."

The relationship between impact of climate change on health to spiritual well-being can be translated by when people are sick because of climate change-related conditions their spiritual well-being will be affected, simultaneously (GOV.UK 2013, 2014; Government of Tonga 2015, 2016). So, supporting people spiritually is vital to their overall health.

Discussion

Perceived Impact of Climate Change on Spirituality and Spiritual Adaptation for Five Coastal Areas and Communities in the Tongan Christians Mind

Despite a considerable body of literature (IPCC 2014b; McIver et al. 2014; Nunn et al. 2016; World Health Organisation (WHO) 2015) devoted to climate change impacts and adaptation strategies to coastal areas and communities, there is little discussion on the literature on the effects of climate change impacts on perceived spirituality and spiritual adaptation. To date, the theoretical and empirical work on impact on livelihood, health and well-being capital has shown limited sensitivity to the impact on perceived spirituality. The current study sought to redress this deficiency in literature.

The paper sought to capture the perceived impact of climate change on spiritual well-being in Tongan Christians aged 15–75 in five coastal areas and communities in Tongatapu, Tonga. A large majority of people perceived that their spiritual well-being was negatively affected by climate change. The study found that women's and men's spiritual well-being was affected by climate change, mainly because they value church significantly, were worried by the impacts of climate change on them or because of the physical impact caused by increasing sea level rise, heavy rainfall, extreme weather events, temperature change (hot and cold), drought, and seasonal shift (for example overlapping of seasonal changes from wet to dry season and vice versa in Tonga), compared to those who were unaffected.

The quantitative and qualitative findings in this study highlighted the significant differences between people who perceived their spiritual well-being was affected (n = 451) and those unaffected (n = 9) ($\chi^2(1) = 424.704$, $p < 0.005$), (see Tables 1 and 2). This is a very important information for Tonga because it shows that the vast majority of the population is affected spiritually by climate change so spiritual adaptation is needed. One female participant stated explicitly how climate change had affected her and her family members: "Dirtiness is a represent of evil spirits and if that evilness invaded your body, your spiritual well-being has as well become dirty and so unhappy." This excerpt illustrated severe affliction to the sense of spiritual well-being because people visualised and conceptualised climate change and dirtiness using the insight of evil spirit (as in laumālie ‘uli). In the same way, when that evil spirit invaded their bodies, as the participant stated, it will have caused them to be unhappy, and so affected their spiritual well-being as well.

The link of dirtiness with physical well-being affected and worried is via the insights of evil spirits (laumālie ‘uli) as conceptualised from the Bible (Bible Society of the South Pacific 1966). In the study, participants discussed dirtiness as a Tongan metaphor of evil spirits, and when this demonic possession in the form of evilness invaded their body, it affected their mental state of health, their physical health and thus lead to madness (puke faka-tevolo), anxiety, worried and/or depressive disorder (puke faka-‘avanga), which eventually would have affected them spiritually. This was concordant statistically, which shows that those whose physical well-being was affected ($p = 0.039$) and worried ($p = 0.035$) are most likely to be affected spiritually (see Tables 3 and 4). Most significantly, as people’s general state of unease increased, their physical and/or mental status would be reflected in their spiritual well-being as well (see Figs. 3 and 4). And since climate change being a very significant source of worry in coastal communities in Tonga because of sea level rise and people perceiving climate change to be responsible for several physical illness, it is therefore to be expected that it also impacts negatively on people’s spiritual well-being.

Perceived impacts of climate change on physical well-being ($p = 0.039$) and worry about impact of climate change ($p = 0.035$) emerged as major predictors of negative impact on spiritual well-being, whereas gender ($p = 0.45$) and mental well-being affected ($p = 0.64$) did not (Michaelis 2011). As one woman from Popua stated when she linked her worries and physical well-being affected to God’s punishment: “To me, this impact of climate change is more like a form of punishment to teach us a lesson to change our lives because we know that Tonga is so different now.” Clearly, this excerpt alone suggests that the participant was spiritually affected because she used words like “punishment”, “teach”, “lesson”, “change”, “lives” to express her fear of God and climate change (Roser-Renouf et al. 2016; USGCRP 2016). And, because of the church she belonged to (FWC) and her perception from the Bible (Bible Society of the South Pacific 1966), it designated her responses to refer to her God for peace and comfort, since no one in Tonga believe that climate change does not exist.

This interpretation of the Bible by the Tongan Christians who advocates climate change are different from the same Christians in Australia, USA, UK and Europe who are climate change denier’s and sceptical, although they all read the same book (Ipsos MORI 2014; Leiserowitz et al. 2015; Morrison et al. 2015; Nunn et al. 2016) but people in Tonga see and experienced climate change impacts everyday. Since these impacts have led Tongans to reference the protection and guidance of God from the Bible (Bible Society of the South Pacific 1966), therefore they perceived that the Holy Scripture can be used to manage climate change adaptation and disasters management as well (Government of Tonga 2015).

A female participant from Kanokupolu explained explicitly how climate change affected her physical well-being in the form of sickness by referring to asthma, shortness of breath, viral illness like flu, common cold, coughing and rhinorrhoea: "The only time that I usually get sick is when there is climate change like a little bit hotter" has "got me sick so easily like flu viral illness, asthma, and shortness of breath, flu, common cold, coughing and rhinorrhoea." From this passage, it showed that because the participant was ill due to climate change such as temperature change variability, it affected her well-being physically in the state of human health and physical well-being and people merging climate change and climate variability is a sign of worry that climate change cause to the people. For the general population, this may mean that they need to be well cared during the wet season (Australian Bureau of Meteorology and Commonwealth Scientific and Industrial Research Organisation (CSIRO) 2014; Tonga Meteorological Service 2015, 2016) in Tonga.

Another female participant from Manuka, explained how climate change was affecting her spiritual well-being because it has affected her schedule to go to church to communicate to God to improve her spiritual well-being. This is such an important well-being impact in the context of Tonga because people who prioritised God in the form of church in the study, may mean that if their programme to attend church activities was affected, consequently it may have influenced their spiritual well-being negatively. As a female participant stated explicitly: "I have attended all the church services, although, there will be climate change like rainfall and so on. Sometimes, however, I missed it due to some extreme natural events, but when I stayed home, I never go to sleep, rather, I keep on reading the scriptures. I don't feel completely spiritually, fully occupied on that day, because I wanted to go to church and praise God by singing." But because she did not go to church, it makes her worried and affected her spiritual well-being.

This is important because, from this excerpt, it is clearly indicated that this woman's priority is by going and attending church services and like everyone else who have faith and trusted in God (Bible Society of the South Pacific 1966). As expressed and argued by a woman from Kanokupolu: Because climate change is nature, and "nature is a true representation of God himself", therefore, worshiping and praying to their Christian God, in her spiritual faith as a Christian, is the only way to help Tonga solve this problem, since King George Tupou I dedicated Tonga to God for protection in 1839 that "God and Tonga are my inheritance" (Government of Tonga 2015, 2016; Sekona 2014).

More interestingly, when she also claimed that our human intelligence cannot solve the impacts of climate change, rather that only the Christian God can, has underscore how devoted some Tongan Christians are, even referring to themselves as "God's children". As a result of this paper supporting people's who perceived climate change is affecting their spiritual well-being negatively, it can then be used to support a policy proposal for the Government of Tonga and the National Council of Churches to scale up CCA. This can be done through preaching, fasting and a national prayer initiative, to pray to their Christian God to reverse the impacts of climate change and natural hazards from affecting people's spirituality. In doing so, these concrete actions could give people a feeling of actively respond to what is written in the Scriptures and make them feel less worried.

Then, promote integration of spiritual adaptation into their roles in their families, churches and communities. In doing so, this would help Church and State achieving a better Resilient Tonga by 2035 (Government of Tonga 2016) and beyond, thus developing a spiritual coping mechanism that offers a unique role in promoting happiness and peacefulness for people who perceived their spiritual well-being was adversely affected. This can be achieved by looking into the impact of climate change on spirituality for the Pacific or other island states beyond the Pacific (Luetz 2017) in order to address the spiritual needs of the people in the Pacific to inform a more holistic regional adaptation plan that can more auspiciously meet the Sustainable Development Goals (SDGs) and Sendai Framework for Disaster Risk Reduction (DRR). Since spiritual well-being is vital for the people of Tonga, it needs to be included in the CCA strategies. However, these strategies need to target the identified predictors, by improving the physical well-being of the people, reducing the physical impacts of climate change and reduce their worries. First, this could be done through CCA project. Second, this may be done in relation with Pastors and churches to reassure them and give them hope by empowering and providing them with information so that they can be part of CCA project and then actively participating in actions that support the teaching of the scriptures.

Conclusion

In conclusion, the main lessons from this paper can be summarised as follows:

- Climate change had affected people's spiritual well-being, negatively;
- Climate change was viewed as punishment from God, but this result in a positive approach to well-being adaptation in order to please God;
- People in Tonga use the Holy Scripture to adapt to climate change;
- Spiritual adaptation in the form of praying or spiritual counselling and rehabilitation should be focused on people who are worried, stressed, and feel anxiety or fear of climate change and whose physical well-being was affected;
- People faith in God has helped them adapt to climate change.

Based on these lessons learnt, the paper recommended that this will be useful to both the State and Church actors—where people in the communities must include in the adaptation strategies, to ensure that the approach is holistic. By doing so, people will be implementing it in the sense that it is the will of God, thus making them more responsible and feel that they own it, since they believe it is what the scriptures in the Bible wanted them to do so. This is how CCA should be incorporated into the government national adaptation action plan and church agenda's, and then use church to co-implement it with them, concurrently meeting both people spiritual needs as well as the requirements and tents of science, in order to achieve Resilient Tonga by 2035 and beyond (Government of Tonga 2016).

For the future, in order to provide a better understanding about how spirituality improves quality of life and heal people who are affected by climate change, further study is still needed in this area for Tonga and the Pacific. This could be a research:

1. To investigate if improving people's spirituality would help them healed their sickness and health (e.g. helping in healing people who are living with malignant cancer, HIV/AIDS or NCD). And if so, does improving spiritual well-being promote peacefulness in them?
2. To find out if the Church and State need to work together closely in addressing spiritual well-being or not. This is an important factor in people's lives to improve overall happiness, peacefulness and enduring their world of religiosity/spirituality in Tonga.
3. To investigate how Christian God's may help Tonga address impacts of climate change and natural hazard on people spiritual well-being, thus co-benefits other human factors as well.

Acknowledgements The authors would like to thank the people of Kanokupolu, 'Ahau, Tukutonga, Popua and Manuka. Also, we want to acknowledge PaCE-SD (Pacific Centre for Environment and Sustainable Development) for supervising this project and the Research Office at USP (University of the South Pacific), for providing financial assistance for this work under its Ph.D. Regional Scholarship Fund, No. 6F004-3107-70766-00. The study was approved by PaCE-SD and funded by the Research Office at USP. Finally, we want to thank the EU PacTVET Project for funding Peni's trip to Apia, Samoa to present this paper in this World Symposium on Climate Change Impact and Adaptation Strategies in Coastal Communities.

References

Australian Bureau of Meteorology and Commonwealth Scientific and Industrial Research Organisation (CSIRO). (2014). *Climate variability, extremes and change in the western tropical Pacific: New science and updated country reports (Vol. 2)*. Melbourne, Australia: Pacific-Australia Climate Change Science and Adaptation Planning Program Technical Report, Australian Bureau of Meteorology and Commonwealth Scientific and Industrial Research Organisation.

Barnes, J. I., MacGregor, J., & Alberts, M. (2012). Expected climate change impacts on land and natural resource use in Namibia: Exploring economically efficient responses. *Pastoralism: Research, Policy and Practice*, 2(1), 1–23.

Bazeley, P. (2007). *Qualitative data analysis with Nvivo* (2nd ed.). London: SAGE.

Bible Society of the South Pacific. (1966). *Koe Tohi Tapu Katoa [The Holy Bible in Tongan]*. Suva: The Bible Society in the South Pacific.

Creswell, J. W. (2013). *Qualitative inquiry and research design: Choosing among five approaches*. United State of America: SAGE.

Creswell, J. W. (2014). *Research design*. Los Angeles: SAGE.

Creswell, J. W., & Plano Clark, V. L. (2007). *Designing and conducting mixed methods research*. United State of America: SAGE.

Creswell, J. W., & Plano Clark, V. L. (2011). *Designing and conducting mixed methods research* (2nd ed.). United State of America: SAGE.

George, D., & Mallery, P. (2016). *Ibm spss statistics 23 step by step—A simple guide and reference* (14th ed.). New York: Routledge Taylor & Francis Group.
GOV.UK. (2013). *Wellbeing and health*. UK: Department of Health and Cabinet Office.
GOV.UK. (2014). *The relationship between wellbeing and health*. UK: Department of Health.
Government of Tonga. (2015). *Tonga strategic development framework 2015–2025—A more progressive Tonga: Enhancing our inheritance*. Tonga: Ministry of Finance and National Planning.
Government of Tonga. (2016). *Tonga climate change policy: A resilience Tonga by 2035*. Tonga: Government of Tonga.
Havea, P. H., Hemstock, S. L., & Jacot de Combes, H. (2016). Preparing for better livelihoods, health and well-being—A key to climate change adaptation. In W. Leal Filho (Ed.), *Climate change adaptation in Pacific countries: Fostering resilience and improving the quality of life*. Berlin: Springer.
Havea, P. H., Jacot des Combes, H., Hemstock, S. L., Gee, K., Han, D., Khan, M. G. M., & Moore, T. (2017). *Climate change impact on livelihood, health and well-being: A mixed method approach (Ph.D. thesis)*. The Unversity of the South Pacific, Fiji.
IPCC. (2014a). *Climate Change 2014: Impacts, adaptation, and vulnerability. Part A: Global and sectoral aspects. Contribution of Working Group II to the Fifth Assessment Report of the Intergovernmental Panel on Climate Change*. Cambridge: Cambridge University Press, 1132 pp.
IPCC. (2014b). *Climate Change 2014: Impacts, adaptation, and vulnerability. Part B: Regional aspects. Contribution of Working Group II to the Fifth Assessment Report of the Intergovernmental Panel on Climate Change*. Cambridge: Cambridge University Press, 688 pp.
IPCC. (2014c). *Climate Change 2014: Mitigation of climate change. Contribution of Working Group III to the Fifth Assessment Report of the Intergovernmental Panel on Climate Change*. Cambridge: Cambridge University Press.
Ipsos MORI. (2014). *Global trends 2014: Environment*. Global Trend: UK. Retrieved Dec 28, 2016 from http://www.ipsosglobaltrends.com/environment.html.
Lātūkefu, S. (1975). *The Tongan constitution: A brief history to celebrate its centenary*. Nuku'alofa, Tonga: Tonga Traditions Committee.
Lātūkefu, S. (2014). *Church and state in Tonga: The Wesleyan methodist missionaries and political development, 1822–1875*. Australia: University of Queensland Press.
Leiserowitz, A., Maibach, E., Roser-Renouf, C., Feinberg, G., & Rosenthal, S. (2015). *Climate change in the American Christian mind: March 2015*. New Haven, CT: Yale Project on Climate Change Communication: Yale University and George Mason University.
Luetz, J. M. (2017). Climate change and migration in the Maldives: Some lessons for policy makers. In W. Leal Filho (Ed.), *Climate Change Adaptation in Pacific Countries: Fostering Resilience and Improving the Quality of Life*. Berlin: Springer.
Matangi Tonga Online. (2001). *Tongans make big investment in education*. Matangi Tonga Online: Tonga. Retrieved May 31, 2016 from http://matangitonga.to/2001/01/30/tongans-make-big-investment-education.
McIver, L., Woodward, A., Davies, S., Tibwe, T., & Iddings, S. (2014). Assessment of the health impacts of climate change in Kiribati. *International Journal of Environmental Research and Public Health, 11*(5), 5224–5240.
Michaelis, L. (2011). Climate change and spirituality. In L. Bouckaert & L. Zsolnai (Eds.), *Handbook of spirituality and business* (pp. 236–244). London: Palgrave Macmillan UK.
Miles, M. B., Huberman, A. M., & Saldaña, J. (2014). *Qualitative data analysis: A methods sourcebook* (3rd ed.). London: SAGE.
Ministry of Environment and Climate Change, & National Emergency Management Office. (2010). *Tonga: Joint national action plan on climate change adaptation and disaster risk management 2010–2015*. Kingdom of Tonga: Nuku'alofa.
Morrison, M., Duncan, R., & Parton, K. (2015). Religion does matter for climate change attitudes and behavior. *PLoS ONE, 10*(8), e0134868. doi:10.1371/journal.pone.0134868.

Nunn, P. D., Aalbersberg, W., Lata, S., & Gwilliam, M. (2014). Beyond the core: Community governance for climate-change adaptation in peripheral parts of Pacific Island Countries. *Regional Environmental Change, 14,* 221–235. doi:10.1007/s10113-013-0486-7.

Nunn, P. D., Mulgrew, K., Scott-Parker, B., Hine, D. W., Marks, A. D. G., Mahar, D., et al. (2016). Spirituality and attitudes towards nature in the pacific islands: insights for enabling climate-change adaptation. *Climatic Change, 136*(3), 477–493.

Ozkan, B. C. (2004). Using NVivo to analyze qualitative classroom data on constructivist learning environments. *The Qualitative Report, 9*(4), 589–603.

Peck, R., & Devore, J. L. (2012). *Statistics: The exploration and analysis of data* (7th ed.). USA: Brooks/Cole Cengage Learning.

Roser-Renouf, C., Maibach, E., Leiserowitz, A., & Rosenthal, S. (2016). *Global warming, god, and the "End Times"*. New Haven, CT (Yale Program on Climate Change Communication): Yale University and George Mason University.

Ryan, B. F., Joiner, B. L., & Cryer, J. (2013). *Minitab handbook*. United State of America: Cengage Learning.

Sekona, H. K. (2014). *Tonga—a land dedicated to god*. The Church of Jesus Christ of Latter-Day Saints: Tonga. Retrieved Mar 7, 2017 from https://www.lds.org/ensign/2014/08/tonga-a-land-dedicated-to-god?lang=eng.

Springer. (2012). *Advanced statistical methods for the analysis of large data-sets*. Heidelberg: Springer.

Statistic Department of Tonga. (2016). *Tonga 2011 census of population and housing: Volume 1—basic tables and administrative report*. Tonga: Tonga Department of Statistics.

Sullivan, M. (2013). *Statistics: Informed decisions using data* (4th ed.). New York: Pearson.

Tonga Department of Statistic. (2008). *Tonga 2006 census of population and housing, volume 1: Administrative report and basic tables* (Vol. 1). Nuku'alofa: Tonga Department of Statistics.

Tonga Department of Statistic. (2011). *Tonga national population and housing census 2011: Preliminary results*. Nuku'alofa: Statistic Department of Tonga.

Tonga Department of Statistic. (2013a). *Tonga 2011 census of population and housing, volume 2: Analytical report*. Nuku'alofa: Government of Tonga.

Tonga Department of Statistic. (2013b). *Tonga 2011 census of population and housing: Volume 1—basic tables and administrative report*. Nuku'alofa: Tonga Department of Statistics.

Tonga Department of Statistic. (2016). *Tonga—National Population and Housing Census 2011*. Tonga: Tonga Department of Statistics.

Tonga Department of Statistic, & Secretariat of the Pacific Community (SPC). (2008). *Tonga 2006 census of population and housing, volume 2: Analytical report* (Vol. 2). New Caledonia: SPC.

Tonga Meteorological Service. (2015). *Climate update for Tonga*. Government of Tonga: Nuku'alofa.

Tonga Meteorological Service. (2016). *Climate summary of Tonga*. Tonga: Government of Tonga. Retrieved Jun 20, 2016 from http://www.met.gov.to/index_files/climate_summary_tonga.pdf.

USGCRP. (2016). *The impacts of climate change on human health in the United States: A scientific assessment*. In A. Crimmins, J. Balbus, J. L. Gamble, C. B. Beard, J. E. Bell, D. Dodgen, R. J. Eisen, N. Fann, M. D. Hawkins, S. C. Herring, L. Jantarasami, D. M. Mills, S. Saha, M. C. Sarofim, J. Trtanj, & L. Ziska (Eds.). Washington, DC: U.S. Global Change Research Program, 132 pp.

Woodward, A., Hales, S., & Weinstein, P. (1998). Climate change and human health in the Asia Pacific region: Who will be most vulnerable? *Climate Research, 11,* 31–38.

World Health Organisation (WHO). (2011). *Public health and environment (PHE): Percentage of deaths and DALYs attributable to the environment, 2004 deaths*. Geneva: WHO. Retrieved Jun 26, 2013 from http://gamapserver.who.int/gho/interactive_charts/phe/total_percentage/atlas.html.

World Health Organisation (WHO). (2015). *Human health and climate change in Pacific Island countries*. Geneva: World Health Organisation.

Author Biographies

Mr. Peni Hausia Havea is a Ph.D. candidate in climate change who holds a Master degree in Public Health majoring in International Health from the University of Melbourne, Australia. He is a finishing Ph.D. candidate from the University of the South Pacific.

Dr. Sarah L. Hemstock holds a Ph.D. in bioenergy systems modeling and is an author and adviser to the Alofa Tuvalu "Small Is Beautiful" project—recognised by UNESCO as one of its "Decade of Achievement Projects". Over the past 20 years, she has worked on community and domestic bioenergy schemes and agricultural extension projects with communities in Sub-Saharan Africa, Europe, India, South America, and the Pacific Islands Region. She also worked on national and regional energy. In 2010, she was made Government of Tuvalu Honorary Ambassador—Officer for Environmental Science. She was the Team Leader for EU-GCCA Project at PaCE-SD. Currently, she is the Team Leader for the European Union PacTVET Project at the Economic Development Division, The Pacific Community; Visiting Fellow at Nottingham Trent University and Adjunct Fellow at the University of the South Pacific.

Dr. Helene Jacot Des Combes obtained her Ph.D. on "reconstruction of the paleoceanography and paleoproductivity variations in the NW Indian Ocean during the last 300 kyr: the geochemical response compared to the biological record", at the University of Lille, France, in 1998. From 2000 to 2008, she worked as a scientist at the Alfred Wegener Institute for Polar and Marine Research in Bremerhaven, Germany. She joined PaCE-SD as a Research Fellow. Currently, she is a Senior Lecturer in Climate Change Adaptation at USP. She is currently developing a training needs and gap analysis on TVET in 15 Pacific island countries for the 6 million Euro European Union Pacific Technical Vocational Education and Training on Sustainable Energy and Climate Change Adaptation Project (EU PacTVET). She is the Course Coordinator for the Postgraduate Diploma course on disaster risk reduction and also delivers guest lectures on the ocean response to climate change.

Dr. Johannes Luetz is Senior Lecturer, Postgraduate Coordinator and Research Chair at CHC Higher Education in Brisbane, Australia. Prior to joining CHC he was a Lecturer at the University of New South Wales (UNSW) in Sydney in the School of Social Sciences, where he also earned his Ph.D. in Environmental Policy and Management. He has worked extensively with World Vision International on research projects raising awareness of the growing effects of climate change on poor and vulnerable communities in Asia, Africa and Latin America.

Ecosystem-Based Adaptation (EbA) for Coastal Resilience Against Water Related Disasters in Bangladesh

M. Mustafa Saroar

Abstract Since the publication of the Millennium Ecosystem Assessment report (MEA in Ecosystems and human well-being: wetlands and water synthesis. World Resources Institute, Washington, DC, 2005), ecosystem-based adaptation (EbA) has appeared as a popular strategy for adaptation to the impacts of climate change. In Bangladesh, people have been adapting to climatic disasters in various ways for long. There exist valuable narratives about many ways of adaptation, yet only sporadic attempts have been observed to analyse the effectiveness of EbA. In fact, there is a clear lack of research that addresses the effectiveness of EbA against water related disasters and the challenges the EbA encounters. This chapter fills these gaps in knowledge in three ways. Firstly, it presents an overview of the vulnerability of coastal social-ecological systems to water related disasters against which increased adoption of EbA has been observed in coastal Bangladesh. Secondly, it examines the status and role of EbA in three distinct coastal regions. Thirdly, it presents an array of factors that put limits on the adoption and functions of EbA in coastal Bangladesh. Finally, it offers some policy suggestions to enhance the effectiveness of EbA. This chapter heavily draws on a review of most recent published and other grey literature on EbA. Most commonly adopted EbAs are mangrove afforestation, wetland restoration, tropical green-belt, hydroponic agriculture, and integrated agriculture and aquaculture. These EbAs build resilience through buffering water related disasters, controlling erosion, recycling nutrients, and offering other co-benefits. The factors that put the limit on the effectiveness of EbA are broadly biophysical, socio-economic, technological, and institutional. In line with the suggestions, initiatives could be taken to make EbA more effective.

M. M. Saroar (✉)
Department of Urban and Regional Planning, Khulna University of Engineering & Technology (KUET), Khulna 9203, Bangladesh
e-mail: saroar.mustafa@yahoo.com

W. Leal Filho (ed.), *Climate Change Impacts and Adaptation Strategies for Coastal Communities*, Climate Change Management,
https://doi.org/10.1007/978-3-319-70703-7_10

Introduction

Anthropogenic climate change has posed a formidable challenge for humanity. Impacts of climate change are already visible in some parts of the globe; others would encounter the impacts before the middle of this century (IPCC 2007). Humanity has only two ways to address the issues of climate change. Firstly, mitigation measures to minimize the release of greenhouse gases, particularly CO_2; this is mostly done by industrialized countries. Secondly, adaptation measures to minimize the impacts of climate change, which are adopted by countries in the global south. If they fail to adapt to the impacts of climate change, it would severely affect the well-being of people at risk. Coastal inhabitants in Bangladesh, one of the countries highly vulnerable to global climate change, are at high risk due to their increased exposure to accelerated sea level rise (SLR), coastal floods, cyclones, surges, droughts and increased salinization (World Bank 2010; CCC 2016). People in the coast have been coping and adapting to many of these water related disasters by employing hard engineering measures. However, recently increased adoption of ecosystem based adaptation (EbA), which offers multiple co-benefits, has been observed as an important adaptation strategy to deals with the impacts of climate change (Pramova et al. 2012; Huq et al. 2013; Reid and Alam 2014; Saroar et al. 2017).

In ecosystem based adaptation (EbA) the services provided by the ecosystem and the functions performed by rich biodiversity are utilized to buffer the worst impacts of climate change (Vignola et al. 2009; Pramova et al. 2012; Dasgupta et al 2014). Most of the EbAs are locally driven and cost effective. They demand less technical knowledge and offer multiple co-benefits; therefore, the pace of adoption of EbAs is very high. However, the effectiveness of EbA measures is constrained by numerous factors (Rahman 2014; Reid and Alam 2014). A growing body of literature provides good narratives on many EbAs, yet there is a lack of research that systematically examines the role of EbAs and the factors influencing the effectiveness of EbAs in coastal Bangladesh. This chapter fills these knowledge gaps in three ways. Firstly, it presents an overview of the vulnerability of coastal social-ecological systems to water related disasters against which increased adoption of EbAs have been observed in coastal Bangladesh. Secondly, it examines the status and role of a suite of EbAs that are widely adopted against water related disasters in the coastal landscape. Thirdly, it presents an array of factors that put limits on the adoption and function of EbAs in coastal Bangladesh. Finally, it offers some policy suggestions to harness the full potential of EbA against the impacts of water related disasters in the coastal setting. However, the author at the very onset acknowledges the limitations of this research findings. This chapter heavily draws on the review of most recent published and other grey literature on EbA. Only a few examples are drawn from an empirical research conducted in three sites of South-western coastal region (i.e. Khulna, Bagerhat and Satkhira district) through nine focus group discussion sessions in the later part of 2015. The findings could be generalized for the entire coast of Bangladesh if empirical investigation could be expanded in all the three coastal zones. As this paper has very limited empirical insight for the entire coastal zone of Bangladesh, the findings need to be generalized with caution.

Vulnerability of Coastal Social-Ecological System in Bangladesh

Bangladesh, a tropical low-lying country, is located at the interface of two different environments; the Himalaya is in the north and the Bay of Bengal is in the south. The coastal zone, with an area of 47,211 sq. km (about one-third of the country's landmass) is located at the lower edge of Himalayan glacier melted Ganges-Brahmaputra delta system and facing the Bay of Bengal (World Bank 2010; Akter et al. 2015; Saroar et al. 2015). The coast line extends over 700 km and most parts are only about one meter above the mean sea level. The entire coastal zone is divided into three coastal regions such as south-western (moderately stable delta), south-central (highly active and unstable delta) and the south-eastern (mostly stable delta) region (see Fig. 1). The whole coastal zone covers 147 Sub-district (out of total 460 Sub-districts) located in 19 Districts (out of total 64). A Sub-district is the lowest level of the spatial unit for which physical development plan is prepared by a specialized institution in Bangladesh. A total 48 Sub-districts under the jurisdiction of 12 Districts are located on the exposed coast. An exposed coast is the area that faces the shoreline of Bay of Bengal which is the entry point of most of the water related disasters in Bangladesh (Islam and Ahmed 2004; Dastagir 2015) (Fig. 1).

Coastal social-ecological system in Bangladesh comprises diverse social and ecological features, such as human settlements, land ownership, agriculture, forestry, fisheries, mangroves (both natural and planted), newly accreted chars (river islands), social and physical infrastructures, and urban centres (Dasgupta et al. 2014; Akter et al. 2015). Agriculture practices cover the largest 59% of total land use followed by shrimp, fisheries and wetlands (13%), mangrove (11%) and others including urban centres (17%) (Alam et al. 2013; Saroar and Routray 2013). Almost 80–90% people depend on both agriculture and fisheries directly, or partly in different seasons for livelihoods and food security (Saroar and Routray 2013; Saroar and Leal Filho 2016). Fisheries are the major concentration of social and institutional activities ever increasing due to population growth and subsistence demand for food and profit oriented business (i.e. shrimp farming) in coastal areas. Newly accreted Char land (river islands) management is largely influencing social and ecological factors of the area due to its potential and multiple uses for afforestation, livelihood and human settlement (Saroar and Routray 2010; Akter et al. 2015). Mangroves, one of the valuable functional ecosystems serves protection against disasters and produce benefits to coastal communities. The traditional natural resource oriented livelihood strategies adopted by coastal residents are highly climate sensitive (Saroar et al. 2015).

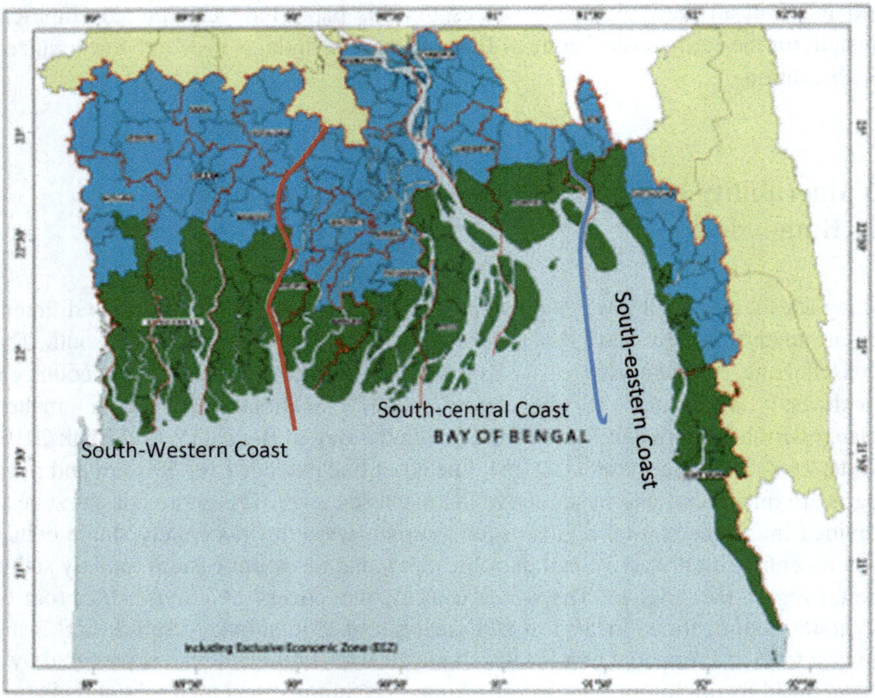

Fig. 1 Coastal zone of Bangladesh. Blue shade indicates interior coast and deep green shade indicates exposed coast. Area between read and blue line is the South-central region; left side of the red line is South-western region and right side of the blue line is the South-eastern region (Adapted from FAO 2007; Islam 2007)

The diversity of coastal resources has different scales of functional capacity and roles to benefit coastal communities for livelihood and food security. Over the time and space, coastal livelihoods have been vulnerable to extreme events in different ways (Saroar and Routray 2013; Reid and Alam 2014). The morphological change caused land erosion hazard and social displacement of coastal communities. Tropical cyclone and storm surge already affected three coastal regions of the country with the massive loss of life and property (Akter et al. 2015). Salinity intrusion already contributed to change of coastal ecosystem beyond threshold into undesirable states (Iftekhar and Takama 2008). Coastal flooding increase risks to livelihoods by sudden submersion of agriculture lands inside the coastal embankment. Available projections show that impacts of many of these already occurring water related disasters might be more severe in the context of climate change (World Bank 2010; Dasgupta et al. 2014; CCC 2016).

The most plausible scenario of climate projection shows that due to global climate change, Bangladesh might experience 15% increase in precipitation during wet season, 1.5 °C rise in summer temperature, and 88 cm rise in sea level along

Bangladesh coast by the end of this century (GOB 2005). Numerous studies have indicated that climate change induced extreme events, such as coastal floods, cyclones, surges, salinity intrusion, and droughts would be more frequent (World Bank 2010; Dasgupta et al. 2014; Dastagir 2015; CCC 2016). Already, coastal zone of Bangladesh is hit by a severe tropical cyclone in every 3 to 4 years with wind storm surges up to seven metres high (GOB 2009; World Bank 2010). Super cyclone Sidr and Aila ravaged the entire coastal landscape in 2007 and 2009 respectively. Once in every 4 to 5 years, the country is visited by a severe flood that inundates more than two-third of the country's landmass (GOB 2009; World Bank 2010). Analysis of 30-year trends of SLR shows that Bangladesh coast has already experienced 6–21 mm/year rise in sea level (CCC 2016). A quarter of Bangladesh's coastline would be inundated before the middle of this century as surge height might increase 13 to 46% for SLR of 30 cm and 100 cm (Ali 1999). Around 13% more land (469,000 ha) will be under regular flooding if sea level rises 62 cm by 2080. Another 16% (551,500 ha) land will be added to water logged condition if rainfall synchronizes with a sea level rise of 62 cm (IWM and CEGIS 2007). Anticipated SLR will severely affect the mangrove forest Sundarbans (World Bank 2010). For a 1 m rise in sea level, the Sunderbans mangrove forest that extended over 6000 sq. km area and supports a very rich and diverse flora and fauna (e.g. 400 species of fishes, 270 species of birds, and over 300 species of plants), is likely to be lost (Agrawala et al. 2003; Islam et al. 2013; CCC 2016).

Water related disasters will severely impact the coastal agriculture. Already about one million hectares of soil in coastal Bangladesh are affected by varying degrees of salinity (CCC 2007). For a 30 cm SLR, which is most likely to occur before 2050, the net decline of rice production is estimated to be 0.5 million metric tons per year (Baten et al. 2015). The presently cultivated rice varieties may not be able to withstand increased salinity (Ahmed 2010). Projection further shows that a 4°C rise in temperature will reduce rice yield by 17% and wheat yield by about 60% in Bangladesh (Mainuddin et al. 2011). All these might have serious repercussion on future food security and livelihood as agriculture is and continue to be the most dominant occupation in coastal Bangladesh. Therefore, Bangladesh is ranked as the most vulnerable to tropical cyclones (Dastagir 2015); the 2nd most vulnerable by death toll (Kreft et al. 2017), and the 6th world's most overall vulnerable to climate change (Kreft et al. 2017).

Adaptation has its limit, yet coastal people in Bangladesh need to continue adaptation to most of the water related disasters cited above. Coastal adaptation can be understood as the adjustment of ecological, social, or economic systems to actual or expected climatic stimuli and their effects or impacts. To help people develop a good portfolio of adaptation, Bangladesh Government has made a significant stride in policy response at multiple levels and scales. At the national levels, it has developed the National Adaptation Program of Action (NAPA) more than a decade ago in 2005, which is basically a blue print for future adaptive responses (GOB 2005). Based on the NAPA, Bangladesh has developed Bangladesh Climate Change Strategy and Action Plan (BCCSAP) in 2009. The BCCSAP has outlined

44 programs under six thematic areas including food and nutritional security, health, coastal infrastructure, and comprehensive disaster management for building a resilient community through adoption of these programs in the short, medium and long-term (GOB 2009).

To support the adaptive responses as outlined in NAPA 2005 and BCCSAP 2009, Bangladesh Government from its own fund created a 100 million US$ Climate Change Trust Fund (CCTF). Another parallel window that supports the adaptation is the multi-donors' Climate Change Resilience Fund (CCRF). Bangladesh had a long tradition of implementation of engineering and technology driven various adaptation initiatives, such as construction and maintenance of coastal roads, cyclone-shelters, public ways, bridges, culverts, rural marketplaces etc. However, since the adoption of Bangladesh Climate Change Strategy and Action Plan in 2009, Bangladesh has been promoting various ecosystem based adaptations (EbAs) as a strategy to build resilience against the challenges posed by climate change (Saroar et al. 2017). Various forms of EbAs, such as restoration of coastal wetlands, afforestation of mangroves, coastal green-belt, hydroponic agriculture, and integrated agriculture-aquaculture have been initiated to build resilience against the water related disasters. In the following section status and the role of major EbAs are examined.

Building Resilience Through EbA in Coastal Bangladesh

Healthy ecosystem provides valuable ecosystem services (MEA 2005; Munang et al. 2013). Ecosystem services are the benefits humans derive from ecosystem either directly (e.g. food and clean water), or indirectly (e.g. flood and erosion control). A well-managed ecosystem can absorb shocks and stresses associated with climate change. Therefore, resilience of coastal community against water related disasters heavily depends on how best the coastal community could adopt ecosystem-based adaptation (EbA) i.e. make use of healthy ecosystem services and biodiversity for building resilience (Dietz et al. 2003; World Bank 2009; Pramova et al. 2012; Munang et al. 2013; Saroar et al. 2017).

Ecosystem-based adaptation (EbA) has generally been defined as the use of biodiversity and ecosystem services as part of an overall adaptation strategy to help people adapt to the adverse effects of climate change (CBD 2009). EbA refers to the use of natural resources through conservation and enhancing the resilience of ecosystem to buffer the worst impacts of climate changes on the ecosystem and human interest (Vignola et al. 2009). EbA offers social, environmental, ecological and economic benefits (Colls et al. 2009; World Bank 2009; Pramova et al. 2012; Duarte et al. 2013; Munang et al. 2013). In the following couple of sections, important EbAs which have been implemented in the coastal landscape in Bangladesh as part of broader adaptation strategy to build coastal resilience are examined.

Mangrove Afforestation and Restoration in South-Central Coast

In line with the BCCSAP 2009 and the NAPA 2005, the 'Community Based Adaptation to Climate Change through Coastal Afforestation (CBACC-CF)' project has been implemented in the south-central coastal region as first priority EbA project. Initially, mangrove has been planted in 9100 ha (91 sq. km) of coastal public land in Hatiya of Noakhali, Char Fasson of Bhola and Barguna Sadar of Barguna district which is later expanded to another 6000 ha newly accreted land. This EbA project has enriched plant density per unit area thus would contribute to reducing wind velocity, tidal surges and other climatic events and ultimately would increase the resilience of social-ecological system through new land stabilization, improving soil quality and protection against air and water induced soil erosions (Iftekhar and Takama 2008; Nandy et al. 2013; Cabo Bujan and Al Hussainy 2016). Co-benefits of this project is huge. The project provided income opportunities through cash for work to more than 12,000 coastal people in afforestation interventions for nursery bed preparation, seedling raising, plantation and maintenance (Cabo Bujan and Al Hussainy 2016). About 3400 coastal households are currently involved (under co-management) as long-term beneficiaries of the strip plantation in four project sites (Ahammad et al 2013).

Several studies including CBACC-CF impact evaluation report have indicated that the CBACC-CF project as a soft EbA measure has enhanced the resilience of coastal ecosystems through increasing forest coverage and creating livelihood diversification among vulnerable communities (Nandy and Ahammad 2012; Cabo Bujan and Al Hussainy 2016). In fact, different livelihoods types including agriculture and fish cultivation, tree plantation and duck rearing are integrated into the CBACC-CF project to ensure co-benefits of the CBACC-CF project in coastal areas. From the experience of CBACC-CF project, it is expected that implementation and maintenance of large-scale mangrove afforestation program along the coastline will work as the first line defence against the cyclones, surges and intrusion of salt water (Uddin 2013; Sajjaduzzaman et al. 2015). This will give impetus to initiate privately/community managed various localized adaptive responses for enhancing coastal resilience.

Sundarbans Mangrove Protection in South-Western Coast

The Sundarbans, the largest mangrove forest in the world and a UNESCO World Heritage site, is located at the lower edge of Ganges-Brahmaputra delta system. It spreads over 10,000 sq. km along the northern edge of Indian Ocean (Bay of Bengal) of which about 6000 sq. km is in Bangladesh (south-western coastal region) (GOB 2011). A well-functioning ecosystem in Sundarbans is crucial to support the livelihood of the coastal communities living adjacent to this forest

(Choudhury 2007). Because, Sundarbans provides the first-line defence and the man-made earthen embankments along the rivers adjacent to the Sundarbans provide the second-line of defence against water related disaster, such as cyclones, surges, salinity intrusion, and coastal flooding (Islam and Ahmed 2004; Duarte et al. 2013). Maintenance of healthy ecosystem of Sundarbans and the embankments are crucial for the well-being of about one million people living adjacent to this forests. However, over the three decades or so both ecosystems of Sundarbans and the embankments have been experiencing damaging effects due to climatic and non-climatic stressors (World Bank 2010; Saroar and Routray 2013; Dasgupta et al. 2014). Clearance of mangrove for saline-shrimp and crop production, illegal logging and unsustainable extraction of Sundarbans' resources, such as fish, crab, honey, and wax have caused unprecedented damage to the Sundarbans mangroves and its rich biodiversity (Saroar and Routray 2010; Sajjaduzzaman et al. 2015). On the other hand, the embankments and the ecologically sensitive areas adjacent to the embankments are also increasingly under threat due to human pressure (e.g. i.e. illegal settlements on embankments) and natural forces (e.g. cyclones, surges) (Uddin 2013; Sajjaduzzaman et al. 2015; Reid 2016). Drawing on the experience of Sundarbans Environmental and Livelihoods Security Project (SEALS), and Sustainable Development and Biodiversity Conservation in Coastal (Protection) Forests (SDBC-Sundarbans) project, the Collaborative REDD + IFM Sundarbans Project (CRISP) has been initiated to build the resilience of the coastal social-ecological system and to earn carbon credits (GOB 2011). Although CRISP is designed to have both mitigation and adaptation benefits, in the following section CRISP is examined only from the perspective of adaptation.

The CRISP project has been implemented in 210 villages that are located within the 10-km wide reference region of the Sundarbans Reserve Forest of Khulna, Bagerhat and Satkhira district. In line with the Bangladesh Climate Change Strategy and Action Plan 2009, the main goal of this EbA was set out, which is an enhancement of coastal resilience. The EbA strategy includes climate change adaptation planning for building community and ecosystem resilience, protection of the Sundarbans mangroves and the adjacent embankments, and capacity building and training for improved biodiversity conservation (GOB 2011; Reid 2016). The project is implemented by building a partnership with community-based co-management groups. About 500,000 people from 125,000 households would be directly benefited from this EbA activities. So far progress indicates, sustainable livelihood development strategy provides alternative and sustainable livelihoods to a growing number of people living in 210 villages in the Sundarbans landscape zone, which is believed to shift biotic pressure away from the Sundarbans mangrove forest (GOB 2011). Due to the environmentally friendly implementation process, this project is not only creating new income opportunities for the local population but is also promoting biodiversity conservation (Duarte et al. 2013). The local population now knows how to plant the dykes and their surroundings with trees to stabilize them and prevent air and water induced erosion and breaching from surges.

Tropical Evergreen Forest Restoration in South-Eastern Coast

The south-eastern coast of Bangladesh is located at the bottom hill areas of Chittagong and Cox's Bazar district along the Bay of Bengal. This region receives about 300 cm rainfall annually and experiences significant flooding, caused mostly by overland flows (runoff water) from the upper catchments. The entire area is vulnerable to tropical cyclones and surges as well. This coastal region is hard hit by severe cyclones in 1964, 1991, and 1996 (GOB 2012). Tropical evergreen and semi-evergreen forests have been playing a significant role to moderate the coastal water related disasters. However, due to increased pressure from new settlers, overland flows of debris from the upper catchment, and water and air induced erosion, the ecosystem in the south-eastern coast have been experiencing significant damage over the last couple of decades (Choudhury 2007). Moreover, as the local residents used to heavily depend on the extraction of forest resources and shifting cultivation inside the forested landscape, maintaining ecosystem health was really a challenging task.

Drawing on the experience of integrated protected area co-management (IPAC) project, the restoration of wet evergreen forest project has been initiated along the south-eastern coast (GOB 2012). The project area covers over 33,000 ha of forest land in six protected areas such as Sitakunda Reserved Forests (SKRF), Dudpukuria-Dhopachari Wildlife Sanctuary (DDWS), Teknaf Wildlife Sanctuary (TWS), Inani National Park (INP), Fasiakhali Wildlife Sanctuary (FWS), and Medhakachapia National Park (MNP). Among the six sites of the project, the SKRF and DDWS are located in Chittagong district; the remaining four are located in Cox's Bazar district. Under this project, a 5-km band around each site is considered as direct impact zone. In the entire impact zone in which over 50,000 households live; 70% of them are extremely poor and 60% are landless (GOB 2012). A significant number of them were largely dependent upon shifting agriculture, illegal harvesting of forest produces, salt cultivation, and fishing.

The main goal of the community co-managed tropical evergreen forest restoration project is to restore vegetative cover through EbA measures such as planting, or human-assisted natural regeneration of woody vegetation. Although the project delivers significant mitigation benefit, here only benefits from EbA are examined. Within five years of initiation of the project, many patches of evergreen and semi-evergreen forests have been developed in degraded forest areas due to reduced biotic pressure (Sajjaduzzaman et al. 2015). Biotic pressure is reduced as the development of conservation-linked value chain has helped to lower the dependency of local residents on this forest resources. The restored forests in two sites (e.g. TWS and INP) have contributed in regulating water flows, checking soil erosion and protecting watersheds, water bodies and coasts (Uddin 2013). On the other hand, in almost all sites reducing the dependency of local residents on forests by providing avenues of alternate income (alongside co-management benefits) has helped conserving biodiversity and ecosystem health.

Wetland Restoration in South-Western Coast

The south-western coast is characterized by the presence of vast tract of the coastal flood plains, natural canals, and mangrove swamp. A significant portion of these wetlands is basically *khas-land*, owned by government departments such as the Department of Land and the Department of Forest. These wetlands buffer the impacts of water related disasters. However, over the last three decades or so about 20,000 ha of the coastal flood plains and mangrove swamps in Khulna, Bagerhat, and Satkhira district of south-western coast have been cleared by influential people to culture salt water shrimp. Hundreds of natural canals, which were under a common pool resource regime, and used for irrigation and subsistence fishing, have been degraded both due to over exploitation and natural forces (Hossain et al. 2013; Akter et al. 2015; Saroar et al. 2017). Due to extensive alteration of these wetlands, water related disasters, particularly high tides, cyclones, and surges cause severe damage to the coastal social-ecological system (Saroar et al. 2015). This has far-reaching implications for coastal ecosystem and biodiversity, livelihood and food security, and settlements and infrastructures.

In the BCCSAP of 2009 high emphasis is given for restoration of coastal wetlands as priority EbA measure. Restoration and rehabilitation of coastal wetlands, such as coastal flood plains, canals, tidal marshes, and mangrove swamps, could provide numerous ecological benefits including nutrient recycling, soil binding, flood control, and livelihood and food security (Choudhury 2007; Uddin 2013; Reid 2016). Following the provisions laid down in the *Jalmahal* (Wetland) policy 2009 of Bangladesh and its subsequent revisions, various EbA measures have been employed by government agencies, local NGOs, and the community people to restore the wetland characteristics. Community managed wetlands and canals restoration activities, such as re-excavation, raising the height of dykes and dyke planting have increased the water retention capacity and water flow of wetlands (Saroar et al. 2017). This EbA already has delivered a range of social-ecological benefits to local people including micro-irrigation, fish culture, soil binding, nutrient recycling, flood and erosion control which help building community resilience against water related disasters.

Ecosystem-Based Climate Smart Agriculture and Aquaculture

Ecosystem-based climate smart agriculture is a new form of adaptive agriculture that has been proven to maintain/improve productivity by adapting to climate variability and change and offers mitigation benefits. This adaptive agriculture can be practised at various scales; from plot to farm to landscape scale (Ahammad et al 2013; Saroar and Leal Filho 2016). Coastal agriculture is very sensitive to water related disasters and climate extreme events, such as excessive/erratic rainfall, severe floods, cyclones, surges and salinity intrusion (Reid 2016). Different suits of EbAs are adopted in coastal agriculture to enhance resilience against each of these water related disasters and climate extreme events.

Hydroponic farming or floating garden is a special kind of EbA which provides alternative agriculture practice to people for producing vegetables in the water-logged conditions through "Hydroponic" system. For vast areas in the south-western coastal districts which have been experiencing seasonal (or permanent) water-logging due to congestions, encroachment and siltation of natural drainage systems, the hydroponic farming has been appeared as a climate smart solution (Reid 2016; Saroar et al. 2017). The hydroponic farming system involves (a) making floating mats of water hyacinth and other aquatic vegetation available in these wetlands in early monsoon, and (b) laying down organic manure (cow dung and other materials) to make beds for growing vegetables. The practice mostly replaces the soil based agriculture which is entirely not possible in this waterlogged condition (Rahman and Islam 2013). By adopting this climate smart technique, farmers usually produce short-rotation vegetable varieties on the floating gardens, which are important for household foods, nutrition, and alternative income generation (Nishat and Mukherjee 2013; Saroar et al. 2017). With more land likely to be submerged due to increased rainfall, flooding, and sea level rise as a result of climate change, more people would be forced to adopt this hydroponic agriculture.

In coastal Bangladesh, almost one million hectors of agricultural land are affected by varying degrees of salinity. Depending on the pattern of rainfall and salinity gradients, farmers employ various ecosystem based adaptive techniques in salt-land agriculture (PDO-ICZMP 2004). In the medium (salinity 4–12 ds/m) and high (salinity >12 ds/m) saline areas, farmers have started growing *Amon rice* (rain-fed wet season rice) during monsoon (May-July) when salinity level goes down due to high precipitation. In the same saline soil area, the land parcels that were earlier left vacant during the dry season (December–February) are now used for dry season's rice such as BRRI Dhan-47 and BINA-8, because of their resilience to salinity and drier condition (CBACC-CF 2012; CSISA 2012). Some other non-cereal crops and vegetables, such as oil seeds, soybean, bean, bitter gourd, sweet gourd, chilli, ground nut, and water melon are also adopted by farmers as these non-cereal crops withstand against high salinity and high moisture stress condition during the dry season (CSISA 2012; UNDP 2012). In fact, the old practice of abandoning farm land in the entire dry season has been reduced; this helps increasing food security.

Coastal lands on the sea/river side of the protective embankment (outside the earthen embankment/polder) are highly exposed to tidal inundation, and highly saline. These land parcels are no more left vacant much of the year. In three coastal districts such as Barguna, Bhola and Noakhali, the beneficiaries of Ditch-and-Dyke for Forest and Livelihood model of EbA have started to grow rice-fish and fruits from this waste land (Ahammad et al. 2013; Nandy et al. 2013). Under this model, the beneficiary farmers by excavating ditches and raising dykes with the excavated soil, create a series of ditches and dykes from flat land. The ditches retain monsoon rainwater (freshwater) in sufficient volume to maintain water levels almost year-round, making them suitable for fish culture (Cabo Bujan and Al Hussainy 2016). The dykes are raised to a level that keeps them free from tidal inundation and helps protect the scheme areas from salinity intrusion. This ditch-and-dyke scheme has been initiated on *khas lands*, which are mostly under the control of forest department, and are then provided to landless beneficiary households (Ahammad et al. 2013; Rahman and Islam 2013; Rahman 2014). The beneficiaries can use the raised dykes for agriculture, horticulture, and forestry, for short, medium and long-term benefits (Cabo Bujan and Al Hussainy 2016). The freshwater from the ditches is used to irrigate the crops and trees on the dyke. Farmers are advised to use organic manure in dyke-cropping, including regular mulching to minimize the effects of soil salinity. Common vegetables cultivated include sweet gourd, pumpkin, bitter gourd, eggplant, tomatoes, beans, spinach and amaranth.

Apart from the above scheme, both at farm and plot level, there has been developed privately managed integrated aquaculture and agriculture practice. Integrated aquaculture and agriculture is a form of EbA where farmers use their rice-fields for combined cultivation of paddy, shrimp and finfish in different seasons (Reid and Alam 2014; Saroar et al. 2017). The paddy cultivation begins with monsoon rain when salinity is about to leach out with freshwater inflow (around June onwards). At that time, there is a particular system locally people adopting (locally called *gher*/pond) to raise their surrounding lands in a dyke form to retain fresh water which is simultaneously used for *Aman* rice (wet season rice) and fresh water shrimp '*Galda*' production (Alam et al 2013; Rahman 2014). Freshwater reduces the salinity level and restores natural quality of the pond where additional organic matters are enriched in soil and planktons available for food of shrimp (Hossain et al. 2013). The integrated aquaculture and agriculture provide twofold household income at a time as people usually harvest both paddy and shrimp between October and November.

Challenges to Adoption of EbAs for Coastal Resilience

From the foregoing sections, it is evident that EbAs have tremendous potential for building coastal resilience against water related disasters in coastal Bangladesh. However, effective adoption of EbA is constrained by numerous factors including bio-physical, socio-economic, technological, and institutional and governance related factors which are examined in some details.

Bio-Physical Factors

Coastal areas of Bangladesh share some common characteristics, such as low elevation, deltaic coast, continuous erosion and accretion/sedimentation (Saroar and Routray 2013; Akter et al. 2015), yet their level of exposure and sensitivity to most of the water related disasters differ significantly (Saroar and Routray 2013; Dastagir

2015). Any robust plan for EbA must consider the differential vulnerability of spatial units for which the EbA plan need to prepare. Unfortunately, there is a clear lack in (spatial) planning unit-wise vulnerability profile, which acts as a barrier to designing and implementation of robust EbA plan. For instance, in the south-western coastal region, vulnerability profiles of places located on the exposed coast are different than the places located in the interior parts of the coast (Dastagir 2015). Again, the vulnerability among places on the exposed coast which are protected by polders is different than places protected by mangroves. Therefore, EbAs plan must be vulnerability specific to be robust and effective.

Socio-Economic Factors

Involvement of diverse stakeholders who often have conflicting interests in planning, designing, and implementation of EbA program/project are keys to success (Vignola et al. 2009; Ahammad et al. 2013). Furthermore, often there is the need for multi-sectoral integration in the EbA processes. It is often the case that involvement of stakeholders is not difficult but managing diverse groups of stakeholders is really a challenging task which influences the success of most of the EbAs. For instance, canal restoration project in south-western coastal region, especially in Bagerhat and Satkhira district, helps local farming community and the artisanal fishers but the salt water shrimp farmers who are very influential recaptured the restored canals for preserving salt water; this has seriously undermined the community's effort to make EbA works for building resilient community. Furthermore, many EbAs projects that operate at the landscape level, such as restoration of mangroves in government's unused land or rehabilitation of network of canals are often designed without due regards to the needs of the local community (Saroar et al. 2017). This kind of projects are hardly owned by the local community, and thus their sustainability is difficult to ensure (Tompkins et al. 2008). Finally, local knowledge and capacity gaps of the actors including community and external institutions must be addressed for promoting new EbAs.

Institutional and Governance Factors

The policymakers have a significant role to eliminate the challenges in policy design of EbA. Most of the EbAs, no matter whether they operate at landscape, farm or plot level, cut across various sectoral policies. Thus, any EbA whether it is mangrove afforestation, restoration of coastal wetland, or rehabilitation of degraded canals to be successful requires support and cooperation from multiple agencies and institutions which are often difficult to ensure on a sustained basis (Ahammad et al. 2013; Nandy et al. 2013; Doswald et al. 2014; Reid 2016). It is partly because there is an inherent lack of policy coordination; agencies operating under different ministries often perceive the same thing very differently and prioritize their actions over others. For instance, water development board which is the nationwide lead agency for water infrastructures prefers replacing the existing earthen dykes with hard engineering infrastructures for the protection of south-western coast (Saroar et al. 2017). But the forest department prefers EbAs approach, such as plantation of mangroves, coastal green belt, and restoration of vegetative coverage (e.g. wet evergreen forest) along the coastline; this is indicated in their new forestry master plan. Mangrove centred EbAs such as protection of coastal mangroves is seen by the Department of Disaster Management as an institutional measure (i.e. EbA) to moderate the risk of water related disasters whereas the Department of Environment counts this as mitigation effort to earn carbon credit through sequestration of carbon. This different conceptualization of same EbA by various departments/agencies of Government influence the work of their functional units operating at local level, which eventually influences the outcome of the EbAs (Saroar et al. 2017).

Technological Factors

Assessment of the sensitivity of coastal ecosystems due to change in bio-physical conditions provides the trend of vulnerability and uncertainty to be taken account in EbA centred adaptation policy formulation (Reid 2016). Effective knowledge on mangrove changes and design of influential policy for reducing land degradation in coastal areas are either less or not evident in Bangladesh (Nandy and Ahammad 2012; Rahman and Islam 2013). To popularize EbA for building resilience against water related disasters at multiple levels and scales, EbA centred policy supported by evidence based research is needed (Doswald et al. 2014). Although assessment of the role of ecosystem services at local level is significant stage of EbA formulation for any specific social-ecological system, however studying specific relation of climatic change impacts on land use and natural resource at local level may not strong enough for scientific projection and policy formulation for entire coastal region of Bangladesh (Ahammad et al. 2013; Alam et al. 2013). Bangladesh NAPA 2005 and BCCSAP 2009 have already recognized this and incorporated some observed scientific evidence in EbA interventions.

Finally, it could be argued that a significant number of barriers exists in operationalizing and institutionalizing of EbA in coastal Bangladesh. Hence, new form of institutional adaptation interface would be required to harness the full potential of EbA in coastal Bangladesh.

Conclusions

This chapter presents an analytical discussion over the various form of EbAs that are adopted to build the resilience of coastal social-ecological systems against the impacts of water related disasters in Bangladesh. Although there are variations among adopted EbAs in three distinct coastal regions of Bangladesh, the commonly adopted EbA includes mangrove afforestation, coastal green belt, restoration of coastal wetlands and natural canals, hydroponic agriculture, and integrated farming. While some of the EbAs have shown tremendous potential, the effectiveness of many others seems constrained by numerous factors including bio-physical, socio-economic, institutional/governance and technological. Some specific issues have been influencing the outcome of most of the EbAs in the entire coastal region of Bangladesh. These issues are: top down approach of project design, lack of inter-agency coordination, a tradeoff between long term gain and immediate investment, and poor understanding of the changing dynamics of the coastal social-ecological system. Moreover, few EbAs are very narrowly focused which fail to integrate conservation and development (Saroar et al. 2017). For instance, there are EbA projects that are designed as a standalone project to stabilize coastal land and protect mangrove forest without an integrated understanding of the ecological and social process that determine the livelihood security in the coastal areas (Ahammad et al. 2013; Uddin 2013). As EbA adoption is constrained by several institutional and policy challenges there is a need for fresh initiatives to overcome these challenges. Accordingly, this chapter came up with the following suggestions.

- A flexible social–institutional interface over the current rigid institutional system, which would foster evolving coastal policies and institutional framework to ensure engagement and management of diverse stakeholders, need to develop (PDO-ICZMP 2004).
- As coastal ecosystems are traditionally managed under fragmented institutional responses for land, water and mangrove forest, an integrated institutional structure could be developed to ensure protective and productive land uses in coastal ecosystems (Reid and Alam 2014).
- Increase resilience and reduce the vulnerability of coastal social-ecological system against the vulnerability of water related disasters by incorporating evidence based ecosystem management activities while designing EbAs (Saroar et al. 2017).
- Some effective/model EbAs could be designed for building resilience of the social-ecological system in each of the three coastal regions by incorporating place-specific vulnerability in integrated natural resource management of coastal wetlands, mangroves, agriculture and aquaculture.
- Climate change knowledge and policy impetus for EbA measures are required to be developed in close consultation with local communities and other actors in order to understand local processes of adaptation and ecosystem management (Saroar et al. 2017).

Although studying specific relation of climatic change impacts on land use and natural resource at local level are not strong enough for EbA centered policy formulation of entire coastal region (Ahammad et al. 2013), still there are reasons to be optimistic about taking initiatives as cited above in regards to effective role of EbAs to build resilience of coastal social-ecological systems against the impacts of water related disasters.

References

Agrawala, S., Ota, T., Ahmed, A. U., Smith, J., & van Aalst, M. (2003). *Development and climate change in Bangladesh: focus on coastal flooding and the Sundarbans*. France: OECD Publications Paris.

Ahammad A. U. (Ed.). (2010). *Reducing vulnerability to climate change: The pioneering example of community based adaptation in Bangladesh*. Dhaka, Bangladesh: Centre for Global Change (CGC) and CARE Bangladesh.

Ahammad R., Nandy P., & Husnain P. (2013). Unlocking ecosystem based adaptation opportunities in coastal Bangladesh. *Journal of Coast Conservation, 17*, 833–840. doi:10.1007/s11852-013-0284-x.

Akter, J., Popescu, I., Sarker, M. H., & Roelvink, D. J. (2015). Evolution of the bengal delta and its prevailing processes. *Journal of Coastal Research, 32*(5), 1212–1226.

Alam, M., Ahammad, R., Nandy, P., & Rahman, S. (2013). Coastal livelihood adaptation in changing climate: Bangladesh experience of napa priority project implementation. In R. Shaw, F. Mallick, & A. Islam (Eds.), *Climate change adaptation actions in Bangladesh* (pp. 253–276). Tokyo, Japan: Springer.

Ali, A. (1999). Climate change impacts and adaptation assessment in Bangladesh. *Climate Research, 12*, 109–116.

Baten, M. A., Seal, L., & Lisa, K. S. (2015). Salinity intrusion in interior coast of bangladesh: Challenges to agriculture in south-central coastal zone. *American Journal of Climate Change, 4* (3), 248–262.

Cabo Bujan, J. A., & Al Hussainy, N. (2016). *Terminal evaluation report: Community based adaptation to climate change through coastal afforestation in Bangladesh (CBACC-CF)*. Dhaka, Bangladesh: UNDP.

CBACC-CF (Community Based Adaptation to Climate Change through Coastal Afforestation). (2012). *Training module on rice production technology* (CBACC-CF Project). Dhaka, Bangladesh: Department of Agriculture Extension Component, Khamarbari.

CBD. (2009). Connecting Biodiversity and Climate Change Mitigation and Adaptation: Report of the Second Ad Hoc Technical Expert Group on Biodiversity and Climate Change. (CBD Technical Series No. 41). Montreal, Canada: Secretariat of the Convention on Biological Diversity.

CCC (Climate Change Cell). (2007). Climate change and Bangladesh. Department of Environment, Ministry of Environment and Forests, Government of the People's Republic of Bangladesh, Dhaka.

CCC (Climate Change Cell). (2016). Assessment of Sea Level Rise on Bangladesh Coast through Trend Analysis. Climate Change Cell (CCC), Department of Environment, Ministry of Environment and Forests, Government of the People's Republic of Bangladesh, Dhaka.

Choudhury, J. (2007). Forests and forest management practices in Bangladesh: The question of sustainability. *International Forestry Review, 9*(2), 627–640.

Colls, A., Ash, N., & Ikkala, N. (2009). *Ecosystem-based adaptation: A natural response to climate change*. Gland, Switzerland: IUCN.

CSISA. (2012). *Farmers manual on improved rice farming technologies*. IRRI (in Bangla): Cereal Systems Initiative for South Asia in Bangladesh.

Dasgupta S., Huq M., Khan Z. H., Murshed M., Ahmed Z., Mukherjee N., et al. (2014). Cyclones in a changing climate: The case of Bangladesh. *Climate and Development, 6*(2): 96–110. doi:10.1080/17565529.2013.868335.

Dastagir M. R. (2015). Modeling recent climate change induced extreme events in Bangladesh: A review. *Weather and Climate Extremes, 7*, 49–60. doi:10.1016/j.wace.2014.10.003.

Dietz, T., Ostrom, E., & Stern, P. (2003). The struggle to govern the commons. *Science, 302*, 1907–1912.

Doswald N., Munroe R., Roe D., Giuliani A., Castelli I., Stephens J., et al. (2014). Effectiveness of ecosystem-based approaches for adaptation: Review of the evidence-base. *Climate and Development*. doi:10.1080/17565529.2013.867247.

Duarte, C. M., Losada, I. J., Hendriks, I. E., Mazarrasa, I., & Marbà, N. (2013). The role of coastal plant communities for climate change mitigation and adaptation. *Nature Climate Change, 3,* 961–968.

FAO. (2007). Proceedings of the workshop on coastal area planning and management in Asian tsunami-affected countries. Workshop held in 27–29 September 2006, Bangkok, Thailand.

GOB. (2005). National Adaptation Programme of Action (NAPA). Final report. Ministry of Environment and Forest, Government of the People's Republic of Bangladesh, Dhaka, Bangladesh.

GOB. (2009). Bangladesh Climate Change Strategy and Action Plan 2009. Ministry of Environment and Forests, Government of the People's Republic of Bangladesh, Dhaka, Bangladesh.

GOB. (2011). *Project Concept Note: Collaborative REDD+IFM Sundarbans Project (CRISP)*. Government of Bangladesh, Dhaka, Bangladesh: Forest Department, Ministry of Environment and Forests.

GOB. (2012). *Project concept note: Bangladesh REDD+ARR Protected Areas Project (BRAPAP)*. Government of Bangladesh, Dhaka, Bangladesh: Forest Department, Ministry of Environment and Forests.

Hossain M. S., Hein L., Rip F. I., & Dearing J. A. (2013). Integrating ecosystem services and climate change responses in coastal wetlands development plans for Bangladesh. *Mitigation and Adaptation strategies for global Change, 20*(2): 241–261. doi:10.1007/s11027-013-9489-4.

Huq, N., Renaud, F., & Sebesvari, Z. (2013). *Ecosystem based adaptation (EBA) to climate change integrating actions to sustainable adaptation*. Bonn, Germany: United Nations University, Institute for Environment and Human Security.

Iftekhar, M. S., & Takama, T. (2008). Perceptions of biodiversity, environmental services, and conservation of planted mangroves: A case study on Nijhum Dwip Island, Bangladesh. *Wetlands Ecology and Management, 16,* 119–137.

IPCC. (2007). *Climate change 2007: Impacts, adaptation and vulnerability*. Contribution of Working Group II to the Fourth Assessment Report of the Intergovernmental Panel on Climate Change. Cambridge University Press, Cambridge, England.

Islam M. R. (2007). Pre-and post-tsunami coastal planning and land-use policies and issues in Bangladesh. In Broadhead J. S., Leslie R. N. (Eds.), *Proceedings of the Workshop on Coastal Area Planning and Management in Asian Tsunami-affected Countries*. Bangkok, Thailand: Food and Agricultural Organization of the United Nations.

Islam M. R., & Ahmed M. (2004). Living in the coast. Problems, opportunities and challenges. Water Resources Planning Organization (WARPO). Ministry of Water Resources, Government of the People's Republic of Bangladesh.

Islam, A., Shaw, R., & Mallick, F. (2013). National Adaptation Programme of Action. In R. Shaw, F. Mallick, & A. Islam (Eds.), *Climate change adaptation actions in Bangladesh* (pp. 93–106). Tokyo, Japan: Springer.

IWM (Institute of Water Modelling) & CEGIS (Center for Environmental and Geographic Information Services). (2007). Investigating the Impact of Relative Sea-Level Rise on Coastal Communities and their Livelihoods in Bangladesh: Final Report, June 2007. IWM/CEGIS/ Government of Bangladesh, Dhaka, Bangladesh.

Kreft, S., Eckstein, A., & Melchior, I. (2017). *Global climate risk index 2017 who suffers most from extreme weather events? Weather-related loss events in 2015 and 1996 to 2015*. Berlin, Germany: Germanwatch e.V.

Mainuddin K., Rahman A., Islam N., & Quasem S. (2011). Planning and costing agriculture's adaptation to climate change in the salinity-prone cropping system of Bangladesh. *International Institute for Environment and Development (IIED)*, London, UK.

MEA (Millennium Ecosystem Assessment). (2005). *Ecosystems and human well-being: wetlands and water synthesis*. Washington, DC: World Resources Institute.

Munang, R., Thiaw, I., Alverson, K., Mumba, M., Liu, J., & Rivington, M. (2013). Climate change and ecosystem-based adaptation: A new pragmatic approach to buffering climate change impacts. *Current Opinion in Environmental Sustainability, 5*(1), 67–71.

Nandy, P., & Ahammad, R. (2012). Navigating mangrove resilience through the ecosystem based adaptation: lessons from Bangladesh. In D. J. Macintosh, R. Mahindapala, & M. Markopoulos (Eds.), *Sharing lessons on mangrove restoration* (pp. 243–254). Bangkok, Thailand: Mangroves for the Future and Gland/IUCN.

Nandy, P., Ahammad, R., Alam, M., & Islam, A. (2013). Coastal ecosystem based adaptation: Bangladesh experience. In R. Shaw, F. Mallick, & A. Islam (Eds.), *Climate change adaptation actions in Bangladesh* (pp. 277–303). Japan: Springer Tokyo.

Nishat, A., & Mukherjee, N. (2013). Sea level rise and its impacts in coastal areas of Bangladesh. In R. Shaw, F. Mallick, & A. Islam (Eds.), *Climate change adaptation actions in Bangladesh* (pp. 43–50). Japan: Springer Tokyo.

Pramova, E., Locatelli, B., Brockhaus, M., & Fohlmeister, S. (2012). Ecosystem services in the national adaptation programmes of action. *Climate Policy, 12*(4), 1–17. doi:10.1080/14693062.14692011.14647848.

PDO-ICZMP. (2004). *Living in the coast: People and livelihoods*. Project Development Office-Integrated Coastal Zone Management Plan (PDO-ICZMP), Water Resources Planning Organization, Government of Bangladesh.

Rahman, M. (2014). *Framing ecosystem-based adaptation to climate change: Applicability in the coast of Bangladesh*. Dhaka, Bangladesh: IUCN.

Rahman, M. M., & Islam, A. (2013). Adaptation technologies in practice and future potentials in Bangladesh. In R. Shaw, F. Mallick, & A. Islam (Eds.), *Climate change adaptation actions in Bangladesh* (pp. 305–330). Tokyo, Japan: Springer.

Reid H. (2016). Ecosystem- and community-based adaptation: Learning from community-based natural resource management, *Climate and Development, 8*(1): 4–9, doi:10.1080/17565529.2015.1034233.

Reid H., & Alam S. S. (2014). *Ecosystem based Approaches to Adaptation: Evidence from two sites in Bangladesh*. (IIED Working Paper). London, UK: IIED.

Sajjaduzzaman, M., Muhammed, N., & Koike, M. (2015). Mangrove plantation destruction in Noakhali Coastal Forests of Bangladesh: A case study on causes, consequences, and model prescription to halt deforestation. *International Journal of Agriculture and Biology, 7*(5), 732–734.

Saroar, M. M., & Routray, J. K. (2010). In situ adaptation against sea level rise (SLR) in Bangladesh: Does awareness matter? *International Journal of Climate Change Strategies and Management, 2*(3), 321–345.

Saroar M. M., Routray J. K. (2013). 'Climate Refugee' is Not a Hoax. But We can Avoid it. Empirical Evidence from the Bangladesh Coast. In Schmidt-Thomé P, Klein J (Eds.) *Climate change adaptation in practice: From strategy development to implementation*. Oxford, UK: Wiley. doi:10.1002/9781118548165.ch21.

Saroar, M. M., Routry, J. K., & Leal Filho, W. (2015). Livelihood vulnerability and displacement in coastal Bangladesh: Understanding the nexus. In W. Leal Filho (Ed.), *Climate change in the Asia-pacific region* (pp. 9–31). Switzerland: Springer.

Saroar, M. M., & Leal Filho, W. (2016). Adaptation through climate smart agriculture: Status and determinants in coastal Bangladesh. In W. Leal Filho, H. Musa, G. Cavan, et al. (Eds.), *Climate change adaptation Resilience and Hazards* (pp. 157–178). Switzerland: Springer.

Saroar, M. M., Rahman, M. M., Bahauddin, K. M., & Rahman, M. A. (2017). Ecosystem-based adaptation: Opportunities and challenges in coastal Bangladesh. In S. Huq et al. (Eds.), *Efforts to address climate change in Bangladesh*. Heidelberg, Germany: Springer Verlag.

Tompkins, E. L., Brown, K., & Few, R. (2008). Scenario-based stakeholder engagement: Incorporating stakeholder's preferences into coastal planning for climate change. *Journal of Environmental Management, 88,* 1580–1592.

Uddin, M. (2013). Status of coastal plantations and its impact on land stabilization, soil PH, and salinity at Nolchira range of Hatiya Island, Bangladesh. *IOSR Journal of Agriculture and Veterinary Science, 3*(4), 57–61.

UNDP. (2012). *Mid-term evaluation of community based adaptation to climate change through coastal afforestation in Bangladesh*. Dhaka, Bangladesh: UNDP.

Vignola, R., Locatelli, B., Martinez, C., & Imbach, P. (2009). Ecosystem-based adaptation to climate change: What role for policy makers, society and scientists? *Mitigation and Adaptation Strategies for Global Change, 14,* 691–696.

World Bank. (2009). *Convenient solutions to an inconvenient truth: Ecosystem-based approaches to climate change*. Washington, DC: The World Bank.

World Bank. (2010). *Economics of adaptation to climate change: Bangladesh country study*. Washington, DC: The World Bank.

Part II
Case Studies on Climate Change Adaptation in Coastal Areas

Black Pearl Farming as an Adaptation to Coastal Climate Change

Mattlan Zackhras, Payton Deeks and Simon Ellis

Abstract Namdrik Atoll is a low-lying atoll approximately 270 miles from Majuro, the capital of the Republic of Marshall Islands, linked only by infrequent boat and plane access. The population of some 500 people consists primarily of subsistence farmers and fishers with limited social or economic safety nets. The population is already facing impacts of climate change, particularly the salination of ground water, king tides, sea level rise, and storm surges. With a strong funding commitment from the US Agency for International Development (USAID), the community has recently revived the farming of black-lipped pearl oysters as a way to provide sustainable income to the farmers and the broader coastal community in order to increase resilience to climate change. During the most recent harvest, the farm harvested over 1000 pearls, with a retail value of US$40,000. Profits from the pearl farm will be re-invested back into the community through renovation of residences, procurement of water tanks and pumps, and other community development and climate change adaptation efforts. This paper will describe the strategies and results of the pearl farming efforts as a means of climate change adaptation, as well as the future prognosis.

M. Zackhras
Namdrik Atoll Development Association (NADA), Majuro, Marshall Islands
e-mail: namdrikalele@gmail.com; mattzackhras@gmail.com

P. Deeks (✉)
Pacific-American Climate Fund, Trade and Financial Tower, Unit 1003,
32nd Street Corner 7th Avenue, Bonifacio Global City, Taguig, NCR 1634, Philippines
e-mail: pdeeks@pgrd.org

S. Ellis
Marine and Environmental Research Institute of Pohnpei (MERIP), Kolonia, P.O. Box 1005,
Pohnpei 96941, Federated States of Micronesia

W. Leal Filho (ed.), *Climate Change Impacts and Adaptation Strategies for Coastal Communities*, Climate Change Management,
https://doi.org/10.1007/978-3-319-70703-7_11

Introduction

Namdrik Atoll, Republic of the Marshall Islands, is a small atoll located 270 miles from the capital of Majuro. It has infrequent and limited air and sea transport which restricts most forms of development and limit livelihood options for its 508 inhabitants (RMI 2011).

The atoll has a small freshwater lens, though most of the populace relies on rainwater harvesting for fresh water. There is local production of bananas and taro, and a small copra (the dried meat of coconuts) export industry. While some fishing and consumption of local breadfruit, bananas, and taro takes place, most of the islanders are heavily reliant on imported food and supplies (NADA 2014).

Climate Change Impacts and Adaptation on Namdrik

The climate change impacts on small island developing states such as RMI have been widely covered (see Barnett and Adger (2003); Campbell 2014; Campbell 2015; Nicholls 2011; Weir et al. 2016 for summaries). These include sea level rise, coastal inundation, greater likelihood of drought, ocean acidification, and a higher intensity of storms. Namdrik has one, small freshwater lens which is highly vulnerable to salt water inundation. In recent years, inhabitants have observed more salinization of their taro pits and banana patches, and both crops are sensitive to salt (NADA 2014).

Climate change is a driver of migration (Campbell 2014), though teasing out the climate change driver from others (such as economic, social, political, etc.) is quite difficult. While some studies indicate that many atolls will become uninhabitable due to sea level rise (Constable 2016; Campbell 2014), exactly where or when is not clear. More recently, worldwide low-lying islands which are vulnerable to sea level rise have begun resettlement (de Sherbinin et al. 2011).

The traditional knowledge embedded in a cohesive island community has been an important part of effective and adaptive responses to threats such as climate change (Weir et al. 2016). The NADA pearl farm project looks to draw on and enhance the knowledge and cohesiveness of Namdrik Atoll.

Methods

Pearl Farming on Namdrik and RMI

Pearl farming has generally been regarded as an environmentally friendly development opportunity for remote coastal communities (Lane et al. 2003; Sims 2003). In addition to relieving pressure on other marine resources such as fisheries, the better the resource is managed (e.g., keeping a lagoon clear), the better the product will be (Cartier 2014). It has also been seen as an activity that incorporates a working environment similar to the traditional activities of many Pacific Islanders on outer atolls (Tisdell and Poirine 2000).

Pearl farming is suited to more remote locations as pearls themselves are easy to export in that they don't cost much to ship, are not perishable, and there are developed markets for the product (Sims 1993, 2003; Tisdell and Poirine 2007). The market for black pearls is dominated by producers in Tahiti and the Cook Islands (Tisdell and Poirine 2000; Ponia 2010). These producers often benefit from an economy of scale, with much lower costs per pearl. The global price of pearls is also dominated by these producers, which means when the supply grew, the price decreased. Between 1990 and 1995, the average price of the Tahitian black pearl decreased by almost 75% (Tisdell and Poirine 2000). Since then prices of high quality pearls have remained high or have continued to increase but lower quality pearl prices remain low. Additionally, smaller producers have difficulty differentiating themselves in the market, as a typical customer cannot tell the difference among pearls produced by the same species of oyster. Some authors suggest working to make products unique by specializing in colors (e.g., Micronesian blue), emphasizing quality, or adding value to the pearls by setting them in jewelry (Fassler 2002; Fong et al. 2005; Tisdell and Poirine 2007; Tisdell and Poirine 2008; Cartier et al. 2012).

The farming of pearls from the black-lipped pearl oyster (*Pinctada margaritifera*) has had a checkered past in the Republic of Marshall Islands (RMI). Robert Reimer Enterprises (owner of one of the prominent hotels on Majuro) started a pearl oyster farm on Arno atoll in 1995 which continued through 2005 and on Jaluit atoll from 2001–2007, but struggled to obtain spat in appropriate, consistent numbers throughout its existence. In 1993, a private company Black Pearls Inc. formed a relationship with the Marshall Islands Marine Resources Authority (MIMRA), which was in charge of one of the main hatcheries (and sources of pearl spat) in RMI (Haws and Ellis 2003b; NADA 2016). This joint venture (called Black Pearls of Micronesia) operated in RMI from 1998 to 2003. And while an article in the year 1999 indicated Black Pearls of Micronesia was set for expansion (Anonymous 1999), four years later it closed up shop. Additionally, as of 2003, Haws and Ellis (2003b) reported two pearl oyster spat hatcheries existed in RMI, one at the College of the Marshall Islands (CMI) and one commercial-scale hatchery overseen by MIMRA. However, at the time of writing there are no operating pearl oyster hatcheries in the RMI. A chronic lack of spat, mortalities due to red tide, and lack of government support contributed to the struggle and ultimate demise of these operations (Johnson 2007; NADA 2016).

The history of black pearl farming on Namdrik has to date followed a similar course. In the 1980s, a survey examined the viability of farming on six atolls in RMI, including Namdrik (NADA 2016), but few black lipped pearl oysters were found to naturally occur. In 1985 the Island Council (de facto the governing body of Namdrik) developed the Namdrik Alele Local Government Pearl Culture Ordinance. This ordinance requires the prior approval of the Island Council for any harvesting and marketing of pearl oysters by anyone. Violations of the Ordinance may result in monetary fines and imprisonment (Dashwood 1991). In 1989,

Marshall Islands Marine Resources Authority (MIMRA) conducted another survey, this time finding more oysters (Dashwood 1991).

Dashwood himself deployed 3500 spat collectors (paid for by USAID) and collected almost 1000 pearl oysters from the wild and hung them on subsurface longlines in the lagoon, though the results of that effort are not clear. Throughout the 1990s, various players (i.e., Secretariat of the Pacific Community, Black Pearls Inc.) made various efforts to get pearl farming in Namdrik off the ground, but none sustained.

In 2004, Dr. Maria Haws conducted research on Namdrik and reported that some remnant pearl oysters were still found in the lagoon, along with old equipment. She also noted that the Namdrik lagoon was experiencing eutrophication (NADA 2016). In 2008, based on a request from community leadership in Namdrik, pearl farming was again re-activated with the assistance of MIMRA, along with other technical partners and funding from UNDP/GEF small grants program (NADA 2016). In early 2009, the Namdrik community had 1500 pearl oysters ready to be seeded, having continued to collect spat over the previous years. Namdrik pearl farming received another shot in the arm in 2012, when it linked with a local hatchery and received hundreds of thousands of pearl oyster spat (ABC Radio Australia 2012) though these were likely to have been very small with low survival rates. Under a grant from Seacology, a seeding technician from the Cook Islands, Ms. Apii McLeod, visited Namdrik for four weeks in 2012 to conduct farm and seeding training. Additionally, the grant included the establishment of a marine protected area (MPA) in the lagoon and the construction of an education, surveillance, and monitoring center (now used as the seeding house) along with kayaks used to enforce the MPA (Seacology 2013). Additionally, some members of the Namdrik community have been trained in grading and pricing of pearls, as well as setting pearls in jewelry (NADA 2014). By 2014, there were three small pearl harvests on Namdrik, consisting of 500 pearls valued at around $20,000 (NADA 2014). Most recently, the pearl farm has been provided a grant of more than $375,000 over three years by the US Agency for International Development's Pacific-American Climate Fund (PACAM). The intent of this support is to provide enabling capital and organizational development to allow the pearl farming to become self-sustaining and thus provide an in situ option to adaptation to climate change. A source of funding for Namdrik Atoll will allow its community to better withstand climate-related shocks and thus adapt to climate change.

Limitations of the Work

While many studies have been produced on the pearl farming industry in Tahiti and the Cook Islands, only limited data exist for the Marshall Islands. Further, the records of the Namdrik Atoll pearl farm are located both on Namdrik and on Majuro. They tend to be temporally incomplete, at time relying on individuals' memories or recollections. For example, locations of the farm and oyster chaplets are simply "known" by the farm workers without thorough GPS locations being documented. Other holes in the data stemmed from misplacement of records during employee turnover.

In instances where hard data are captured at the farm level, there are technological challenges which prevent these data from being used by the researchers. Specifically, one computer is used at the farm to log seeding, mortality, and other data. Internet connections to the atoll are frequently down and/or intermittent. Power on the atoll is solar generated, so computer use and internet router use are subject to the vagaries of the weather. In short, the challenges of scrupulous electronic recordkeeping are magnified in an outer atoll environment. This paper therefore is a best attempt to document the history of the farm, the rationale of pearl farming as an adaptation, and discuss future prospects for the Namdrik pearl farm.

Discussion

Outmigration

In the Pacific, movement among atolls or from the periphery to the core of island nations is a longstanding norm and a way for Pacific Islanders to spread environmental and economic risk; adapt to climate change; or seek more opportunities (Connell and King 1990; Barnett and Chamberlain 2010; Campbell and Warrick 2014). Young people will move from the rural to the urban, from the periphery to the core, looking for education, paid work, medical facilities, or entertainment (Heine 1984; Locke 2009; Weir et al. 2016).

Like much of Micronesia, Namdrik Atoll has been affected by outmigration. Young people, looking for jobs or education, leave outer atolls, often for Majuro (the capital of RMI) or beyond. A 2011 RMI national census showed that 4000 people had left the outer islands since the previous census in 1999. In twelve years, roughly ¼ of the outer islands population had migrated out (Johnson 2015). In 1948, the population of Namdrik (spelled Namorik at that time) was 327 people (Williamson and Sabath 1984). When Dashwood visited Namdrik in 1991, the population had increased to 814 people living in 96 households. The population had decreased again by 2016 to 508 people (though notably a similar number of households). The bulk (some 74%) of the population on Namdrik is 18 or younger (IOM 2016).

Migration in the RMI has a couple of characteristics which set it apart from the rest of the Pacific. Under the original Compact of Free Association (1986–2001), the Marshall Islands received roughly $1 billion in aid from the US. Under the renegotiated compact (2004–2024) the Marshall Islands will receive roughly $1.5 billion in direct US assistance (CIA 2017), and RMI's Compact of Free Association allows Marshallese visa-free entry to the US, with Hawaii being a common location (US Department of State 2003). While migration to the US is a legal possibility, it also requires additional resources such as money and social connections, which many do not have (Barnett and Chamberlain 2010; Constable 2016). Additionally, relocations and migration have a sometimes fraught past in the RMI. Forced relocations due to nuclear testing remain a sore point in the collective memory (Johnson 2015).

Migration can have positive or negative effects on the origin island as well as the destination. There are economic, social, cultural, and psychological costs associated with migration such as loss of tradition, language, identity, livelihoods, and community cohesion (Campbell and Warrick 2014). But it can also help home island adaptation by providing remittances, reducing pressure on environmental resources, and creating an opportunity for skills and knowledge transfer if migrants return home (Campbell and Warrick 2014).

In the Pacific, the migration of young people to urban centers has left children and elders on the outer atolls, two of the more vulnerable groups to the impacts of climate change. Atoll residents have historically been resilient in the face of disasters and extreme events in the past (McMillen et al. 2014). Partly this may be due to traditional knowledge and the social capital of a cohesive atoll community (Weir et al. 2016) but also to the reciprocity present among people, communities, and neighboring islands (Connell 1993). While the islands themselves may be vulnerable, there is evidence to suggest that Pacific Island people are resilient (Campbell 2014).

The urban centers of RMI have suffered deleterious effects of migration from the periphery into the core. Majuro and Ebeye were experiencing high population density even in 1984 (Heine) and continue to have some of the highest population densities in the Pacific. This has led to deteriorating sanitation, increased conflict, and informal settlements of communities originating from outer atolls (Cocklin and Keen 2000; Campbell and Warrick 2014). Additionally, some authors lament the loss of traditional values and customs of the outer islands which have been lost in the urban areas (Heine 1984; Royle 1999).

In the Pacific Islands more generally, authors have identified a strong reluctance of Pacific Island people to leave their homes due to a fear of loss of land, sovereignty, and cultural identity. The link between identity and land means people do not readily wish to move (Adger et al. 2011; Barnett and O'Neill 2012; Adger et al. 2013; Rudiak-Gould 2013; Campbell and Warrick 2014; Weir et al. 2016). This reluctance, and attachment to a place may mean that coastal communities in the Pacific will try to adapt in situ to climate change impacts (Adger et al. 2013; Nunn 2013) with pearl farming being one of the adaptation strategies.

Pearl Farming as an Adaptation and Means of Community Development

A strong, self-sustaining Namdrik pearl farm will provide an ecologically-friendly livelihood option for residents of Namdrik. This endeavor will not only create a wage-earning opportunity, but will strengthen the social cohesion of the community by involving members in the farm or peripheral activities such as jewelry making. Income accruing to individuals or the community will also strengthen their resilience to climate change and allow them to withstand climate-related stresses. Studies have shown that households that save money have a greater capacity to cope with and recover from shocks (Davis et al. 2004; Gaillard et al. 2009). Similarly, these savings help households adapt to climate change impacts as people can relocate homes (e.g., away from rising sea levels or away from areas prone to flooding), or reinforce their homes to be more resilient to high wind from typhoons, etc. People living in areas such as atolls have a lower capacity to protect themselves from climate-related threats if they have a fragile livelihood. Those with stronger and more diverse livelihoods are better able to face coastal hazards (Gaillard et al. 2009), so adding an additional income source to the atoll will strengthen the community's ability to adapt.

Further, small islands must work towards financial self-reliance in order to meet the economic costs of adaptation activities, reduce donor dependence, and for survival itself as climate change stressors intensify. The alternative to continued poverty and economic dependence is outmigration, possible future abandonment of the atoll and the loss of the identity of the Namdrik community (NADA 2014). Pearl farming will provide income for daily life on Namdrik and may replace other activities such as agriculture which is suffering from salt water intrusion. It can act as a catalyst for addressing climate change issues, providing environmental education, and be a source of pride for the community.

The more cynical authors in the field have noted that communally-owned and -operated pearl farms have struggled, either due to the politics of a small island, an unmotivated labor force, or theft (Sims 2000), while others have attributed the collapse of many small farms in French Polynesia and the Cook Islands to falling pearl prices (Tisdell and Poirine 2000). The Namdrik pearl farm recognizes these risks (NADA 2014) and has tried to address them as described above.

Haws and Ellis (2003a) note that pearl farming can foment changes which may impact the community. For example, changes in the economic status of individuals associated with the farm. It was noted during a donor visit to Namdrik that the pearl farm (still below the sustainable size) employs 7 people on a part-time basis (Fig. 1). It is estimated that each worker on the pearl farm will provide for a family of 4–10, and additional jobs will be created from the jewelry making efforts (Fig. 2). It is expected that the number of farm workers will double every year, and that community groups will be involved in making chaplets to hang the oysters and also to help clean oysters to prepare for seeding (NADA 2014).

There are also ongoing efforts to include community members in jewelry making. During an interview with community members, those that were aware of the pearl farming operations believed it was already making a lot of money (which it is not, yet) (PACAM 2016b). During the most recent harvest, the farm harvested over 1000 pearls, with a retail value of US$40,000. However, these pearls have not all been sold, so the full revenue has not yet been realized. Communications and transparency will be an important part of a successful endeavor moving forward.

Fig. 1 Farm workers of the Namdrik pearl farm. *Photo* © USAID/Matt Abbott

Fig. 2 Namdrik residents participate in training on setting pearls in jewelry to add value locally. *Photo* © NADA

The Namdrik Pearl Farm aims to provide economic benefits to the community, while also being mindful of social and environmental sustainability (NADA 2016). One of the mechanisms by which it aims to do this is the establishment of a trust fund or through community development programs. This might include residence renovations, water tanks, water pumps, a community center, financial support for Namdrik residents seeking medical treatment off-island, and solar energy for homes (NADA 2014). Once the pearl farm has become profitable, the cooperative intends to create a trust fund or a time deposit account, putting aside funds to be saved for the long term.

Addressing Challenges

Many small farms experience negative cash flow for about five years. Startup costs range from nearly $39,000 (Haws 2002) to over $200,000 for a farm with 25,000 oysters (Fong et al. 2005). Much of the startup costs are fixed costs, which must be borne no matter the size of the operation. The injection of funding from the United States Agency for International Development was intended to kick-start the pearl farm once again, enabling it to move to a point of being profitable. Figure 3 below shows the current location of the pearl farm in the lagoon, as well as the spat collector lines.

Oyster spat. Pearl farms in French Polynesia and the Cook Islands have readily available oyster spat. The largely enclosed lagoons in these areas have limited exchange of water with the open ocean, so spat remains in the lagoon, allowing for easier capture (Friedman et al. 1998). The cost and availability of spat have been an ongoing challenge in RMI (see abovementioned history) and numerous authors have flagged the importance of reliable hatcheries for small-scale aquaculture (Bell 1999). Dashwood (1991) noted that "the long term viability of pearl farming [on Namdrik] will depend almost entirely on the ability to collect seed stock through the implementation of a successful spat collection program." The Namdrik pearl farm initially anticipated accessing spat from the College of Marshall Islands (CMI) hatchery as well as deployed spat collectors. During the course of the project implementation, the CMI hatchery decided to close for one year due to facility maintenance, rendering it unable to provide spat to the Namdrik pearl farm. Additionally, contacts at the Marshall Islands Marine Resources Authority (MIMRA) identified a potential spat-producing facility in Arno Atoll (PACAM 2016a) but this did not prove feasible. The closing of hatcheries delivered a blow to the work schedule of the Namdrik pearl farm. In response, the team has procured and deployed additional spat collectors (Ellis 2016). To date, with USAID support, sixteen additional spat collectors have been deployed since April 2015, with 18 more planned for 2017 (NADA 2017) (see Fig. 4).

Number of oysters. Dashwood (1991) recommended that a pearl farm on Namdrik have 10,000 oysters between 7–15 cm in diameter as the nucleus of a pearl farm. Other authors have indicated that 3000 oysters of a suitable grafting size are needed,

Fig. 3 Map of Namdrik Atoll and pearl farm

working out to be 12,000–15,000 oysters in various stages of culture (Haws 2002). The grafting technician assisting the Namdrik pearl farm (whose family runs a farm in the Cook Islands) suggested that the pearl farm might be able to run sustainably on as few at 10,000 oysters (PACAM 2016b). As of August 2016, the Namdrik pearl farm had a total of 4165 oysters in its inventory (PACAM 2016b).

Grafting costs. Costs associated with seeding (also referred to as grafting, when a nucleus is implanted along with a tissue graft into a pearl oyster) are also estimated to be some 40% of operating expenses (Fong et al. 2005). Trained technicians charge about US$3–$4 for each pearl oyster he or she grafts, or some percentage of the harvest (Haws 2002). The availability (both in terms of numbers and geographically) of seeding technicians has been identified as a limit to growth of the industry (Sims 2000). The pearl farm on Namdrik, with USAID support, has been able to have Apii McLeod visit Namdrik twice to seed 3107 oysters and harvest 1063 pearls (Fig. 5).

Fig. 4 Namdrik pearl farm worker inspects spat collectors. *Photo* © USAID/Matt Abbott

Fig. 5 Grafting technician Apii McLeod traveled to Namdrik from the Cook Islands. *Photo* © USAID/Matt Abbott

To reduce future costs associated with hiring a grafting technician, the Namdrik pearl farm project sent two farm workers who had demonstrated skill and leadership to the Cook Islands to learn both grafting techniques and general pearl farm management. This will allow oysters to be grafted onsite and increase the skills of those involved with the farm. While conventional wisdom dictates that a trained seeding technician needed to seed 10,000–15,000 oysters to become proficient, Nava et al. (2000) found that new technicians were reaching the trained technicians' level of 60–80% nucleus retention after only seeding 7000 oysters. This may further underscore the wisdom of having locally-available grafting technicians. Additionally, having technicians onsite will allow grafting on a continuous basis which is more appropriate for small farms (NADA 2014).

Poor governance. Earlier attempts at a self-sustaining pearl farm on Namdrik failed due to poor management (NADA 2014). Currently, the pearl farm enjoys the support of both the Mayor of Namdrik (Clarence Luther) and its Senator (Mattlan Zachkras) as well as some (but not all) of the local council (PACAM 2016b). However, in an effort to de-politicize the efforts, a Namdrik Pearl Farming Cooperative is being established. With PACAM support, a workshop to review the structure of the proposed cooperative was conducted with the community on Namdrik on July 7, 2016 and subsequently the cooperative's papers were completed. The pearl farm has engaged a lawyer in Majuro to validate the legal status of the papers and help file them with the appropriate authorities.

Environment. As part of a best practice for pearl farming, and in acknowledgement of its reliance on its natural resources, the Namdrik community began efforts to develop a marine protected area in the lagoon in 1994 (NADA 2014). Following a grant from Seacology in 2012–2013, 86 acres (34.8 hectares) were designated as a marine protected area through a local ordinance approved by the local government. A no-take zone was declared around the pearl farm in 2014 (NADA 2016). The pearl farm best management practices (see Haws and Ellis 2003a) encourage a development of a resource management plan, which was completed in 2016.

Conclusion

Namdrik Atoll is threatened by a global issue to which it made little or no contribution. With the goal of enabling Namdrik residents to develop climate-resilient livelihoods in situ, the Namdrik pearl farm has been revived with the support of USAID, community members, and technical expertise.

This most recent effort to revive the Namdrik pearl farm (described in this paper) has addressed risks that have led to previous efforts' demise. These include attracting and collecting spat locally; developing grafting skills locally; and creating value-added with the pearls through jewelry setting.

Looking forward, the pearl farm intends to keep selling pearls (available at the Namdrik Atoll Development Association office in Majuro, Republic of Marshall

Islands). Proceeds from these sales will help to keep the project functioning once the grant funding from USAID finishes.

Additional future steps include outreach to the Namdrik community about the farm and potential livelihood supplements the farm could bring and how this might help residents to stay and thrive on Namdrik in the face of climate change.

References

ABC Radio Australia. (2012). *Marshall Islands developing black pearl industry. Interview with Dr. Maria Haws. Geraldine Coutts, presenter*. February 15, 2012. Available at http://www.radioaustralia.net.au/international/radio/onairhighlights/marshall-islands-developing-black-pearl-industry.

Adger, W. N., Barnett, J., Chapin, F. S., & Ellemor, H. (2011). This must be the place: Underrepresentation of identity and meaning in climate change decision-making. *Global Environmental Politics, 11,* 1–25.

Adger, W., Barnett, J., Brown, K., Marshall, N., & O'Brien, K. (2013). Cultural dimensions of climate change impacts and adaptation. *Nature Climate Change, 3,* 112–117. doi:10.1038/nclimate1666.

Anonymous. (1999). *Black pearl industry continues to expand* (CTSA Regional Notes, 11 (1). Fall 1999, p. 1). Waimanalo, HI 96795: Centre for Tropical and Subtropical Aquaculture, The Oceanic Institute and University of Hawaii.

Barnett, J., & Adger, W. N. (2003). Climate dangers and atoll countries. *Climatic Change, 61,* 321–337.

Barnett, J., & Chamberlain, N. (2010). Migration as climate change adaptation: Implications for the Pacific. In B. Burson (ed.), *Climate change and migration: South Pacific perspectives* (pp. 51–60). Victoria University of Wellington.

Barnett, J., & O'Neill, S. (2012). Islands, resettlement and adaptation (commentary). *Nature Climate Change, 2,* 8–10.

Bell, J. (1999). Aquaculture: A Development Opportunity for Pacific Islands. Information Paper 19, Secretariat of the Pacific Community, Noumea, New Caledonia.

Campbell, J. R. (2014). Climate-change migration in the Pacific. *The Contemporary Pacific, 26,* 1–28. doi:10.1353/cp.2014.0023.

Campbell, J. (2015). Development, global change and traditional food security in Pacific Island countries. *Regional Environmental Change, 15,* 1313–1324. doi:10.1007/s10113-014-0697-6.

Campbell, J., &Warrick, O. (2014). *Climate change and migration issues in the Pacific*. United Nations Economic and Social Commission for Asia and the Pacific. Kadavu House, Level 5, 414 Victoria Parade, Suva, Fiji.

Cartier, L. E. H. (2014). Sustainability and traceability in marine cultured pearl production. *Thesis*, University of Basel.

Cartier, L. E., Krzemnicki, M. S., & Ito, M. (2012). Cultured pearl farming and production in the Federated States of Micronesia. *Gems & Gemology,* 48(2), 108–122. Available at http://www.sustainablepearls.org/fileadmin/Documents/PDF/Cartier_et_al_2012_Micronesian_pearls_G_and_G.pdf.

CIA. (2017). *World Factbook: Republic of Marshall Islands*. https://www.cia.gov/library/publications/the-world-factbook/geos/rm.html. Accessed April 3, 2017.

Cocklin, C., & Keen, M. (2000). Urbanization in the Pacific: Environmental change, vulnerability and human security. *Environmental Conservation, 27*(4), 392–403.

Connell, J. (1993). Climatic change: A new security challenge for the atoll states of the South Pacific. *Journal of Commonwealth & Comparative Politics, 31*(2), 173–192.

Connell, J., & King, R. (1990). Island migration in a changing world. In R. King & J. Connell (Eds.), *Small worlds, global lives: Islands and migration* (pp. 1–26). London: Pinter Publications.

Constable, A. L. (2016). Climate change and migration in the Pacific: Options for Tuvalu and the Marshall Islands. *Regional Environmental Change*, 1–10.

Dashwood, J. (1991). *The Namdrik pearl oyster project*. Noumea, New Caledonia: South Pacific Commission.

Davis, I., Haghebeart, B., & Peppiatt, D. (2004). *Social vulnerability and capacity analysis. Discussion paper and workshop report*. Geneva: ProVention Consortium.

de Sherbinin, A., Castro, M., Gemenne, F., Cernea, M. M., Adamo, S., Fearnside, P. M., et al. (2011). Preparing for resettlement associated with climate change. *Science, 334*(6055), 456–457.

Ellis, S. (2016). "Spat Collection for Pearl Farming" PowerPoint presentation used for training on USAID-funded project. *Marine and Environmental Research Institute of Pohnpei*, Kolonia, Pohnpei, Federated States of Micronesia.

Fassler, C. R. (2002). *Recent development in selected Pacific and Indian Ocean black pearl projects*. Presented at the Pearl Session, World Aquaculture Society's Annual Meeting, Beijing, People's Republic of China, April 24, 2002.

Fong, Q. S. W., Ellis, S., & Haws, M. (2005). Economic feasibility of small-scale black-lipped pearl oyster (*Pinctada margaretifera*) pearl farming in the Central Pacific. *Aquaculture Economics & Management, 9*(3).

Friedman, K. J., Bell, J. D., & Tiroba, G. (1998). Availability of wild spat of the blacklip pearl oyster, *Pinctada margaritifera*, from 'open' reef systems in Solomon Islands. *Aquaculture, 167,* 283–299.

Gaillard, J. C., Maceda, E., Stasiak, E., Le Berre, I., & Espaldon, M. V. (2009). Sustainable livelihoods and people's vulnerability in the face of coastal hazards. *Journal of Coastal Conservation, 13*(2/3), 119–129.

Google Earth. (2017). *Satellite image of Namdrik Atoll*. Accessed March 29, 2017.

Haws, M. (2002). *The basic methods of pearl farming: A layman's manual*. Center for Tropical and Subtropical Aquaculture Publication No. 127.

Haws, M. C. & Ellis, S. C. (2003a). *Best management practices for pearl farming in the Western Pacific*. Pacific Aquaculture and Coastal Resources Center, University of Hawai`i Hilo. Publication 001–2003.

Haws, M. & Ellis, S. (2003b). Pearl farming in Micronesia. *Pearl World, the International Pearling Journal, 12*(3) (July/August/ September 2003).

Hayes, K. (2015). *Map of Namdrik pearl farm drawn based on conversations with farm workers*. Unpublished.

Heine, A. (1984). Urbanization and social change in the marshall islands. *Ambio, 13*(5–6), 313–315.

IOM (International Organization on Migration). (2016). *Namdrik community profile, surveyed on August 7, 2016*. Available at http://www.micronesia.iom.int/publications/category/17-country-state-profile?download=165:namdrik-namdrik-profile&start=60.

Johnson, G. (2007). Pearl farms shuts down. *Marshall Islands Journal*. Available at http://yokwe.net/index.php?module=News&func=display&sid=1832.

Johnson, G. (2015). *Idyllic no more: Pacific island climate, corruption, and development dilemmas*. Self-published.

Lane, C., Oenpepa, C., & Bell, J. (2003). Production and grow-out of the black-lip pearl oyster *Pinctada margaritifera*. *Aquaculture Asia, 8*(1), 5–7.

Locke, J. (2009). Climate change-induced migration in the Pacific Region: Sudden crisis and long-term developments. *Geographical Journal, 175,* 171–180.

McMillen, H. L., Ticktin, T., Friedlander, A., Jupiter, S. D., Thaman, R., Campbell, J., et al. (2014). Small islands, valuable insights: Systems of customary resource use and resilience to climate change in the Pacific. *Ecology and Society*.

NADA. (2014). "*Namdrik Atoll pearl farming: Linking adaptation and livelihoods.*" Final Project Proposal to PACAM. Unpublished.

NADA. (2016). "*Namdrik Atoll Resources Management Plan.*" Deliverable to PACAM. Submitted September 2016.

NADA. (2017). Grantee Quarterly report to PACAM: October 1, 2016 to December 31, 2016. *Namdrik Atoll Development Association*, Majuro, Republic of Marshall Islands.

Nava, M., Arizmendi, E., Farell, S., & McLaurin, D. (2000). Evaluation of success in the seeding of round nuclei in Pteria sterna, a new species in pearl culture. *SPC Pearl Oyster Information Bulleting*. December 14, 2000.

Nicholls, R. J. (2011). *Sea level impacts of climate change*. Presented at: Improving the assessment and valuation of climate change impacts for policy and regulatory analysis: Research on climate change impacts and associated economic damages, January 27–28, 2011. Washington, DC. https://yosemite.epa.gov/ee/epa/eerm.nsf/vwAN/EE-0566-111.pdf/$file/EE-0566-111.pdf.

Nunn, P. (2013). The end of the pacific? Effects of sea level rise on pacific island livelihoods. *Singapore Journal of Tropical Geography, 34,* 143–171.

PACAM (Pacific-American Climate Fund). (2016a). *Personal communication from Lara Studzinski, Regional Fund Manager (Micronesia) during Pacific-American Climate Fund team meeting on April 14, 2016*. Trade and Financial Tower, Unit 1003 32nd Street corner 7th Avenue, Bonifacio Global City, Taguig, NCR Philippines 1634.

PACAM (Pacific-American Climate Fund) (PACAM). (2016b). *Mid-term review report: Namdrik Atoll Pearl Farming, August 2016*. Trade and Financial Tower, Unit 1003 32nd Street corner 7th Avenue, Bonifacio Global City, Taguig, NCR Philippines 1634.

Ponia, B. (2010). *A review of aquaculture in the Pacific Islands 1998–2007: Tracking a decade of progress through official and provisional statistics*. SPC Aquaculture Technical Papers, Secretariat of the Pacific Community, Noumea. ISSN: 1683-7568.

Ramsar Sites Information Service, undated. *Annotated list of wetlands of international importance: Marshall Islands*. Available at https://rsis.ramsar.org/sites/default/files/rsiswp_search/exports/Ramsar-Sites-annotated-summary-Marshall-Islands.pdf?1490083290.

Republic of Marshall Islands. (2011). *The RMI 2011 Census of Population and Housing Summary and Highlights Only*. Economic Policy, Planning, and Statistics Office. Office of the President. Available at https://www.doi.gov/sites/doi.gov/files/migrated/oia/reports/upload/RMI-2011-Census-Summary-Report-on-Population-and-Housing.pdf.

Royle, S. A. (1999). Conservation and heritage in the face of insular urbanization: The Marshall Islands and Kiribati. *Built Environment, 25*(3), 211–221 (Conservation: Experience outside the Industrialized West).

Rudiak-Gould, P (2013). Climate Change and Tradition in a Small Island State: The Rising Tide. Routledge, New York. ISBN-13: 978-0415832496.

Seacology. (2013). Project website. https://www.seacology.org/project/329-marshall-islands/. Accessed March 28, 2017.

Sims, N. A. (1993). Pearl oysters. In A. Wright & L. Hill (Eds.), *Nearshore resources of the South Pacific* (pp. 409–430). Honiara: IPS, Suva/ Forum Fisheries Agency.

Sims, N. A. (2000). *From the editor, pearl oyster information bulletin No. 14 December 2000*. B. P. D5, 98848 Noumea Cedex, New Caledonia: Secretariat of the Pacific Community, Marine Resources Division, Information Section SPC.

Sims, N. A. (2003). *From the editor, pearl oyster information bulletin No. 16 December 2003*. B. P. D5, 98848 Noumea Cedex, New Caledonia: Secretariat of the Pacific Community, Marine Resources Division, Information Section SPC.

Tisdell, C., & Poirine, B. (2000, June). *Socio-economics of pearl culture: Industry changes and comparisons focusing on Australian and French Polynesia* (pp. 30–60). World Aquaculture Magazine.

Tisdell, C., & Poirine, B. (2007). Economics of pearl oyster culture. Working Papers on Economics, Ecology, and the Environment. Working Paper No. 143. Available at https://www.researchgate.net/publication/46471162_Economics_of_Pearl_Oyster_Culture

Tisdell, C., & Poirine, B. (2008). Economics of pearl farming. In P. Southgate & J. Lucas (eds.), *The pearl oyster* (pp. 473–496). Elsevier Science.
US Department of State. (2003). Compact of free association agreement between the United States of America and the Marshall Islands. Available at https://www.state.gov/documents/organization/173999.pdf.
Weir, T., Dovey, L., & Orcherton, D. (2016). Social and cultural issues raised by climate change in Pacific Island countries: An overview. *Regional Environmental Change*, 1–12.
Williamson, I., & Sabath, M. (1984). Small population instability and island settlement patterns. *Human Ecology, 12*(1), 21–34.

Reviving Traditional Croplands to Improve Community Climate Resilience

Bernie Besebes

Abstract Small islands such as Palau are highly vulnerable to Climate Change. Rising tides and extreme weather events are two climate effects that are negatively impacting Palauan communities. These effects coupled with changing demographics and urban development, have resulted in watershed degradation and declines in agriculture and coastal fisheries productivity. Given Palau's dependence on the health of its ecosystems to its economic and social objectives it is essential that Palau address these vulnerabilities, if its communities are to become resilient in the face of a changing climate. To resolve these challenges, Palau is looking to its past to inform the development of solutions that can be applied now and into the future. One solution is to adapt successful traditional soil conservation practices utilized in taro farming to modern farming and storm water management practices. The benefits of implementation of these agro-ecological farming principles in farming and as guidelines for development include increased food, environmental and economic security which are the cornerstones of resilient communities.

Introduction

This project: *Reviving Traditional Croplands to Improve Community Climate Resilience* is one solution for Palau's communities to build resilience to climate change. Activities undertaken will enhance community resiliency by increasing food security and ensuring ecosystem health of rivers and coral reefs so that they continue to provide the provisioning services required by the Palauan people. A key project outcome is improved governance of watersheds. Project outputs include identifying traditional soil conservation best management practices to manage storm water and storm water management plans which will be trial tested during the course of the project time line and institutionalized for continuous use.

B. Besebes (✉)
Palau Conservation Society, 1811, 96940 Koror, Republic of Palau
e-mail: bbesebes@palauconservation.org

W. Leal Filho (ed.), *Climate Change Impacts and Adaptation Strategies for Coastal Communities*, Climate Change Management,
https://doi.org/10.1007/978-3-319-70703-7_12

Successful implementation of this project will result in state governments with the institutional capacity to manage their natural resources in this changing climate by addressing sedimentation in rivers, streams, and coral reefs. It will result in community members who are food secure therefore resilient with the added value of increased income. More importantly, the project will build awareness on climate change and storm water management as well as provide knowledge and guidance documents that will enable community members to become better stewards of their natural resources thereby improving their resilience to climate change impacts. These tools will enable decision makers in the communities to put measures in place to ensure future developments can occur without sacrificing the environment of each state. The end result will be improved community and ecosystem resilience to climate change by means of reviving traditional landscapes and its associated ecological knowledge to increase food, environmental and economic security of communities in Babeldaob, Palau.

Palau and Climate Change

"As an island nation, we are literally in the frontlines of climate change. Never before in the history of our islands and humanity have we faced a bigger and more universal threat. Climate change has negatively impacted our resources, infrastructure, and livelihoods—not once, but several times in recent years. Our continuity and future existence as a people are now at the mercy of climate change." (*Message from His Excellency Tommy E. Remengesau, Jr., President of the Republic of Palau—Palau Climate Change Policy*).

For centuries, the people of Palau have existed and thrived because of their close ties and reliance on the natural environment and natural resources. In recent years, concern is mounting as we are experiencing changes in weather patterns resulting in strong typhoons, extreme drought, rising sea levels, salt water inundation and intrusion, among others. These concerns are reinforced by scientific findings of the Palau International Research Center (PICRC), a local research institution, along with findings of international groups such as the Intergovernmental Panel on Climate Change (IPCC). These findings demonstrate real and immediate impacts of climate change to our livelihoods.

As a Small Island Developing States (SIDS), Palau faces major threats from the effects of climate change such as typhoons, storm surge, drought, and persistent and prolonged rainfall events that cause major damage to community assets, infrastructure and ecosystems. This also adversely affects the sustainable economic development of Palau. In response, Palau has developed a Climate Change Policy Framework to address these and other threats. The policy has three overarching objectives: (1) enhance adaptation and resilience, (2) manage disaster and minimize disaster risk, and (3) mitigate global climate change by working towards low emission development.

Palau is susceptible to ecological disturbances related to global climate change. The effects of climate change on both marine and terrestrial ecosystems impacts the organisms that depend on them. These effects of climate change contribute to increased seawater temperature, increased air temperature, sea level rise, climate extremes, and changes in weather and precipitation patterns. These factors resulting from climate change lead to ecosystem level impacts that can adversely affect a community's social and economic well-being.

Palau's susceptibility to the effects of climate change is clearly demonstrated by past extreme climate events. In 1998, an El Niño event resulted in significant ecosystem damage in both marine and terrestrial environments in Palau. This event created elevated seawater temperatures that contributed to massive coral bleaching and decline of sea life in near shore areas. To date and to some extent, affected areas are still recovering from the event. In March 1998 at the peak of the El Niño, Palau received the lowest amount of rainfall in over 100 years of records. This resulted in depletion of water supplies, crop failures, and uncontrolled wildfires on some islands.

Palau experienced drier than normal conditions in March 2016 with Koror recording 2.68 in. of rainfall which is lower than the minimum threshold of 4–8 in. required to meet monthly water needs. Historical data for the US Affiliated Pacific Islands (USAPI) demonstrated that the months of October 2015 to March 2016 were the driest for Koror out of 65 years of data collected (NOAA State of the Climate: Drought for March 2016).

This led to approximately 80% of Palau experiencing decreased water supply resulting in crop damage, wildfire outbreaks and the shutting down of the Ngerimel Dam, one of the two water sources for the Koror-Airai Water System. Water rationing was implemented for 5 hours each day which was eventually reduced to 3 hours per day due to diminishing water level at the Ngerikiil Dam (Drought Report, Republic of Palau, June 2016).

The Ring of Fire or typhoon belt is a large Pacific Ocean region between 10° and 40° north latitude where typhoons, volcanic eruptions and earthquakes occur. Palau is located 7° north latitude and therefore considered outside of the typhoon belt. However, the country recently experienced two major typhoons. Typhoon Bopha hit Palau in December 2012 and Typhoon Hyan followed 11 months later in November 2013. Both typhoons went on to cause substantial death and destruction in the Philippines. In Palau these typhoons caused significant wind damage to homes and trees, storm surge flooding coastal areas, heavy rains, and alterations in lagoon channels. Typhoon Haiyan caused particularly severe damage in Kayangel state. Nearly all structures in the state were destroyed, the vast majority of trees were toppled by high winds, taro patches were inundated with seawater, and the drinking water supply was contaminated with saltwater (Republic of Palau Fifth National Report 2014).

The following table provides a snapshot of Palau's climate vulnerabilities. It illustrates the various effects of Climate Change, implications and possible impacts on Palau's communities (Table 1).

Table 1 Climate change effect, implications and possible impact on communities

Effect	Implications	Possible impacts on biodiversity
Increased seawater temperature	• Coral bleaching • Decline of fisheries	• Loss of coral species, organisms dependent on corals • Habitat loss • Fish nursery decline
Increased average air and ocean temperature	• Increased energy consumption • More severe weather events • Changes in water quality	• Destruction and alteration of habitat by storms • Coral bleaching • Import of petroleum to power air conditioning
Increase in sea-surface temperature	• Increased frequency and severity of tropical storms and typhoons	• Coral bleaching • Habitat loss due to storms
Sea level rise	• Flooding • Coastal erosion • Salt intrusion in taro fields • Damage to low-lying hamlets and infrastructure	• Loss of terrestrial habitat • Loss of agricultural area • Contamination of freshwater lenses
Climate extremes	• Droughts, storms and floods	• Increased susceptibility to invasive species
Changes in precipitation	• Decreased reliability of water supply	• Wild fires • Agricultural decline

(PCS Grant Agreement)

The Palau Climate Change Policy Framework further highlights Palau's vulnerability to both human-induced and natural disasters, which may be exacerbated by the impacts of climate change. These impacts include:

- Sea level rise of 3.0–6.3 in. by 2030;
- Increased frequency and severity of extreme weather events (particularly rain events), with increasing numbers of extreme rainfall days and rainfall intensity;
- Increased inundation, storm surges, erosion, and other coastal hazards;
- Changes to quantity, quality, and variability in seasonal and annual flows of surface and underground water;
- Adverse effects to subsistence and commercial agriculture and fisheries;
- Reduced food security due to extreme weather events and changes in seasonal weather patterns;

Palau's food production and watershed are predicted to be negatively affected due to the impacts of climate change. Today, increased frequency and severity of rainfall, salt water inundation and intrusion, and droughts affect taro production-based livelihoods which is a sector managed by women. Furthermore, Palau's watersheds are continuously degraded as a result of poorly planned developments which often result in soil erosion and sedimentation into estuaries and coral reefs. Sustainable economic development is negated by this degradation of Palau's terrestrial and marine ecosystems. More importantly, these impacts of climate change greatly threaten the food, environmental, and economic security of Palau.

Palau's Agriculture and Climate Change

Through years of colonization, Palauans have been exposed to, influenced, and now practice agricultural methods developed in and used in other parts of the world including Spain, Japan, US, and more recently, China and the Philippines. These agriculture techniques have become widespread however it is now clear that they are not suitable or sustainable for Palau's soil and climate conditions.

Traditional Palauan farming techniques along with other sustainable farming methods produce erosion rates that fall well below the erosion conservation goal of 11.2 metric tons per hectare per year (5 tons/acre/year). However, unsustainable techniques have been found to result in erosion rates up to 65 times higher than the erosion conservation goal.

Research conducted by the Palau International Coral Reef Center (PICRC) show that "land-based development activities have a direct impact on the amount of sediment that goes into rivers and eventually ends up on coral reefs. The amount of sediments being released into the rivers and reefs on Babeldaob Island, Palau, depended on the degree of development within adjacent watersheds" (Golbuu et al. 2003).

To persist in these unsustainable techniques will continue to result in high rates of erosion therefore continued decrease to agricultural productivity. This not only leads to food insecurity, it also leads to environmental and economic insecurity for the communities of Babeldaob thereby exacerbating the impacts of climate change for these communities.

Mesei and Taro Cultivation

According to the legend of *Iluochel*, taro cultivation has been around since the first people arrived in these islands dating back to approximately 3500 years before Christ. Iluochel, a demi-goddess carved out the first *Mesei* or taro patch in Angaur and continued to develop *Mesei* as she migrated north through Peleliu and unto Babeldaob. It is this belief of divine origin that makes the *Mesei* sacred to the people of Palau. Yet another proverb: "*A Mesei a delal a telid*" literally means "The Taro Patch is the mother of our breath". In other words, the *Mesei* is the essence of our being providing sustenance for Palauan families, taro used in cultural ceremonies, and provides a source of income for most families especially in the rural areas of Babeldaob.

The *Mesei* is a paddy like system in Palau's wetlands where taro is cultivated. This is the most productive and sustainable method of food production in Palau. A domain dominated by women, the skills for taro cultivation is handed down from mother to daughter and traditionally, a source of pride for Palauan women. In most cases, the same area has been farmed in taro for hundreds if not thousands of years.

Within a *Mesei* system are many individual taro patches that share one water source that is irrigated through the taro patches by means of a series of ditches and dikes. This makes taro production a shared community activity as individuals must work together to ensure the "*bong*" or waterways are always clear and functional to ensure fresh water flows to each taro patch. The health of the whole system is dependent on the health of individual taro patches to ensure weed and pests are controlled and not transferred to adjoining patches.

The taro patches are arranged where one is always slightly higher than the other to allow for water irrigation to all patches from a water source upstream. Narrow water channels are built in and maintained to ensure fresh water flows from one patch to another. The in and out flow of water is controlled by each taro cultivator depending on the stage of growth of the taro. This also serves to slow the rate of flow of water traveling through taro patches resulting in sediment particles in the water settling in the taro patches or waterways thereby filtering the water.

In recent studies, the *Mesei* has been proven to help protect the environment from ridge to reef. A research paper titled "*Palau taro fields and mangroves protect the coral reefs by trapping eroded fine sediment*" published by the Palau International Coral Reef Center found that taro fields are able to trap sediments similar to the sediment trapping capacity of mangroves.

> A 4-month long field study was conducted to quantify the sediment accumulation rate for three different types of taro fields and to determine their sediment trapping efficiency. The results showed that the taro fields have the capacity to trap up to 90% of sediments (Koshiba et al. 2013).

Taro cultivation is labor intensive requiring continuous soil conditioning through compost and mulching as well as water control within the taro patch. The method of conditioning the soil is called "*Smalo*", the act of turning over the mud, inserting "*Ramk*" (green compost) and replacing the mud. The mud that is dug up (average of 2 feet depth) is turned upside down to bring the decomposed green to the surface thus providing nutrients to the newly planted taro shoots. Once planted, the "*Dekedek*" (green mulch) is applied to the top of the mud surrounding the taro shoots for several purposes: additional nutrients, hold moisture, and prevent growth of weed. These methods among others lead to the *Mesei* being the most productive and sustainable agricultural system in Palau.

According to Dr. Aurora Del Rosario, a researcher from the Palau Community College—Cooperative Research Extension, the average taro patch size in a *Mesei* system is 12 ft by 12 ft and may yield an average of 50 lb of taro per year. The average number of patches within a *Mesei* is 20 patches. Taro has a very high cultural value and therefore indispensable in cultural exchanges (Del Rosario et al. 2015).

To date, the taro is in high demand because it's a major staple food in the home, used in various traditional ceremonies, and provides a source of income especially for Palauans living in the rural areas of Babeldaob. Lastly, Palau is experiencing a boom in tourism in recent years creating a demand for taro from hotels and restaurants. These are just some of the reasons why the current demand for taro far exceeds the supply.

Changing Demographics and Urbanization

Since Palau's independence in 1994, all kinds of opportunities are becoming available to young Palauan women including higher education and formal employments that turn into careers. These opportunities enable young Palauans to live abroad during their college years and eventually gain employment outside of Palau. Additionally, the Compact of Free Association allows citizens of Palau to legally live and work in the US and its affiliated states. These and other reasons are resulting in out-migration of young Palauans.

In Palau, Koror is the hub of commerce and employment as well as the seat of Palau High School and Palau Community College, the only public high school and higher education institute on island. For these reasons, many young Palauans choose live and work in Koror. This leaves only grandparents and grandchildren in the rural areas of Babeldaob.

This situation is slowly changing due to the completion of the Compact Road in 2008. The Compact Road is a 52 mile road that circles the island of Babeldaob connecting all 10 states to Koror. Part of the benefit package of the Compact of Free Association between Palau and the US, this road was affectionately called "*Bemrei Highway*" or Come Back Home Highway. One of the main purposes of this road was to enable commerce to spread to Babeldaob enabling the return of Babeldaob people to their home states. As predicted, many Palauans are moving back to their home states resulting in land clearing for the building of homes and *Sers* or dry land farms.

Due to changing demographics and urbanization, the communities in Babeldaob face a double-edged issue. On one hand young women are choosing to join the work forces, working 40+ hours per week in an office leaving no time or interest to work the taro patch. This is partly due to the intensive physical labor required in cultivating a taro patch not to mention having to "get down and dirty" as an average taro patch is two to three feet deep of soft, wet mud. Palauan women choosing to have a career and earn a salary results in a growing number of abandoned taro patches which exacerbates the community's vulnerability to climate change impacts as it threatens food and environmental security for these communities.

On the other hand we have people who are financially secure and are now able to move back to Babeldaob. This results in land clearing to build houses and for *Sers*. Most states in Babeldaob do not have clear regulations regarding land development.

In this instance, this leads to soil and sedimentation deposits in the rivers and eventually draining into the reefs. The urbanization of Babeldaob without clear development regulations results in poorly planned developments that leads to damage to the watersheds resulting in reduced quality and quantity of water supply to the community. This further increases the community's vulnerability to climate change impacts.

Traditional Soil Conservation

Traditional soil conservation (TSC) consists of agriculture practices that are designed to make Palau's poor soils productive and manage the climate effects of too much rain and tropical heat on agricultural production. These TSC practices are utilized in the cultivation of taro. To condition Palau's poor soils taro farmers use a constant source of organic matter in the form of fresh leaf bulk called "*Ramk*" pushed down into the bottom layer of the soft mud and leaf cover called "*Dekedek*" which is layered on top of the surface of the wet paddy after the taro is planted. Every time taro is planted utilizing this process, bringing up mud from the bottom of the patch that consists of organic matter and pushing down a fresh batch of leaf bulk, the soil becomes more fertile and therefore productive. This cyclic process is how the *Mesei* or taro farm is maintained. There is little soil erosion in this system and maintaining an organic matter layer helps to maintain soil fertility, provides nutrients, and controls weeds. The quality of the taro depends on the amount of water in the paddy. Too much or too little water yields soggy tubers that are considered low grade and at times inedible. The women who work the farms control the water flow and maintain a healthy amount for best quality crops. TSC measures also incorporate agro-forestry practices around taro farms. Trees such as the Pongamia pinnata are nitrogen fixers and are usually found planted along the dykes separating taro farms. The leaves of the pongamia tree are especially good for use as leaf bulk in the taro patch because as they decompose they release nitrogen into the soil which is then absorbed by the taro plants. Additionally the plants around the *Mesei* provide shade and hold moisture in the ground during the drier seasons. The health of the taro farm is dependent on the function of water, trees, shrubs, and grasses which ensure proper fertilization, crop production, and quality of taro corms. (PCS Grant Agreement)

According to the IPCC Report 2014, more than 70% of agriculture is rain fed. This suggests that agriculture, food security, and nutrition are all highly sensitive to changes in rainfall associated with climate change. Adaptation outcomes focusing on ensuring food security under a changing climate could have the most direct benefits on livelihoods, which have multiple benefits for food security, including enhancing food production, access to markets and resources, and reduced disaster risk. Effective adaptation of cropping can help ensure food production thereby contribute to food security and sustainable livelihoods in developing countries, by enhancing current climate risk management (IPCC Report 2014).

The Resulting Crisis

Sediment arising from erosion on agricultural land lacking any conservation practice is particularly problematic because farming is a key driver of food security. To date, *Sers* or dry land farms, unlike taro farms are contributing to Babeldaob's sediment budget. *Sers* utilize introduced farming practices that aren't always compatible with Palau's soil profile. Additionally, a vast majority of *Sers* occurs on slopes that are cleared and planted with crops such as tapioca, sweet potatoes, other root crops and vegetables. Most *Sers* occur on slopes because Babeldaob Island is characterized by hilly terrain.

The challenge of farming on slopes is exacerbated by Babeldaob's highly acidic soils making large-scale agriculture unfeasible. In highly acidic soils, nutrients dissolve quickly and leach away leaving the soil unsuitable for large-scale farming. Most dry land farmers use fertilizers to increase soil fertility even though it can be quite costly. Additionally, widespread open burning is a common but unsustainable farming practice that releases nutrients resulting in unproductive crops. This is because open burning depletes soil organic matter, kills microorganisms, and leads to long-term land degradation.

Land degradation coupled with increased rainfall events contributes to erosion and sedimentation in estuarine and coral reef systems leading to declines in water quality and productivity of reefs resulting in environmental and food insecurity. As well, sustainable economic development for Palau is greatly dependent on the health of its environment. In this connection, the implementations of best management practices that are Palau-specific and based on traditional knowledge are necessary in order to secure the food, environment, and economic security of the rural communities of Palau. Sustainable economic development will be enabled for these climate resilient communities as agriculture, tourism and other industries will further develop and thrive.

Project Background

Reviving Traditional Croplands to Improve Community Climate Resilience is a project funded by USAID and is being implemented by the Palau Conservation Society (PCS). The timeline for this project is from August 2016 to June 2018. This project focuses on three of the 10 states of Babeldaob, Palau, namely Melekeok, Ngarchelong and Ngaremlengui.

These three project sites were selected based on climate related impact to community landscapes and the community and local government's existing capacity to respond and address those impacts. The village of Ngerubesang in the state of Melekeok was selected because this village is located along the shoreline and taro farms are susceptible to salt water intrusion and inundation, making them vulnerable to rising sea levels associated with climate change. The village of Imeong in Ngaremlengui State was selected as a project site because a number of taro farms in this village are susceptible to flooding making them vulnerable to projected intense rainy events associated with Palau's climate change profile. The villages of Ngebei and Ngriil in Ngarchelong State were selected because they have taro farms that have consistently been cultivated in the traditional manner and can serve as a control site for project field investigations. Finally, all the three states are rural communities with very little development and minimal institutional structures to manage natural resources or to address impacts of climate change on those resources and as such would benefit from the lessons in storm water management best practices work arising from this project.

Project activities are composed of strategies and actions to "*enhance adaptation and resilience*" (objective 1, Palau Climate Change Policy) with specific emphasis focused on the development of climate change-resilient agriculture and watershed management strategies and interventions. In other words, we are looking to the past to inform the development of solutions that can be applied now and into the future.

Palauans rely heavily on ecosystem services and goods for their livelihoods and cultural practices therefore ecosystem degradation and its associated impact on water quality and biodiversity poses a major threat to communities as it leads to Palau's food, environmental and economic insecurity. This project will address the climate associated drivers of food, environmental and economic insecurity. However, project efforts will be focused on mitigating inadequate agriculture and watershed management and institutional weaknesses of state governments to effectively manage natural resources necessary for a resilient community.

These challenges will be addressed by utilizing a mixture of traditional and modern resource management in an effort to improve watershed management and agriculture in Babeldaob. The major objective of this project is to increase the adaptive capacity of Palau's rural communities by strengthening community and ecosystem resilience to the impacts of climate change. Achieving this objective comprise actions to revive traditional landscapes into productive agricultural landscapes. This includes efforts to document associated traditional ecological knowledge of soil and water conservation derived from a long tradition of taro farming into best practices that can be integrated into existing watershed and resource management regimes. Reviving traditional croplands and sustainably managing watersheds are key actions in Palau's efforts to increase food, environmental and economic security of the rural communities in Babeldaob, Palau.

Additionally, the objectives of this project are aligned with both Palau's Climate Change Policy Framework and Palau's agriculture and aquaculture policy. Specific action of this project will enable Palau to meet a number of objectives in both policies.

Project Goal

The overall goal of this project is to improve community and ecosystem resilience to climate change by reviving traditional landscapes and its associated ecological knowledge to increase food, environmental and economic security in Melekeok, Ngarchelong, and Ngaremlengui states of Babeldaob Island. Enhancing community resiliency by increasing food security and ensuring ecosystem health of rivers and coral reefs will result in continued provisioning services required by the Palauan people. Project outcomes such as mainstreaming of the taro farmer/state partnership have built in sustainability features because there is interest and means for taro production to occur. The project will also document traditional soil conservation methods, trial test methods on *Sers*, and work with project partners to produce guidance and knowledge documents in an effort to provide reliable information to enable sound decision and policy making for communities and decision makers of the states of Babeldaob. Finally, these tools and guidance will allow community members to become better stewards of their natural resources thereby improving their resilience to climate change impacts.

Project Components

This project will utilize an Ecosystem-based Adaptation (EbA) approach via three interrelated strategies or components.

Component 1: Improve existing institutional and personal capacities of local or state governments to manage their watersheds via storm water management.

Component 2: Implement priority actions to promote traditional wetland taro farming. These actions will focus on reviving neglected taro patches through the creation of community partnerships and optimizing dry land farming using traditional soil conservation measures.

Component 3: Strengthen knowledge management of Climate Change across all local sectors so that communities are empowered to take local action to address global climate change impacts.

The overall project rationale asserts that effective water regime management in both agriculture production and watershed management must be incorporated in the natural resource management in Palau. As well, this water regime management must utilize traditional soil conservation measures. Traditional soil conservation measures can be used to slow down the rate of water flow to ensure that crops get the water and nutrients they need and used to reduce runoff to ensure that sediment drops from the water column so that it does not enter rivers, streams and the coral reefs. Effective management of water resources through agriculture irrigation or storm water management results in enhanced watershed management. Achieving effective watershed management improves the health of both terrestrial and marine resources, enhances community development and leads to community resiliency.

A key project outcome is improved governance of watersheds. Project outputs include best practices to manage storm water and storm water management plan which will then be implemented during the project life time and then institutionalized for continuous use. Additionally, the Environmental Quality Protection Board (EQPB) as a technical partner and project implementer has indicated that it will institutionalize the traditional soil conservation measures as best practices for erosion control for farming and other related development at the national level.

Activity Narrative

Implementation of this project's three components and achievement of its associated objectives provides the foundational building blocks of community sustainability and resiliency in Palau because it ensures that communities are able to develop their economies and society without compromising the health of their natural resources.

- **Objective 1** applies Palau's traditional ecological knowledge of soil conservation to address the problems of sedimentation that is pervasive throughout Babeldaob. These traditional soil conservation measures will be identified and documented during the implementation of component 2 activities of this project. Implementation of storm water management measures utilizing locally relevant best management practices such as the traditional soil conservation measures identified in component 2 are a low technological and low cost measure to ensure that land based activities do not result in erosion and sedimentation. These measures have a high sustainability factor because they are cost effective and easily applied and do not require any engineering knowledge as well as high budgets.
- **Objective 2** identifies and documents traditional soil conservation measures that are utilized in taro farming which can be modified for dry land farming to ensure soil productivity and minimize runoff. These modified practices can then be utilized to reduce runoff from farms, road construction and building development. Additionally, component 2 has the added benefit of increasing food production in the communities as a result of utilizing traditional soil conservation measures to revive non-productive farms.
- **Objective 3** is intended to capture the lessons learned from implementation of components 1 and 2. The overall project rationale asserts that natural resource management in Palau needs to incorporate effective water regime management in both agriculture production and watershed management and that this water regime management is one that utilizes traditional soil conservation measures. Traditional soil conservation measures can be used to slow down the rate of water flow to ensure that crops get the water and nutrients they need and used to

reduce runoff to ensure that sediment drops from the water column so that it does not enter rivers, streams and the coral reefs. Effective management of water resources through agriculture irrigation or storm water management results in enhanced watershed management. Achieving effective watershed management improves the health of both terrestrial and marine resources, enhances community development and leads to community resiliency.

Conclusion

Palau identified its main vulnerabilities due to climate change in a National Communication to the United Nations Framework Convention on Climate Change (UNFCC). These vulnerabilities are related to: increased drought and storm activity; extreme high tides; sea level rise; coastal erosion; habitat fragmentation; sea surface temperature rise; and coral bleaching. Palau has also identified a number of potential adaptation actions in the areas of water and agriculture, among others. Adaptation measures for water include the following: improved management and maintenance of existing water supply systems is a high priority; centralized water treatment in urban centers; catchment protection and conservation; and drought and flood preparedness strategies. For agriculture, measures include the following: identify and document the uses, potential uses and preferred growing environment for trees and plant species in order to better enable selection of species suited to a particular physical environment; introduction of salt-tolerant root crops for use in low-lying areas; breeding more drought resistant cultivars and crops for use in drought prone upland areas; introduction of alternative cultivation practices such as use of irrigation and raised-bed systems; improved soil and water conservation practices; promote use of Agroforestry; preservation and dissemination of traditional knowledge; and diversification of subsistence crops (Palau Climate Change Profile V2—June 2013).

The goal of this project reflects Palau's adaptation measures. Activities undertaken in this project will enhance community resiliency by increasing food security and ensuring ecosystem health of rivers and coral reefs so that they continue to provide the provisioning services required by the Palauan people. A key project outcome is improved governance of watersheds. Project outputs include best practices to manage storm water and storm water management plan which will then be implemented during the course of the project life time and then institutionalized for continuous use. Additionally, the EQPB as a project partner and implementer has indicated that it will institutionalize the traditional soil conservation measures as best practices for erosion control for farming and other related development at the national level.

Successful implementation of this project will result in state governments with the institutional capacity to manage their natural resources in this changing climate by addressing sedimentation in rivers, streams, and coral reefs. It will result in community members who are food secure therefore resilient with the added value of increased income. More importantly, the project will build awareness on climate change and storm water management as well as provide knowledge and guidance documents that will enable community members to become better stewards of their natural resources thereby improving their resilience to climate change impacts. These tools will enable decision makers in the communities to put measures in place to ensure future developments can occur without sacrificing the environment of each state. The end result will be improved community and ecosystem resilience to climate change by means of reviving traditional landscapes and its associated ecological knowledge to increase food, environmental and economic security of communities in Babeldaob, Palau.

Sustainable economic development for Palau is greatly dependent on the health of its environment. In this connection, the implementations of best management practices that are Palau-specific and based on traditional knowledge are necessary in order for tourism, agriculture and other industries to develop and thrive. This is the basis of this project.

References

Climate Change Profile, Republic of Palau Version 2 (June 2013), p. 15.

Del Rosario., Aurora, G. (2015). Taro Production in Palau, College of Micronesia land grant programs.

Drought Report, Republic of Palau. (June 2016).

Golbuu, Y., Victor, S., Wolanski, E., & Richmond, R. H. (2003). Trapping of fine sediment in a semi-enclosed bay, Palau, Micronesia. *Estuarine, Coastal and Shelf Science, 57,* 941–949.

Koshiba, S., Besebes, M., Soaladaob, K., Isechal, A. L., Victor, S., & Golbuu, Y. (2013). Palau's taro fields and mangroves protect the coral reefs by trapping eroded fine sediment. *Wetlands Ecology and Management*. Palau International Coral Reef Center.

McLean, R., Agard, J., Briguglio, L., Lefale, P., Payer, R., & Sem, G. (2014). *IPCC Report on Climate and Small Islands.*

NOAA National Centers for Environmental Information, State of the Climate: Drought for March 2016 (published online April 2016).

Palau Climate Change Policy: For Climate and Disaster Resilient Low Emissions Development. Government of Palau 2015, p. 3.

PCS Grant Agreement, USAID Pacific-American Climate Fund Agreement No. PACAM-16-0017.

Republic of Palau Fifth National Report to the Convention on Biological Diversity. Office of the Environment, Response and Coordination (OERC), 2014.

Coping with Future Coastal Floods in Denmark—Advancing the Use of Global Frameworks

Martin Jebens and Carlo Sorensen

Abstract The main aim of Disaster Risk Management and Climate Change Adaptation is to lower the risk for the population and the society at large. Risk assessments constitute an important part of flood risk management and their quality is crucial to well-informed decision making. This requires an in-depth understanding of the society and its vulnerabilities. Often attention to the flood risk and vulnerability in developed countries is absent due to the assumption that society can cope with disaster; For Denmark, a mixed methods' research inquiry reveals that this is not always the case. In a critique of current Danish approaches to deal with Disaster Risk Management and Climate Change Adaptation including coordination and planning, the paper proposes a new pathway for coping with the risks of coastal floods: Global frameworks like the Hyogo and Sendai tailored to suit Danish conditions may serve to mainstream Disaster Risk Management and Climate Change Adaptation and provide for a holistic, sustainable and more consistent national approach. In addition, such an approach will advance coordination and produce more elaborate guidelines and uniform progress report mechanisms to local and national actors when dealing with flood risks and adaptation.

Introduction

In the early morning of 14 November 1872, the disastrous effect of the worst coastal floods impacting Denmark in recent time became evident. When waters receded the storm had killed more than 250 people around the Southwestern Baltic Sea;

M. Jebens (✉)
Danish Coastal Authority, Coast and Climate, Lemvig, Denmark
e-mail: mje@kyst.dk

C. Sorensen
Technical University of Denmark, DTU Space, Kongens Lyngby, Denmark

W. Leal Filho (ed.), *Climate Change Impacts and Adaptation Strategies for Coastal Communities*, Climate Change Management,
https://doi.org/10.1007/978-3-319-70703-7_13

destroyed more than 3000 houses, and made 15,000 people homeless (Dahlberg et al. 2017). The event revealed a lack in society to mitigate large flood events, insufficient technologies to support mitigation, and a lack of coastal resilience in the early days of industrialization. The aftermath of the 1872 event brought about change through the first Coastal Act passed by the Danish Parliament (1874), and larger dikes were constructed in some of the affected areas. There is little recollection of the 1872 event today and, in fact, Denmark last experienced fatalities related to coastal floods in 1923. Despite numerous flood events over the past century, the risk has gradually increased in coastal communities and urban areas. Also, there is a lack of planning measures to reduce the current risks and adapt to effects of climate change. Thus the level of risk awareness among key stakeholders and particularly the civil society is low (Sorensen and Jebens 2015).

Two minor storm surges 25–27 December 2016 and 4–5 January 2017 impacted different coastal areas of Denmark. Despite their inferiority to the 1872 event, approximately evaluated at 50 year return water levels as opposed to a 1000 year event in 1872 (Sorensen et al. 2017), both decision makers and the press responded strongly to the situation and put society in a state of alert. Seen in this perspective alone, the insufficient understanding in society of the current and future vulnerabilities with climate change is remarkable.

In Denmark and globally, the challenges faced from rising mean sea levels and more extreme storm surges are immense, and are exacerbated by the ongoing developments and urbanization taking place mainly in coastal areas (e.g. Grinsted et al. 2015; IPCC 2012, 2013; Nicholls 1995). Coordination and planning is fundamental to implement appropriate Disaster Risk Management (DRM) and Climate Change Adaptation (CCA) measures. Building on prior work of the authors, the paper investigates the level of structural planning in Denmark to address vulnerabilities and impacts from climate change in coastal areas. The past Hyogo Framework for Action 2005–2015 (UN 2005) and the current Sendai Framework for Disaster Risk Reduction (UN 2015a) together with the Paris Agreement (UNFCCC 2015) may offer ways forward to build a holistic and sustainable system which can incorporate and reflect Danish conditions. The frameworks are implemented and used in less developed countries to improve planning and coordination but often have a small impact in more developed countries. Objectives of the paper thus are: can Denmark decrease its risk of flooding due to climate change by implementing internationally acknowledged frameworks, and, how may these frameworks be coherently applied?

The research uses a mixed methods' approach where a quantitative review of existing work on Danish adaptation to coastal flood hazards which e.g., includes an assessment of national progress reports on the implementation of the Hyogo Framework for Action is combined with qualitative interviews. Information based on interviews with key stakeholders among emergency managers is included to discuss pathways ahead for the CCA and risk reduction in Denmark that extend between local and global level methodologies and frameworks.

Denmark—Geographical Setting

Denmark is a low-lying North-European country (Land area c. 43,000 km^2; Population c. 5.7 million) which borders the Baltic Sea and the North Sea. Denmark has no large rivers (at a European scale) and the current flood hazards from cloudbursts or heavy precipitation are generally low. Natural hazards which may have a profound impact on the Danish society are storms and storm surges/ coastal floods. Location dependent surge levels range between two and five meters above normal water level along the 7300 km long and diverse coastline. The 10 largest cities are located at the coast and 80% of the population lives within three kilometers from the sea. The country has many unprotected coasts but also large stretches with erosion protection (sand nourishments, groins, revetments etc.) and flood protection measures (sea walls, artificial dunes, dikes etc.). Many coasts are flood prone with additional areas, often in relation to towns, potentially becoming at risk over the next century with sea level rise. The maps in Fig. 1 show differences in areas vulnerable to coastal floods today and in 2115 with sea level rise. DRM and CCA are inter-connected and measures to prevent one should also have the ability to lower the impact from the other. In other words, mitigation and adaptation measures must be integrated, coordinated and well planned to avoid risk transfer along the Danish coastline.

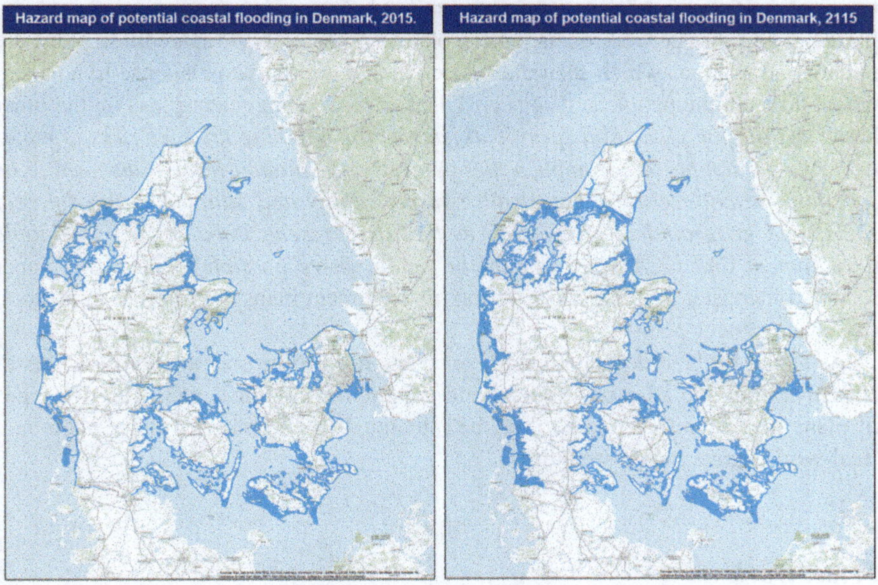

Fig. 1 Maps of Denmark with low-lying areas liable to coastal floods shown in blue today (left) and in 2115 from extreme sea level statistics and the RCP 8.5 scenario (DMI 2014; IPCC 2013; Sorensen et al. 2013)

Background

The section provides a brief overview of definitions used and introduces the rationale behind this paper by building on prior research, which point to the lack of coordination and planning in Denmark to mitigate flooding and adapt to climate change in coastal regions. This is followed by an account of the Danish governance structure and of the current work status on DRM and CCA in relation to coastal floods that include the national engagement in international treaties and frameworks.

Definitions

To distinguish between concepts and different types of plans the following terms and definitions are applied according to the United Nations Office for Disaster Risk Reduction (UNISDR; unisdr.org/we/inform/terminology): Disaster management is understood as "*The organization, planning and application of measures preparing for, responding to and recovering from disasters*" and further "*it focuses on creating and implementing preparedness and other plans to decrease the impact of disasters and build back better.*"

It is reasonable to assume that climate change is an event giving rise to a global lack in coping and thus to treat climate change as a slow-onset disaster and use terminology from the disaster management field. This paper discusses preparedness planning and disaster risk management plans, respectively. Preparedness plans are understood as plans which secure an effective and appropriate response to a hazard which threaten the people and society. Disaster risk management plans on the other hand "*set out the goals and specific objectives for reducing disaster risks together with related actions to accomplish these objectives. National-level plans need to be specific to each level of administrative responsibility and adapted to the different social and geographical circumstances that are present. Linkages to sustainable development and CCA plans should be made where possible*". Since this paper focuses on coastal flooding, disaster risk management plans are referred to as flood risk management plans.

As stated by UNISDR, national-level plans need to be specific across levels of administration and adapted to social and geographic circumstances. The two types of plans may therefore relate to, or, be integrated in a number of other plans in a 'real-world' governance structure.

Prior Work

When acting upon disaster and climate change the work must be organized and relate to the hazards, vulnerability and hence the risks faced in society from both events. Current storm surges are rapid-onset disasters, whereas climate change is a slow onset-disaster affecting the risks ahead. Here, risk assessments constitute an important part and the quality of such assessments is crucial to informed decision making. This requires an in-depth understanding of the history and vulnerability of the society in focus. In order to create holistic and sustainable adaptation and mitigation measures that target population needs, coordination and planning at different levels is called upon which ideally should build on already established frameworks. In a number of papers the authors have identified main concerns in the way the Danish society addresses DRM, CCA, and coastal flood risk. Sorensen and Jebens (2015) pointed to planning and coordination needs at all levels in order to improve coastal CCA in Denmark based on a municipal level study. In addition, the uncertainty about which position and role the national government will take in relation to future implementation and advancement of DRM and CCA in Denmark is still pertinent. Sorensen and Jebens (2015) identified complex legislation, too, as a potential hindrance to advance public awareness and coordination between DRM and CCA. Jebens and Sorensen (2016) point to new policies, legislative changes, preparedness and capacity building as means to advance coordination between DRM and CCA. Furthermore, a lack of coordination and planning was identified to hamper a shift towards non-structural measures for CCA and risk reduction (Jebens et al. 2016). Figure 2 provides an overview of vulnerabilities in a Danish flood risk and climate change context.

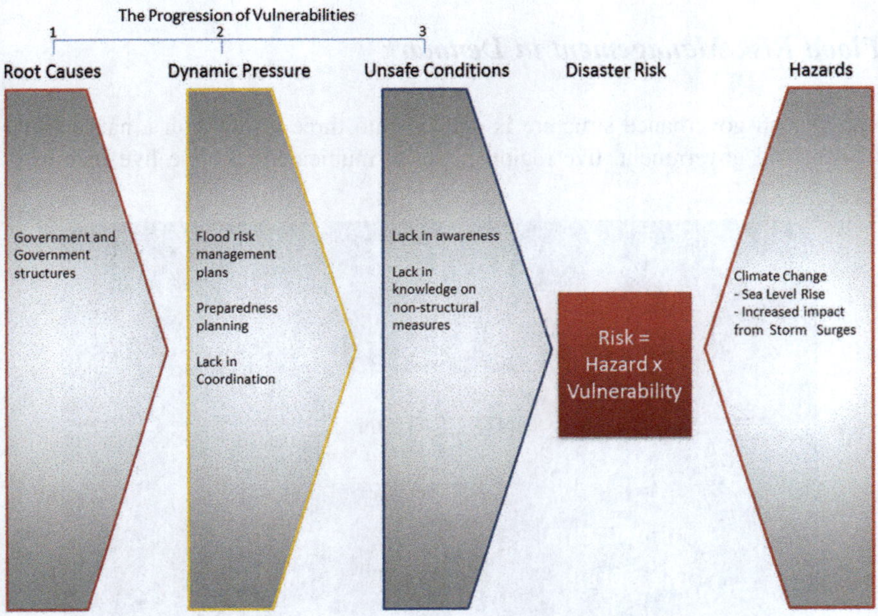

Fig. 2 The Crunch Model after Wisner et al. (2004). The graphic shows the progression of vulnerabilities in Denmark due to climate change and flood hazards (modified from Sorensen and Jebens 2015)

A detailed understanding of a flood risk situation calls for assessments (ACAPS 2014). Risk assessments provide the fundamental basis for CCA and DRM and are essential to how decisions are made on an informed basis in multiple sectors by quantifying the potential losses and hereby improving the resilience to extreme weather. The selection of mitigation or adaptation measures depends on the assessment of current and future risks. This can lead to development of policies and actions which will be required to work at different time-scales and have to take into account future risks as a result of changes in social, economic and environmental conditions (GFDRR 2016).

The assessment is also the basis for defining the measures needed for adaptation to future flooding. In Flood Risk Management (FRM) the adaptive measures can be divided into four groups (Davis 2015; UN 2015a): structural measures, non-structural measures, semi-structural measures and environmental measures as seen in Fig. 3. The measures should be used in an integrated approach to mitigate and adapt to future flooding and will lead to more flexible ways to protect the population. Structural measures are physical constructions to reduce the impact from a hazard. In this context this includes dikes, groins and breakwaters as well as CCA initiatives (engineering new buildings, retrofitting, strengthening of infrastructure and critical infrastructure). Non-structural measures do not involve physical constructions but reduce impacts by implementing FRM and preparedness, assessments and planning, legislation, land use and public awareness.

Flood Risk Management in Denmark

The Danish governance structure is divided into three levels with a national parliament and government, five regions, and 98 municipalities. The five regions do

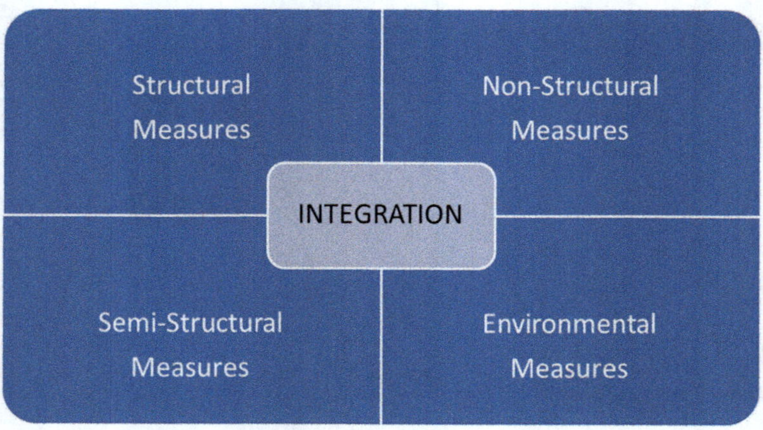

Fig. 3 The measures which should be used in an integrated approach to adapt to future flooding can be divided into four types: structural, non-structural, semi-structural and environmental measures

not possess formal obligations in relation to the overall FRM but may act on a voluntary basis to engage in and facilitate initiatives with their respective municipalities. One such example is the 'Coast to Coast Climate Challenge (2017–2022)' (c2ccc.eu) partnership project involving municipalities and water sector utility companies in the Central Denmark Region (CDR) as well as universities. With a total budget of approximately 13.5 million US$, the CDR-led initiative is, by far, the largest in Denmark with a regional scope to: formulate a common long term strategy; to coordinate and implement municipal solutions and activities, and to identify and improve resources and capacities amongst citizens, municipalities, and the water sector.

The national government and its ministries have over the years participated in and reported on United Nations Frameworks and Commissions like the 'Brundtland Report' (UN 1987) and the Assessment Report cycles of the Intergovernmental Panel on Climate Change (e.g. IPCC 2013). Denmark has ratified the 'Paris Agreement' (UNFCCC 2015) and committed to the 'Sendai Framework of Disaster Risk Reduction' (and participated in the Hyogo framework) (UN 2005; UN 2015a), and 'Sustainable Development Goals' (ICSU 2015; UN 2015b).

In relation to the IPCC-work and according to provisions, national reports have been published in English and Danish addressing both local and global issues related to climate change and including translations and summaries of international work. Also, independent organizations and universities have spawned numerous publications on climate change and sustainable development over the past three decades. In 2004, the need to implement CCA in all future planning was explicitly addressed (Miljostyrelsen 2004) and in 2008 a strategy for CCA in Denmark was published (Danish Government 2008). The strategy has very few concrete recommendations but does point to the need of ad hoc adaptation and cross-sectoral coordination. Much of the initiated work, especially regarding the cross-ministerial focus and collaboration on CCA, lost momentum immediately succeeding the COP15 meeting in Copenhagen in 2009 (UNFCCC 2009).

In 2011, the Danish government put forward legislative acts for the municipalities to individually make climate adaptation plans by late 2013 and incorporate these in their municipal planning. The main focus is on cloudburst events and the sewer systems and the plans do not deal with CCA in a broader sense, however (cf. Jebens et al. 2016). Based on a governmental decision, these plans are not likely to be reevaluated or followed up and therefore may have a limited impact. In relation to coastal floods and adaptation, the Danish Coastal Authority (DCA) under the Ministry of Environment and Food has published guidelines for coastal climate adaptation (Sorensen and Sorensen 2012), has led a 'coastal analysis' (Ministry of Environment and Food 2016) and has developed tools for decision making. Although touching upon planning dimensions, the DCA work is mainly focused on hazards, vulnerability and risk mapping.

Parallel to the municipal adaptation plans, the DCA is implementing the EU Floods Directive (EU 2007). The implementation by EU member states is divided into six year plan periods containing appointments of risk areas; hazard, vulnerability and risk mapping, and the implementation of risk management plans. In the first plan period (2009–2015), nine coastal flood risk areas were appointed that involve 21 municipalities. Whereas the DCA was in charge of the appointment of risk areas and production of maps, the municipalities are in charge of the risk management plans and of carrying out risk reduction initiatives. The municipal level work for Denmark is unique in relation to the European nations' implementation of the Directive. Jebens et al. (2016) provide an overview of the national implementation of the Directive. Whereas some guidance on sea level rise and uncertainties is brought into the DRM and CCA work, Denmark has not politically decided upon any scenarios or projections to plan from. This yields a somewhat 'messy' picture e.g., between neighboring municipalities using different numbers and timeframes for FRM. The work in relation to the Directive and the municipal adaptation plans has raised CCA issues on the political agenda both locally and nationally, however.

The Danish Emergency Management Agency (DEMA) is a governmental agency under the Ministry of Defense. DEMA's mission is "to lower the adverse impact from accidents and disasters on the society and to mitigate harm to people and damage to property and the environment" (DEMA 2015). DEMA has an operational, supervisory and regulatory function on a national and municipal level towards emergency management and preparedness. DEMA provides guidance to national and local organizations on how to develop emergency and preparedness plans and is consulted when local emergency managements develop preparedness plans. In addition, it implements various international legislation and frameworks including the Sendai and its forerunner the Hyogo (UN 2005; 2015a). DEMA therefore supposedly has the national in-depth overview of the status of implementation of emergency and preparedness plans as well as of international disaster risk reduction (DRR) framework implementation.

In addition to the national government and administration, several funding institutions, the insurance sector, NGOs and civil society organizations are active on the national scene; either pursuing their own or common agendas, calling upon the national government to take more responsibility and action, or both. The national coastal act states that it is the house owners'/citizens' own responsibility to protect themselves against flooding. Whether acting as an individual land owner or in a house-owner's association within a coastal community it is, in other words, those who benefit must pay. Irrespective of the rightfulness of citizens to take responsibility of their own situation, this has led to a buck-passing situation in many cases instead of creating a "let's work together and coordinate attitude" in civil society. Table 1 provides an overview of the levels of governance and of main actors in relation to flood risk management in Denmark.

Table 1 Main structure and responsibilities in flood risk reduction in Denmark (not exhaustive)

Level of governance	Actors	Legislation/ Frameworks	Responsibility
Global	UN (NGOs etc.)	International frameworks and collaborations	
European	EU Commission EU Parliament	Floods Directive Water Framework	
National	National Government Various Ministries National Emergency Man. Units Interest organizations for e.g.: Regions, Municipalities, Utility Sector, Insurance, Funding, Academia.	Implement UN frameworks etc. Implement EU Directives Danish legislation Preparedness and emergency planning	Incorporation into Danish law Implementation of Danish law
Regional	5 Regions:	Voluntarily in DRM & CCA	Health (hospitals)
Local	98 Municipalities Local Emergency Man. Units Utility companies	Municipal Level Planning Acts incl. Climate Adaptation Plans	Running the municipalities Local emergencies Securing functional water supply and storm drainage etc.
'Sub-local'	Coastal communities House owners' associations Citizens, House/land owners Companies etc.	Local association rules	Owners/Citizens are responsible for and must pay for their own protection

Methods

This work builds on a mixed method approach. It combines quantitative research by re-viewing literature and qualitative research by conducting semi-structured interviews. Previous work (Jebens and Sorensen 2016; Jebens et al. 2016; Sorensen and Jebens 2015) identified needs for improving flood risk management and CCA in the coastal zone among civil society and decision makers. To support and expand the knowledge on coordination and planning two key informants with an in-depth knowledge about national implementation of international frameworks were interviewed for the presented work. The informants were specifically selected from their core knowledge on the Hyogo Framework implementation and on preparedness planning in general. A need to include further key informants could be argued. In the context of the presented paper, the key informants do, however, provide core knowledge about emergency management and national implementation of the Hyogo and Sendai frameworks. This provides information about the overall national status of implementation of the frameworks but does not allow for a detailed analysis of the Danish engagement in international programs and frameworks at large. Even though there are a number of disadvantages in doing a single case study (comparative study can be more subjective and have higher external validity) a single case study was identified as the relevant approach to examine the current state and need for improved planning in the field of DRM and CCA in Denmark. This research has a specific focus on Danish flood risk management and the level of planning and is analyzed and discussed in relation to disaster risk science and global frameworks.

The flood risk management plans of the EU Floods Directive are also a part of national legislation, and initiatives mentioned in the plans must be implemented. It is not known whether the local emergency management units have received these plans or if coordination takes place between municipalities and emergency management. DEMA has not received the plans, and the follow-up of plans within the municipalities is often not clear due to a lack of ownership and responsibility as pointed out by Jebens et al. (2016). Likewise, work in relation to national reports under the Hyogo framework is unclear. In order to analyze these issues further, semi-structured interviews were held with two employees at DEMA. Interviews were held at telephone meetings and informants have been anonymized in the results. In interviews Informant 1 primarily relates to national level work, and Informant 2 relates to the international dimension. The interview focus is strictly scientific and the informants have accepted in advance to provide information for use in the presented work. Interview summaries were carefully translated from Danish to English, and the liability of potential misinterpretation and translation errors rests solely with the authors.[1] In addition, the Danish use of and progress reports on the implementation the Hyogo Framework for Action is analyzed.

Results

Results from the two semi-structured interviews are presented separately and are followed by a brief analysis of Danish work in relation to the Hyogo framework.

Interview Results—Informant 1

The interview focused on the development and existence of flood risk management planning and preparedness planning in Denmark. Based on an in-depth experience

[1]The scientific data use fulfils the Danish Ethical legislation regarding the legal and ethical handling of personal data. The Data Protection Act implementing Directive 95/46 was passed on in Law no. 429 of 31/05/2000. The Act faithfully transposes the provisions of the EC Directive into Danish law. The University (DTU) follows the Danish Statutory Order on Security on Data processing of Personal data, no. 528 of 15/06/2000 when processing Personal Data.

and knowledge in the field of disaster and emergency management the informant was asked questions in relation to:

1. The use of preparedness planning,
2. How the system of preparedness planning works in Denmark, and
3. The need for flood risk management plans now and in the future to adapt for climate change.

Summary of replies: Since the end of the 'cold war' in the late 1980s the disaster and preparedness planning has changed considerably. During the cold war the planning had an extremely high level of detail towards the continued service of the society, but after the cold war the level of detail has diminished.

Today planning is made on two levels; municipal and national. The municipal level is responsible for maintaining all its critical functions and to prevent loss of life and other losses. In the case of large events in which the local emergency management cannot cope with the situation, the national emergency (DEMA) can be called for. In the national planning, all ministries have to plan for their areas of responsibility and there are plans for national crisis management. Local emergency managements and DEMA do not have a national plan regarding flooding.

Municipalities must make preparedness plans. It is up to the municipalities and local emergency management to make a risk dimensioned plan which targets the most important risks the society is facing on a municipal level. The important risks are mostly defined as fire, terror, and pollution. A few municipalities and local emergency managements have made a preparedness plan for flood events. All preparedness plans go through a consultation period during which DEMA can comment. DEMA has no legal rights to alter the plans but can only give recommendations on how to improve them.

The local emergency managements and DEMA is often retroactive, even though the emergency management act states that risk reduction has to be included. Local emergency managements often purchase mitigating measures e.g., pumps after a flood event without knowing how they can be integrated in the response. Only after larger events a real effort to behave pro-actively is identified. There is a need to improve both local and national flood risk management plans identifying where the hazard will strike and which vulnerabilities will be impacted.

Strengthening the education of local emergency managers and on site commanders towards a better understanding of climate change could strengthen the preparedness level. This is not included in the current education.

Interview Results—Informant 2

The interview focused on the Danish implementation of the Hyogo Framework for Action, lessons learned, and the potential use in Denmark. Questions based on the informants' in-depth knowledge about these matters related to:

1. The applicability of the framework in a Danish context and which organization in Denmark who had the responsibility of implementing the Hyogo Framework for Action; and who has the current responsibility of implementing the Sendai Framework for Disaster Risk Reduction.
2. Where the data used in the final Hyogo Framework for Action progress report came from and how this data input was summarized.

Summary of replies: The Hyogo Framework for Action was under the jurisdiction of the Ministry of Foreign Affairs of Denmark (www.um.dk) for approximately 8½ years without being implemented nationally. After this period it was transferred to the Ministry of Defence and DEMA. In contrast, Scandinavian partners did a lot of work to implement the framework including development of national platforms and progress reports. Currently it has to be decided which ministry will be in charge of the implementation of Sendai Framework for Disaster Risk Reduction. DEMA is capable to act as a focal point but cannot perform the task alone. Because of the short timeframe in which DEMA tried to implement [the Hyogo Framework for Action] there was no time to establish a national platform.

DEMA has a more retroactive approach and therefore there is less emphasis on risk reduction. Seen in the perspective of limited resources the costs and benefits of implementing a large framework should be discussed, since it is complex and involves coordination of a very diverse group of stakeholders.

The lack of focus on these frameworks could be due to the general assumption that Denmark has a well-functioning risk reduction system, and that the frameworks have been developed for the purpose of less developed countries. Due to the focus on transparency and corruption it is not the entire framework Denmark can use.

DEMA did not have time to go into detail with this very complex field and the [national] value of it has been limited. Input to the progress report was mainly given by municipalities and the private sector and to a lesser degree by governmental institutions. Grades are an average reflection on how the conditions are regarding the different hazards.

Reporting on the Hyogo Framework for Action

During the period 2005–2015 in which the Hyogo Framework for Action was in place, Denmark only reported on the last progress period 2013–2015 (HFA 2014). Here, the Danish progress reporting 2013–2015 is evaluated to identify whether it targets and includes the current state of FRM and preparedness planning. As the national reporting embraces all identified hazards, an explicit and differentiated assessment of the current state of FRM is difficult to carry out. Some of the reporting benchmarks do not mention FRM. For instance, disaster risk is not always taken into account in public investment and planning decisions for flooding prevention; there are no strategies and plans for flooding mitigation; there is no standard assessment method; there is no coordinating lead institution for DRR for

Table 2 Challenges for Areas 1–3 defined in the Hyogo Framework for Action and the outlook statement given for Denmark by DEMA (HFA 2014)

Future Challenges Area 1	Future Challenges Area 2	Future Challenges Area 3
There is a lack for all organizations to understand that they have a part to play in DRR	There is a lack of DRR and preparedness planning in parts of the central administration defined by an audit in 2014 of the Danish emergency preparedness. Lack of resources is the central challenge to many organizations	Climate change is felt in Denmark mainly due to water management
Future Outlook Statement Area 1	**Future Outlook Statement Area 2**	**Future Outlook Statement Area 3**
Cloudbursts have created a momentum for stronger and better coordination in the DRR sector	A revision of the Emergency Management Act making clear requirements for planning by the central administration has been made. Could create focus on planning in relevant sectors and organizations	There is a need for an integrated DRR approach which could improve adaptation strategies

coastal floods, and there is no national multi-hazard risk assessment with a common methodology available to inform planning and development decisions. Despite all of the above is absent for FRM, several of the benchmarks have been answered with a 'yes' in the progress reporting since these parameters are in place in other sectors.

The progress reporting points to a number of overall challenges and outlook statements of importance to this study as seen in Table 2. Overall challenge Area 1 refers to a better integration of DRR and sustainable development including planning. Overall challenge Area 2 refers to development and strengthening of capacities which systematically can build resilience. Overall challenge Area 3 refers to incorporation of DRR into all aspects of the DRM cycle including emergency preparedness.

Coastal floods have the potential to cause devastating economic impacts on the Danish society. However, this hazard is often forgotten in the national and municipal planning and in the emergency management. The local emergency managements primarily respond to flood events and pay less attention to preparedness planning, and flood risk management plans are only made by municipalities appointed under the EU Floods Directive. According to DEMA (2017), the Danish ministries "are responsible for planning within their respective areas of responsibility. Their tasks are to maintain the functions of the Government and the public administration, producing necessary legislation and providing guidance to regional and municipal authorities".

The analysis points to a careless approach to the implementation of the Hyogo Framework, where Denmark only reported on the progress of implementation of the

Hyogo Framework for Action for the last period 2013–2015. Will carelessness also characterize Danish implementation and reporting of the Sendai Framework for Disaster Risk Reduction, too? One factor supporting this proposition is that stakeholders believe that the Danish society is well equipped to cope with and adapt to disasters. Furthermore, the opinion is that the frameworks mainly target the needs of less developed countries. In addition, it is not clear yet whether DEMA, the Ministry of Foreign Affairs or another public entity will take or be given responsibility to implement the Sendai Framework.

Discussion

The Hyogo Framework for Action progress report ideally describes the overall state of DRR initiatives embracing a variety of hazards which Denmark is prone to. There are no specific details on current or future flooding hazards. The report generally overestimates the actual planning level for coping with and adapting to coastal flooding. Furthermore, the process of preparing the final progress report was influenced by confusion as to which governmental organization had the responsibility, and where DEMA eventually was handed over the responsibility at a late stage. The lack of an entity in charge of national flood risk management is considered a main concern, and this has to be solved if coherent plans are to be made. The local emergency managements together with the national emergency management do not have a national plan in relation to floods due to their retroactive approach. In addition, emergency management does not coordinate efforts with e.g., the DCA. In the progress report, DEMA pointed to the complexity and that a high degree of coordination is needed. The gain from introducing the frameworks and create a national platform should therefore be evaluated. An overall challenge of Action point 2 on preparedness planning and DRR in Denmark in the Hyogo progress report is a lack in coordination and planning to which a national platform is a good tool for improvement. Danish experiences working with the frameworks abroad almost certainly exist and may serve as a starting point. During implementation Denmark could learn from best practices from the long list of nations which use the Hyogo or the Sendai frameworks to bring coordination and planning onto a national level. However, the Danish flood risk management sector has to realize and acknowledge the beneficial use of the frameworks to improve coping mechanisms and to decrease the risk due to climate change. With planning we will be able to manage the risk and not always fall back to manage the disaster.

As mentioned, the international frameworks often embrace all hazards. There is thus a need to establish a system which can plan, do, monitor and evaluate the different initiatives on both a national and a local scale. For FRM the need for national planning may be appropriate since only a few municipalities have the resources and competencies to coordinate this work and to secure that initiative taken in one municipality do not transfer risks to the neighboring areas.

Since the Sendai Framework for Disaster Risk Reduction together with CCA and sustainable development contain cross-cutting issues which give guidelines on coordination and planning, it is believed that these frameworks could be used in Denmark.

Flood risk assessments in Denmark have a focus on the collection of quantitative data and tangible losses, and they do not include qualitative data from stakeholders or intangible losses. In addition, planning is often absent or is developed to different levels between areas. This calls for a more systematic approach for assessments which includes both qualitative and quantitative data collection. To identify vulnerabilities and capacities e.g., the participatory vulnerability and capacity assessment method (VCA) was developed by the IFRC (2006). Today, methods have been developed by numerous INGOs and NGOs and offer a comprehensive set of tools which identify and assess the current capacities and vulnerabilities in communities. Tools include baseline studies using primary and secondary data, interviews and focus group discussions hereby incorporating both qualitative and quantitative data. As in other assessments there are challenges in conducting them. First, they are generally resource intensive and need to be implemented in a large number of communities (van Aalst et al. 2008) and secondly, there are challenges in incorporating CCA into the process (IFRC 2012; van Aalst et al. 2008). A way to incorporate CCA in the vulnerability and capacity assessment is to invent or improve tools to assess long term trends. The assessments are important to define the combination of structural, non-structural, semi-structural and environmental measures in order to cope with future climate changes.

International Frameworks and Standards

Denmark needs to develop a customized framework which is systematic and defines FRM on all levels. The Sendai Framework for Disaster Risk Reduction, the Paris Agreement, and the Sustainable Development Goals are all highly developed frameworks which together could form a new pathway for CCA in coastal communities in Denmark. The driver in all of these frameworks is to decrease the risk for the life of humans by developing capacity and increase resilience and hereby creating a better future for the society at local and national levels. This makes risk assessments the driver in identifying appropriate mitigation and adaptation measures; hence it is important to define the combination of structural, non-structural, semi-structural and environmental measures in order to cope with future climate changes.

The lack of coordination between DRM and CCA has recently received attention in international development and particularly in the Pacific region (Glasser and Burkot 2016). During the same time there has been a continued global increase in economic losses from disasters which implies that there is a need for increased management so that investments will become risk informed.

The Paris Agreement (UNFCCC 2015) and the 2030 Agenda for Sustainable Development (UN 2015b) could together reduce the risk related to climate change. The Paris Agreement argues the need to implement new approaches to manage climate change risks. The new focus should be on comprehensive management and DRR instead of focusing on disaster loss. A way forward to manage this paradigm shift is to build it on the Sendai Framework for Disaster Risk Reduction 2015–2030 (UN 2015a) and focus on resilience building, risk assessment and adaptation planning. The 2030 Agenda for Sustainable Development and the Sustainable Development Goals (SDGs) offer a way forward to decrease climate impacts among coastal communities. Important examples are Goal 11 which aims at making cities and human settlements inclusive, safe, resilient, and sustainable; Goal 13 which calls for urgent action to combat climate change and its impacts; and Goal 14 which focuses on conservation and sustainable use of the oceans, seas and marine resources for sustainable development.

If implemented, Danish FRM should build on these agreements and frameworks to improve the protection of coastal communities. In the Pacific region a framework for resilience has been developed which targets an integrated approach to address climate change and disaster risk management (SPC 2016). The framework uses the Paris Agreement, the Sendai Framework for Disaster Risk Reduction and the SDGs to create an integrated approach to decrease risks in the region. The framework provides guidance on how CCA and DRM are understood as development challenges and gives a number of priority actions to build resilience and identify vulnerabilities. The work is coordinated, implemented and monitored by the Pacific Resilience Partnership. In Denmark, as well as in the Pacific disaster risk is for the most part concerned with increasing severity of storms and sea level rise. The Pacific approach could therefore be beneficial in a Danish context too, although the threats from sea level rise are not yet as imminent in Europe.

There are some weak links between CCA, DRR and SDGs which need to be improved before they can be coordinated. First, the Paris Agreement is weakly linked to DRR. A way forward would be to strengthen implementation of local and national adaptation planning hereby linking CCA and DRR. Secondly, on all levels a greater collaboration among management and stakeholders is needed to merge CCA, SDGs and DRR (Shaw et al. 2016). This integration is not just related to policies and planning but also by 'doing', 'monitoring' and 'evaluating'. Thirdly, national focal points should mainstream and monitor indicators established in the three frameworks.

Implementation calls for an approach building e.g. on the 'Plan, Do, Check, Act' (PDCA) principles which are also required by the Sendai Framework for Disaster Risk Reduction (Fig. 4).

The Danish flood risk management plans developed by the municipalities as part of the EU Floods Directive have been criticized for not including how measures will be implemented, monitored and evaluated (Jebens et al. 2016). Therefore, to fulfill the guidelines from international frameworks and to improve Danish flood risk management plans a systematic approach like the PDCA principles is needed. The standards developed by the International Organization for Standardization

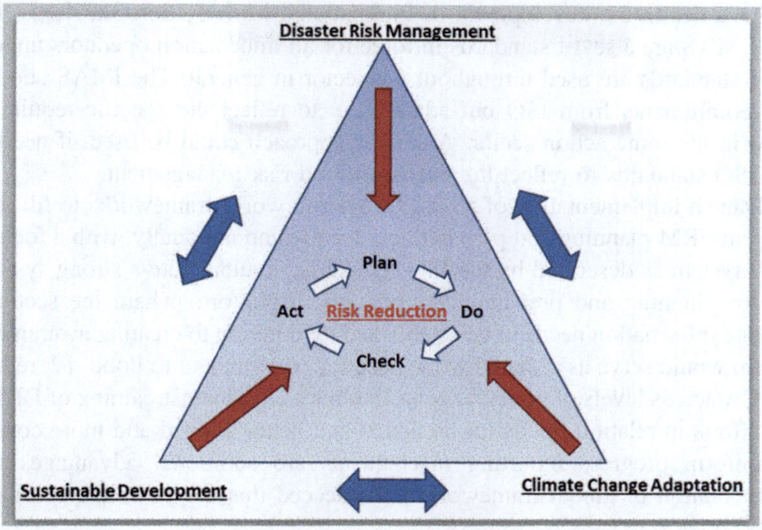

Fig. 4 An illustration of the need to integrate DRM, CCA and SDG´s and to move away from managing the disaster to managing disaster risk reduction. It underlines the need for introducing the PDCA principle in risk reduction work potentially using already known standards, or, to adapt these standards to reflect and target the specific challenges in integrated flood risk management (coastal flooding, river flooding, cloud burst, and groundwater) and CCA. The standards should be the basis for developing a national plan to create holistic and sustainable solutions, taking into account regional differences and hereby avoiding transferring risks. In the Danish case this has to be done by assessing the needs of society. It is important to keep in mind that planning itself does not lead to a decrease in risk

(ISO) are used globally and ensure that product and services are safe, reliable and of good quality and the concept build on the PDCA cycle. This system is widely used in Denmark and the capacity is already in place for implementation.

The development of international standards cannot take place in isolation. Of value to the ISO system is its multi-stakeholder approach, which builds on recommendations and contributions from a broad spectrum including governments, academia, research, INGOs and NGOs, and the private sector. One example is within water management where ISO cooperates with e.g., WHO, WMO, and FAO (ISO 2017). This could potentially lead to more hands-on recommendations in existing or new ISO developments to target the threat of natural hazards and different aspects of climate change.

The ISO systems already target the environment (Environmental management—ISO 14000) and have a strong focus on quality management and its principles (ISO 9000 and ISO 9001) and risk management (ISO 31000). In addition, the ISO system works with sustainable development in communities (ISO 37101) and SDGs (ISO 26000). Also, ISO has published guidelines on tools for addressing climate change. To decrease the landmine hazard, the mine clearance sector is currently including the ISO series in the development of International Mine Action Standards (IMAS 2017). IMAS are a set of standards in force for all mine action operators under UN but the standards are used throughout the sector in general. The IMAS adopts the main requirements from ISO but adapt them to reflect the specific requirements needed in the mine action sector. A similar approach could be used, if needed, to adapt ISO standards to reflect the needs of flood risk management.

A Danish implementation of global 'proven to work' frameworks to fill in gaps in current FRM planning and preparedness locally and nationally, with a focus on a PDCA system as described by the ISO standards, would create a strong systematic basis for planning and preparedness. For this, a platform where the sectors can exchange information needs to be established. In addition to creating awareness, the platform would serve as a shared workspace for commitment to flood risk reduction and CCA across levels of governance in Denmark. The mainstreaming of DRM and CCA efforts in relation floods for all actors and better aligned and more consistent and uniform progress reporting mechanisms are additional advantages to the implementation of global frameworks. To succeed this has to be made a national priority, however.

Conclusions

Denmark is at risk from coastal flooding and the risk will increase in the future due to climate change and continued economic development in coastal areas. Flood risk management therefore needs to take both current and future conditions into account. Flood risk management plans and preparedness plans are generally not at hand although a few Danish municipalities and local emergency managements have made first attempts of producing plans. In order to advance Disaster Risk Management and Climate Change Adaptation to coastal flooding, risk management plans need to build on a common understanding or framework, they must be coordinated, and they must be aligned and shared between actors within the water sector and across levels of governance.

Denmark only reported on the last progress period (2013–2015) of the Hyogo Framework for Action (2005–2015) and the national engagement in the Sendai Framework for Disaster Risk Reduction (2015–2030) is still uncertain. The paper suggests that these two global frameworks together with the Paris Agreement and the Sustainable Development Goals may serve as a pathway to tailor a Danish national framework for flood risk management. Here, Denmark may prosper from existing experiences e.g., in the Pacific region in its implementation of a framework and in the establishment of a shared platform to better deal and cope with current and future flood risks and to improve planning, preparedness, and awareness in the public at large. Furthermore, this will guide Denmark towards a holistic and more sustainable national approach to Climate Change Adaptation.

Acknowledgements The authors wish to thank the anonymous reviewers for comments and suggestions to improve the manuscript and Ms. Alexandra Strand Holm for fruitful discussions. Funding for the research presented is kindly provided by Innovation Fund Denmark (Grant no. 1355-00193).

References

ACAPS. (2014). Humanitarian needs assessment: The good enough guide. The Assessment Capacities Project (ACAPS), Emergency Capacity Building Project (ECB). Rugby, UK: Practical Action Publishing.

Dahlberg, R., Albris, K., & Jebens, M. (2017). The Baltic Storm Surge in November 1872. Urban processes, gender vulnerability and scientific transformation. In D. Simonton & H. Salmi (Eds.), *Catastrophe, gender and urban experience, 1648–1920*. New York & London: Routledge Research in Gender and History. 177–197.

Danish Government. (2008). *Strategi for tilpasning til klimaændringer i Danmark*. (March 2008) Danish Government. The Danish Energy Agency. Schultz, Valby, Denmark.

Danish Parliament. (1874). *Lov om diger til beskyttelse mod oversvømmelse fra havet. Act no. 53 of 10 April 1874*. Danish Parliament, Copenhagen, Denmark.

Davis, I. (2015). Introduction. In I. Davis (Ed.), *Disaster Risk Management in Asia and the Pacific*. New York & London: Routledge. 3–32.

DEMA. (2015). *Crisis management in Denmark*. Birkerød, Denmark: Danish Emergency Management Agency.

DEMA. (2017). Danish Emergency Management Homepage. http://brs.dk/eng/emergency_management/Pages/emergency_management.aspx. Accessed 19 Mar 2017.

DMI. (2014). *Fremtidige Klimaforandringer i Danmark*. Danish Meteorological Institute, Copenhagen, Denmark, Danmarks Klimacenter Rapport no. 6. 35p.

EU. (2007). DIRECTIVE 2007/60/EC of the European Parliament and of the council of 23 October 2007 on the assessment and management of flood risks. *Official Journal of the European Union*, L 288/27 (6/11/2007).

GFDRR. (2016). *The making of a riskier future: How our decisions are shaping future disaster risk* (p. 143p). Washington, D.C.: U.S.A., Global Facility for Disaster Reduction and Recovery report.

Glasser, R., & Burkot, C. (2016). From managing disasters to managing disaster risk: An interview with Robert Glasser. http://devpolicy.org/managing-disasters-managing-disaster-risk-interview-robert-glasser-20161128/. Accessed 19 Mar 2017.

Grinsted, A., Jevrejeva, S., Riva, R. E., & Dahl-Jensen, D. (2015). Sea level rise projections for northern Europe under RCP8.5. *Climatic Research, 64,* 15–23. doi:10.3354/cr01309.

HFA. (2014). *Denmark: National progress report on the implementation of the Hyogo framework for action (2013–2015)*. Danish Emergency Management Agency, Birkerød, Denmark, Hyogo Framework for Action (UNISDR). (http://preventionweb.net/go/41422. Accessed 19 Mar 2017.

ICSU. (2015). *Review of the sustainable development goals: The science perspective*. France, Paris: International Council for Science (ICSU).

IFRC. (2006): *What is VCA? An introduction to vulnerability and capacity assessment*. International Federation of Red Cross and Red Crescent Societies, Geneva, Switzerland, Handbook on VCA. 51p.

IFRC. (2012). *How can climate change be considered in Vulnerability and Capacity Assessments? A summary for practitioners*. International Federation of Red Cross and Red Crescent Societies, The Hague, Netherlands, Red Cross/Red Crescent Climate Centre Report. 15p.

IMAS. (2017). International mine action standards homepage. https://www.mineactionstandards.org/. Accessed 19 Mar 2017.

IPCC. (2012). Managing the risks of extreme events and disasters to advance climate change adaptation. In C. B. Field., V. Barros., T. F. Stocker., D. Qin., D. J. Dokken., K. L. Ebi., M. D. Mastrandrea., K. J. Mach., G.-K. Plattner., S. K. Allen., M. Tignor., & P. M. Midgley. (Eds.), *A special report of working groups I and II of the Intergovernmental panel on climate change*. (Cambridge, UK; New York, NY: Cambridge University Press), 582.

IPCC. (2013). Climate change 2013: The physical science basis. In T. F. Stocker., D. Qin., G. -K. Plattner., M. Tignor., S. K. Allen., J. Boschung., A. Nauels., Y. Xia., V. Bex., & P. M. Midgley. (Eds.), *Contribution of working group I to the fifth assessment report of the Intergovernmental panel on climate change* (Cambridge, UK; New York, NY: Cambridge University Press), 1535.

ISO. (2017). ISO & water. Global solutions to global challenges. International Organization for Standardization, Geneva, Switzerland. https://www.iso.org/files/live/sites/isoorg/files/archive/pdf/en/iso_and_water.pdf. Accessed 19 March 2017.

Jebens, M. & Sorensen, C. (2016). Advancing coordination between DRM and CCA in integrated flood risk management. In *Proc. 6th International Disaster and Risk Conference 2016 (IDRC 2016)* 292–295 (Davos, Switzerland, 2016).

Jebens, M., Sorensen, C., & Piontkowitz, T. (2016). Danish risk management plans of the EU floods directive. *E3S Web of Conferences*, 7, [23005]. doi: 10.1051/e3sconf/20160723005.

Miljøstyrelsen. (2004). Tilpasning til Fremtidens Klima. Miljøstyrelsen, Copenhagen, Denmark, Miljøtema, 28.

Ministry of Environment and Food. (2016). *Kystanalyse*. Ministry of Environment and Food of Denmark, Copenhagen, Denmark.

Nicholls, R. J. (1995). Coastal megacities and climate change. *Journal of Geography, 37,* 369–379.

Shaw, R., Prabhakar, S., & Chiba, Y. (2016). *SDGs, DRR and CCA: potential for strengthening inter-linkages*. Institute for Global Environmental Strategies, Kanagawa, Japan, Policy Brief, Number 34, July 2016. p. 12.

Sorensen, C., & Jebens, M. (2015). How can awareness in civil society and in governance be raised? Reducing risks from coastal hazards. *Proc. 24th New South Wales Coastal Conference*, Forster 11–13 Nov 2015, NSW, Australia.

Sorensen, C., Madsen, H. T., & Knudsen, S. B. (2013). *Hojvandsstatistikker 2012* (p. 158p). Lemvig, Denmark: Danish Coastal Authority.

Sorensen, C., & Sorensen, P. (2012). *Guidelines for klimatilpasning i kystområder* (p. 43p). Lemvig, Denmark: Danish Coastal Authority.

Sorensen, C., Sorensen, P., Jürgensen, C., Jorgensen, N-E., Jebens, M., & Knudsen, P. (2017). 1872. In *19. Danske Havforskermøde. Program & presentations* (p. 108).

SPC. (2016). Framework for resilient development in the pacific: An Integrated approach to address climate change and Disaster Risk Management (FRDP) 2017–2030. Pacific Community (SPC), Suva, Fiji, Voluntary Guidelines for the Pacific Island Region (11 September 2016). 38p. http://www.pacificdisaster.net/dox/FRDP_2016_Resilient_Dev_pacific.pdf.

UNFCCC. (2009). COP15 Homepage, United Nations framework convention on climate change. http://unfccc.int/meetings/copenhagen_dec_2009/meeting/6295.php.

UNFCCC. (2015). *Adoption of the Paris Agreement, 21st Conference of the Parties*. Paris: United Nations Framework Convention on Climate Change.

UN. (1987). Our Common Future—Brundtland Report. Oxford, UK: Oxford University Press. 204p.

UN. (2005). Hyogo framework for action 2005–2015: Building the resilience of nations and communities to disasters. United Nations, New York.

UN. (2015a). Sendai framework for Disaster Risk Reduction 2015–2030 (resolution 69/283, annex II), United Nations Office for Disaster Risk Reduction (UNISDR). 32p.

UN. (2015b). Transforming our world: The 2030 Agenda for Sustainable Development (A/RES/70/1), United Nations.

Van Aalst, M. K., Cannon, T., & Burton, T. (2008). Community level adaptation to climate change: The potential role of participatory community risk assessment. *Global Environmental Change, 18,* 165–179. doi:10.1016/j.gloenvcha.2007.06.002.

Wisner, B., Blaikie, P., Cannon, T., & Davis, I. (2004). *At Risk. Natural hazards, people's vulnerability and disasters* (2nd Edition), New York & London: Routledge.

Local Knowledge Co-production, Emergent Climate Adaptation Publics and Regional Experimentalist Governance: An Institutional Design Case Study

Nicole Lisa Klenk, James Ian MacLellan, Kim Reeder and Dragos Flueraru

Abstract In this paper, we ask how are climate hazards, local knowledge, affects and political arrangements assembled and generated by co-productive adaptation planning? Additionally, we ask how climate change preparedness comes to articulate and embody social imaginaries of the future and the governance arrangements these call forth. Using a paradigmatic case study from New Brunswick, Canada, we ask how the material dynamics of climate change impacts and local knowledge co-production in community-based adaptation planning are constitutive of the formation of "climate adaptation publics"? Our chapter argues that current governance arrangements are not adequate to the task of empowering and coordinating emerging climate adaptation publics, and keeping different levels of climate adaptation decision-making transparent, adaptive and accountable. We propose an institutional design based on an experimentalist form of regional adaptation governance to support climate adaptation publics and derive insights from this case study to inform regional adaptation governance more generally. This article also makes a theoretical contribution to non-extractive conceptions of local knowledge mobilization in climate change adaptation governance.

N. L. Klenk (✉) · J. I. MacLellan · D. Flueraru
Department of Physical and Environmental Sciences, University of Toronto, 1265 Military Trail, Toronto, ON M1C1A4, Canada
e-mail: Nicole.klenk@utoronto.ca

K. Reeder
Climate Change Research Collaborative, 5524 Route 127, New Brunswick, NB E5B 0B7, Canada

W. Leal Filho (ed.), *Climate Change Impacts and Adaptation Strategies for Coastal Communities*, Climate Change Management,
https://doi.org/10.1007/978-3-319-70703-7_14

Introduction

Climate change impacts are forecast on the basis of global scale scientific models, yet adaptation to projected changes will have to be informed by the particular knowledge of the communities where consequences are felt. Knowledge co-production recognizes this essential, mutual relationship in seeking not only to account for the expertise of academics and researchers, but to include local expertise and knowledge in climate adaptation. However, the effective mobilization of local knowledge through community-engaged planning entails more than simply assembling an academic and community team to inform local plans, it requires rethinking how such co-productive processes are experiments in climate governance (Rojas 2015).

In this article the authors ask what climate impacts, local knowledge, affects and political arrangements are assembled and generated by adaptation planning? Additionally, the authors are interested in how climate change preparedness comes to articulate and embody social imaginaries of the future and the governance arrangements these call forth. Using a case study of a community-based adaptation planning process in 5 coastal communities of the province of New Brunswick, Canada, the authors demonstrate how and why the material dynamics of climate change impacts and local knowledge co-production in community-based adaptation planning can be understood as constitutive of the formation of "climate adaptation publics". Drawing upon the notion of 'material participation' developed by Marres (2012), this chapter argues that attending to the specificity and contingency of social-material co-production of local knowledge gives rise to specific governance challenges. Current governance arrangements are not adequate to the task of empowering and coordinating emerging climate adaptation publics, and keeping different levels of climate adaptation decision-making transparent, adaptive and accountable.

While the focus of this chapter is on a case study of local knowledge co-production in New Brunswick, Canada, the insights derived from this case study and its implications for adaptation governance are not limited to this location, but speak to adaptation governance more generally. This case study was carefully selected to exemplify the complexities of mobilizing local knowledge in climate change adaptation governance. Focused on the climate hazards of sea level rise and extreme weather events, such as repeated severe flooding, the communities at the core of this case study exemplify many of the governance challenges reported in the literature on regional climate governance. In addition to presenting a paradigmatic case (Flyvbjerg 2006), this chapter also makes a theoretical contribution to non-extractive conceptions of local knowledge mobilization and the implications of a co-productive perspective for multi-level and regional adaptation governance (Jacobs et al. 2016; Dannevig and Aall 2015; Termeer et al. 2011; Cloutier et al. 2015; Antonson et al. 2016; Hanssen et al. 2013; Klenk et al. 2017).

In the following sections, the authors provide a short explanation of their theoretical perspective, then present results of the case study analysis and discuss how the process of local knowledge co-production created new assemblages of material conditions, maps, and social imaginaries that generated new collective identities, in turn given rise to new climate adaptation publics. The authors argues that these publics are attempting to reconfigure local, regional and provincial socio-material and political arrangements to address climate impacts, giving rise to specific governance challenges. To conclude, the authors discuss the governance implications of understanding climate change adaptation planning as the formation of experimental Deweyan publics.

Theoretical Background

Local and Regional Adaptation Governance

Locally based adaptation-planning faces a number of governance challenges identified within the literature (Antonson et al. 2016). Biesbroek et al. (2011) have characterized governance challenges to adaptation as follows: (1) conflicting timescales, (2) substantive, strategic and institutional uncertainty, (3) institutional crowdedness and institutional void, (4) institutional fragmentation, (5) lack of awareness and communication, (6) motives and willingness to act, and (7) lack of resources. Conflicting timescales are the source of significant barriers to adaptation because of projected long-term changes and imperfect knowledge of the timing and magnitude of climate impacts, which is difficult to integrate into the short-term nature of decision-making, policies and political cycles. Uncertainty can also emerge through the quality, availability, legitimacy and credibility of knowledge used in decision-making (Zegwaard et al. 2015). While adaptation planning and implementation has largely had a focus at the national scale, the past decade has seen emphasis on locally based adaptation initiatives (IPCC 2014). In addition to the realization that adaptation strategies at the national scale have largely failed to produce tangible adaptation action, research suggests that the nature of adaptation is inherently "context-specific" (Measham et al. 2011).

Local and regional governance systems are therefore thought to be the appropriate institutions to address these impacts (Mukheibir et al. 2013; Hanssen et al. 2013; Termeer et al. 2011; Dannevig and Aall 2015; Jacobs et al. 2016; Antonson et al. 2016). In a multi-level governance perspective, local institutions play a crucial role in local knowledge mobilization, setting priorities for action, building adaptive capacity, strengthening social capital within the community and in international networks of climate change governance, and in the implementation of climate change policies (Urwin and Jordan 2008). However, climate change planning represents an added financial burden to municipal governments already short on resources to provide the services under their jurisdiction. Such fiscal challenges and the recognition that climate change impacts spillover local political boundaries, have led to calls for regional governance arrangements (Termeer et al. 2011;

Antonson et al. 2016; Dannevig and Aall 2015). Previous studies have highlighted the capacity of regional institutions to support local level adaptation and foster integrated land-use planning that mitigate climate risks (Hanssen et al. 2013; Jacobs et al. 2016; Dannevig and Aall 2015), however the relationship between local level decision-making and regional governance institutions have been found to be fraught with tensions (Storbjörk and Hedrén 2011; Nilsson et al. 2012; Antonson et al. 2016). While studies on regional adaptation governance have begun to examine these tensions, there are few studies that examine institutional architectures that address the challenges associated with local–regional governance, with some exceptions (Jacobs et al. 2016). A key design element of regional adaptation governance arrangements is integrating different sources of knowledge: local, experiential, traditional and scientific (Measham et al. 2011; Lieske et al. 2015; Dannevig and Aall 2015; Cloutier et al. 2015).

Local Knowledge

While the term "local knowledge" originally referred to indigenous ways of knowing (Agrawal 1995; Ellen et al. 2000), couched in terms such as "traditional ecological knowledge" or "indigenous knowledge", in this article the authors refer to Cruikshank's (2005 p. 9) definition of local knowledge as "tacit knowledge embodied in life experiences and reproduced in everyday behavior and speech", including but not restricted to indigenous knowledge (see also Turnbull 2003; Abbott and Wilson 2014). A recent critical review of local knowledge in global climate change research suggest that it largely refers to perceptions of environmental change, practices associated with local livelihoods, traditional indigenous knowledge, and the "adaptive" capacity of local communities to respond to environmental change (Klenk et al. 2017). Recent reviews (Naess 2013; Reyes-García et al. 2016) treat local knowledge as an object that can be extracted from its local circumstances of production, circulated and aggregated without questioning its "localness". The objectification of local knowledge is closely linked to a misunderstanding of, and perpetuation of, a conception of science as removed from its social context, leading to a focus on the extraction of local knowledge without due consideration to how this process is co-productive of local knowledge and social orders (Turnbull 2003).

In science and technology studies, co-production refers to the intimate and inseparable processes of knowledge production and how we organize our societies and governance arrangements (Turnbull 2003; Latour 2004; Jasanoff 2004). Co-production involves the encounter between different local knowledges and, as Law (2012, p. 157) notes, "assemblages of these practices do realities…realities are inseparable from the patterning juxtaposition of practices." This performative perspective suggests that realities are inseparable from the production of knowledge claims, and since knowledge production practices are done in multiple ways, in principle those realities—or the balance between them—could be different

(Law 2012, p. 156). By understanding 'local knowledge' mobilization as a process of co-production, rather than a system of extraction, the procedural or performative aspects of local knowledge mobilization come into greater focus. What previous, paradigmatic case studies (Whatmore and Landström 2011) demonstrate is that embedded in the process of local knowledge co-production in climate change planning are fresh proposals for individual and collective reorientation in the face of climate change. We conceptualize this process in terms of the formation of climate adaptation publics.

Climate Adaptation Publics

Local knowledge co-production refers not only to extractive knowledge mobilization (Lezaun and Soneryd 2007), but to the material conditions of knowledge production (Cruikshank 2005; Donaldson et al. 2013). In the process of climate change adaptation planning, local knowledge co-production assembles human and non-human actors around a future they co-construct and prepare for, but over which they have little or no control—resulting in the formation of publics in the Dewean sense (Marres 2012; Donaldson et al. 2013; Sabel 2012). For Dewey (1927), a public often begins its life as inchoate, unorganized, unresourced, and unrecognized. Publics emerge in circumstances when governance structures are defective or unprepared to respond adequately to an issue.

The formation of a public is intrinsically connected to the (re)organization of socio-material governance arrangements. Hence, every new public represents a *political experiment* by attempting to secure the necessary involvement of human and non-human actors in the process of the domestication of scientific, technological, natural entities in the governance of society (Marres 2012). In other words, such political experiments serve the role of producing new knowledge by means of exemplification and demonstration, but also consist in the composition of a common world, where new issues, entities and their relations, are accommodated or alienated (Latour 2004). These processes are experimental in the sense that materials, entities and relations are assembled deliberately and guided by a means-ends logic that integrates experiences and evidence that reflexively approach dominant regimes of evidence (Callon et al. 2011; Marres 2012; Whatmore and Landström 2011; Tironi et al. 2014). In this process, collective objectives, identities and social imaginaries are generated both to test a path of action and to unveil desired future social, material and political conditions (Callon et al. 2011). Bellamy and Lezaun (2015) and Donaldson et al. (2013) call attention to the performative dimension of public engagement and public formation, which are fraught with risks, such as the materializing of particular technological futures that reinforce current inadequate governance regimes, and hence they call for a more reflexive approach that seeks to open rather than foreclose policy options, and seeks to enmesh particular imaginaries of the future within broader societal, political and economic discourses.

Case Study Context

Climate Change Decision-Making in New Brunswick

Municipalities in Canada are empowered by provincial statutes—hence the nature and extent of their jurisdictional responsibilities are imposed by a higher level of government. The main role of municipalities is to provide services such as police protection, emergency measures, solid waste management, transportation planning, land-use planning, and water management among others, depending on the capacity of the municipality. In New Brunswick, municipalities are subject to the department of the Environment and Local Government, which oversees and addresses issues within land-use planning and management, zoning development and waste management and ensure that municipalities comply with environmental legislation.

Climate change planning is the purview of the New Brunswick Climate Change Secretariat (forthwith simply named the Climate Secretariat), situated within the Department of the Environmental and Local Government. The Climate Secretariat develops, implements and oversees climate change mitigation and adaptation strategies and policies in consort with multiple stakeholders and with other provincial departments. Funding for adaptation action largely comes from the Regional Adaptation Collaborative (RAC) and the Environmental Trust Fund (ETF). The RAC is a federally funded 30 million dollar program, which funded the Atlantic Climate Adaptation Solutions Association (ACASA) designed to help people in Canadian Maritime provinces to adapt to climate change (Lieske et al. 2015). The Environmental Trust Fund is a provincial program under the Department of Environment and Local Government that delivers funding for projects aimed at protecting, preserving or enhancing the natural environment.

Case Study Description

Charlotte County, situated in the southwestern region of the Atlantic province of New Brunswick, is the drainage basin for much of the water received in the province, with three (3) major river systems, being the Magaguadavic, Digdeguash and St. Croix Rivers. It is also situated close to the fourth major drainage basin in New Brunswick, namely the Saint John River. In the last several years, severe weather events have stricken Charlotte County (Signer et al. 2014). On November 5, 2010, the Town of St. Andrews experienced strong winds accompanied by an extreme high tide and 45 mm of rain. This caused a storm surge, which inundated the coastal homes on low-lying, commercial and residential streets. On December 13, 2010 a rainfall event of 166.4 mm hit coastal communities and a similar event was repeated on July 26 2013, with a 163 mm of rain. In addition, in late December 2013, a series of intense storm events comprised of freezing rain, ice pellets, extreme wind chill temperatures, and snow storms impacted Charlotte County for an extended period. The first and most intense storm hit on December 22, 2013 and lasted for 24–36 h. It was estimated that 50,000 residences were without power across the province of New Brunswick (Signer et al. 2014). The region was also impacted by tropical hurricane Arthur and others since this time. Together, these events have caused health impacts, physical and infrastructure damage, loss of household savings, temporary loss of services resulting in economic disruption, and environmental damage (Signer et al. 2014). Similar events across the province have led the Government of New Brunswick to prioritize flood risk prevention in its climate change adaptation strategy (Province of New Brunswick 2014a, b).

In 2013, Kim Reeder, at the time the Executive Director of the non-governmental organization The St. Croix Estuary Project Inc., and Donald Killorn, Executive Director for Eastern Charlotte Waterways Inc., and their intern Kristie Signer organized the Charlotte County Community Vulnerability Assessment (CCCVA) with the support of the New Brunswick Environmental Trust Fund. They sought to enable Charlotte County communities to share knowledge and concerns relative to climate change, as well as to develop and share information on such topics as socioeconomic systems, sea-level rise, and inland flooding (Signer et al. 2014). They organized working groups (6–8 individuals) from five coastal municipalities (St-Stephen, St-George, St-Andrews, Black's Harbour and Grand Manan). Beginning in the fall of 2013, Reeder and her colleagues organized 30 working group meetings with participating municipalities. Meeting on a bi-weekly basis, the working groups took part in an interactive community mapping exercise to identify physical, social, economic, and environmental climate vulnerabilities according to a five-step process (Fig. 1).

The CCCVA process worked to determine which community elements are most sensitive to environmental and climatic changes and to start the development of efforts that focus on building community resilience (Signer et al. 2014). A number of tools were used to co-produce local knowledge, including flooding scenarios (using LiDAR data and IPCC scenarios to create scenarios of future sea-level rise for coastal Charlotte County), wet-areas mapping to better understand inland

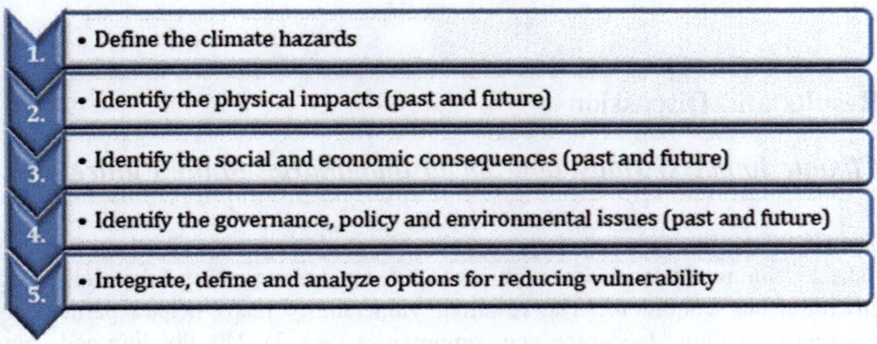

Fig. 1 Stakeholder engagement process used in the CCCVA. *Source* with permission: Signer et al. (2014)

flooding issues (on the basis of provincial digital elevation data), and participatory vulnerability mapping. The latter maps consisted of a series of numbered points representing social, environmental and economic vulnerabilities identified during working group meetings at each of the municipalities, subsequently overlaid on a community map. The working groups also identified priorities for action, based on the mapping exercises, previous discussions, expert information, and their personal experiences.

Case Study Methods

The first author collaborated with Reeder and her colleagues to design and analyze a 'pre and post-CCCVA' survey of participants' perceptions of the process, and recommendations for action. She also conducted in-depth qualitative interviews (n = 41) with working group participants to elicit their personal narratives of recent extreme weather events, flood response and recovery, perceptions of how environmental change is affecting them and their communities, and what were their expectations from the CCCVA process. Interviews were conducted halfway through the CCCVA process (November 2013) and lasted approximately one (1) hour. Interviews were recorded and transcribed for accuracy. The second author was instrumental in conveying socio-economic information to inform working group discussions.

Our analysis of the working groups and interviews is set within an inductive and interpretive research tradition in social science and humanities (Gomm et al. 2000) that seeks to generate knowledge from immersion within a particular context. Efforts were made to draw out themes from interview transcripts following a grounded theory approach where themes emerge from an iterative coding process. The analysis is also informed by the authors' notes taken during working group meetings. The results and discussion presented in the next section draw from the interviews, participant observations, and the final CCCVA report (Signer et al. 2014).

Results and Discussion

Climate Impacts, Maps, and Social Imaginaries of the Future

The CCCVA process invited experts trained in producing maps of inland flooding risks and future sea level rise scenarios to inform working groups about changing environmental conditions. The resulting vulnerability maps helped participants imagine their future landscape and communities (Fig. 2). The flooding and community vulnerability maps were essential to the working group discussions, yet for

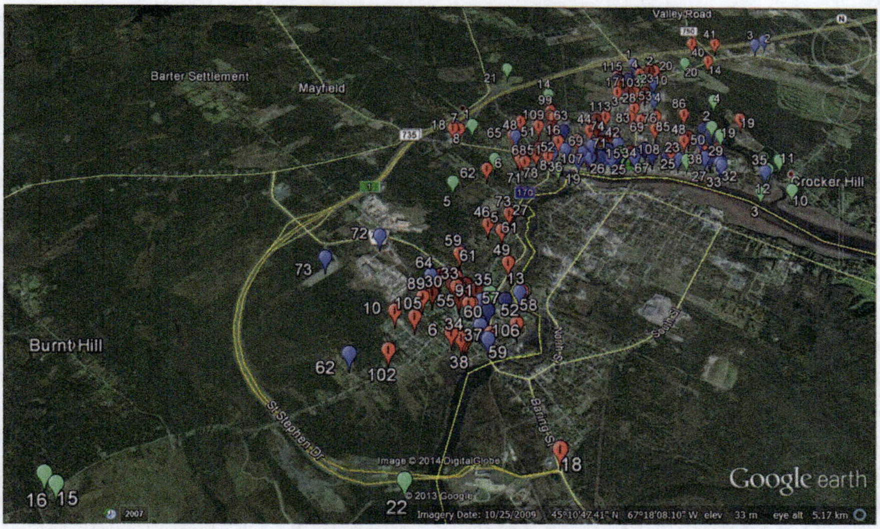

Fig. 2 Results of the community mapping exercise for St. Stephen. Indicators: blue, socio-economic impacts; green, environmental impacts; red, physical and infrastructure impacts. *Source* with permission: Signer et al. (2014)

some participants, these maps merely quantified and rendered legible something they already knew—through recent lived experience. Participants were aware that sea levels would rise due to climate change and that extreme weather events such as those they experienced, were going to become more frequent. The maps, while useful for constructing likely future scenarios, could not tell them precisely when sea level rise will happen, nor when the next severe weather event would occur. Nevertheless, producing this baseline mapping data was a central aspiration of the CCCVA process, as the maps were meant to enable participants to visualize which places within of their communities were most at risk from flooding and future sea level rise. The encounter between the lived experience, concerns and aspirations of participants and a scientific approach to vulnerability assessment was insightful, especially with regards what aspects of local knowledge were integrated into the vulnerability maps, and which were not.

Interviews with participants suggest that some of the local problems with infrastructure, for example ineffective storm water conduits and culverts, were easy to point to as "vulnerable locations" on maps. Yet for some participants, identifying these "vulnerable spots" was not sufficient. Stories participants told during working group meetings about poorly designed or ineffective infrastructure, how their businesses and homes were impacted during the recent floods, and what solutions they thought might mitigate future flood impacts, expressed the continued frustration that some participants felt with local and provincial governance arrangements. These stories intertwined memories of recent extreme weather events, local knowledge associated with livelihoods, household and community life, and affective relationships between individuals and their changing, sometimes "disastrous environment" (see also Tironi 2015). Mapping social and economic vulnerabilities as dots on a community map was meant to capture some of the more intangible vulnerabilities participants felt.

The stories participants told during working group meetings and the interviews, are not merely an emotive layer separate and independent from local knowledge, inconsequential for adaptation planning. These lived experiences provide both the social and emotive content of local knowledge, vital to understanding the nature of the vulnerability participants were asked to identify in their communities, and with which they identified (also see Abbott and Wilson 2014). While placing a numbered dot to represent a particularly "vulnerable spot" on a community map was informative from a Cartesian perspective, it could not adequately express how this vulnerability was tied to individuals' patterns of local transportation, livelihoods, and the distrust some participants felt towards those who were in charge of maintaining or updating infrastructure—important elements of local knowledge as embodied experience. Indeed, stories of recent floods and extreme weather events recounted by participants are material stories, shaped by a history of relationships, between people, place, decision-makers, and environmental change. These stories can be lost, or dismissed, if they are solely represented by a dot on a community vulnerability map. As Julie Cruikshank noted (Cruikshank 2005 p. 359): "Narrative recollections and memories about history, tradition and life experience have to be appreciated in their totality, rather than fragmented into data, if we are to learn anything from them." The resulting maps are two-dimensional: they emphasize quantities (inland flooding risks tied to elevation, projected sea level rise), and the distribution of these and other risks across Cartesian points. The stories of the recent floods, and individual life histories revealed in the interviews, speak to more than such quantifiable risks. As Smith et al (2013) point out in their study of river basin planning and spatial planning in Scotland, reliance on spatial planning maps may serve to downplay the need for broader discussion and dialogue between community members and decision-makers.

Moreover, sea level rise scenarios produced in the CCCVA were particularly revealing in how local knowledge co-production in adaptation planning can be productive of unintended risks and vulnerabilities due to the nature of the participatory materials produced—supporting Marres's (2012) argument that material participation is contingent and is normatively ambiguous. During the working group meetings concerns emerged about the community vulnerability and flooding maps would be made available to local citizens, how they might be used by real estate agents, potentially negatively impacting the value of residential and commercial buildings identified at risk from future sea level rise. It is important to note that these coastal communities are situated in a rural, economically depressed region of the province, and characterized increasingly by an aging demographic with low and/or fixed incomes (Finn 2008; Signer et al. 2014). The CCCVA maps identified homes and businesses that are at risk given projected climate change impacts—buildings that may represent the owners' only financial assets. The normative register of these maps thus shifted from a "public good", informing local climate change adaptation planning, to a new vulnerability—an unintended outcome of the CCCVA process, which revealed the contingencies of such co-productive experiments and their impacts on social imaginaries of the future.

The Emergence of Climate Adaptation Publics

During the CCCVA, working groups came to recognize their shared experiences with recent extreme weather events and articulated common concerns about climate change. The blending of stories of past extreme weather events and associated damages, responses, recovery, and the CCCVA mapping exercises fostered the emergence of new collective identities. The CCVA can be characterized by a movement away from the aggregation of data points on a map to an assemblage of flood waters, sea level rise, individual and community vulnerabilities and aspirations. In other words, the CCCVA did more than add together individual experiences and perceptions of environmental change—the extraction of local knowledge—it represented the deliberative crafting of common concerns and aspirations, the co-production of locally grounded knowledge, resulting in recommendations for collective action (Callon et al. 2011). As others have noted, flood events can be politically disruptive (Zegwaard et al. 2015), generating publics around them (Donaldson et al. 2013), and creating "windows of opportunity" to trigger policy change, reorganization, innovation and widen the range of actors involved in governance arrangements.

The CCCVA working groups suggested that a top priority for their communities was to craft locally specific but regionally coordinated emergency response plans. Working groups came to identify with a collective vision of 'preparedness'. This does not come as a surprise, as vulnerability assessments and climate change adaptation planning come from a long history of all hazard planning (Lakoff and Collier 2010). Technologies of planning make use of imaginative scenarios to articulate potential future risks that cannot be predicted with accuracy, but whose eventuality is not in question. In turn, participatory planning for preparedness assembles material objects (i.e., maps in our case) to visualize who needs protection and why, and how such measures transform the relationship between climate hazards, households, businesses and decision-makers. Yet, the CCCVA is more than an imaginative rehearsal of how to respond to climate hazards, it is also generative various and contingent normative registers of these communitities' social imaginaries of the future. Working groups' recognition that they need to be better prepared for climate impact was enmeshed with aspirations for greater social cohesion, seeking to ensure that the most vulnerable individuals within their communities are protected, while at the same time recognizing that the knowledge they produced might create new vulnerabilities for some community members.

Drawing upon John Dewey's pragmatism, Marres (2012) argues that the material dynamics of problematization and issue formation must be understood as constitutive of the very process of a public's formation. Whether through technologies, such as carbon monitoring devices installed in ecohomes (Marres 2012), or in relation to computer simulations in floor risk management (Whatmore and Landström 2011), assemblages of humans and non-humans emerge as a public when they are affected by an issue, such as climate hazards, but have no way to resolve it. From a Deweyan perspective, the CCCVA generated new climate adaptation publics: "The public consists of all those who are affected by the indirect consequences of transactions to such an extent that it is deemed necessary to have those consequences systematically cared for" (Dewey 1927: 48). In our case, working groups recognized that ecologically, politically, economically, socially—the effects of and responses to climate hazards are interconnected at different scales, from local to global.

For instance, the CCCVA revealed how participating communities felt vulnerable to decisions made upstream from their communities. Working groups observed that local is not enough—both in terms of local knowledge—hence their interest in mobilizing the expertise of climate scientists in the CCCVA—but also because local decision-making cannot work independently from the larger governance regime within which it is embedded and controlled. For example, St-Stephen working group participants identified governance challenges in planning for adaptation with regards to the control of the flow of the St-Croix river by private dam owners. Another example, is the control of water flow in the Bonny River in St George area by the private company Irving. Other governance issues relate to land-use plans allowing the building of new housing developments in flood plains; and, communities already suffering from inadequate storm water infrastructure. Participants realized their communities' need for governance arrangements that are responsive to the complex social, economic and environmental vulnerabilities associated with locally felt climate hazards. Interviewees were generally dissatisfied with current adaptation governance arrangements, both at the local and provincial scale. With regards to the province's responsibilities, participants suggested that their participation in the CCCVA working group was necessary if any adaptation plans were to be implemented. In addition, participants suggested that the province should set requirements for municipalities to put adaptation plans in place.

In summary, CCCVA participants made recommendations for how to improve the coordination of shared responsibilities between different levels of government in New Brunswick, including the need to recognize the environmental, social and economic needs and aspirations of localities and the need to more closely link locally-relevant action and regional planning processes (Signer et al. 2014). Moreover, shared ownership of impacted areas entails the need for cooperation between the municipalities and the provincial government to manage municipal infrastructure. Participants referred to the control of water flow through the St-Croix and Bonny rivers as important aspects of mitigating flood risk reduction in the area and thus critical in discussions about adaptation planning. Indeed, interviewees pointed to the need for multilevel (or multi-jurisdictional) governance of climate change adaptation planning, including the participation of the operators of the privately owned dams on the St. Croix River at Forest City, Vanceboro, Grand Falls and Milltown, communities upstream of the participants of the CCCVA. Such demands for regional coordination have been noted by others as both necessary and a key challenge for the interplay between local-regional climate adaptation governance (Fidelman et al. 2013; Antonson et al. 2016).

Governance Implications

The province of New Brunswick is facing a number of challenges to local governance, including urbanization, an ageing demographic and population decline (Finn 2008). Moreover, the province's economic path is shifting away from a rural natural resource economy to a service-based economic in urban centers (Finn 2008). Taken together, these trends are creating new challenges for rural municipalities with a shrinking residential tax basis on which they are partially dependent to pay for the services they are mandated to deliver. Additional governance challenges include the, mismatch between the large number of municipalities and Local Service Districts (LSDs are areas unincorporated into municipalities and are serviced by the province) relative to the tax base and population distribution, and the challenge of having 35% of the province's population living in LSDs without representations by local elected officials. In response to these challenges, and the lack of regional cooperation and cost-sharing between communities for the provision of services, the province created Regional Service Commissions (RSCs) in 2008. Their mandate is to develop a regional emergency plan, coordinate regional land-use planning and manage several services, services that were previously the responsibility of local decision-makers. Some RSCs have included climate change adaptation planning within their mandate, particularly in regions that have well resourced planners and a climate change adaptation plan already in motion.

New Brunswick communities face similar constraints to local and regional adaptation governance reported in studies in other countries, including competing and contradictory agendas at the local level (Mukheibir et al. 2013); public risk perceptions and attitudes towards risks (Measham et al. 2011; Adger et al. 2009; Lieske et al. 2015); event-driven and short term policy horizons (Naess et al. 2005; Measham et al. 2011; Amundsen et al. 2010; Adger et al. 2009; Dannevig and Aall 2015) and institutional constraints involving inadequate structures, processes, and distribution of responsibility across decision-making levels, and a lack in local capacity (Naess et al. 2005; Measham et al. 2011; Amundsen et al. 2010; Adger et al. 2009). Proposals for more reflexive and responsive adaptation governance arrangements suggest that these reflect the multi-level, polycentric nature of the adaptation problem, and encourage knowledge generation, sharing, and learning from a variety of sources (Emerson and Gerlak 2014; Fossum 2012). From this case study, the authors would argue that it should also strive to sustain and empower climate adaptation publics.

Recently proposed governance models can help us think about how institutional arrangements may encourage local knowledge co-production and sustain the action of emerging climate adaptation publics. Wyborn (2015) has proposed co-productive governance as a model to account for how the encounter between the cognitive, normative, material and social dimensions of knowledge systems that shape the interactions between knowledge and power, science and governance. In Wyborn's model, public participation targets "stakeholders", whose identities are already articulated around particular interests and demands. In contrast, Tironi (2015) also enrolls the idiom of co-production in his case study of climate change adaptation in Peru, but in his study it is precisely the nature of participants, the formation and marginalization of emerging climate adaptation publics that is in question. His work highlights the idea that Braun and Schultz (2010, p. 406) have aptly articulated, namely that publics then are "never immediately given but inevitably the outcome of processes of naming and framing, staging, selection and priority setting, attribution, interpellation, categorization and classification". Indeed, disaster recovery and adaptation planning are political experiments that are generative (Tironi et al. 2014). The precariousness of public formation, how it can be encouraged to organize itself, coordinate and publicize its actions, and be held accountable for its actions, requires a more focused understanding of the institutional arrangements that facilitate such political experiments (Overdevest et al. 2010; Sabel 2012; Bellamy and Lezaun 2015).

An experimentalist approach to governance may be a more helpful model to understand and critically assess the challenges of supporting local knowledge co-production, and the emergence and empowerment of climate adaptation publics. Defined as a recursive process of provisional goal-setting, and revision based on learning from the comparison of alternative approaches to advancing them in different contexts (Sabel and Zeitlin 2011), it encompasses an architecture that brings together multiple stakeholders in a multi-level decision-making process. From its pragmatic roots in the Deweyan tradition, experimentalist governance treats assumptions, knowledge and practice with skepticism and sees policies and practices as contingent, provisional and constantly subject to revision (Sabel and Zeitlin 2011). Higher levels of decision-making articulate broad and provisional framework goals such as "climate change adaptation", the implementation of which is at the discretion of local decision-makers. Local actor groups must in turn report their performance and participate in a peer-review process in which decisions and outcomes from different localities are compared. If local units fail to meet assessment standards and criteria, they are required to make necessary adjustments in consultation with other localities. Alternatively, this may result in the redefinition of original goals and assessment standards and criteria (Sabel and Zeitlin 2011).

Transparency, accountability and participatory democracy are pursued through the comparison of locally crafted approaches to advancing broad goals in different contexts. By instituting an iterative and comparative process to publicize and hold to account local efforts to advance broad goals, this experimental governance model supports the joint-production of knowledge and goals (Sabel and Zeitlin 2011). This governance model is functional rather than structural, and therefore the model can take a variety of institutional forms. As of yet, experimentalist governance has not been applied to climate change adaptation planning. Nevertheless, its focus on empowerment of local publics within an iterative, power-sharing, learning-by-doing process can conceptually lend itself well to addressing a number of the specific barriers to adaptation planning and implementation in New Brunswick. Unfortunately, while Sabel and Zeitlin (2011) effectively highlight the functional elements of the process, they do not adequately address its' implementation and sustainability.

The authors contend that the nascent foundations of experimental governance model are already in place within the Environmental Trust Fund (ETF) program, as evidenced by the CCCVA. This funding framework simply needs to be extended in a manner consistent with Sabel and Zeitlin's (2011) governance model. As already discussed, the ETF program was developed to provide competitive funding for short term, stand-alone municipal and environmental projects aimed at protecting, preserving and enhancing the Province's natural environment. Funding sources are stable insofar as the program is not sourced from government operating budgets, but is derived directly from the province's recycling program. As originally defined then, the Environmental Trust Fund is an opportunistic mechanism used to address environmental issues with short-term solutions, implemented by unspecified partners. Since its original conception though, this program has evolved to support more strategic, complex, socio-environmental projects.

To support this change in strategic direction, government departments have become more proactive in encouraging the participation of outside investigators, as well as facilitating the formation of alliances across government departments. Thus, not only are appropriate external partners encouraged to partake in the process, but the nuanced relationships that exist within the Government of New Brunswick must be negotiated, as must the relationships that exist between Provincial and Federal governments, as well as between government, academia and the business community. Much of this is implicit in the process of partnership formation, but as Beck has pointed out "a different computer, a different specialist, a different institute [results in] a different 'reality'" (Beck 1992: 56). In other words, partnership choices have implications for knowledge co-production and governance, and care must be taken to systematically address these variances (MacLellan 2008).

Increasingly, the favored approach for integrating these varied functional elements is place-based case studies. Conventional wisdom holds that case studies are not useful in generating broad insights or generalizations (i.e. generating insights about governance systems), but Flyvbjerg (2006) contends the opposite: formal generalization is overvalued as a source of scientific development, whereas "the force of example" is underestimated. Case studies provide a practical focus for dealing with highly complex, concrete problems which facilitates the testing of theories, the application of data, tools and methods, the emergence of shared learned experience, and as our case study demonstrates—the emergence of climate adaptation publics. In Flyvbjerg's typology (2006), the heuristic value of the CCVA case study appears to have been its' information orientation Flyvbjerg (2006), or rather its' ability to maximize the utility of information generated from the unusual circumstances associated with multiple flooding events within the region, as well as the possibility of sea level rise.

The motivation for the various partners to partake in the CCCVA process was clearly dependent upon their specific objectives: municipalities were primarily concerned with developing Emergency response plans and partners within the Government of New Brunswick used the opportunity to fund Lidar collection for unmapped regions of the province, to promote a system hydrological mapping, as well as extend their mapping of sea level rise along the coast. What was unexpected were the implications of this case study for the emergence of local governance variants, as has already been shown. Given this broad indication of the multiple, integrated benefits form this case study, the question for us becomes how to adjust the ETF funding mechanism to meet the functional requirements of Sabel and Zeitlin (2011) experimental governance model?

By strategically developing a coherent, iterative, linked, system of individual case studies that is sourced through the ETF mechanism, a broad range of methods and protocols can be tested in a manner that minimizes institutional risk, maximizes both the potential for innovation and application (MacLellan 2013), and helps continually redefine climate adaptation publics. The ETF process already provides a number of functional elements of Sabel and Zeitlin's (2011) approach: 'higher levels of decision-making' are clearly articulated which are then implemented at the 'discretion of local decision-makers'. The necessity for local actors to report their 'performance' is also provided within the ETF process, as well as an evaluative component that is tied to the possibility of future funding. A knowledge sharing platform appears to be a natural extension of this system, adding the missing element of institutional memory that Sabel and Zeitlin (2011) do not address. What is less evident is: (i) the nature of the comparative 'peer review process;' and (ii) the potential redefinition of original goals and assessment standards and criteria.

In terms of this evaluative component, the authors would suggest an approach that brings local climate adaptation publics together to set broad framework goals and compare their own contextualized plans (and outcomes) to achieve those goals. The climate adaptation publics would then adjust their plans based on what has worked for their peers, or collectively readjust framework goals. To address issues that spill out of particular localities, it would seem germane therefore to allocate the role of coordination, mediation and communication to a regional body (Antonson et al. 2016; Dannevig and Aall 2015; Jacobs et al. 2016), namely RSCs. This would entail RSCs and municipalities working as peers and collaborating across jurisdictional divides. An experimentalist governance arrangement would not curtail the power of local government in making decisions to mitigate climate change impacts in the way unilateral regional planning does.

Pragmatist institutions such as experimentalist governance arrangements function to systematically provoke doubt about their own assumptions and practices, they treat alternative courses of action as incomplete and corrigible; and produce an ongoing, reciprocal readjustment of ends and means through the comparison of locally crafted approaches to advancing common general aims (Sabel and Zeitlin 2011). Indeed, drawing from a Deweyan perspective on the formation of publics, this model of experimentalist governance recognizes that in the context of global environmental change, climate adaptation public formation is intrinsically linked to how climate change impacts are locally manifested and produced through specific social and material encounters. What types of local knowledge re mobilized depends on how issues are framed (Vink et al. 2013), hence the importance of reflexive governance processes that enable political dialogue involving local communities.

Conclusion

Our case study suggests that in participatory climate change adaptation planning, such as the CCCVA, local knowledge is not extracted from participants, but co-produced by encounters between different knowledge, with different emotive and normative qualities moderated by objects and environments. The CCCVA utilized a community level advisory and engagement process to identify locations, groups, and processes that are most susceptible to climate change hazards, based on individuals lived experience and scientific projections. Numerous maps with annotations on locally specific areas of vulnerability were co-produced with scientists, which in turn gave rise to new vulnerabilities associated with the mapping process itself. In addition to these maps, the CCVA process allowed participants to voice their lived experience—connecting the dots, both literally and figuratively, between how individuals felt about and perceived climate change impacts and what they imagined their future might be, and what role they could play in reducing their vulnerability to climate hazards (Signer et al. 2014).

Drawing upon recent scholarship on knowledge local co-production and material participation the authors show how the CCCVA represents the formation of emerging climate adaptation publics, which has important implications for adaptation governance. The authors propose an experimentalist approach to regional governance to empower climate adaptation publics to act upon locally framed issues and concerns, publicizing these frames and (in) actions widely across localities, and at the same time institute a regional coordination mechanism that can ensure accountability, learning and memory. These are design elements that transcend the particular challenges facing the coastal communities studied in this chapter, indeed they speak more broadly to the need to design climate adaptation governance arrangements that seek to empower and sustain emergent climate adaptation publics.

Experimentalist governance is a novel governance model that takes a different approach to the integration of knowledge and decision-making. Conceptually it encompasses a multi-scale architecture that brings together multiple stakeholders (public, private) in a polycentric decision-making process. Experimentalist governance also recognizes that given the increasing number of, and multifaceted aspects of climate hazards, emerging climate adaptation publics not only need to organize themselves to effectively craft locally appropriate solutions, obtain resources and institutional support, but decision-making bodies at all scales also need to be held to account for how they mobilize local knowledge, in its multifaceted forms (also see

Nagy et al. 2009). As this case study of the CCCVA suggests, local knowledge co-production is not enough—climate adaptation publics are multi-scalar social, material and political assemblages that require multi-scalar, experimental apparatus to empower and sustain them.

Acknowledgements This research was supported by a Canadian Social Science and Humanities Research Council Partnership Development Grant # 605512 (2013–2015): "Living with Climate Change: Mapping Experiences and Adaptation in the Global South and North to X and X. The authors are also thankful for the financial support of the New Brunswick Government Environmental Trust Fund granted to Kim Reeder for the CCCVA study.

References

Abbott, D., & Wilson, G. (2014). Climate change: Lived experience, policy and public action. *International Journal of Climate Change Strategies and Management, 6*(1), 5–18.

Adger, W. N., Dessai, S., Goulden, M., et al. (2009). Are there social limits to adaptation to climate change? *Climatic Change, 93,* 335–354.

Agrawal, A. (1995). Dismantling the divide between indigenous and scientific knowledge. *Development and change, 26,* 413–439.

Amundsen, H., Berglund, F., & Westskog, H. (2010). Overcoming barriers to climate change adaptation: A question of multilevel governance? *Environment and Planning C: Government and Policy, 28,* 276–289.

Antonson, H., Isaksson, K., Storbjörk, S., & Hjerpe, M. (2016). Negotiating climate change responses: Regional and local perspectives on transport and coastal zone planning in South Sweden. *Land Use Policy, 52,* 297–305.

Beck, U. (1992). *Risk society: Toward a new modernity*. London: SAGE Publications.

Bellamy, R., & Lezaun, J. (2015). Crafting a public for geoengineering. *Public Understanding of Science, 19,* 403–419.

Biesbroek, R., Klostermann, J., Termeer, C., & Kabat, P. (2011). Barriers to climate change adaptation in the Netherlands. *Climate Law, 2,* 181–199.

Braun, K., & Schultz, S. (2010). "…a certain amount of engineering involved:" Constructing the public in participatory governance arrangements. *Public Understanding of Science, 14,* 299–441.

Callon, M., Lascoumes, P., & Barthe, Y. (2011). Acting in an uncertain world: An essay on technical democracy. Cambridge, MA: The MIT Press.

Cloutier, G., Joerin, F., Bubois, C., Labarth, M., Legay, C., & Viens, D. (2015). Planning adaptation based on local actors' knowledge and participation: A climate governance experiment. *Climatic Policy, 15,* 458–474.

Cruikshank, J. (2005). *Do Glaciers Listen? Local Knowledge, Colonial Encounters & Social Imagination*. Vancouver and Seattle: UBC Press and University of Washington Press.

Dannevig, H., & Aall, C. (2015). The regional level as boundary organization? An analysis of climate change adaptation governance in Norway. *Environmental Science & Policy, 54,* 168–175.

Dewey, J. (1927). *The public and its problems*. New York: Henry Holt.

Donaldson, A., Lane, S., Ward, N., & Whatmore, S. (2013). Overflowing with issues: Following the political trajectories of flooding. *Environment and Planning C: Government and Policy, 31,* 603–618.

Ellen, R. F., Parkes, P., & Bicker, A. (2000). *Indigenous environmental knowledge and its transformations: Critical anthropological perspectives*. Amsterdam: Harwood Academic.

Emerson, K., & Gerlak, A. K. (2014). Adaptation in Collaborative Governance Regimes. *Environmental Management, 54,* 768–781.

Fidelman, P., Leitch, A. M., & Nelson, D. R. (2013). Unpacking multi-level adaptation to climate change in Great Barrier Reef Australia. *Global Environmental Change, 23,* 800–812.

Finn, J. G. (2008). Building Stronger Local Governments and Regions: An Action Plan for the Future of Local Governance in New Brunswick. *Report of the Commissioner on the Future of Local Governance,* Canada: Government of New Brunswick.

Flyvbjerg, B. (2006). Five misunderstandings about case-study research. *Qualitative Inquiry, 12* (2), 219–245.

Fossum, J. E. (2012). Reflections on experimentalist governance. *Regulation & Governance, 6,* 394–400.

Gomm, R., Hammersley, M., & Foster, P. (Eds.). (2000). *Case Study Method.* London, UK: Sage Publications.

Hanssen, G. S., Mydske, P. K., & Dahle, E. (2013). Multi-level coordination of climate change adaptation: By national hierarchical steering or by regional network governance? *Local Environment, 18,* 869–887.

IPCC. (2014). Summary for policymakers. In C. B. Field., V. R. Barros., D. J. Dokken., & et al. (Eds.) *Climate Change 2014: Impacts, Adaptation, and Vulnerability. Part A: Global and Sectoral Aspects.* Contribution of Working Group II to the Fifth Assessment Report of the Intergovernmental Panel on Climate Change. Cambridge, United Kingdom and New York, NY, USA: Cambridge University Press. pp. 1–32.

Jacobs, B., Lee, C., Watson, S., et al. (2016). Adaptation planning process and government adaptation architecture support regional action on climate change in New South Wales, Australia. In W. Leal Filho (Ed.), *Innovation in Climate Change Adaptation.* Berlin: Series in Climate Change Management, Springer.

Jasanoff, S. (2004). *States of knowledge: The Co-production of science and social order.* London: Routledge.

Klenk, N. L., Fiume, A., Meehan, K., & Gibbes, C. (2017). Local knowledge in climate adaptation research: Moving knowledge frameworks from extraction to co-production. *Wiley Interdisciplinary Reviews: Climate Change.* (in press).

Lakoff, A., & Collier, S. J. (2010). Infrastructure and the event: The political technology of preparedness. In B. Braun & S. J. Whatmore (Eds.), *Political Matter: Technoscience, Democracy, and Public Life* (pp. 243–266). Minneapolis: University of Minnesota Press.

Latour, B. (2004). *The politics of nature: How to bring the sciences into democracy.* Cambridge, USA: Harvard University Press.

Law, J. (2012). Collateral realities. In P. Baert & F. D. Rubio (Eds.), *Politics of knowledge* (pp. 156–178). Abingdon: Routledge.

Lezaun, J., & Soneryd, L. (2007). Consulting citizens: Technologies of elicitation and the mobility of publics. *Public Understanding of Science, 16,* 279–297.

Lieske, D. J., Roness, L. A., Phillips, E. A., & Fox, M. J. (2015). Climate change adaptation challenges facing New Brunswick coastal communities: A review of the problems and a synthesis of solutions suggested by regional adaptation research. *Revue D'Études Sur Le Nouveau Brunswick, 6,* 32–54.

MacLellan, J. I. (2008). Brokering the Local Global Dialectic. In A. Fenech & J. I. MacLellan (Eds.), *Linking Climate and Impact Models to Decision and Policy Making Environment.* Toronto: Canada.

MacLellan, J. I. (2013). Research Scoping Framework and Exercise: Potential 2014–2015 ETF Research Projects. In J. I. MacLellan & D. Floyd (Eds.), *The New Brunswick Climate Change Research Collaborative ETF 130240 Final Report.* Canada: New Brunswick Climate Change Research Collaborative, University of New Brunswick, New Brunswick.

Marres, N. (2012). *Material Participation: Technology, The Environment and Everyday Publics.* London: Palgrave Macmillan.

Measham, T. G., Preston, B. L., Smith, T. F., et al. (2011). Adapting to climate change through local municipal planning: Barriers and challenges. *Mitigation and Adaptation Strategies for Global Change, 16,* 889–909.

Mukheibir, P., Kuruppu, N., & Gero, A. (2013). Overcoming cross-scale challenges to climate change adaptation for local government: A focus on Australia. *Climatic Change, 121,* 271–283.

Naess, L., Bang, G., Eriksen, S., & Vevatne, J. (2005). Institutional adaptation to climate change: Flood response at the municipal level in Norway. *Global Environmental Change, 15,* 125–138.

Naess, L. O. (2013). The role of local knowledge in adaptation to climate change. *Wiley Interdisciplinary Reviews: Climate Change, 4,* 99–106.

Nagy, G., Seijo, L., Verocai, J., & Bidegain, M. (2009). Stakeholders' climate perception and adaptation in coastal Uruguay. *International Journal of Climate Change Strategies and Management, 6*(1), 63–84.

Nilsson, A. E., Gerger Swartling, Å., & Eckerberg, K. (2012). Knowledge for local climate change adaptation in Sweden: Challenges of multilevel governance. *Local Environment, 17*(6–7), 751–767.

Overdevest, C., Bleicher, A., & Gross, M. (2010). The experimental turn in environmental sociology: Pragmatism and new forms of governance. In M. Gross & H. Heinrichs (Eds.), *Environmental Sociology: European Perspectives and Interdisciplinary Challenges* (pp. 279–322). Berlin: Springer.

Province of New Brunswick. (2014a). New Brunswick Climate Action Plan 2014–2020. Retrieved from the Department of Environment and Local Government http://www2.gnb.ca/content/dam/gnb/Departments/env/pdf/Climate-Climatiques/ClimateChangeActionPlan2014-2020.pdf.

Province of New Brunswick. (2014b). New Brunswick's Flood Risk Reduction Strategy. Retrieved from the Department of Environment and Local Government http://www2.gnb.ca/content/dam/gnb/Departments/env/pdf/Publications/NBFloodRiskReductionStrategy.pdf.

Reyes-García, V., Fernández-Llamazares, A., Guèze, M., et al. (2016). Local indicators of climate change: The potential contribution of local knowledge to climate research. *Wiley Interdisciplinary Reviews: Climate Change, 7,* 109–124.

Rojas, D. (2015). Environmental management and open-air experiments in Brazilian Amazonia. *Geoforum, 66,* 136–145.

Sabel, C. (2012). Dewey, democracy, and democratic experimentalism. *Contemporary Pragmatism, 9,* 35–55.

Sabel, C., & Zeitlin, J. (2011). Experimentalist Governance. In D. Levi-Faur (Ed.), *The Oxford Handbook of Governance*. Oxford, UK: Oxford University Press.

Signer, K., Reeder, K., & Killorn, D. (2014). Community vulnerability assessment of climate change and variability impacts in Charlotte County, New Brunswick; Eastern Charlotte Waterways Inc., Canada: New Brunswick.

Smith, H. M., Wall, G., & Blackstock, K. L. (2013). The role of map-based environmental information supporting integration between river basin planning and spatial planning. *Environmental Science & Policy, 30,* 81–89.

Storbjörk, S., & Hedrén, J. (2011). Institutional capacity-building for targeting sea-level rise in the climate adaptation of Swedish coastal zone management. Lessons from Coastby. *Ocean & coastal management, 54*(3), 265–273.

Termeer, C., Dewulf, A., van Rijswick, H., et al. (2011). The regional governance of climate adaptation: A framework for developing legitimate, effective, and resilient governance arrangements. *Climatic Law, 2,* 159–179.

Tironi, M. (2015). Disastrous publics: Counter-enactments in participatory experiments. *Science, Technology, & Human Values, 40,* 564–587.

Tironi, M., Rodríguez-Giralt, I., & Guggenheim, M. (2014). *Disasters and Politics. Materials, Experiments, Preparedness*. Chichester, UK: Wiley Blackwell.

Turnbull, D. (2003). *Masons, Tricksters and Cartographers*. London: Routledge.

Urwin, K., & Jordan, A. (2008). Does public policy support or undermine climate change adaptation? Exploring policy interplay across different scales of governance. *Global Environmental Change, 18,* 180–191.

Vink, M. J., Boezeman, D., Dewulf, A., & Termeer, C. J. A. M. (2013). Changing climate, changing frames. Dutch water policy frame developments in the context of a rise and fall of attention to climate change. *Environmental Science & Policy, 30,* 90–101.

Whatmore, S. J., & Landström, C. (2011). Flood apprentices: An exercise in making things public. *Economy and Society, 40,* 582–610.

Wyborn, C. (2015). Co-productive governance: A relational framework for adaptive governance. *Global Environmental Change, 30,* 56–67.

Zegwaard, A., Petersen, A. C., & Wester, P. (2015). Climate change and ontological politics in the Dutch Delta. *Climatic Change, 132,* 433–444.

Economic Resiliency and Food Security in the Marshall Islands Through *Polydactylus sexfilis* Aquaculture

Kathleen Hicks and Ryan Murashige

Abstract Climate change in the Pacific puts pressure on one of the most basic human needs: food. In the Marshall Islands, pressures on the coastal ecosystem have increased due to a high volume of fishing. In the context of climate change, the economic livelihood and day-to-day food procurement of residents of Pacific islands are threatened due to stress on the coral reef ecosystem and specifically the fish populations that provide these livelihoods. The RMI aquaculture fisheries project is an experiment in one option to address the vulnerability of the fishing sector to climate change. The impacts of the RMI aquaculture fisheries project are three-fold. First, when enacted on a large scale, aquaculture will provide a sustainable, exportable fishing business to offer an alternative to purse seine tuna fishing, increasing the economic resiliency of the Marshall Islands. Second, when enacted on a small scale, will offer a secure source of animal protein that is locally produced. Thirdly, there will be less dependence on expensive imported agricultural inputs by instead using locally sourced by-catch for fish feed. This paper outlines the basic principles, necessary infrastructure, and results from October 2015 to January 2017 of hatchery and nursery operations, feed production, and open-ocean cage grow-out utilized by the Aquaculture Technologies of the Marshall Islands (ATMI). The ATMI program can be used as a pilot program to inform others in the Pacific interested in cage aquaculture as a climate change adaptation strategy for coastal communities.

K. Hicks (✉) · R. Murashige
Aquaculture Technologies of the Marshall Islands, 3378, Majuro,
MH 96960, Marshall Islands
e-mail: khicksd12@gmail.com

R. Murashige
e-mail: hukilau11@gmail.com

W. Leal Filho (ed.), *Climate Change Impacts and Adaptation Strategies for Coastal Communities*, Climate Change Management,
https://doi.org/10.1007/978-3-319-70703-7_15

Introduction

Pacific Islands and atoll nations in particular face many severe upcoming challenges due to climate change. Climate change increases the vulnerability of islands like the Marshall Islands. Atoll nations struggle to find resources for adaptive measures due to many factors, including economic vulnerability and unsustainable development (Barnett and Adger 2003).

Vulnerability is a result partly due to few coping options and low adaptive capacity, involving factors such as financial and natural resources and the networks that manage them (Brooks 2003). Economic development and new opportunities to manage the coastal resources will become increasingly important as climate change progresses. New technologies and new connections within the community will need to be formed to adapt to the increasing severity of climate change effects. (Tompkins and Adger 2004).

One key area in which climate change will take its toll in the Marshall Islands is its fishing industry. Ocean temperatures rise and conditions become more acidic from increased atmospheric carbon dioxide, both of which are detrimental to coral reef ecosystems and to the human societies dependent on them. Severe weather can damage coastal infrastructure and as many in the RMI depend on fishing as a livelihood, many stand to be affected by climate change and climate events. The struggling private sector involved with fishing as an economic activity is vulnerable because few are educated on the options for coping with the hazards of climate change (Brooks 2003). New techniques and the ability to adjust the species targeted and harvest methods will be important in increasing their adaptive capacity (Badjeck et al. 2010) New options are also important as human fishing activities cause noticeable decreases in fish populations in comparison to unpopulated islands in the RMI (Houk and Musburger 2013). As population density increases on islands like Majuro in the RMI, fishing pressure and ways to relieve that pressure will become an even greater factor in food security. This is why it was necessary to test the viability of aquaculture in the coastal community of Majuro. Aquaculture, or farming of aquatic plant and animal species, is a possible adaptation strategy to reduce the stress from secondary effects of habitat loss from ocean acidification (Cooley and Doney 2009).

Marshall Islands and Climate Change

The Marshall Islands economy is struggling. Private sector development is slow and the RMI has few exportable resources (Chutaro 2005). Even in industries that are alive and well, the private sector in the RMI sees little of the benefits. The Marshall Islands fishing industry is predominated by foreign vessels, with license fees and income related to transshipment activities as the main contribution to the RMI economy from the tuna fishery. Much of the income from the tuna fishery goes to the government and not the private sector (Moss 2007). Majuro's unemployment is approximately 30.8% with some areas as high as 47%. Even if employed, it is not uncommon for Marshallese working in the fishing sector to receive below minimum wage. Fertility rates in the Marshall Islands are some of the highest in the world (Chutaro 2005). Such economic stresses run the risk of crippling the RMI's ability to manage its response to the pressures on natural resources and food security that are promised by climate change predictions.

As an atoll nation, the RMI faces an increasing risk of inundation threatening land-based programs. It is evident from recent decades that sea level rise will increase the severity of inundations in the Pacific, particularly when coupled with El Niño/Southern Oscillation (Hoeke et al. 2013). Land-based agricultural activities face the risk of stress from saltwater inundation and drought events as climate change progresses (Barnett and Adger 2003). Ocean-based food sources are important to offset land-based agriculture for food security. Coastal communities must experiment with adaptation measures that allow them put more of the food burden on the ocean.

The RMI depends on the health of coral reef systems and a thriving fishery for economic well-being, recreational fishing activities, and for some, daily sustenance. Though productive, the tuna fishery is predominated by foreign vessels, management policies encourage unsustainable overfishing, and climate shifts are expected to add even more pressure to these resources (Moss 2007). Reef fishing is more common for private sector fishermen in the Marshall Islands. However, the coastal resources in the Marshall Islands are becoming more and more at risk. Corals face pressures from local anthropogenic factors as well as the rising temperatures of the atoll oceans. The worst recorded RMI coral bleaching event occurred in late 2014 and Majuro was particularly affected, with some groups of corals showing bleaching in more than half of the population (Fellenius 2014) As corals are affected by global warming, the reef fish can be affected both by the loss of their food source as well as a decrease in topological complexity. Numbers and diversity of fish can decrease. Even fish not directly feeding off of coral are affected because of loss of habitat or loss of predators. This can cause the value of fishermen's catches to decrease (Pratchett et al. 2008). The coral reef ecosystem and Marshall Islands fisheries may also face difficulties from increasing acidification, causing decline in coral reef habitat (Cooley and Doney 2009).

Polydactylus sexfilis Aquaculture

As current fishing methods leave the RMI vulnerable to effects of climate change on their fisheries, experimenting with aquaculture is important in order to evaluate other options. In order to decide which species to farm, species that have had success in similar coastal areas in the region were considered. *Polydactylus sexfilis* (moi) aquaculture has a well-established history in Hawaii. In nature, moi adults are found around sandy or rocky bottoms and feed on small crustaceans or other invertebrates. Their larvae, however, may use coral reef structures as shelter after metamorphosis (Callan et al. 2012). Moi have a regular spawning schedule, a good feed conversion ratio, high quality flesh with a good demand in markets both North America, Asia, and the Pacific (Callan et al. 2012). Aquaculture in Hawaii has been enacted both on land and open-ocean. Open-ocean grow-out aquaculture does not require pumping of large volumes of seawater to land-based facilities. Moi were primarily chosen as a suitable test for exploring aquaculture in the RMI because of how well established their aquaculture practices are. An added benefit was that this fish is not typically hunted, so if successful it could increase the species available in the local market and decrease pressure on other, more heavily fished and at risk species in RMI coastal areas.

The RMI Aquaculture Fisheries Project

ATMI, a private commercial venture, brought aquaculture practices and technologies to the Marshall Islands in 2013. A hatchery/nursery facility, feed manufacturing facility, and cage grow-out facility were constructed. The results of this venture will inform others in the region of the viability of moi aquaculture. Data from the project were collected to see if moi was feasible to produce in commercial quantities. Acceptance of the technique by the community is also key in realizing the adaptive capacity of the community to decrease its vulnerability (Brooks 2003). If well-received as an alternative sustainable activity in the local fishing sector, moi aquaculture can be recommended as an adaptation strategy to decrease the vulnerability of Pacific coastal communities. ATMI's hatchery facility is located on land in Ajeltake, Majuro. This allows for better control of the temperature and water quality that the moi larvae and broodstock are exposed to. Fingerling moi are transferred from the hatchery to the grow-out cage location at Drirej. Ocean cages can be surface or submerged, but the location and weather conditions in the Majuro lagoon allow ATMI to use surface cage structures. Through the RMI Aquaculture Fisheries Project (funded by a PACAM grant from April 2015–March 2018) ATMI is putting more resources towards training for both expertise and experience for prospective employees as well as disseminating best practices to interested parties. Fishing is a popular traditional practice so launching new production methods makes sense both in the context of harnessing traditional practices and increasing opportunities in the global economy for assisting the RMI's adaption to climate change (Barnett and Adger 2003).

Methods

Staff Recruitment and Development

Training knowledgeable employees has been a key focus of ATMI under the PACAM funded RMI Aquaculture Fisheries Project. Trainings are offered periodically in each area of aquaculture production: hatchery/nursery operations, feed mill, and cage grow-out. Trainings allow for more people to be exposed to the new technology of aquaculture, including women who traditionally have low involvement in fishery efforts. Trainings provide for community connections as people from different areas on Majuro and from different islands are in touch with each other. ATMI is also committed to development of the local work force by promoting employees from within for supervisor and management positions. Additional empowerment of women through our trainings and empowerment of individuals through merit-based promotion of local employees will help contribute to the RMI's resilience to climate change (Tompkins and Adger 2004). High risk groups and the general population will be taught new adaptation strategies to be used in climate-sensitive economic activities.

Hatchery/Nursery

ATMI's hatchery and nursery practices are both local and sustainable. The facility is a land-based facility. Broodstock, hatchery, and nursery tanks are located in the indoor facility. Broodstock are obtained from wild fish on the neighboring Arno atoll and raised at the facility until they are mature enough to spawn. Ocean water is pumped into the hatchery facility for the broodstock, larval rearing tanks, and nursery tanks. Eggs are collected by hand after monthly broodstock spawning and hatched overnight in a separate hatchery room, where they are raised for approximately one month before being moved to the nursery tanks. Newly hatched larvae are fed from live feed grown in the hatchery facility. The fish are in the nursery tanks for another month and then transferred to the cage grow-out facility upon reaching fingerling size. Along with providing fish for commercial sale, aquaculture technologies have the potential to be used for building up wild stocks of fish (Leber et al. 1998).

Open-Ocean Cage Grow-Out

ATMI's cage grow out practices are very sustainable. After transfer, fish take approximately 7 months to reach harvestable size. The site selected is sandy bottom and at a depth of 80–100 feet. This ensures that any sedimentation or eutrophication that may occur would not stress a coral reef structure. These are both combatted, however, by the steady current which consistently removes wastes and provides fresh oxygenated water to the growing fish. Cage system anchors are buried in the sand and do not destroy any coral habitat during installation. Cage maintenance requires predominantly just manpower, without using chemicals. Good feeding practices are followed to prevent pollution; fish are monitored by snorkelers and divers in the water to ensure they are well-fed, but not overfed, and as much as possible pellets are used as they do not spread oils to the water. Daily oxygen monitoring, quarterly ammonia monitoring, and benthic monitoring twice a year of the site ensure that the site is not harming the environment.

Feed Production

Feed production for ATMI's cage grow-out site relies to a large extent on locally-sourced protein ingredients. The only feed ingredient that must be imported is cassava powder. Starch from local breadfruit would cost $8.00 per pound, which is prohibitively expensive at the scale of production. Furthermore, local breadfruit farming would not be able to supply the volume of starch needed. The source of protein in the feed is bycatch obtained locally from the tuna fishing vessels that dock in the Majuro Lagoon, serving a dual purpose of being readily available at low cost to ATMI as well as reducing waste from the tuna boats, the main fishing industry. Fishmeal used in feed is a by-product of local tuna plant Pan Pacific Foods. Thus the feed ingredients do not require any extra pressure on fishing resources because they are making use of what would otherwise go to waste.

Limitations

Both the feed mill and hatchery facilities require an initial investment in infrastructure. The feed mill requires machinery, a facility with a sewage system, and cold storage for ingredients and the finished feed. The hatchery facility requires infrastructure for water circulation and oxygen supply for broodstock, larvae, nursery fish, and live feed. A backup power supply is required in the event of a power failure.

ATMI has only been able to devote resources to researching the feasibility of one fish species for commercial aquaculture. Moi was chosen because of success in similar environments. Moi broodstock, once domesticated, lay eggs monthly. However, broodstock fresh from the wild (as required for responsible genetic practices in aquaculture) follow the more natural spawning pattern of only spawning in the summer months. Larger broodstock are sensitive and require a lot of attention to detail in their care, in everything from their feeding to ensuring a steady supply of fresh water pumped into their tank. Experience of employees caring for the broodstock is therefore very important, and errors can cause delays in production.

Gear must be procured for the cage site such as scuba gear, tanks, and cleaning equipment to clean the cage nets. Boats and gasoline for transportation to the site must be provided. Buoys and cage nets must be imported and installed.

Results

As of January 2016, weekly harvest records show that ATMI has harvested a grand total of 20,224.97 pounds of fish since the first harvest in October 2015. Tests harvests were repeated in December of 2015. More frequent harvests began after that in April 2016 and moved to a more regular weekly schedule in the second half of May. Total monthly harvests are tallied in Fig. 1a. The spread in local popularity of moi fish coupled with successful negotiations and shipments with the international market has allowed for exponential growth in ATMI's moi trade, visible in the graph in Fig. 1b.

(a)

Total Moi Harvested	
Month	Pounds
October	50
November	0
December	441
January	0
February	0
March	0
April	1089.4
May	662.5
June	362
July	1300
August	1505
September	1258.9
October	1549.6
November	2432.8
December	3763.62
January	5,810.15
Total	20224.97

(b)

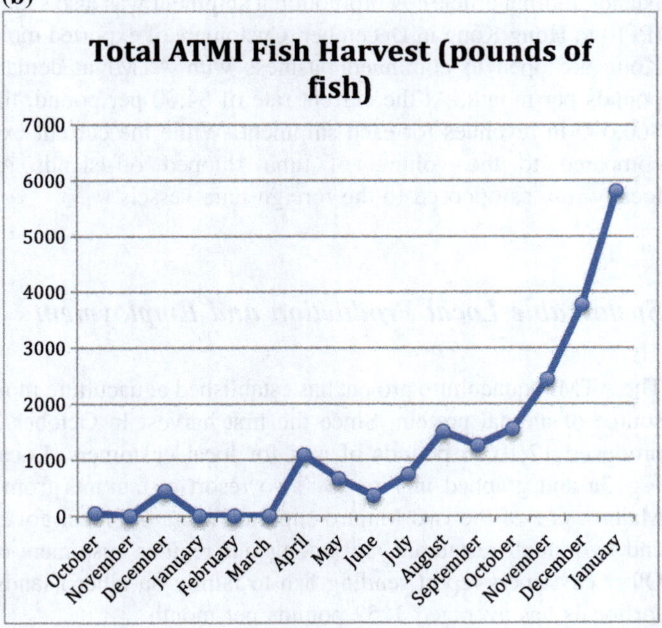

Fig. 1 **a** Charts the monthly totals of fish harvested from October 2015 through January 2017. Test harvests occurred in October and November 2015. In April, May, and June, harvests were intermittent. Regular Wednesday harvests began in July 2016 and continue to the date of the authoring of this paper. **b** Displays graphically the increase over time of the total monthly harvests that were possible as a result of the ATMI aquaculture program

The cage grow-out site currently has six net enclosures installed. Over the time period included in Fig. 1, only the first 2.5 cages of moi produced in Majuro were harvested, translating to 6.4 months per cage. ATMI plans to add another 6 cage enclosures in 2017, resulting in 12 enclosures eventually harvested at the rate of one cage per month. Thus the moi production in Majuro has great potential for growth. The results of this study therefore represent preliminary findings. Further insight into the full extent of production of the site would be gained by collecting data after all 12 cages are active.

Sustainable Exports

The results of harvests and exports from ATMI's aquaculture project show that it is a sustainable, exportable alternative to tuna fishing. Importantly, by raising fish from eggs, the wild stock of moi is not affected even with a high volume of production. ATMI's exports have increased rapidly in recent months, with the volume shipped in January alone totaling more than in all previous months together (see Figs. 2a, b). ATMI has two main commercial exporters: Kendall Micronesian Shipping Agency (KMI) and Marshall Islands Fishing Venture (MIFV) (Marshall Islands Journal article). A promotional shipment was also sent by Pan Pacific Foods (PPF) to Hong Kong in December. Customers of exported moi in Taiwan and Hong Kong are open to continued business with ATMI at demand of almost 15,000 pounds per month. At the current rate of $4.00 per pound, this would translate to $60,000 in revenues for each shipment. While the current export volume is small compared to the volume of tuna shipped off-island, ATMI is completely locally-run, as opposed to the foreign tuna vessels.

Sustainable Local Production and Employment

The ATMI aquaculture project has established aquaculture moi fish as a viable local source of animal protein. Since the first harvest in October 2015 the project has produced 12,102.6 pounds of moi for local customers. Local sales are tallied in Fig. 3a and graphed in Fig. 3b. Two resort restaurants from the major hotels on Majuro, one of the two Majuro supermarkets, one local government organization, and two small restaurant are among our regular customers ordering every week. Other customers report sending fish to family on other islands. Since May harvest for locals has averaged 1252 pounds per month.

Additional local benefits come from the cage structures. Cages act as a fish aggregating device. Species attracted to the area include remora (*Echeneis maucrates*), mackerel, various tuna species, various mullet species (*Crenimugil crenilabis* and *Liza vaigiensis*), bluefin trevally (*Caranx melampygus*), trumpet emperor

(a)

Exported Moi	
Month	Pounds
October	0
November	0
December	0
January	0
February	0
March	0
April	742.4
May	0
June	0
July	559
August	0
September	0
October	0
November	571.2
December	1876.62
January	4,373.15
Total	8,122.37

(b)

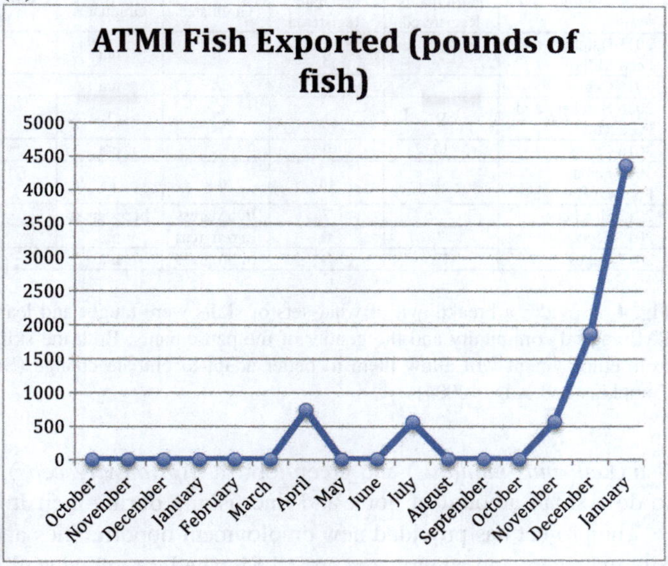

Fig. 2 **a** Charts the monthly totals of fish exported internationally from October 2015 to January 2017. Fish were exported via PPI, KMI, and MIFV to Hong Kong, Taiwan, and Hawaii, respectively. **b** Displays graphically the increase of commercial harvests achieved each month

(a)

Local Moi Harvest	
Month	Pounds
October	50
November	0
December	441
January	0
February	0
March	0
April	347
May	662.5
June	362
July	741
August	1505
September	1258.9
October	1549.6
November	1861.6
December	1887
January	1,437.00
Total	12,102.6

(b)

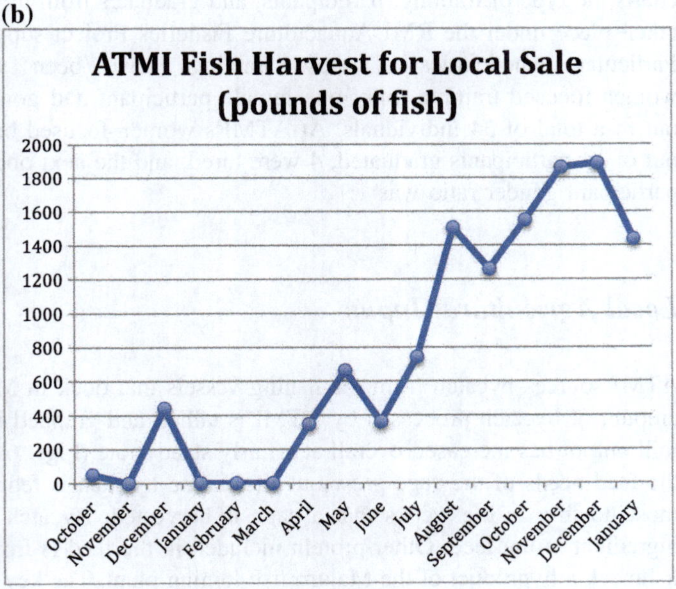

Fig. 3 **a** Charts the monthly totals of fish sold locally on Majuro. Increased awareness of moi fish has assisted with increased sales. ATMI has dependable weekly harvests to provide fish for local customers. Many customers order every week such as restaurants and supermarkets. **b** Displays graphically the overall increase in local fish sales

Training	Male Participants Registered	Female Participants Registered	Male Graduates	Female Graduates	Males Subsequently Hired	Females Subsequently Hired
2015 Hatchery and Feed Mill	5	0	4	n/a	4	n/a
2015 Cages	7	0	4	n/a	4	n/a
2016 Hatchery and Feed Mill	9	1	7	1	7	1
2016 Cages	12	0	9	n/a	9	n/a
2016 Female-Focused Hatchery	0	14	n/a	13	n/a	4
2017 Hatchery	7	7	In progress	In progress		
2017 Feed Mill	7	0	In progress	n/a		
2017 Cages	12	0	In progress	n/a		

Fig. 4 Provides a breakdown of what sets of skills were taught and learned by members of the RMI coastal community and the gender of the participants. Both the skills and economic capital from employment will allow them to better adapt to climate change (Barnett and Adger 2003; Tompkins and Adger 2004)

fish (*Lethrinus miniatus*) and green jobfish (*Aprion virescens*). Employees are able to do a small amount of hook and line fishing during their lunch hour.

The project has provided new employment opportunities and staff development. Site managers and supervisors are all Marshallese, office staff is 60% Marshallese. All staff, managers, and supervisors have been trained by ATMI since it is the first commercial fish aquaculture endeavor in the RMI. Besides aquaculture-specific skills, participants have reported that some of the strengths of the training were extra instruction in aspects of math, measuring, and English language. Figure 4 charts the type of training, participants, and graduates from the trainings that have taken place under the RMI Aquaculture Fisheries Project supported by PACAM. Particular beneficiaries of these trainings have been women. Before a women-focused training, only one female participant had gone through trainings out of a total of 34 individuals. At ATMI's women-focused hatchery training, 13 out of 14 participants graduated, 4 were hired, and the next open hatchery training participant gender ratio was 1:1.

Local Agricultural Inputs

ATMI sources bycatch from the fishing vessels that dock in Majuro Lagoon. The amount of bycatch processed by ATMI is tallied and graphed in Figs. 5a, b. Feed mill output has increased overall at a fairly steady rate (Figs. 6a, b), in response to the feed needs of the cage grow-out site. Some bycatch is fed directly to the fish, most notably in the weeks that a cage is harvested. Bycatch is also used as an ingredient in the feed. Other protein included in the feed is from locally produced fishmeal, a byproduct of the Majuro fish loining plant. The key accomplishment of

(a)

ATMI Bycatch Production	
Month	Kilograms
October	0
November	0
December	2402
January	3745
February	9500
March	8480
April	13200
May	4300
June	3940
July	3887.4
August	34493.7
September	4678.4
October	4350
November	20657.7
December	27652.2

(b)

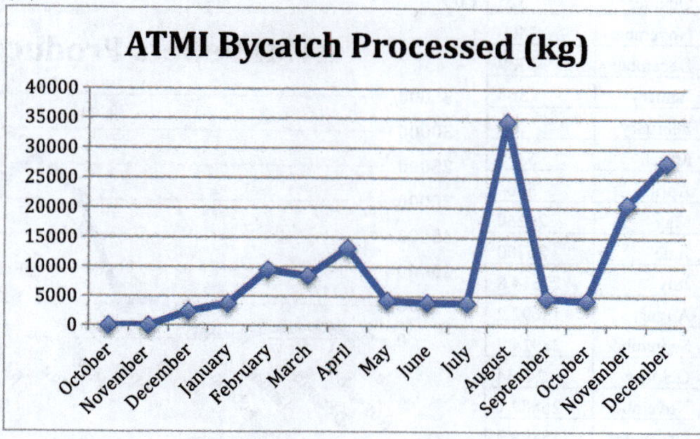

Fig. 5 **a** Charts the amount of ground bycatch produced by ATMI's feed mill from October 2015 until December 2016. **b** Graphically represents the amount of bycatch processed by ATMI's feed mill

(a)

ATMI Total Production	
Month	Kilograms
October	325
November	1400
December	4131
January	4113
February	10250
March	13870
April	18464
May	24550
June	30240
July	26502.2
August	48185.9
September	33602.6
October	29084
November	47544.8
December	56551.1

(b)

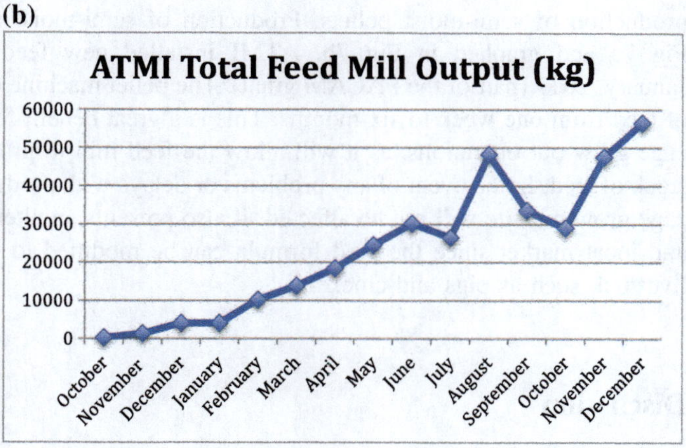

Fig. 6 **a** Charts the data collected for the amount of total feed produced by ATMI's feed mill from October 2015 until December 2016. No figures on the production of dry feed were available. **b** Graphically represents the increase in overall feed mill output. Note that the feed production increased in response to the need for feed from the cage grow-out site

(a)

ATMI Pellet Production	
Month	Kilograms
October	325
November	1400
December	1729
January	368
February	750
March	5390
April	5264
May	20250
June	26300
July	22614.8
August	13692.2
September	28924.2
October	24734
November	26887.1
December	28898.9

(b)

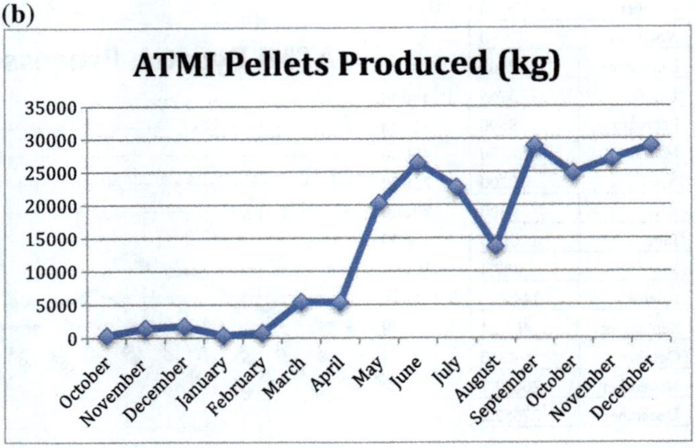

Fig. 7 **a** Charts data collected for the amount of semi-moist pellets produced by ATMI's feed mill from October 2015 until December 2016. No figures on the production of dry feed were available. **b** Graphically represents the increase in production of semi-moist pellets over

the feed mill is that has been able to supply 100% of the feed needs of the cage grow-out site; no feed for the cages has had to be purchased from an outside source.

To date, the only data available about formulated feed production is on the production of semi-moist pellets. Production of semi-moist pellets is charted in Fig. 7a and graphed in Fig. 7b. ATMI installed new feed mill equipment in January, 2017 (part of the PACAM grant). The pellet machine extends the shelf life of feed from one week to six months. This is a great benefit for the security of the cage grow-out operations, as it will allow the feed mill to produce a large reserve stock of feed. In the event of any problems or delays with production, feeding at the cage grow-out site will not be affected. It also presents an alternate revenue stream and local market since the feed formula can be modified to make feed for other livestock such as pigs and chickens.

Discussion

Since the establishment of ATMI, there has not been expansion to other islands. ATMI offers a source of both feed and fingerlings for those interested. Advertising efforts may need to be increased in order to reach possible interested parties. Though less populated, the outer islands of the RMI exhibit other risk factors of climate change vulnerability (Barnett and Adger 2003) and therefore they are an important audience for the benefits of aquaculture technologies. As production stabilizes, ATMI will do a closer cost analysis to more exactly determine profit margins. A closer analysis on the feed conversion ratio can help better plan for feed needs as production expands.

ATMI is still in the process of scaling up its operations. Full capacity has yet to be reached at the cage grow-out site and with feed production. Other parties in the region starting hatchery operations should pay particular attention to the maturation of their broodstock and the design of their hatchery facility, particularly in securing sources of sea water in case of any equipment or power malfunction. After fish production has been maximized, the possibilities of the feed mill facility in providing food to other aquaculture facilities in the region can be further explored.

An additional benefit of the feed mill technology used by ATMI is the option to use different feed ingredients to produce feed for different species. ATMI has the potential to diversify feed production to market to chickens, pigs, or other species of fish. This ability to diversify is an important aspect in increasing the resiliency and decreasing the vulnerability of the RMI's aquaculture industry (Badjeck et al. 2010). ATMI will continue to consider alternative sources for the starch powder necessary to make the feed.

The key success of the ATMI aquaculture project has been establishing *Polydactylus sexfilis* as a valuable market species. Though native to the Marshall Islands, this species was not popular for human consumption before the project but now is a well-recognized species. This ability to shift target species is key in facing possible challenges from climate change (Badjeck et al. 2010; Cooley and Doney 2009). ATMI also recognizes the benefit of diversification in experiments with development of broodstock of different species. Current focus for broodstock diversification is rabbitfish. ATMI has even experimented with small-scale production of seaweed. Seaweed production is very simple, as it just requires circulation of ocean water in a sunny location, however it is not currently a popular item for consumption on a large scale in the RMI.

Conclusion

The ATMI aquaculture project has proved that international exports of locally-produced *Polydactylus sexfilis* fish is possible. Moi are indeed a viable species for aquaculture in the Pacific region. Shipments to Taiwan, Hong Kong, and Hawaii have allowed ATMI to experiment with shipping and markets. If current production rates continue to increase, markets in Taiwan and Hong Kong offer the possibility of $60,000 per month in revenues. These new markets and fishing technologies benefit the Marshall Islands' economic resiliency, giving them capital for necessary climate change adaptation. It is likely that other coastal communities would similarly benefit economically from inclusion of aquaculture as a new technique in their fishing industry.

Local fish purchases from supermarkets, restaurants, governmental organizations, and individuals show that ATMI's moi fish are now a widely recognized as a viable local source of fish protein for human consumption. Aquaculture operations at a non-industrial scale would still have a reliable customer base. ATMI's training programs spread this expertise to employees and other interested parties. Together these indicate that moi aquaculture has been accepted by the local community, enhancing adaptability in the use of coastal resources under threat from climate change. The acceptance of moi aquaculture by the local community in Majuro indicates that in similar communities in the region, it is likely that aquaculture technology as an alternative to traditional fishing would be well received.

In feed production, ATMI has uniquely made use of the local commercial tuna industry's waste. In this way, ATMI is helping locals benefit from an industry dominated by foreign vessels. ATMI moi fish are truly a local product. The feed facility can meet 100% of its fishmeal and bycatch needs from local sources, even after allowing for expansion of the feed mill production. This helps ensure the sustainability of moi as a food source and enhances RMI food security. Possibilities for exporting feed as well as moi will add to the economic benefits of this pilot project. The unique use of bycatch as an ingredient if fish feed has the potential to be replicated by other coastal economies involving large-scale tuna fishing by purse-seine vessels.

Overall ATMI's introduction of *Polydactylus sexfilis* aquaculture in the Marshall Islands enhances the ability of the Marshallese fishing industry to increase economic resiliency and food security to reduce their climate change vulnerability, particularly by introducing new aquaculture technologies that are adaptable to the uncertain conditions resulting from global climate change.

Acknowledgements ATMI thanks USAID/PACAM for their generous funding which made the majority of our trainings possible and for their assistance with our program in general.

References

Badjeck, M., Allison, E., Halls, A., & et al. (2010). Impacts of climate variability and change on fishery-based livelihoods. *Marine Policy 34*: 375–383. Oxford: Elsevier.

Barnett, J., & Adger, W. (2003). Climate dangers and atoll countries. *Climatic change*, *61*(3), 321–337. Netherlands: Kluwer Academic Publishers.

Brooks, N. (2003). Vulnerability, risk and adaptation: A conceptual framework. *Tyndall Centre for Climate Change Research Working Paper*, *38*, 1–16. Norwich UK.

Callan, C. K., Laidley, C. W., Ostrowski, A. C., & Molnar, A. (2012). *Pacific Threadfin, Polydactylus sexfilis (Moi), Hatchery Manual*. Waimanalo, HI, USA: Center for Tropical and Subtropical Aquaculture.

Chutaro, B. (2005). *Social and Economic Baseline Survey: Jenrok Village, Majuro (Republic of Marshall Islands). IWP-Pacific Technical Report (International Waters Project), No. 15*. SPREP: Apia, Samoa.

Cooley, S., & Doney, S. (2009). Anticipating ocean acidification's economic consequences for commercial fisheries. *Environmental Research Letters 4*. IOP Publishing. Accessed Feb. 21, 2017. stacks.iop.org/ERL/4/024007.

Fellenius, K. (2014). *Republic of the Marshall Islands Coral Bleaching Report*. Paper for Marshall Islands Coastal Management Advisory and NOAA Coral Watch. College of the Marshall Islands: University of Hawai'I Sea Grant.

Hoeke, R. K., McInnes, K. L., Kruger, J. C., & et al. (2013). Widespread inundation of Pacific islands triggered by distant-source wind-waves. *Global and Planetary Change*, *108*, 128–138. Oxford: Elsevier.

Houk, P., & Musburger, C. (2013). Trophic interactions and ecological stability across coral reefs in the Marshall Islands. *Marine Ecology Progress Series*, *488*, 23–34. Inter Research.

Leber, K., Brennan, N., & Arce, S. (1998). Recruitment patterns of cultured juvenile Pacific threadfin, Polydactylus sexfilis (Polynemidae), released along sandy marine shores in Hawaii. *Bulletin of Marine Science*, *62*(2), 389–408. Miami, FL, USA: Rosenstiel School of Marine & Atmospheric Science.

Moss, R. (2007). Environment and development in the Republic of the Marshall Islands: Linking climate change with sustainable fisheries development. In *Natural resources forum 31*(2), 111–118. Oxford: Blackwell Publishing Ltd.

Pratchett, M., Munday, P., Wilson, S., & et al. (2008). Effects of climate-induced coral bleaching on coral-reef fishes. *Ecological and economic consequences. Oceanography and Marine Biology: An Annual Review*, *46*, 251–296. Abingdon, UK: Taylor & Francis.

Tompkins, E., & Adger, W. (2004). Does adaptive management of natural resources enhance resilience to climate change? *Ecology and Society 9*(2): 10. The Resistance Alliance. Accessed Feb. 23, 2017. http://www.ecologyandsociety.org/vol9/iss2/art10.

Advantages and Challenges of Participatory Management of Customary Coastal Areas in the French Islands of the Pacific

Allenbach Michel, Le Duff Matthieu, Dumas Pascal and Bouteiller Yolaine

Abstract Shoreline erosion affects all Southwest Pacific island countries and territories. This erosion is directly related to global climate change. In this general context, the management of the coastal areas of the French Pacific territories is constrained by the customary status of most of the coastline of the Main Island of New Caledonia, of the Loyalty Islands and of Wallis and Futuna. The French national risk management model is not applicable even though the risks are high, notably because of high population density and a specific geomorphological context. Within the IFRECOR (France) and INTEGRE (European Union) projects, some thoughts on the content of a territorialized prevention and adaptation policy are presented. This policy takes into account concepts of vulnerability and resilience adapted to the local cultural context, but also the local population's needs in an area with limited resources, where pressure is being felt due to aggregate removal, coral reef destruction and defective civil engineering works. Participatory management is a promising approach for raising awareness of the risks associated with climate change in the coastal zone and to find efficient solutions. It brings together expert and local knowledges to provide relevant and acceptable responses for Pacific peoples. Ongoing applications in French territories demonstrate tangible results but also difficulties. These applications may be transposable to a number of island states of the Pacific sharing the Oceanian vision of the shoreline.

A. Michel (✉) · L. D. Matthieu · D. Pascal
University of New Caledonia, Noumea, New Caledonia
e-mail: allenbach@univ-nc.nc

L. D. Matthieu
e-mail: matthieu.leduff@yahoo.fr

D. Pascal
e-mail: pascal.dumas@univ-nc.nc

A. Michel
Labex Corail, Puna'auia, French Polynesia

B. Yolaine
South Pacific Commission, Noumea, New Caledonia
e-mail: yolaineb@spc.int

W. Leal Filho (ed.), *Climate Change Impacts and Adaptation Strategies for Coastal Communities*, Climate Change Management,
https://doi.org/10.1007/978-3-319-70703-7_16

Introduction

Significant shoreline erosion now affects all southwest Pacific island countries and territories. In their 5th report (IPCC 2014), Intergovernmental Panel on Climate Change experts attribute it to global change-related marine transgression that has been affecting the planet since the beginning of the first industrial revolution. The trend is now accelerating with heavy fossil-fuel use and the rising world population.

Pacific countries' soft shorelines are being heavily affected by the consequences of climate change and marine transgression, due to their geomorphology in some cases, i.e. being low islands, and their population's current lifestyles with housing and public infrastructure built along the very coasts that are under threat. The stakes are high for the Pacific island countries and territories (PICTs) and SPREP (Secretariat of the Pacific Regional Environment Programme), through its Pacific Futures Programme, aims to assist Pacific countries and territories organise and cope with the threats and pressures facing the oceanic and island systems. (http://www.environnement.pf/lien-web/programme-regional-oceanien-de-l-environnement-proe-pacific-regional-environment-program).

There has been no trouble reaching the consensus required in Pacific-island decision-making practice on this priority objective and, while the Southwest Pacific islands are not the only ones in jeopardy, PICTs are particularly exposed to natural disasters. They are now striving to have international climate refugee status recognised (Le Monde 2008; Gemenne 2009). Despite facing a legal conundrum, the move has the support of all Pacific countries who work together at every opportunity in an attempt to make the major industrial powers admit responsibility for the problems now heavily affecting Pacific coastlines. A major step forward was made at COP 21 in 2015, when an international agreement was reached to keep the global temperature rise to between 1.5 and 2 °C by 2100.

The French Southwest Pacific territories are, of course, affected by coastline and coastal-area management issues, as their shores have felt the brunt of global change to varying degrees. They have increasingly been joining forces with Pacific island countries (Oceania 21 2015) as part of a bid to improve regional integration. Locally, discussions and initiatives are underway in each territory to set up monitoring mechanisms and regulate risk management. These efforts supplement actions taken by France in all its overseas territories and by Europe through programmes specifically designed for the Pacific. In the framework of the IFRECOR (France) and INTEGRE (European Union) projects, a reflection on what a territorialized prevention and adaptation policy could be will be presented. This policy would be based on taking into account, on the one hand, the concepts of vulnerability and resilience adapted to the local cultural context, and on the other hand, the local population needs in an area with limited resources, and which is also under pressure, notably due to aggregates removal, coral reefs destruction, and defective civil works.

The management of the coastal area of the French Pacific territories is constrained by the customary status of most of the coastline of the main Island of New

Caledonia, of the Loyalty Islands and of Wallis and Futuna. The territories discussed in this article and the broad principles of their legal frameworks will be described with a focus on their powers in environmental and coastal management matters. The French national integrated coastal area and risk management model will then be briefly presented. This French national model of risk management is not applicable even though the risks are elevated, notably because of high population density and specific geomorphological context. It will be explained why they are inapplicable and ideas for overcoming the issues encountered will be proposed and discussed.

Participatory management is a promising approach.The value of participatory coastal management in the French territories of the south-west Pacific came from observation. Noted were the great difficulties encountered by coastal management specialists in raising awareness of the risks involved and choosing the best options to protect the coastline. By closely associating local populations with research and achievements, it is possible to understand the real motivations and issues specific to the Pacific peoples in the coastal space. When these motivations and concerns are not integrated into coastal management, the implementation of regulations and environmentally-friendly development options can be difficult to achieve. Participatory management allows the exchange of ideas between expert and local knowledge stores and contributes positively to greater efficiency in the fight against the effects of climate change on the coastal zone.

Two experimental projects are underway in New Caledonia at Saint-Joseph tribe in northern Ouvea (Loyalty Islands Province) and Ouara tribe on Ouen Island (South Province). One of the limitations to the conclusions that will be reached concerns the relatively short time available for participatory management practice. The first of these experiments began three years ago and is due to conclude at the end of 2017. The second one is more recent (one year) and will also end on the same date. The project on Wallis and Futuna is still being defined with chiefdoms and technical services. There is thus a potential bias on the observations made. The number of cases studied is limited. They concern coastal areas where the damage caused by the effects of climate change is significant - heavy erosion of the coastal dune in the first case and an area subject to frequent submergence in the second. The fact that the problems are proven certainly facilitates the establishment of a dialogue to try to solve the dysfunctions. It has not been tested for the moment in areas where the problems ahead are not yet visibly expressed. It is therefore impossible to say whether the method can be efficient for prevention. It has only been tested in a crisis management situation.

French Southwest Pacific Territories

Broad Physical Features

The territories are New Caledonia, with its main island and the Loyalty Islands, and Wallis & Futuna. Fig. 1 provides the geographical location of these territories that

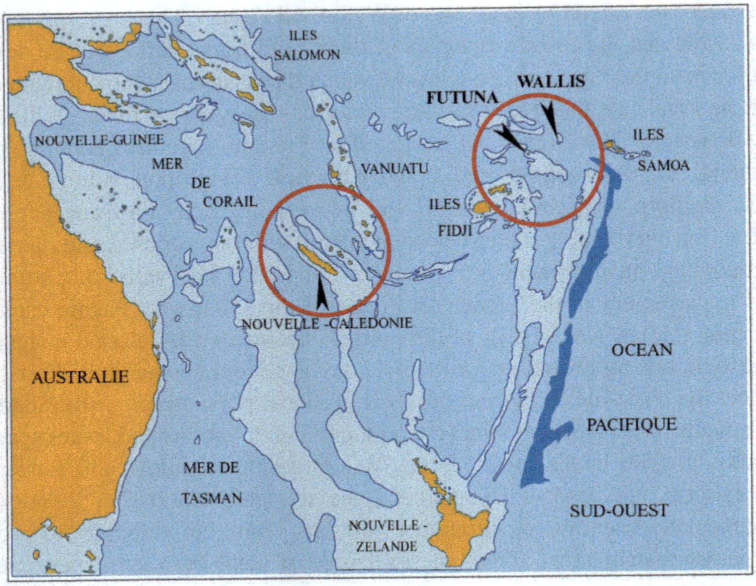

Fig. 1 Geographical location of the French Southwest Pacific territories

differ significantly in physical terms. New Caledonia and the Loyalty Islands lie approximately 1500 km off Australia's east coast, across the Coral Sea. In terms of latitude, they are located near the Tropic of Capricorn. Wallis & Futuna are located closer to the Equator on the 15th parallel south and some 2500 km to the east of New Caledonia.

The different latitudes make for different climates, i.e. subequatorial in Wallis & Futuna and temperate-tropical oceanic in New Caledonia and the Loyalty Islands (Météo France 1995). The islands are geologically different too, with Wallis & Futuna being small volcanic islands (274 sq. km) as are the Loyalties (1980 sq. km for all four islands in the group), although the latter are exclusively made up of carbonate rock at the surface, which is uncommon in Wallis & Futuna, other than on fringing and lagoon reefs. New Caledonia (18,575 km^2) is a complex combination of volcanic-sedimentary formations and ocean floors inherited from the Australian continent. The formations have had a long geologic history (Cluzel 2006) marked by subduction followed by obduction since the ocean basin opened, making it a major world geological attraction.

The New Caledonian mainland is surrounded by a huge lagoon bordered by a barrier reef protecting it from ocean hydrodynamics, as is Wallis Island. Neither the Loyalties nor the Horn Group (Futuna and Alofi Islands) are, however, and so, are fully exposed to ocean hydrodynamics.

The Loyalty Islands rise to 138 m (Mare) and Wallis to 151 m and could be described as moderately low islands, while Futuna at 524 m and the New Caledonian mainland, with two peaks at nearly 1700 m, are high islands. These altitudes are less alarming than on the region's low islands, such as in the Tuvalu group, where the atolls are less than 2.5 m above sea level on average. The low-lying coastal areas on the main islands and offshore lagoon islands in New Caledonia and Wallis are, nevertheless, sensitive and prone to erosion, often exacerbated by human activity.

Population Figures and the Different Legal Frameworks

The population is also diverse with an amazing kaleidoscope of ethnic groups and skin colours found in the territories. Following several successive waves of migration, the Melanesians of New Caledonia and the Loyalty Islands and the Polynesians of Wallis & Futuna have mixed considerably in Noumea with its transient population, although they do remain homogeneous and attached to their ancestral ways in isolated areas that have always been far from the main maritime trading routes, such as Wallis & Futuna and the Loyalty Islands. According to currently accepted anthropological theories, the territories' peoples are descended from a common Austronesian lineage originally hailing from Asia (Latouche 2017), which is undoubtedly one of the reasons why a common "oceanic vision of coastal areas" is shared by the first nations of the region, as will be discussed later in this article.

Officially, New Caledonia has a population of 268,767 (2014 census, https://www.insee.fr/fr/statistiques/1560282), i.e. a density of 14 pop/sq. km with 18,297 in the Loyalty Islands (density: 9 pop/sq. km). Wallis & Futuna's 2013 census (https://www.insee.fr/fr/statistiques/2121453) recorded a population of 9019 on Wallis (or Uvea in the local language) and 3848 on Futuna, i.e. densities of nearly 100 pop/sq. km. New Caledonia's population is growing, while Wallis and Futuna's is shrinking. In both territories, most of the population lives on the coastal fringes, near the shoreline. Anthropogenic pressure differs, of course, from one location to another. In New Caledonia it is high in the Noumea area, but less so elsewhere on the mainland and Loyalty Islands, due to the low population density (Dumas 2004; Pestana 2016). On Wallis and Futuna, it is significant, owing to higher population densities.

For a long time, both territories had a common political status, but their situations began to differ more and more after the 1988 Matignon and 1998 Noumea Accords, which paved the way for a self-determination referendum to be held in 2018 in New Caledonia and the Loyalty Islands. The result will determine the territory's future political status, namely either independence in association with France or full independence or broad-ranging autonomy. For the time being, the territory is an overseas country (OC) with very extensive *sui-generis* autonomy under the Noumea Accord.

In 1959, Wallis & Futuna asked to enter the French Republic by referendum and were initially administered in large part from New Caledonia after choosing overseas territory status in 1961 (Lotti 2008). This has no longer been the case since the Noumea Accord and the group is now an overseas entity (or *collectivité* in French) that is increasingly forging its own destiny rather than being the New Caledonian dependency it once was. These islands have their own specific features (Allenbach 2008) that distinguish them from all other French overseas territories. Their political framework places powerful traditional authorities (two kingdoms on Futuna and one on Wallis Island) alongside the executive (French central government) and legislative (Territorial Assembly) arms of government. The territory is now a kind of modern theocracy combined with a traditional, aristocratic social structure. It is administered by a representative of the French Republic under an arrangement that is a far cry from French centralist Jacobinical principles with a legal framework that upholds both the Roman Catholic religion and customary authority (Lotti 2008).

Because these provisions are so far removed from the rules of play in mainland France and owing to current changes in the territories' legal frameworks, coastal environmental management powers, insofar as they previously existed on the New Caledonian mainland, changed hands to fall under the purview of provincial administrations, with each province having its own policies. The public maritime domain is recognised in the Southern Province as well as in the Northern Province, though with a few restrictions there (Bodmer 2011). The concept is less clearly defined in the Loyalty Islands Province, where land tenure is strictly traditional and the Pacific island view of coastal areas pervasive (Herrenschmidt et Clua 2006). On custom lands in the territory, as well as on Wallis & Futuna, the shoreline that backs onto traditional coastal land belongs to the clans (New Caledonia and Loyalties) and villages (Wallis and Futuna) who own that part of the shoreline. The fact that the environment codes drafted for the Loyalty Islands Province and Wallis & Futuna are inapplicable and that there is no public maritime domain there are currently major obstacles to introducing territorial coastal-area risk-management policies to cope with global change. Solutions, therefore, need to be found that reflect the reality of Pacific-island coastal-land-tenure notions.

National Rules

A brief explanation would be required at this stage of the applicable rules in mainland France for coastal areas, which are national standards independent of the above special-status arrangements. Nationally, there are two basic legal concepts, namely the public maritime domain and the Coast Act. The natural maritime domain (https://www.legifrance.gouv.fr/—General Public Property Code) is the land, sea floor and underlying bedrock located between the high-water mark, other than during unseasonal weather, and the territorial water boundary. Authority over this area is exercised by central government through the maritime prefect. For the reasons stated above, this notion cannot be relied upon for integrated coastal management.

The Coast Act of 10 July 1975 (https://www.legifrance.gouv.fr/—Town Planning Code) on developing, protecting and using the coast is a law that aims at regulating coastal development to protect it from property speculation and ensure the public has free access to seaside footpaths. It instituted the Sea and Lake Shore Conservatory for turning a third of the French coastline into sanctuaries by 2050 (Brigand 1995). The Act does not, however, apply to the territories dealt with here for obvious reasons to do with special status and, more specifically, the special status of custom lands and cannot, therefore, be relied on as is.

The coastal areas of French Southwest Pacific territories face a variety of challenges ranging from the erosion that is nibbling away at ever more beaches to powerful earthquakes and tsunamis, particularly in the Loyalty Islands and Futuna with their specific geological features (Pelletier et Pillet 2004; Sahal et al. 2010; Lamarche et al. 2013; Le Duff et al. 2016). The civil defence issues are considerable. In mainland France, the coastal hazard management model is built around a central pillar, namely Natural Disaster Mitigation Plans (PPRs). PPRs (https://www.georisques.gouv.fr/…/plan-de-prevention-des-risques-naturels) are town-planning documents regulating land use based on the natural disasters that affect them. Regulatory measures range from construction bans to restricted planning permission. PPRs, like all other town-planning and land-use documents, are based on a fundamental principle, namely the existence of a land register (Le Duff et al. 2014).

In pursuance of the Noumea Accord's devolution provisions, authority over civil defence has just been transferred to New Caledonia and all measures must now be taken locally, based on locally applicable regulations. As a result, the national model for coastal hazards cannot be directly transposed to the territory, due to special-status traditional land tenure arrangements (Pontier 2010; Le Duff et al. 2014). There is no land register and land is deemed inalienable, non-transferable and held in absolute ownership. Other channels must, therefore, be found that identify strategic points for proposing a management model that would be acceptable to the communities living in custom areas. The model must be defined with the involvement of the communities and an imported model must not be imposed. It must be supported by effective participatory management of the relevant areas.

Contributions from National and European Initiatives

Some strategic points that could contribute to defining such a management model include a few key general initiatives. Based on the awareness of the hazards that may affect coastal infrastructure and human beings as a result of global change, policies have been introduced at various levels (European, national, regional and local) to quantify them, devise adaptation strategies and design management models that are acceptable to communities. Two examples discussed below are taken from French Southwest Pacific territories:

IFRECOR National Action Plan (French Coral Reef Initiative) 2016-2021

One example of national initiatives for attempting to adapt to global change is the IFRECOR National Action Plan (French Coral Reef Initiative) 2016–2021 that covers both territories dealt with in this paper and which both have major coral reef formations on their coasts. IFRECOR (http://www.ifrecor.com) is a national policy instrument focusing on coral reefs and their associated ecosystems in French oversea territories. Since 2000, this program is designed and implemented through five-year plans at local, national and international levels. All the IFRECOR actions contribute to a better understanding and management of French coral reefs spread over the three oceans. Six strategic axis are defined within IFRECOR framework:

1. Planning of IFRECOR actions
2. Reducing negative impacts of human activities and promote sustainable development;
3. The development of research, monitoring and decision support tools;
4. Information, training and education;
5. The development of regulation policies and financial tools;
6. The development of local and regional cooperation.

The National Action Plan IFRECOR for 2016–2021 set up a TIT (transverse thematic area of interest) on "Adapting to climate change" in 2011. Its long-term goal is to benefit from the IFRECOR actions and recommendations to raise awareness of economic actors and policy makers on reef environments and associated ecosystems, with the view to contribute to the design of appropriate plans for overseas regions. It is also to proposing specific technical, scientific measures for socio-economic development. It also aims to encourage French overseas communities to define a strategy for adapting to climate change taking into account the reef environments. Carried out by Wallis and Futuna Department of Environment (South Pacific) with the support of a S2C consultant, University of New Caledonia (UNC) and the National observatory on the effects of global warming (ONERC) various products are made and visible on scientific interoperable portal developed within the University of New Caledonia (http://portail-scientifique.univ-nc.nc). It provides access to data (http://servlet.univ-nc.nc/series/ifrecor) compiled in the different French overseas territories on indicators identified to be relevant to assess climate change in reef environment (sea surface temperature, acidification, coastal development, state of health of the reefs).

Local authorities and their departments in the territories are beginning to take ownership of these initiatives, because they provide them with communication materials for the affected coastal communities in order to heighten their awareness of the hazards. In both territories, sociological and anthropological studies (Worliczek 2013) show that communities have little understanding of global change-related hazards for the shoreline. They cannot see them and, therefore, do not believe the phenomena described by scientists are real. Environmental change is partly related to natural cycles and partly to local human error. Providing material based on the figures and involving communities are useful ways of heightening awareness, as detailed below.

Integre Litto-Ouvea Programme Driven by Spc Overview

INTEGRE (Pacific Territories Initiative for Regional Management of the Environment) is a sustainable development project covering four European Pacific overseas countries and territories (OCTs), namely New Caledonia, French Polynesia, Wallis & Futuna and Pitcairn. Funded by the European Union, it aims at improving the management and sustainable development of the environment for the good of communities and to strengthen regional co-operation in sustainable development. Activities funded by INTEGRE (https://integre.spc.int) include the Litto-Ouvea programme aimed at studying the effects of erosion on the shoreline in and around Lekine Faiava Reserve, Mouli District, in Ouvea Island's south (Le Duff et al. 2017).

The study began in 2015 and will end in 2017, i.e. this year, with the following objectives:

- Help improve understanding of the physical and human processes at work on Loyalty Island coastal areas
- Help gain recognition for and showcase traditional knowledge
- Develop and apply a cross-sector analysis methodology based on field trials with the aim of defining operational management tools
- Help integrate global warming issues into New Caledonian public policy

These objectives are directly related to the core issues of this paper, i.e. presenting participatory management and then discussing its pros and cons. The adopted approach involves:

- quantifying the hazards and their development combining field work, statistical processing with meteorological and oceanographic databases and diachronic analysis based on satellite and aerial images and maps; and
- understanding the relationships between communities and their coastal areas and the importance of natural disasters in coastal narratives and uses, and in the spatial and social organisation of local communities based on field studies, interviews, term analysis, story compilations, genealogical discourses, tales and legends involving natural phenomena affecting coastal areas.

The field work was carried out by conducting participatory monitoring on the sites and involved all the local stakeholders, i.e. elected representatives, traditional leaders, government departments and associations (Ibid). The aim was to foster knowledge sharing and encourage Loyalty Islanders to acquire scientific skills and culture. IFRECOR climate-change observatory data made a contribution at this stage in the work by demonstrating how real the phenomena were and how they were spreading out to many overseas locations. Generally speaking, the sharing afforded by the participatory approach heightened community awareness of the environmental issues affecting the coast, namely erosion, submersion and natural disasters. More than simply sharing the data, the exercise effectively involved the Ouvea communities in acquiring data about their own territory. Actively involving the communities and the resulting understanding they acquired of the phenomena is undeniably a major plus in the island's integrated coastal management.

Field Operations

The participatory coastal monitoring work carried out on Ouvea consisted of involving the communities and environmentally-aware associations in measuring the geomorphological changes occurring on beaches over time using a simple technique, namely the Emery method (Emery 1961) for obtaining readings that were sufficiently precise for the measurements' end purpose. They were supplemented by measurements taken to higher scientific standards by DGPS and drone surveys (Cohen et al. 2016) commonly applied in coastal geomorphology research and the results were excellent (Le Duff et al. 2017). The work encouraged the locals who took the readings to own the findings and the skills needed to measure the extent of erosion on their shoreline. As a result, these resource persons now understand the facts and figures and can lobby decision-makers to advocate for their coastal development needs. Having a facilitator, an INTEGRE-funded Ph.D. student, on hand locally helped them understand the phenomena and encouraged them to respond quickly by correcting human error that was aiding and abetting erosion and undermining the shoreline. As a result, two polders jutting out from the shore and hampering sedimentary drift were urgently destroyed, while this sort of decision usually requires consensus that takes a long time to achieve. This was the outcome of observations made by the local residents themselves and confirmed by both scientific experts and the readings taken by the community that was directly affected by coastal erosion.

Discussion

Integrated coastal zone and hazard management in French Southwest Pacific territories is complicated by the special legal frameworks that govern them and by local developments. A French centralist model that directly transposes rules from coastal areas in mainland France to the Pacific islands would be unacceptable to the communities.

Pacific islanders view coastal management from a very different perspective. In the Pacific's dominant traditions and customs, the coast belongs to the coastal people (Leblic 1991). The rules governing traditional land tenure in the Loyalty Islands, Wallis & Futuna and parts of the New Caledonian mainland require solutions that address the reality of these special-status interests that can, in certain cases legitimately oppose the common interest when it is not understood or owned by the communities involved. The sea is part of the Pacific islander's persona. For the people of all 22 Pacific island countries and territories and, therefore, the French territories in the ocean's southwest, the sea is primarily the link that unites them in the migrations by which they gradually settled the islands through time. As a means of transport and source of food, the sea forged Pacific islanders' identity as much as the land they settled on did. Along with ancestor worship, the sea and land are the mainstays of Pacific society (Worliczek 2013). These social pillars are still very much alive today and transgressing them without consensus, by imposing imported solutions, ignores this identity and is unacceptable (Allenbach et al. 2016).

The unruly human settlement of coastal areas in Wallis and Futuna offers a salutary lesson. It has generated serious dysfunctions that have exacerbated natural patterns undermining the coastline. No action can be taken or solution applied to the issues without first addressing the way local communities perceive their coastal areas. Throughout the Pacific, be it in Melanesia or Polynesia, traditional maritime law differs widely from western law. The sea is treated like land (Leblic 1991; Herrenschmidt 2004). The right to build on or take sand from the shore, collect shells on the reef flats or fish in the lagoon depend on land rights. It is, therefore, governed by custom, a set of unwritten rules deposited in the memories of traditional leaders. Each island has its own rules, but the spirit is the same. On Wallis & Futuna, this traditional law entitles communities that have the resources to do so, to reclaim the sea adjacent to their family land, develop the coast in defiance of building standards and quarry beaches already undermined by oceanic transgression. Private interests trump collective interests when the community has not been made sufficiently aware of the hazards and owned them.

Awareness efforts via the media or by visiting scientists have little effect, however. At best, their message is not understood, but it is usually contested or sometimes even deemed a tall story (Worliczek 2013). Participatory management that aims to provide the locals with the skills and equipment they need to monitor, quantify and understand the phenomena through two-way sharing of scientific and traditional knowledge is a more effective means of communicating and heightening awareness. The European-funded (INTEGRE) experiment with Loyalty Islanders on Ouvea has borne fruit. The associations involved in the partnership are active and getting the message across. The experiment conducted on the island has had a favourable reception outside the Loyalty Island Province and been transposed to Ouen Island, also through INTEGRE, at the request of the Ouara tribe.

In the Ouara tribal area on Ouen Island, the traditional context is similar to Ouvea's, although public maritime domain management and the locally applicable environment code are different to the Loyalty Island Province's. Their Pacific island concept of coastal and maritime areas is very close to the Lekine tribe's on Ouvea and Vailala villages on Wallis. The protocols applied on Ouvea were implemented on Ouen Island with the stated aim of reproducing a system that is currently working in the Loyalty Islands Province. It must be stated, however, that after a year of trialling, the expected momentum failed to materialise. The Southern Province Department of Surveys scientists did their job, but the tribe failed to truly own the tasks that were required of them. The reason for this relative failure, at least at the current stage of the activity, could be due to the fact that there is no facilitator to regularly assist the community in the work required. It could also be due to the lack of funding for the readings to be taken on Ouen Island, unlike Ouvea where they were funded. The cost involved is very small, but then so are incomes in the area and whatever money can be made goes a long way. This, as well as the commitment of the associations, could be one of the reasons why the momentum is currently so positive on Ouvea, while the experiment has met with failure on Ouen Island.

When the facilitation work and dedicated INTEGRE funding draw to a close, it will be possible to gauge whether the local communities in the New Caledonian trials have been made sufficiently aware and taken ownership of this participatory coastal management technique for monitoring geomorphological changes in eroded shorelines. Whether or not the participatory management activities started on Ouvea continue will determine whether or not funding needs to be pursued for technical support to this type of environmental monitoring, whether voluntary or paid civic action should be used or whether paid labour can be resorted to. Scientific programmes are temporary and cannot be used to gather long series of environmental data and yet this information is vital for analysing the many heavily affected but resource-deprived Pacific island locations. The communities living there must take ownership of data acquisition tasks and work hand in hand with scientists and their local and regional government departments and international bodies to convince aid donors to help them mitigate and adapt to the global change that is affecting their shores. The trials underway on Ouvea and Ouen Islands are intended to be models that can later be transposed throughout the Pacific, affected as it is by global change, but lacking the major resources required to cope with it.

Conclusion and Prospects

The possibility of sensitizing the Pacific populations to the risks and problems caused by climate change in the coastal space has been tested through a participatory management approach. Three geographic projects that are at different stages are currently being carried out in the French territories of the Pacific under the INTEGRE and IFRECOR programs.

In the territories concerned, the impact of climate change on the coastal zone is major. The slow, but real rise in sea level, facilitates the erosion of the beaches and dunes where the infrastructures of the tribes and villages studied are located. Anthropogenic errors (faulty coastal development, aggregate extraction) add to the adverse trends induced by climate change. The inhabitants of the sites concerned note the facts, but most often, attribute them to natural cycles. They are asking for immediate help to stop erosion, marine submergence or saltwater intrusion into coastal fresh water bodies, but they do not project themselves into the future. Moving to secure sites inland, the removal of structures that amplify the crisis and the halting of disruptive practices must be understood and accepted by communities in order to be implemented. The participatory management approach makes it possible to associate the populations more closely with the studies carried out on the threatened coastlines and to explain what is at stake. These findings are easily accepted. The exchange of local and technical knowledge enables efficient decision-making, a pre-requisite to consensus-based acceptance on customary land.

The results obtained show, however, that the approach must be sustainable in order to be well understood. It must rely on the presence of resource persons accompanying the inhabitants and associations wishing to invest in the approach and using specific dedicated funding. This makes it possible to be as close as possible to the events that impact the coastline and to reactive positively to these events. In particular, it enables people to become aware of the harsh reality of long-term and amplified impacts of climate change on coastal lands traditionally occupied by Pacific Islanders. The temporary solutions implemented are more easily accepted as well as the principles of adaptation and the inevitable fallback to secure spaces in the medium term.

In New Caledonia, experiments are being carried out at other sites and extended at current sites. The technical services of the communities who participated in the work on the experimental sites of Ouvea and Ouen Island are in favor of the approach. This approach will therefore be able to be tested in other physical and socio-cultural environments to clarify its efficiency and possible general applicability in island states of the Pacific sharing the Oceanian vision of the shoreline.

References

Accord de Nouméa. JORF n°121 du 27 mai 1998, p. 8039.

Accord de Matignon. www.mncparis.fr/uploads/accords-de-matignon_1.pdf.

Allenbach, M., Le Duff, M., Dumas, P., & Cohen, O. (2016). Gestion intégrée, risques côtiers et aménagements sur les littoraux coutumiers océaniens français. Les apports croisés de la gestion participative et des méthodes classiques. Conference: Journées Nationales Génie Côtier—Génie Civil, Ed. (pp. 127–136). Paralia. doi:10.5150/jngcgc.2016.015.

Allenbach M., & Hoibian, T. (2008). Enjeux, spécificités et perspectives de la gestion intégrée du littoral dans l'espace insulaire français du sud-ouest Pacifique, *European Journal of Environmental and civil Engineering, 12*—N°. 1–2/2008, 15–34.

Bodmer, D. (2011). *Le littoral de la Province Nord en Nouvelle-Calédonie: Quel développement durable?* (p. 570). Thèse de doctorat en Géographie: Université Michel de Montaigne-Bordeaux III.

Brigand, L. (1995). D'île en île: l'archipel du Conservatoire du littoral. Conservatoire du littoral, 1995.

Cluzel, D. (2006). Synthèse Géologique de la Nouvelle Calédonie et de sa Zone Economique Exclusive, Institut des sciences de la terre d'Orléans—Rapport EXTRAPLAC; https://wwz.ifremer.fr/ncal/Geosciences-Marines/Publications.

Cohen, O., Dumas, P., Le Duff, M., & Allenbach, M. (2016). Études du rivage par photogrammétrie aéroportée à très haute résolution spatiale. Exemple à Ouvéa, Nouvelle-Calédonie, Ed Paralia, pp. 361–370. DOI:10.5150/jngcgc.2016.041.

COP, 21. (2015). http://www.cop21paris.org/about/cop2.

Domaine public maritime naturel. https://www.legifrance.gouv.fr/—Code général de la propriété des personnes publiques.

Dumas, P. (2004). Caractérisation des littoraux insulaires: approche géographique par télédétection et SIG pour une gestion intégrée. Application en Nouvelle-Calédonie. Ph.D. thesis in Geography, 404 p. University of Orléans.

Emery K-O. (1961). A Simple Method of Measuring Beach Profiles. *Limnology and Oceanography*, *6*(1), 90–93. doi:10.4319/lo.1961.6.1.0090 http://www.aslo.org/lo/toc/vol_6/issue_1/0090.pdf.

Gemenne, F. (2009). *Géopolitique du changement climatique*. Armand Colin, 256 p.

IPCC, (2014). Summary for Policymakers. In Edenhofer, O., R. Pichs- Madruga, Y. Sokona, E. Farahani, S. Kadner, K. Seyboth, A. Adler, I. Baum, S. Brunner, P. Eickemeier, B. Kriemann, J. Savolainen, S. Schlömer, C. von Stechow, T. Zwickel & J.C. Minx (Eds.), *Climate Change 2014: Mitigation of Climate Change. Contribution of Working Group III to the Fifth Assessment Report of the Intergovernmental Panel on Climate Change* (32 p). Cambridge, United Kingdom and New York, NY, USA: Cambridge University Press.

Herrenschmidt, JB et Clua, E. (2006). Vers une gestion plus "culturelle" des milieux naturels en Océanie? *En ligne]:* http://www.c3ed.uvsq.fr/cdgecorev/fr/pdf/t7/Clua.pdf, 2006.

Herrenschmidt, JB. (2004). Territoires coutumiers et projet de développement en Mélanésie du Sud (Iles Loyauté, Vanuatu, Fidji), PhD thesis in Geography, 656 p. University of Paris IV Sorbonne.

IFRECOR. http://www.ifrecor.com.

INTEGRE. https://integre.spc.int.

Lamarche, G., J. Mountjoy, S. Popinet, B. Pelletier, J. Goff, J. Bind et S. Woelz (2013). L'aléa tsunami à Wallis et Futuna, modélisation numérique et inventaire des tsunamis. Rapport préparé pour l'Administration Supérieur du Territoire de Wallis et Futuna, 86 p. Niwa, leading environnmental science.

Latouche, Austronésiens, *Encyclopædia Universalis* [en ligne], consulté le 22 mars 2017. URL: http://www.universalis.fr/encyclopedie/austronesiens/.

Leblic I. (1991). Représentations du foncier en Nouvelle-Calédonie et identité culturelle kanak. Evolution de ces représentations liée à la revendication identitaire et aux processus de développement économique. 3e journées scientifiques de la Société d'écologie humaine, Perceptions et représentations de l'environnement, Novembre 1991, 25 p. Aix-en-Provence, France.

Le Duff, M., Dumas, P., Cohen, O., & Allenbach, M. (2017). Coastal Erosion Monitoring on Ouvea Island (New Caledonia): Involving the Local Community in Climate Change Adaptation. HandBook on climate change adaptation, Ed. Springer, pp. 255–268. doi:10.1007/978-3-319-50094-2_15.

Le Duff, M., Dumas, P., Sabinot, C., & Allenbach, M. (2016). Le risque tsunami en Nouvelle-Calédonie: Évolutions des facteurs de vulnérabilités et de résiliences à Lifou en territoire coutumier kanak. *VertigO—la revue électronique en sciences de l'environnement* [En ligne], Volume 16 numéro 3| décembre 2016, mis en ligne le 20 décembre 2016, consulté le 02 mai 2017. URL: http://vertigo.revues.org/17951; doi: 10.4000/vertigo.17951.

Le Duff, M., Dumas, P., & Allenbach, M. (2014). Quelle gouvernance des risques littoraux aux îles Loyauté (Nouvelle-Calédonie), dans le cadre du réchauffement climatique? *Actes Du Colloque Cocorisco, Brest, Juin, 2014,* 429–436.

Le Monde. (2008). http://www.lemonde.fr/asie-pacifique/article/2008/06/09/les-naufrages-de-l-archipel-de-tuvalu_1055642_3216.html.

Loi Littoral. https://www.legifrance.gouv.fr/—Code de l'urbanisme.

Lotti, A. (2008). *Le statut de 1961 à Wallis et Futuna: genèse de trois monarchies républicaines (1961–1991).* Thèse de doctorat en histoire contemporaine. Bordeaux 3. ISBN: 978-2-296-13871-1.

Météo France. (1995). Atlas climatique de Nouvelle-Calédonie, décembre 1995, Nouméa.
Oceania, 21. (2015). http://www.lemonde.fr/planete/article/2015/05/01/oceania-21-les-iles-du-pacifique-plaident-pour-une-revolution-internationale.
Observatoire des indicateurs des changements climatiques de l'IFRECOR. http://servlet.univ-nc.nc/series/ifrecor.
Pelletier B., et Pillet R. (2004). Tectonique active, tsunamis et sismicité en Nouvelle-Calédonie. Notes techniques sciences de la terre, géologie, géophysique, revue n 28, 24 p.
Pestaña G. (2016). Du spectre du dépeuplement à celui de l'indépendance, *Espace populations sociétés* [En ligne], 2015/3–2016/1 | 2016, mis en ligne le 20 mars 2016, consulté le 02 mai 2017. URL: http://eps.revues.org/6197; doi: 10.4000/eps.6197.
Portail scientifique de la Nouvelle-Calédonie. http://portail-scientifique.univ-nc.nc.
Pontier, JM. (2010). Les conditions du transfert de compétences en matière de sécurité civile. Les assises de la sécurité civile, juillet 2012, Nouméa.
PPR. https://www.georisques.gouv.fr/…/plan-de-prevention-des-risques-naturels.
PROE. http://www.environnement.pf/lien-web/programme-regional-oceanien-de-l-environnement-proe-pacific-regional-environment-program.
Insee (2014). Recensement Nouvelle-Calédonie 2014. https://www.insee.fr/fr/statistiques/1560282.
Insee (2013). Recensement Wallis et Futuna 2013. https://www.insee.fr/fr/statistiques/2121453.
Sahal, A., B. Pelletier, J. Chatelier, F. Lavigne et F. Schindelé. (2010). A catalog of tsunamis in New Caledonia from 28 March 1875 to 30 September 2009. CR Géoscience, Tome 342, fascicule n 6, Ed. Elsevier, pp. 434–447. doi:10.1016/j.crte.2010.01.013.
Worliczek E. (2013). La vision de l'espace littoral sur l'île Wallis et l'atoll Rangiroa dans le contexte du changement climatique. Une analyse anthropologique de la perception des populations locales. Thèse en Anthropologie, Université de la Nouvelle-Calédonie, 500 p.

Courts as Decision-Makers on Sea Level Rise Adaptation Measures: Lessons from New Zealand

Catherine Iorns Magallanes, Vanessa James and Thomas Stuart

Abstract New Zealand has an extensive coastline, and most of its cities, significant infrastructure and population are located in low-lying coastal areas. Sea level rise is already impacting on coastal communities, and a sea level rise of one metre would have a devastating impact on the country. Cohesive and appropriate adaptation measures are urgently needed at both central and local government level to reduce the impact on communities, protect against potential loss of life, and limit economic losses. However, responsibility for climate change adaptation measures in New Zealand has been devolved to local government without any significant direction or guidance from central government. This has produced climate change adaptation policies that vary significantly across the country and leaves councils vulnerable to legal action. This paper addresses legal challenges to sea level rise adaptation measures in New Zealand. It adopts a case study approach, discussing four decisions in cases where a precautionary sea level rise adaptation measure taken by a territorial authority was challenged by a holder of property rights in the coastal area. The paper explores the broader lessons they offer for decision-making on climate adaptation measures. While the decisions are grounded in New Zealand's particular legal framework for resource management, they offer broader insight into the types of issues that arise in adaptation decision-making and how to overcome them. It suggests some lessons that are applicable beyond New Zealand—lessons that better enable appropriate sea level rise adaptation measures and thereby increase the resilience of coastal communities to climate change.

Introduction

A key component of strategic adaptation to climate change in coastal communities is the establishment and use of appropriate planning processes, activities and measures. Such processes, activities and measures have been much studied,

---

C. Iorns Magallanes (✉) · V. James · T. Stuart
Victoria University of Wellington, Wellington, New Zealand
e-mail: Catherine.iorns@vuw.ac.nz

W. Leal Filho (ed.), *Climate Change Impacts and Adaptation Strategies for Coastal Communities*, Climate Change Management,
https://doi.org/10.1007/978-3-319-70703-7_17

including the decision-making frameworks around them. Yet such decision-making processes sit within a legal framework, and more attention needs to be paid to this framework. Globally, courts are important decision-makers on climate change adaptation measures, often making final decisions on adoption of appropriate measures. However, while litigation risk may be identified as a factor for consideration in decision-making, research on adaptation decision-making has largely overlooked the role of courts and judges.

This paper addresses legal challenges to coastal climate adaptation measures in New Zealand. Responsibility for climate change adaptation measures in New Zealand has been devolved to local government without any significant direction or guidance from central government. This leaves councils vulnerable to legal action and has produced climate change adaptation policies that vary significantly across the country. Various legal challenges to council decisions have occurred and illustrate the issues faced by local councils, the concerns of communities, and the responses by the Courts.

The results of these challenges are varied, with some measures being upheld by courts and others being denied. While expert evidence underpins the planning processes, the use of expert evidence does not always serve to protect decisions from being overturned by courts. Further, results have not been predictable; different courts emphasise different reasons for the results, including illustrating varying approaches to risk and caution. This demonstrates the need for more consistency and clarity in the legal frameworks to assist the planning process, and thereby assist decision-making on climate adaptation measures.

The paper illustrates some features of the New Zealand court decisions on sea level rise and explores the broader lessons they offer for decision-making on climate adaptation measures. While the decisions are grounded in New Zealand's particular legal framework for resource management, they offer broader insight into the types of issues that arise in adaptation decision-making and how to overcome them. The aim is to find lessons that are applicable beyond New Zealand—lessons that better enable appropriate sea level rise adaptation measures and thereby increase the resilience of coastal communities to climate change.

The paper first discusses the New Zealand framework within which courts are decision-makers on climate adaptation measures, and summarises some aspects of such decisions. It then uses a case study approach to discuss four individual decisions on climate adaptation measures, to illustrate features that give rise to lessons. Two cases are from 2010, before a new framework for coastal management was adopted; two cases arise after that framework coastal policy statement was adopted. The conclusion then summarises features identified in the case studies and lessons that they suggest.

In taking a case study approach, this paper does not provide a comprehensive analysis of the evolving jurisprudence on climate adaptation in New Zealand. Nor does it discuss the legal framework empowering territorial authorities to factor climate adaptation into decision making. Instead, the focus is on a small number of cases chosen for their innovative and future-focussed approach, and for the way the court responded to claims that precautionary council action was improper in each case. The events which led to each of the judicial proceedings will be considered, but no investigatory analysis of those events will be conducted. Instead, this article will explain some of the ways that Courts have played their part in adapting New Zealand to a warmer climate.

New Zealand Decision-Making on Sea Level Rise Adaptation Measures

Statutory Framework

Decisions on coastal climate adaptation measures in New Zealand are largely made by local and regional authorities—usually called councils—pursuant to the Resource Management Act 1991 (RMA). The RMA is an all-encompassing environmental, land and resource use management statute that provides for the development of long-term plans regarding the types of land and resource uses that will be allowed, as well as for the issuing of immediate, individual water and land use permits (consents), including house building consents. It is local and regional authorities that may make policy decisions on sea level inundation lines, and how they should be notified to New Zealanders—such as a decision to include sea level rise information on official Land Information Memoranda ('LIM') for certain properties in an area (these Memoranda are not legal titles but record the official information held by a council about a property, including its history and established future risks).

The national-level central government may issue policy directions which all local authorities and relevant decision-makers must abide by. Of particular relevance to sea level rise adaptation, the central government has issued New Zealand Coastal Policy Statements (NZCPS) which govern use and protection of the coast. To date, two such policy statements have been issued: 1994 and 2010 (the 2010 statement supersedes that of 1994). These coastal policy statements contain a number of objectives and policies directly relating to key aspects of the coastal environment. Such aspects include: understanding of coastal processes; protection of natural character, landscape values and wild or scenic areas; protection of coastal species, habitats and ecosystems; and identification of coastal hazards and risks. While the 1994 Coastal Policy Statement (Department of Conservation 1994) included six policies in relation to coastal hazards, they were specifically reviewed for the 2010 Coastal Policy Statement (Department of Conservation 2010) in light of the increased understanding of the likely effects of climate change and the responses that would be required to address them (Orchard 2011, p. 33). Accordingly, the 2010 Coastal Policy Statement introduces as a key issue the continuing coastal erosion exacerbated by climate change (Department of Conservation 2010, p. 5).

While New Zealand has a population of less than 5 million, a gross domestic product of only NZD 260 billion (Statistics New Zealand 2016), and a government spend of NZD 50 billion, it has the ninth longest coastline in the world. A sea level rise of one metre, as anticipated by 2115, would encroach significantly along the 15,000 km coast (Ministry of Fisheries 2011). It would have significant and costly effects on low-lying towns and cities, estuary settlements, beachfront homes, and other coastal infrastructure. The majority of New Zealand's major cities (including Auckland, Wellington, Christchurch and Dunedin) are situated in coastal areas that will be affected by rising sea levels. Significant commercial and residential development exists in low-lying areas and both Christchurch and Wellington are at severe risk if sea levels rose by one metre. Good city planning involves preparing for the threat of climate change and, in particular, rising sea levels. However, the mandate for such planning is often unclear and short-term considerations regularly get in the way. It is not politically palpable to prepare for the worst, particularly when considerable uncertainty remains around the degree of harm that is likely and when it will occur.

In order to incentivise territorial authorities to take precautionary action, centralised government direction is necessary. Although the New Zealand government has not been particularly forthcoming in this regard, the New Zealand Coastal Policy Statement 2010 has provided a modest foundation for adaptation policy, at least in coastal areas. The 2010 Coastal Policy Statement requires territorial authorities to identify possible hazards arising in the next 100 years and to plan accordingly. While not particularly empowering, it has given territorial authorities justification for precautionary and preventative action in the coastal zone. Objective Five of the Statement is of particular relevance to climate adaptation. It requires authorities to ensure that coastal hazard risks arising from climate change are considered in their plans and factored into their decision making. They must limit development in at-risk coastal areas, develop managed retreat plans, and restore natural defences to coastal hazards (Department of Conservation 2010, p. 10). All such actions are of critical importance in ensuring that development along the New Zealand coastline is responsive to, and ready for, a changing climate.

However, despite the direction provided by Objective Five, planning for the threats posed by rising sea levels, extreme weather events, and other climatic changes is still not widespread among local authorities. While the 2010 Coastal Policy Statement has provided a foundation for some plan changes and consent decisions on climate adaptation in the coastal area, many councils appear reluctant to anger ratepayers or uncertain as to how best to address potential coastal hazards. Coastal property owners affected today feel they are bearing the cost for wider social benefits tomorrow and are likely to object to any effort taken to prepare the coastline for rising sea levels (Boston 2017).

Those councils that have taken action in response to the threat of rising sea levels have met with considerable resistance from private property owners. The four cases examined in this paper exemplify this. In each case, a precautionary action taken by a territorial authority was challenged by a holder of property rights in the coastal area. While such litigation is not desirable for councils, perhaps it is inevitable in the early stages of any new policy direction. Further, it has not been as damaging as councils might have expected. While councils must be sure to ground precautionary action in robust science and take proportionate responses, the courts in New Zealand appear willing to recognise a broadly worded national policy statement as a strong mandate for preventative council action. It is suggested that national guidance documents could be introduced elsewhere to generate similar results.

New Zealand Courts as Decision Makers

Council decisions on coastal climate adaptation measures are often challenged in New Zealand courts, in respect of long-term planning decisions as well as more immediate consents. Legal challenges will typically go first to the Environment Court, which is comprised of legally trained Environment Court judges and other specialist hearing commissioners. Some matters may be able to be appealed to the generalist courts above the Environment Court—the High Court, Court of Appeal and the Supreme Court. Judges in these courts are generalist in that they may be allocated a case on any topic that comes before the court—from tax and corporate law, to criminal law to constitutional, administrative and environmental law topics. Such judges may have no specific training in environmental law, nor scientific method, let alone climate science; and there are no specialist commissioners appointed to decide cases with them. Yet decisions of these higher, generalist courts are binding on courts below, including on the Environment Court. It is thus noticeable at the outset that a very real question is whether judges are receiving the appropriate scientific information to enable them to decide cases on climate adaptation measures, especially in light of the speed at which climate risk profiles are changing and the need to provide guidance to councils in this area.

On the specific topic of adaptation to expected sea level rise, there have already been at least 19 court challenges to council decisions. Interestingly, while expert evidence underpins councils' planning processes, it does not always serve to protect councils' decisions.

Summarising the results of these cases, some legal challenges to protective measures have been upheld by the courts and the relevant adaptation measures rejected, even where the measures were based on scientists' reports and on forecasts of climate change [*Weir v Kapiti Coast District Council* (2015), *Coopers Beachfront Estate Management Company Ltd v Far North District Council* (2006)]. Conversely, other similar council decisions have been upheld [*Southern Environmental Association (Wellington) Inc v Wellington City Council* (2010), *Gallagher v Tasman District Council* (2014)].

In a number of cases, a principle of voluntary assumption of risk has been applied. This has been used to uphold councils' decisions [e.g., *Mahanga E Tu Inc v Hawkes Bay Regional Council* (2014), *Waterfront Watch Inc v Wellington Regional Council* (2009)], or to uphold but add conditions [*Otago Regional Council v Dunedin City Council & B S & R G Holt* (2010)]. Yet the same principle has also been used to overturn councils' decisions not to allow such conditions [e.g., *Hemi v Waikato District Council* (2010)]. In yet other cases, voluntary assumption of risk has been insufficient and consents have been denied in seemingly similar circumstances [e.g., *Buckley v South Wairarapa District Council* (2008), *Department of Building and Housing, Determination* (*concerning Storm Ltd and the Hastings District Council*) (2007)].

The cases illustrate varying approaches to risk and caution, with different factors being given prominence. Such divergence may well demonstrate an appropriate flexibility, yet individual human preferences are often given a greater weighting than climate risks. Further investigation is needed as to the consistency of these factors and whether courts are adopting an appropriate approach in terms of the relevant climate science. They certainly illustrate a lack of clarity on the key drivers councils should use to underpin their planning processes. Stronger central government guidance would better equip both councils and courts to make decisions on climate adaptation measures.

It is in light of this background that this paper now turns to examine four case studies of litigation faced by particular local decision-making authorities, two before and two after the 2010 Coastal Policy statement.

Houses in Coastal Inundation Zones: Voluntary Assumption of Risk

In 2010, while still under the 1994 Coastal Policy Statement, the Environment Court considered two cases involving granting of consent under the Resource Management Act 1991 (RMA) to build houses in coastal inundation zones. The houses required consent as they were non-complying activities under their local councils' district plans.

A Hemi v Waikato District Council

In the first case, Mr. Hemi applied to the Waikato District Council for a resource consent to build a family home on (indigenous Maori) ancestral land owned by his family trust at Whaingaroa/Raglan on the west coast of the North Island of New Zealand. The site was described as "on the landward side of an in-filled boulder spit. The crest of the spit lowers towards the east where a lagoon occurs, into which waves enter." (*Hemi v Waikato District Council* (2010), paras. 1 & 2). (See Appendix 1 for a map.) Mr. Hemi had originally applied for resource consent in 2005, and the Council declined his application on the basis that the risk of coastal inundation was unacceptable. Mr. Hemi appealed, and Waikato District Council then supported his application, but it was opposed by one of the neighbouring landowners and was thus heard by the court.

The court identified risks of inundation from the ocean and the lagoon (coastal inundation), and of lagoon edge erosion (*Hemi v Waikato District Council* (2010), para. 44). It noted that these could result in the house being damaged, becoming unusable and having to be removed to protect the natural character of the environment, and that personal safety could be compromised during an exceptional storm (para. 45). (The definition of 'environment' in the RMA includes "ecosystems and their constituent parts, including people and communities" (RMA 1991, s2)).

The coastal inundation risk was assessed by four experts. The first, Mr. Dahm, was engaged by the Council. The second, Dr. Shand, was engaged by Mr. Hemi to address concerns in raised in Mr. Dahm's report (*Hemi v Waikato District Council* (2010), para. 46). The Council then engaged Dr. Mead to review Dr. Shand and Mr. Dahm's assessments, and the neighbouring landowner opposing the application engaged Professor Manning to review Mr. Dahm, Dr. Shand, and Dr. Mead's reports. Ultimately these experts worked together to advise the court, concluding that sea level rise values between 1.5 and 2 m could not be ruled out, but that an estimate of 0.5 to 1.5 m should be adopted. From this, they suggested that the upper estimate should apply to "expensive investments with high public welfare and benefit and no hazard adaptation options" (para. 57), and the lower estimate "should apply to investments of limited value where personal safety is not an issue and viable adaptation options are available" (para. 57). Mr. Hemi's house was assessed as being on the lower end and the sea level rise figures accepted were 0.6–0.9 m. This meant that the potential water level would be below the floor level of the basement and the risk was therefore minor. Similarly, the potential effect of lagoon edge erosion was determined to be of low probability and impact (para. 71).

In assessing the overall risk of coastal hazard, the court followed the decision in a previous case, *Waterfront Watch Inc v Wellington Regional Council* (2009), stating "that there is an element of 'voluntary assumption of risk' by people who choose to live near the coast in situations such as this" (*Hemi v Waikato District Council* (2010), para. 77) and the court must decide if the risk is acceptable, rather than requiring the avoidance of risk (para. 77). In these cases it was considered acceptable.

Finally, as a non-complying activity, the court could only grant consent if it found that the application "represent[ed] unusual qualities or [was] a true exception" (*Hemi v Waikato District Council* (2010), para. 96). It decided that, because of Mr. Hemi's historical and cultural links to the land, and strong support from his *hapu* (a division of a Maori tribe), the proposal did have special characteristics, and consent was granted.

B Otago Regional Council v Dunedin City Council & B S & R G Holt

In the second case, Mr. and Mrs. Holt applied to Dunedin City Council for land use consent to build a house on their rurally zoned 4048 m^2 property in Karitane

Village, 35 kilometres north of Dunedin in New Zealand's South Island. The site is across the Coast Road from the Karitane Estuary and is close to sea level, with the lower areas of land covered in salt marsh species (*Otago Regional Council v Dunedin City Council & B S & R G Holt* (2010), para. 10). (See Appendix 2 for a map.) The Holts proposed to build a pole house with a minimum floor level of 3.7 m above mean sea level, with the garage at a lower level (paras. 7 & 36).

The Dunedin City Council (DCC) granted the consent, but the Otago Regional Council (ORC) appealed the decision, leaving it to be determined by the court. The court found that "the adverse effects of the proposal that may be more than minor relate[d] solely to the amenity and safety of people who may be on the site" (*Otago Regional Council v Dunedin City Council & B S & R G Holt* (2010), para. 52). These adverse effects were flooding caused by water spilling over the Coast Road from the Waikouaiti River, inundation, climate change, storm surge, tsunami, and on-site effects of wave action and wave run-up (para. 56). After assessing the probability and effect of flooding from these events, the court determined that the risks to human safety were sufficiently low and the house should be built (para. 75). The Holts were advised to enter into a covenant with DCC stating that they understood and accepted the probabilities and consequences of flooding, that they would not complain or sue the city council for negligence, and that they would obtain a similar covenant from any subsequent purchasers, and a further deed with ORC stating that they would not seek flood protection works (para. 81). In agreeing to assume the risks, the Holts volunteered a condition to the covenant to keep a boat on a derrick on their house to escape from floodwaters if necessary. The court agreed that "[a] condition to that effect should be added" (para. 69).

Analysis

These cases reflect a broader trend in New Zealand of decisions being made based on an analysis of individuals' ability to assume risk, and indicate a preference for human safety as a predominant concern. Determining the level of risk required extensive analysis of the scientific evidence in both cases. In *Hemi*, it was noted that the science of sea level rise is "complex and still under development" (para. 57) and "characterised by potentially large changes and significant uncertainties" (para. 56). The court sought to obtain a consensus amongst the four experts advising on the case, rather than determining to prefer the evidence of one expert over another. In *Holt*, the court arrived at its risk assessment conclusions after extensive modelling and consideration of a variety of scenarios for the identified potential adverse effects.

The creation of a "moral hazard" was also considered in *Holt*, where the house might be sold to a third party who was less informed or accepting of the risk (para. 67). However, this was considered to be addressed by the reference to any subsequent purchasers in the covenant, indicating a preference for voluntary assumption of risk (as described in *Hemi*) rather than protection of third parties.

In *Hemi*, the relevant district plan included a building setback line of 100 m from mean high water springs. This was "said to be an arbitrary number" as "100 m up an 80 m high cliff is quite different to 100 m on a flat sandy coast" (para. 43) and a site-specific assessment was applied. While an individual assessment of the facts of the particular case is the natural duty of the court, setback and hazard lines are commonly used in council planning and therefore it would be appropriate if their underlying rationale were sufficiently robust to not require court intervention. In general, stronger guidance from central to local government could assist council decision-making and potentially reduce court intervention, particularly in cases such as *Hemi* where potential sea level rise was used as a proxy for allocation of level of risk.

Finally, it is interesting that the final decision to grant consent in *Hemi* turned not on these climate hazard considerations but on cultural factors related to indigenous ancestral land. This shows that additional factors may be included in whether to adopt an approach of voluntary assumption of risk, not just the climate hazard.

Challenges to Planning and Policy Measures Post-2010: Varied Results

The two cases examined in the previous section took place just prior to the New Zealand Coastal Policy Statement 2010 coming into force. The 2010 Coastal Policy Statement was designed to enable councils to better manage risks and hazards along New Zealand's coast, including risks and hazards arising from climatic changes (Orchard 2011). It is worthwhile to examine some of the approaches taken, and decisions made, following its introduction to determine whether a new approach is evident or whether it has facilitated the continuation of the status quo.

A Weir v Kapiti Coast District Council

The case of *Weir v Kapiti Coast District Council* (2013) concerns a clash between ratepayers and their local council over action taken in response to a localised assessment of projected sea level rise—The Kapiti Coast Erosion Hazard Assessment 2012 (KCEHA). This assessment, originally conducted in 2008 and revised in 2012, found that 1800 properties along the Kapiti Coast could be partially underwater or suffering from significant erosion by 2112 (Shand 2012). It included a series of maps overlaid with erosion hazard prediction lines (see Appendix 3 for an example), which the Kapiti Coast District Council felt obliged to place on its planning maps until such a time as the report could be factored into the next District Plan (*Weir* (2013), para. 14). The practical effect of this was that the Land Information Memorandum (LIM) over each of these 1800 properties indicated that they were both at risk of erosion over the next 100 years and within erosion hazard lines likely to form the boundary of a "no-build" zone in the upcoming District Plan.

The affected property owners were bewildered by this conclusion, as some of them lived as much as 80 m from the shoreline (Shand 2012). One affected property owner, Doug Palmer, told the local newspaper, the Dominion Post "If I go, New Zealand goes. No question, we are finished. [The prediction line is] so far out it is ridiculous" (Fairfax 2015). He, along with other ratepayers, brought a judicial review claim against the Kapiti Coast District Council before the High Court. Concerned that the inclusion of this information on the LIM, and potentially the District Plan, would seriously devalue their properties and restrict their options for future development, they argued that the Kapiti Coast Erosion Hazard Assessment was irrelevant, based on a flawed methodology, and did not take a sufficiently holistic view of the coastal environment in Kapiti (*Weir v Kapiti Coast District Council* (2013), para. 43).

Kapiti Coast District Council defended their claim on the basis that, as a territorial authority, they were statutorily obliged to incorporate the findings of the report on all relevant LIMs (*Weir* (2013), para. 14). The Local Government Official Information and Meetings Act 1987, s.44A(2) obliges councils to include in a LIM, any information that is "known to the council" and that identifies a "special feature or characteristic of the land concerned, including but not limited to potential erosion … or inundation, or likely presence of hazardous contaminants." The Council argued that the Hazard Assessment prediction lines amounted to information about potential erosion and inundation of the relevant properties, and were therefore caught within the aforementioned section.

The resulting High Court judgment focussed on two legal issues: whether or not the council was obliged to include the conclusions of the Hazard Assessment on the relevant LIMs and, if so, whether the method used to display them was proportionate. It did not focus on the more fundamental and scientific question of whether or not the predictions contained in the Hazard Assessment were sufficiently robust that they amounted to a "reasonable possibility". The judge preferred to put that question to one side for examination by an expert panel (*Weir v Kapiti Coast District Council* (2013), para. 35).

On the question of a council's obligation to include all relevant information on a LIM, the judge took a strict view. He did not accept the argument that the council needed to know the probability of the predicted erosion occurring before it was obliged to include it on the LIM. It was sufficient that the Council knew it to be a "reasonable possibility"—something more than mere speculation (*Weir v Kapiti Coast District Council* (2013), paras. 46 & 64). He also did not accept the proposition that a specific site-by-site analysis was necessary before the assessment could be said to relate to the property. As stated at paragraph 58, the report was "unquestionably about potential erosion as a special feature or characteristic of all coastal land along the Kapiti Coast, and therefore of every individual property fitting that description." The judge was of the view that the purpose of a LIM was to inform, and if necessary warn, potential purchasers about all relevant features of the land. That purpose included "trying to warn the market about the potential local effects of a global phenomenon." (*Weir v Kapiti Coast District Council* (2013), para. 58). He concluded that some reference to the information contained in the Hazard Assessment needed to be included on LIMs (para. 66).

Before examining the second legal question, it is worth reflecting on the consequence of this conclusion. It represents an affirmation by the High Court that a local council in possession of an evidentiary-based assessment of projected future land erosion must, in some way, incorporate that information on a LIM, and subsequently in the District Plan. This is likely to have significant precedential effect for other councils, who, by virtue of Policy 24 of the 2010 Coastal Policy Statement, are obliged to consider all potential coastal hazards that might arise over the next 100 years—including hazards arising from climate change (Department of Conservation 2010, p. 23).

On the second question of how to display the information on a LIM, the judge was critical of the approach taken by Kapiti Coast District Council. At paragraph 69, he empathised with the applicants' concerns that "the reports and, particularly the lines, have the potential to seriously affect the value and marketability of coastal properties in the district." Accordingly, he found that including bold erosion prediction lines on LIM reports and on planning maps, without adequate explanation of the inherent assumptions or methodology, was misleading and inappropriate. While the council has a very broad discretion as to how it represents information on a LIM, the information must "be accurate, state the position fairly, and it must not mislead." The action taken by the Council had fallen outside of these requirements and needed to be altered to ensure that the erosion risk was fairly represented on the LIMs.

The implications of the court's decision are significant. It creates a three-part test, which can be used by councils to determine whether or not they are obliged to disseminate predictions about potential coastal hazards and how they might go about doing so. Councils should ask themselves: whether the prediction relied on represents an evidence-based reasonable possibility (even if that possibility is the worst case scenario); whether the prediction relied on affects some private properties more than others, to the point of becoming a special feature or characteristic; and whether the method used to display that information states it fairly, without being misleading? While this process was tailored to councils' specific statutory obligations, the three requirements of sound science, localised effects and accurate representation of the facts are good guidelines for policy-makers everywhere.

In some instances the sticking point will be the first hurdle of whether or not the scientific prediction is sufficiently well-established to warrant legal intervention. This was the case in *Weir*. As determined by a panel of experts following the interim judgment, the Kapiti Coast Erosion Hazard Assessment was a useful data-set, but was not sufficiently robust to function as the evidentiary basis for building controls (Carley et al. 2014). Although the Hazard Assessment had reached its conclusions on the basis of the best global estimates of sea level rise and had accurately applied those estimates to the topography of the Kapiti Coast, it had ignored localised considerations, such as past evidence of shoreline accretion, which would have offset many of the findings (Carley et al. 2014). This was enough to put the data into question and, with the validity of the "prediction lines" questioned, the ability of the Kapiti Coast District Council to justify their inclusion on the LIMs was compromised—it would not be an accurate representation of the facts. On this basis, and facing mounting public pressure, the Council chose to remove the lines from all LIMs (Cairns 2015). Instead, it adopted the court's suggestion of a measured and proportionate indication of the findings; in this case this was precautionary wording about coastal erosion, as agreed between the parties.

What can be learnt from the case of *Weir v Kapiti Coast District Council* (2015) is that taking action in the face of uncertainty and on the basis of scientific modelling is difficult, particularly when private property owners stand to lose. Nevertheless, where scientific predictions regarding the potential effects of climate change are sufficiently robust to stand up to expert review, they can and must be incorporated into building and planning controls. The first stage of this is inclusion of such information in official records about the property in question.

B Gallagher v Tasman District Council

The case of *Gallagher v Tasman District Council* (2014) ('*Gallagher*') offers an interesting point of comparison to *Weir*. It revolves around a proposed plan change (Plan Change 22 or 'PC22') to the Tasman Resource Management Plan. This plan change sought to impose controls on the subdivision and development of coastal land in the Ruby Bay area, which was deemed to be at risk of coastal erosion and seawater inundation due to climate change (see Tasman District Council 2011). The Tasman District Council considered that future development in the area should, wherever practicable, occur outside of coastal hazard risk areas. They proposed to curtail further subdivision and habitable building construction accordingly. Consistent with Objective Five of the NZ Coastal Policy Statement 2010, the Council proposed to promote residential development on the hills to the northwest of Mapua and Ruby Bay instead (para. 8).

The Gallaghers objected to PC22 insofar as it applied to their property. They owned a 3.2 hectare land block on the coastal plains of Ruby Bay and wanted to subdivide this land for development and sell it as small residential land blocks. However, as their property fell within the potential coastal hazard zone, this was prohibited. The Gallaghers argued that PC22 should not apply to their property as the extent of coastal hazard risk was not sufficiently high to warrant the imposition of building controls and as any risks that did exist could be mitigated by building relocatable houses on elevated building platforms (*Gallagher v Tasman District Council* (2014), paras. 25 & 26). The Council disputed this claim. On the basis of updated flood hazard maps, the Council claimed that there was a high likelihood of severe flooding on the Gallagher property during a significant storm between now and 2115 (para. 116). They also claimed that allowing an exemption for the Gallagher property would be contrary to Objective Five and Policy 25 of the 2010 Coastal Policy Statement (para. 154).

In making its determination on the facts, the court broke its discussion into two key issues: whether or not the projected flood data was sufficiently accurate so as amount to a reasonably possible coastal hazard; and, whether or not an exception to Plan Change 22 could be made on the basis that the Gallaghers would assume any risk associated with the identified coastal hazard.

In respect of the first issue, which was a largely scientific inquiry, expert conferencing was used to put the projected flooding data under the spotlight (*Gallagher v Tasman District Council* (2014), paras. 57 & 58). While the perspectives of the experts differed, the court was able to identify a general consensus that, taking sea level rise over the next 100 years into account, a significant storm surge at high tide would, to some degree, flood the Gallagher property. At paragraph 156 of the judgment, the court expressed its satisfaction that, while some uncertainty remained, "the risk scenario identified…is a sufficiently realistic possibility to justify imposition of the controls proposed in PC22."

In assessing the second issue, the court did not undertake a balancing inquiry, as it may have prior to 2010. Instead, it examined whether or not the Council's inclusion of the property in PC22 was consistent with the relevant guidance document—namely, the 2010 Coastal Policy Statement. On the facts, the court considered that in light of the identified flood risk, the imposition of tighter building controls in the coastal plain of Ruby Bay (including on the applicant's property) was consistent with the purpose of the Coastal Policy Statement. It went towards directly achieving Objective Five, the relocation of future development away from coastal hazard risks. Additionally, it was consistent with Policies 3, 24 and 25 requiring a precautionary approach, the identification of potential coastal hazards and the avoidance of land uses that increase the risk of adverse effects from those hazards (*Gallagher v Tasman District Council* (2014), para. 147). There is no requirement that a particular coastal hazard be particularly severe or likely before these policies are triggered. Accordingly, the court dismissed the applicants' claim, finding, at paragraph 165 of the judgment, that:

> "PC22 in its present form gives effect to [the 2010] NZCPS as it is required to do…The inclusion in PC22 of site specific provisions of the kind proposed for the Gallagher property fails to give effect to the [2010] NZCPS and is directly contrary to the provisions of Objective [Five] and Policy 25 of that document.

This strong language reflects the prominence of the 2010 Coastal Policy Statement in precautionary coastal management. It should encourage other decision-making authorities to take precautionary action in response to climate-change-induced coastal threats. Contrary to the earlier cases, it finds that private property owners should not be permitted to bear the risk of potentially catastrophic coastal hazards.

Analysis

These cases demonstrate that, when a council faces a legal challenge questioning a decision made to avoid or mitigate a potential future costal hazard, the NZ Coastal Policy Statement 2010 can be used to defend their actions. Provided a council's decision to avoid or mitigate a future hazard rests on a sound evidentiary base, and provided the response taken is proportionate to the evidence possessed, acting cautiously in anticipation of a potential climate-change-induced coastal hazard is justified.

Unfortunately for the Kapiti Coast District Council in *Weir*, their defence failed at the first stage of this test—the scientific analysis on which their hazard lines were based did not stand up to scrutiny. Without a sufficiently robust evidentiary base, there was no solid foundation for the Council's actions and their defence fell down. They did everything they could in a way that was consistent with the 2010 Coastal Policy Statement—they obtained expert evidence on the potential for coastal hazards risks, they informed affected parties of the risk identified, and they took action to avoid the realisation of those risks. But, as the scientific predictions were inadequate, their actions were found to be overly cautious and baseless. Kapiti Coast District Council's experience should serve as a warning that, while the 2010 Coastal Policy Statement requires councils to take a precautionary approach to coastal hazard risks, there must be more than mere speculation about such risks. A degree of reasonable possibility is required (*Weir v Kapiti Coast District Council* (2015), para. 17).

Gallagher presents an interesting contrast to *Weir*, due to the wealth of coastal hazard information available and the consensus among scientists regarding its accuracy. Thanks to robust expert conferencing, the court was able to put the evidence of potential coastal hazard risk under the spotlight, gain insight into its methodology, hear opposing perspectives and clarify the analysis. The net result was that the court could be satisfied that there was a reasonable possibility that the potential coastal hazard risk identified (namely, flooding and seawater inundation) would occur. The Tasman District Council's plan change had a sound evidentiary base and, with that initial hurdle overcome, there were little grounds to question its wisdom. A council acting cautiously in response to good information about a potential coastal risk has a strong defence in the 2010 Coastal Policy Statement.

The court recognised this, steering clear of the risk-sharing analysis which had previously dominated court decisions about potential future coastal hazard risk. Instead, it acknowledged that the 2010 Coastal Policy Statement places the responsibility and burden for managing coastal hazard risk squarely on councils, and that they must be given scope to manage this risk as they see fit. The question is not whether or not the owner of a property can safely assume the risk themselves, but whether or not a council's action in response to the risk is reasonable and consistent with the provisions of the 2010 Coastal Policy Statement (*Gallagher v Tasman District Council* (2014), para. 180).

These cases demonstrate the critical importance of national guidance documents such as the NZ Coastal Policy Statement 2010 in responding to long-term risks. They not only assist territorial authorities in their decision-making but also provide powerful bases for precautionary action. Since the enactment of the 2010 Coastal Policy Statement, councils have been able to point to clear objectives and principles justifying precautionary action in respect of projected coastal hazards. The 2010 Coastal Policy Statement thus ought to be welcomed as a strong pillar underscoring precautionary council decisions. One can only hope that its potential will be fully realised and translated into other policy spheres above and beyond the coastal environment.

Conclusion

Responding to sea level rise requires coordinated solutions, extensive information gathering, significant financial input, and anticipatory policy-making (Boston 2017), particularly given the requirement to use a planning horizon of at least 100 years. Just the few cases focused on in this paper evidence several lessons for authorities that make decisions on sea level rise adaptation measures.

Territorial authorities must first ensure that any adaptation actions taken are grounded in robust science. They need good information about the risks and uncertainties associated with climate change, and must be prepared to assess a range of adaptation options and to justify the one chosen (Flood & Lawrence 2016). The Kapiti Coast District Council appears to have learnt this lesson the hard way, while the Tasman District Council can be commended for its measured and evidence-based response to a re-zoning issue.

It is recommended that future inundation lines be publicly recorded, including against official private and public property information. Future planning by way of determining appropriate activity zones is also required, taking into account likely sea level rise.

The cases also illustrate some lessons for judges as decision-makers on such options. In relation to the science, the approach taken by the courts to expert evidence in *Hemi*, *Weir* (2013) and *Gallagher* was extremely useful. Despite it being an adversarial system, the court required all the experts to identify areas of consensus and, in *Gallagher*, required an expert conference to be convened. Yet the courts still had to decide what to do with disagreement over scientific evidence, and to appropriately assess the risks. For example, as in *Holt*, not only does the court have to decide whether or not to issue a building consent, it has to decide on conditions to impose on any consent issued. This suggests that research is needed into whether the courts are adopting appropriate responses to changing climate risk profiles and thereby assisting the relevant authorities with their long-term adaptive planning and enabling deliberative and anticipatory approaches (McKim 2016; Lawrence 2015). Existing work has already shown that ambiguity surrounding the understanding of climate change adaptation is compounded by differences in interpretation of risks by courts (McKim 2016; Lawrence 2015).

Linked with an assessment of the content of judicial decisions as appropriate responses to climate adaptation is the issue of whether judges themselves have access to the scientific information they need in order to make informed decisions about climate change adaptation. If they do not have appropriate information, especially in light of fast changing climate risk profiles, assessing what type of information they need and in what format becomes critical to climate adaptation decision-making.

Finally, in relation to all levels of decision-making, the cases show that central government guidance is extremely helpful; moreover, it can be expected that challenges will be prevented if the guidance is clear and clearly followed. In New Zealand, the 2010 NZ Coastal Policy Statement has partially fulfilled this role, but perhaps as the knowledge base develops, such guidance could be expanded and made more useful.

In summary, the case studies examined in this paper provides the following lessons:

- *Local authorities*:
- Adaptation actions by local authorities must be underpinned by robust science
- Future planning should determine appropriate activity zones
- *Courts*:
- A review of the approaches by the courts is required to ensure that court decisions represent appropriate responses to changing climate risk profiles
- Consideration is needed as to whether judges have sufficient access to robust scientific information to enable them to make informed decisions about climate change adaptation
- *Central government*:
- Stronger central government guidance is needed to support local authorities to develop appropriate climate change adaptation measures
- Future inundation lines need to be publicly recorded, including against official property information (whether at local or central level).

The paper demonstrates that a more cohesive approach by central government is required to support the future development of appropriate climate change adaptation measures in New Zealand. In particular, further work is required to support local government and the courts to deliver appropriate and consistent decision-making across New Zealand to reduce the impact of sea level rise on local communities and the economy, and to protect against potential loss of life. The New Zealand Coastal Policy Statement 2010 could be significantly more directive and robust to assist with these aims, and thus become a useful model for other countries to consider.

Appendix 1: The *Hemi* House Site

From Google maps—https://www.google.co.nz/maps/place/Karioi/@-37.8241015,174.795483,431m/data=!3m1!1e3!4m5!3m4!1s0x6d1329d519efd905:0x96161aad18419689!8m2!3d-37.865!4d174.8025

Map Data: Google, MapData Sciences Pty Ltd., PSMA
Imagery: CNES /Airbus

Appendix 2: The *Holt* House Site

From Google maps: www.google.co.nz/maps.

https://www.google.co.nz/maps/place/96+Stornoway+St,+Karitane+9471/@-45.6378235,170.6519138,17z/data=!3m1!4b1!4m5!3m4!1s0xa8295ad82b666cf7:0x55a2090f60b8305!8m2!3d-45.6378235!4d170.6541025

Map Data: Google

Imagery: Google, TerraMetrics

Appendix 3: The *Weir* SLR Lines

Shand, R. (2012) *Kapiti Coast Erosion Hazard Assessment—2012 Update*. Wanganui, New Zealand: Coastal Systems Limited. Also available from: http://coastalsystems.co.nz/kapiti-erosion-assessments/

References

Boston, J. (2017). *Safeguarding the Future: Governing in an Uncertain World*. Wellington, NZ: Bridget Williams Books.

Buckley v South Wairarapa District Council. (2008). NZEnvC 4.

Cairns, L. (2015). *Kapiti Residents Joint Christchurch's Erosion Debate*. Available from www.stuff.co.nz.

Carley, J., Komar, P., Kench, P., & Davies, R. (2014). *Coastal Erosion Hazard Assessment for the Kāpiti Coast: Review of the Science and Assessments Undertaken for the Proposed Kāpiti Coast District Plan 2012*. Wellington, New Zealand.

Coopers Beachfront Estate Management Company Ltd v Far North District Council. (2006). NZEnvC 06.

Department of Building and Housing, Determination (concerning Storm Ltd and the Hastings District Council. (2007). 110.

Department of Conservation. (1994). *New Zealand Coastal Policy Statement*. New Zealand: Wellington.

Department of Conservation. (2010). *New Zealand Coastal Policy Statement*. New Zealand: Wellington.

Fairfax Media. (2015). *How Sea Level Rise Will Affect the Wellington Region and Possible Solutions*. Retrieved from www.stuff.co.nz.

Flood, S., & Lawrence, K. (2016). *Framing conversations around risk and uncertainty*. Prepared for the New Zealand Climate Change Research Institute, Victoria University of Wellington, as part of the 'Climate Change Impacts and Implications for New Zealand to 2100' research program.

Gallagher v Tasman District Council. (2014). NZEnvC 245.

Google, (n.d.), Google map of *Hemi* house. Retrieved 6 May 2017 from https://www.google.co.nz/maps/place/Karioi/@-37.8241015,174.795483,431m/data=!3m1!1e3!4m5!3m4!1s0x6d1329d519efd905:0x96161aad18419689!8m2!3d-37.865!4d174.8025.

Google, (n.d.), Google map of *Holt* house. Retrieved 6 May 2017 from https://www.google.co.nz/maps/place/96+Stornoway+St,+Karitane+9471/@-45.6378235,170.6519138,17z/data=!3m1!4b1!4m5!3m4!1s0xa8295ad82b666cf7:0x55a2090f60b8305!8m2!3d-45.6378235!4d170.6541025.

Hemi v Waikato District Council. (2010). NZEnvC 120.

Local Government Official Information and Meetings Act. (1987).

Lawrence, J. (2015). *The adequacy of institutional frameworks and practices for climate change adaptation decision making*. PHD Thesis, Victoria University of Wellington. Retrieved from http://researcharchive.vuw.ac.nz/handle/10063/4707.

Mahanga E Tu Inc v Hawkes Bay Regional Council. (2014). NZEnvC 248.

McKim, L. (2016). *A systematic review of recent research: Implications for policy and management, and tools to support adaptation decision making in New Zealand*. Prepared for the New Zealand Climate Change Research Institute, Victoria University of Wellington, as part of the 'Climate Change Impacts and Implications for New Zealand to 2100' research programme.

Ministry of Fisheries. (2011). *Coastal marine habitats and marine protected areas in the New Zealand Territorial Sea: A broad scale gap analysis*. Available from www.doc.govt.nz.

Orchard, S. (2011). *Implications of the New Zealand Coastal Policy Statement for New Zealand Communities* (Rep.). Wellington, New Zealand: Environment and Conservation Organisations of New Zealand.

Otago Regional Council v Dunedin City Council & B S & R G Holt. (2010). NZEnvC 120 (2010). NZRMA 263.

Resource Management Act. (1991).

Shand, R. (2012). *Kapiti Coast Erosion Hazard Assessment—2012 Update*. Wanganui, New Zealand: Coastal.

Southern Environmental Association (Wellington) Inc v Wellington City Council. (2010). NZEnvC 196.

Statistics New Zealand. 'Gross Domestic Product: December (2016) quarter' Liz MacPherson, Government Statistician, ISSN 1178–0290, 16 March 2017. Retrieved from: http://www.stats.govt.nz/browse_for_stats/economic_indicators/GDP/GrossDomesticProduct_HOTPDec16qtr.aspx.

Tasman District Council. (2011). *Change 22: Mapua and Ruby Bay Development*. Retrieved from www.tasman.govt.nz.

Waterfront Watch Inc v Wellington Regional Council. (2009). W043/09.

Weir v Kapiti Coast District Council (Interim Judgment). (2013). NZHC 3522 19 December 2013 15 NZCPR 28.

Weir v Kapiti Coast District Council (Final Judgment). (2015). NZHC 43.

Mapping of Benthic Habitats in Komave, Coral Coast Using WorldView-2 Satellite Imagery

Roselyn Naidu, Frank Muller-Karger and Mathew McCarthy

Abstract Human uses or resources of coral reef ecosystems have synergistic effects that compound environmental pressures, often with negative results on ecosystem services. For example, sea-surface temperatures in many tropical regions have increased by almost 1 °C over the past 100 years, and in the area of Fiji are currently increasing at approximately 2 °C per century. This is combined with increased fishing pressures and increased pollution from populated land areas. Resources in shallow coastal ecosystems may be conserved for sustainable use through the design and implementation of resilient networks of marine conservation or marine protected areas (MPAs). This requires accurate spatial data on the distribution and extent of habitats (coral reef, seagrass and mangroves) at scales which match conservation planning decisions by policymakers. Habitat maps improve management by guiding sampling strategies, mapping resources in a way that facilitates assessment of status and change, involving local communities, and identifying conservation areas that can be fixed or rotated to promote sustainable use. Social and ecological vulnerability to disasters and outcomes of any particular extreme event are influenced by buildup or weakening of resilience, both before and after disaster occurs. The findings of the present study help communities consider options in planning management strategies including MPA networks that can assist the communities adapt to changing climate. To support the design of a resilient MPA network for the Komave region, a high-resolution habitat map was generated based on WorldView-2 imagery, literature review and ground-truth data.

R. Naidu (✉)
School of Environment, The University of Auckland, City Campus, Auckland, New Zealand
e-mail: roselynnaidu87@gmail.com

F. Muller-Karger · M. McCarthy
College of Marine Science, University of South Florida, St. Petersburg, FL, USA

W. Leal Filho (ed.), *Climate Change Impacts and Adaptation Strategies for Coastal Communities*, Climate Change Management,
https://doi.org/10.1007/978-3-319-70703-7_18

Introduction

Coastal areas worldwide are suffering significant, perceptible, and quick degradation in water quality, biodiversity, and various other critical parameters as a direct result of climate change (Lovejoy and Hannah 2005). Especially hard-hit are tropical island environments in the Pacific Ocean, such as Fiji, where climate change threatens to aggravate other environmental problems (Hurrell and Trenberth 1994; Hoegh-Guldberg et al. 2007). Reef islands are small islands, composed of mainly unconsolidated Holocene sands and gravels that have been deposited on the rim of coral atolls and on other reef platforms during the past few thousand years (McLean and Stoddart 1978; Woodroffe et al. 1999; Woodroffe and Morrison 2001). Coral reefs and their connected habitats are recognized as an extremely important resource in the Pacific (Guinotte et al. 2003).

While coral reef species have survived substantial climate and sea-level change events in the geological past, this time human impacts are undermining reef resilience (Hughes et al. 2003). Fiji has a tropical marine climate, with average daily temperatures at Suva ranging from 20.4–26.5 °C in July to 23.8–31.0 °C in February (Fiji Meteorological Service 2017). In the lee of the mountains, however, the daytime temperatures often rise 1–2 °C above those on the windward sides or on the smaller islands. Also, the humidity on the lee side tends to be somewhat lower. Average sea surface temperatures range from 26 °C in June to October, to 29 °C in February. December to April is just below the February maximum with an average temperature of 28 °C (Fiji Meteorological Service 2017). Water temperature in Fiji regularly reaches the upper thermal limit for coral. Corals in such locations exist within 1 or 2 °C of their upper thermal limit during summer months. When corals are exposed to higher than normal seawater temperatures, during the warm season, their algal symbiosis is disrupted, leading to the phenomenon known as bleaching (Baker et al. 2008; Spalding and Brown 2015). This physiological symptom of stress is followed by large-scale coral mortality if high water temperatures are sustained. Since the 1980s, rising sea surface temperatures owing to global warming have caused unprecedented mass bleaching of corals, including three pan-tropical events in 1998, 2010 and 2015–2016 (Obura and Mangubhai 2001; Heron et al. 2016). During 2015–2016 bleaching event, record temperatures triggered a pan-tropical episode of coral bleaching, the third global-scale event since mass bleaching was first documented in the 1980s (Hughes et al. 2017).

The morphology and dynamics of reefs are a function of the climatic situation, reflecting wave and wind energy, and the occurrence of tropical disturbances such as cyclones (McLean and Woodroffe 1994; Perry and Smithers 2011). There have been quite a few assessments of the physical vulnerability of low-lying coral atoll nations to sea level rise (Nunn and Mimura 1997; Woodroffe 2008; Webb and Kench 2010; Biribo and Woodroffe 2013) where the reefs in the Pacific region were classified as highly vulnerable to extreme events and environmental changes (Roy and Connell 1991; McCarty et al. 2001; Morgan and Kench 2016). Several past shoreline studies indicate that most islands are morphologically stable (Dawson and Smithers 2010; Ford and Kench 2014; Kench et al. 2015; McLean and Kench

2015), while number of studies suggest that reef islands are just as resilient to sea level rise and can become accustomed to the climate changes geographically (Kench and Cowell 2001; Kench et al. 2005, 2015). Despite this trend, some islands have eroded or completely disappeared (Kench et al. 2005). Tropical cyclones in Fiji occur in the period of November to April, with greatest frequency during January and February (Fiji Meteorological Service 2017). On average, some 10–15 cyclones per decade affect some parts of Fiji, and 2–4 exert severe damage. The dominant north-west to south-east tracks of these cyclones gives some increased risk of damage in outlying north-west island groups. There has been an apparent increase in the incidence of cyclones in Fiji since the early 1990s. The projected changes in climate have been predicted to increase the frequency of intensity cyclones (Knutson et al. 2010). An increase in severe cyclones threatens to lead to the degradation of vulnerable reef structures with reduced chances of recovery (Gardner et al. 2005).

According to Hughes et al. (2007), the removal of herbivores from the reefs puts stress on the coral communities and leads to eventual growth of macro algae causing their decline. There is a substantial body of literature that demonstrates how remote sensing imagery is useful in providing timely and up-to-date information for benthic and coral reef mapping and monitoring (Eakin et al. 2010; Goodman et al. 2013; Maglione 2014; Martin et al. 2016). Therefore, it is important to understand the distribution of coral reef resources in planning fisheries management options.

The purpose of this study is to explore the utility of multispectral imagery to identify and classify coral reefs; marine benthic habitats and seagrass in Komave, Coral Coast (Latitude 18°13′15.32″S and Longitude 177°45′31.82″E).Such comprehensive maps of marine resources provide crucial information for local fisheries management (Crowder and Norse 2008) and biodiversity conservation that will help the communities develop local adaptation strategies to combat climate change and increase their resilience to the impacts (Hatziyanni 2015).

The WorldView-2 multispectral sensor are amongst the recently launched sensors, which provides the highest spatial and spectral resolution of multispectral satellite imagery (DigitalGlobe 2009). The sensor has an 8 band multispectral resolution with 6 bands in the visible spectrum and 2 near-infrared bands at a 2 m spatial resolution. This increased spectral resolution (8 bands), improves the accuracy of bathymetric mapping applications (Collin and Hench 2012) and the discrimination of marine habitats such as corals (Botha et al. 2013), as compared to other sensors such as QuickBird and IKONOS (4 bands). Satellite imagery and in situ observation are the basis for more complex forecasting models and ecosystem-based management (Sherman et al. 2011; Lyons et al. 2012; Röckmann et al. 2012). Without maps created from the integration of these tools, it is difficult to conduct research and evidence based planning and decision making. Such maps can be used to plan the placement of Marine Protected Areas (MPAs). There is a growing network of locally managed Marine Protected Areas or Tabu areas initiated by Fiji Locally Managed Marine Areas Project (LMMA). The study promotes the use of conservation education to highlight the advantages of voluntarily established marine reserves, such as increased fish catches and tourist revenue, to local communities (Rowlands et al. 2005; White et al. 2010).

It is difficult for small Marine Protected Areas (MPAs) used in remoteness to support marine species populations, which are adequately large to sustain themselves, while to create one single large MPA of such sufficient size is also difficult in consideration of economic, social and political constraints in many regions (Laffoley et al. 2008: Friedlander et al. 2016). The networks may improve ecological resilience against the risk of localized disasters, climate change, failures in management o)r other hazards, and thus better help sustainability of populations than single sites (NRC 2001). Thus, networks of MPAs of different amounts and scales are established to help to reduce human impacts without compromising fisheries benefits and marine conservation, and provide necessary significant spatial linkages to preserve ecosystem processes (PISCO 2011). For planning and establishing MPA networks, a map of marine biogeographic regions can help to identify unique and representative areas to achieve conservation goals (Roberts et al. 2003). Similarly, based on the identification of the existed MPAs and their networks, their adequacy in each classified ecosystem could be assessed for the further MPA network planning in the insufficiently protected areas.

Due to the benefits of mapping biogeography for planning MPAs networks, the need for a globally, detailed, and comprehensive consistent map of marine biogeography has been acknowledged for decades in marine conservation by various regional, national, and global legal frameworks and planning commitments, including the convention of Biological Diversity's Programme for Work on Protected Areas, the Nature Conservancy, World Wildlife Fund, and United Nations (Spalding et al. 2007).

Habitat Description

Coral reef benthic habitat maps and maps of coastal land resources provide critical information needed for the management of coastal ecosystems and are used in numerous research and monitoring activities, for example, coastal development, fisheries and other resource uses, coral reef resiliency, connectivity, sea-level change, climate change, and ocean acidification (Miller 2010). Such maps facilitate describing the physical environment in coral reefs, identify connectivity to relevant land-based and marine threats, and set a baseline reference for change detection analysis and monitoring (Jennings 1998; Hedley et al. 2016). The data support a wide range of applications in such areas as agriculture, forestry, geology, resource management, geography, mapping, water quality, coastal studies, and in global change. This comparison process can reveal land-cover changes that occur slowly and subtly, or quickly and devastatingly.

A habitat classification scheme allows grouping of habitat types based on common ecological or geomorphological characteristics (Mumby and Harborne 1999). There are a variety of marine benthic habitat characterization schemes around the world. Here, we considered the knowledge of the area of local experts involved in the previous fieldwork activities Tawake (2000), and the characteristics in the images, to develop a generalized habitat classification scheme. Immersed aquatic vegetation (e.g. seagrasses and algae) are vital to coastal ecosystem health and resilience. For example, Zostera marina and other seagrass species provide key ecosystem services such as sediment retention, carbon cycling and providing physical stability to coastlines by preventing against wave and current action (Hemming and Duarte 2000; Mateo et al. 2003). The habitat map derived using satellite imagery and field work as part of this study provides the basis for consensual decision-making and conflict resolution among stakeholders of the Fijian coral reefs located near Komave. This will hopefully lead to informed decision making and implementation of environment management plans for identified areas.

Scientific Questions

Globally, fisheries industries in coral reef areas reach the value of $400 billion dollars annually (Ichiro 2009). Such valuation is not yet available for reefs of Fiji, yet these ecosystems support the economy and health of communities throughout the Indo-Pacific Ocean.

Based on the objectives described as above, the following research questions are formulated:

1. To what extent can coral, mangrove and benthic habitats of Votua and Komave be classified and mapped using easily accessible satellite imagery (e.g. images from the WorldView-2 and Landsat satellites)?
   Importance of research question: Clear delineation of important coastal ecosystems will help improve monitoring and management efforts.
2. Do the latest advances in satellite sensors provide any benefit to local and regional coastal ecosystem classification accuracy?
   Importance of research question: New sensors on-board satellites (such as Landsat 8 and Sentinel-2 with new spectral bands for coastal applications, and the WorldView-2 satellite with two-meter spatial resolution) are better suited for coastal applications including more accurate ecosystem classifications.
3. What are the benefits of object-oriented habitat digitization using GIS combined with satellite observations over standard satellite imagery pixel based classification techniques?
   Importance of research question: More accurate classification of particular ecosystems can be achieved with baseline information to develop an automatic satellite classification system. The result is a geospatial tool that can be used by different stakeholders to track the health of the resource.

Methods

Study Site

The study was conducted in Cakaubalavu Reef, Cakau Lekaleka Reef, Vatumalawa Reef and Nalumu Reef in the south western part of Viti Levu (Fig. 1). This area is prominently known as the Coral Coast due to its reef systems and beautiful beaches which have allured tourists for many decades. Geographically, the study site was situated between 177° 47′ 48″–177° 50′ 59″ east longitude and 18° 14′ 50″–18° 16′ 11″ south latitude.

The WorldView-2 image was used to map the study site. WorldView-2 provides a more appropriate resolution to measure the small-scale dynamics (2 m × 2 m pixel resolution) (Reshitnyk et al. 2014; Hoang et al. 2016; Martin et al. 2016). The image was taken on the 24th of January, 2013. The image was acquired through the Institute of Marine Remote Sensing, University of South Florida. The data was delivered as a “LV1B”. This is the least processed of the World View-2 imagery products which are radiometrically and sensor corrected, but not projected to a plane using a map projection (Cheng and Chaapel 2010).

The image was radiometrically calibrated to produce radiance values by multiplying the metadata based gain with the pixel value and adding the offset with

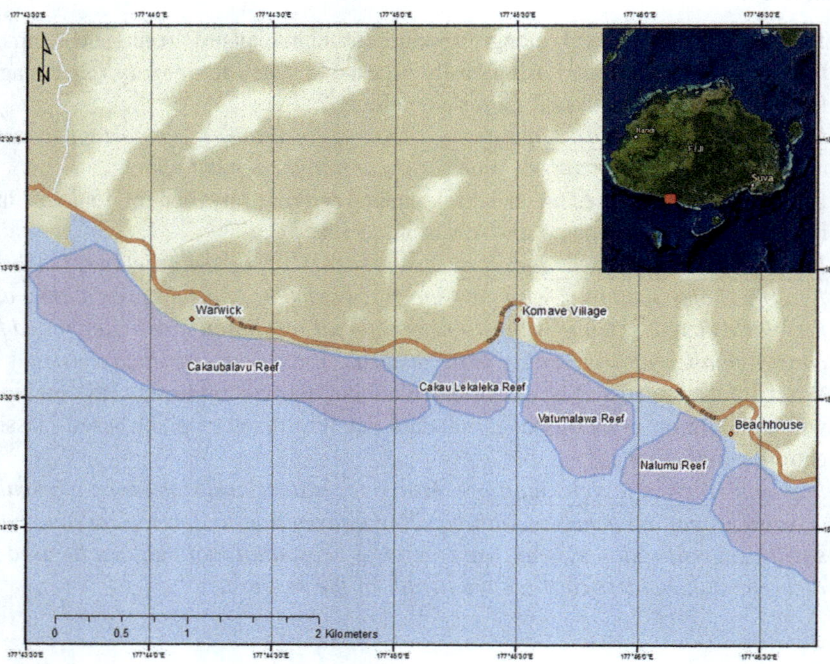

Fig. 1 Shows the locality map of the reefs near Komave, Coral Coast, Fiji

ENVI's WorldView-2 Radiance tool (DigitalGlobe 2014). Radiometric corrections were performed to reduce the atmospheric effects in the images to enable derivation of the water leaving radiance in each pixel using Fast Line-of-sight Atmospheric Analysis of Hypercubes (FLAASH). FLAASH corrects at-surface reflection of objects hence enhancing the reflectance properties of the benthic cover types in the imagery (Phinn et al. 2011). Geometric corrections were also performed on the imagery using ENVI's RCP Ortho-rectification Workflow tool. This ensured that the ground reference points (GRPs) were correctly aligned with the imagery. Since the study site was small, a region of interest was created and the study area was cropped out of the entire imagery for analysis. A water mask was also constructed to remove areas of "white water" from the reef crest.

Ground-truthed data of the shallow water benthic habitats were collected between July and November, 2016. The geographical positioning of the field sampling points was collected using a Trimble Pro 6T receiver and Juno 3D with a horizontal accuracy of 1 m. At each ground reference point (GRP), GPS locations were recorded, a photograph was taken, and the habitat class and supplementary notes were taken. To minimize spatial auto-correlation effects, the GRPs were taken at least 20 m apart (McCarthy and Halls 2014). Four benthic classes were defined and labelled as 'Bare', 'Coral', 'Algae' and 'Grass' based on field observations. Out of the 600 GRPs collected, 300 were selected at random and used for training the supervised classification while the remaining 300 were used to carry out the accuracy assessment.

Supervised Classification and Accuracy Assessment

After classification, the data were imported into a geographic information system, ArcGIS 10.3 and randomly split 50%: 50% into training and validation data. Training data were visualized over the WorldView-2 image to select 2 × 2 pixel Regions of Interest (ROIs) (4 pixels per ROI) thus ensuring that pixels represented each substrate class based on the ground-truthed data. Training data ROIs were used to assign a habitat class for conducting a supervised classification with the WorldView-2 imagery. Validation data were processed to ensure that (1) validation data did not overlap with training ROIs and (2) that there was only one validation point per 2 × 2 m pixel (resolution of the WorldView-2 imagery) thus minimizing false inflation of the accuracy assessment (McCarthy et al. 2015). Validation data were used to assess the accuracy of the final WorldView-2 satellite habitat maps.

The supervised method of classification allows automatic categorization of all the pixels in an image into previously defined classes (Vahtmäe et al. 2012). The Maximum Likelihood Classifier (MLC) is one of the most widely used supervised classification methods (Yang et al. 2015) and was also applied in this study. MLC evaluates the relationship between the brightness levels in one band versus the other band (variance and covariance) of the training classes and designates pixels based on the maximum probability of it belonging to each class (McCarthy and Halls 2014). The training GRPs were then used in ENVI for the supervised MLC.

Accuracy assessment was the most important step in the classification process. The main idea was to quantitatively determine how effectively the pixels were grouped in the user defined classes (Ismail and Jusoff 2008). Accuracy of the map was determined by constructing matrices in order to test if any difference exists in the interpretation. Using the Confusion Matrix Using Ground Truth ROI Tool in ENVI, the remaining GRPs were used to construct the matrix after training the supervised classification and the accuracy of the maps were recorded.

Results

The atmospheric correction approach showed clear improvements to the spectral reflectance within the WorldView-2 image. A total of 300 ground truth data points were used to assess the accuracy of the WorldView-2 classifications (approximately 50% of the ground-truth data). The field data were distributed among four benthic habitat classes: coral, bare, algae, grass. The habitat maps resulting from each classification are shown in Fig. 3. Overall, producer's and user's accuracy for all benthic substrate classifications of the WorldView-2 image are reported in Table 1. The overall classification accuracy for the simulated WorldView-2 scene was 71%.

The analysis contained within this paper concentrates on general benthic cover of the reef flat. The created coastal ecosystem map and the map legend were shown in Fig. 2. This ecosystem map shows the spatial distribution and zonation of each thematic category in the study site.

On a per class basis, the use of spectral indices did improve either the user's or the producer's accuracy of certain habitat classes. For example in Table 1, the algae class showed a higher producer's accuracy for the classification with the spectral indices compared to the classification with the user's accuracy (84% compared to 77%). The other classes such as grass (64% compared to 62%), coral (60% compare to 50%) and bare (73% compared to 60%) showed higher user's accuracy for classifications performed with the spectral indices.

The habitat map was created from the shallow (<3.0 m) WorldView-2 map using the spectral variables. This selection was made taking into account several considerations. Firstly, the WorldView-2 classification performed best in the shallow regions. Secondly, the WorldView-2 image has a continuous surface of pixels, which improves the mapping of smaller patches of habitat and habitat edges. The habitat classification (and map accuracies) created from the datasets are shown in Fig. 3. Overall, user's and producer's accuracy are reported and these values represent the classification of the habitat map.

Algal and seagrass habitats were spectrally and spatially different. Though the photosynthetic pigments in algae and seagrass (e.g. chlorophyll, phycoerythrin and fucoxanthin) have different reflectance characteristics, satellite spectral bands are generally unsuitable for distinguishing them (Maritorena et al. 1994; Mumby et al.

Table 1 Confusion matrix results for Project Site at Komave, Coral Coast, Fiji Islands

Class	Producers Accuracy	Users Accuracy
Grass	61.73	64.10
Coral	53.06	60.47
Bare	60.00	72.97
Algae	83.91	76.44
Overall Accuracy	71.35	
Kappa Coefficient	0.5553	

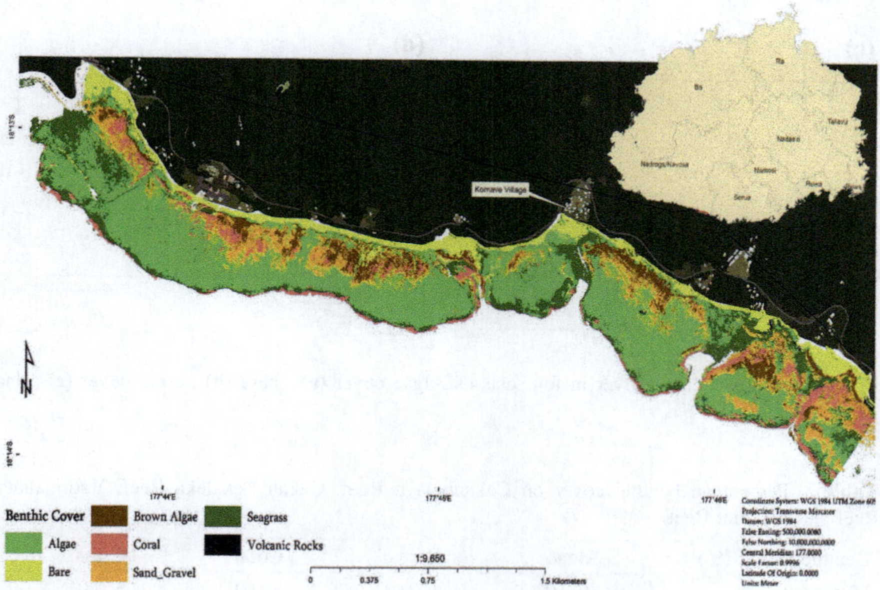

Fig. 2 Map showing the benthic cover of Cakaubalavu Reef, Cakau Lekaleka Reef, Vatumalawa Reef and Nalumu Reef

1997), because at wavelengths of >580 nm penetration of water is poor, preventing the characteristic reflectance minima and maxima of photosynthetic pigments being detected. For example, though most photosynthetic pigments show reflectance minima below 450 nm, the maxima lie between 670 and 700 nm. Where distinguishing minima and maxima exist within the water-penetrating spectrum, satellite bands may be too broad to distinguish them. For example, the reflectance minima for both green and brown algae are below 500 nm and their maxima are 550 and 575 nm, respectively (Maritorena et al. 1994; Mumby et al. 1997).

Algae lead total benthic cover of the Cakaubalavu Reef, Cakau Lekaleka Reef, Vatumalawa Reef and Nalumu Reef, however, it is not distributed evenly (Fig. 3a). Levels of between 0.0015% and 35.13% algal cover were seen (Table 2). Algae tend to be more abundant towards the outer reef zone, as well as on the margins of

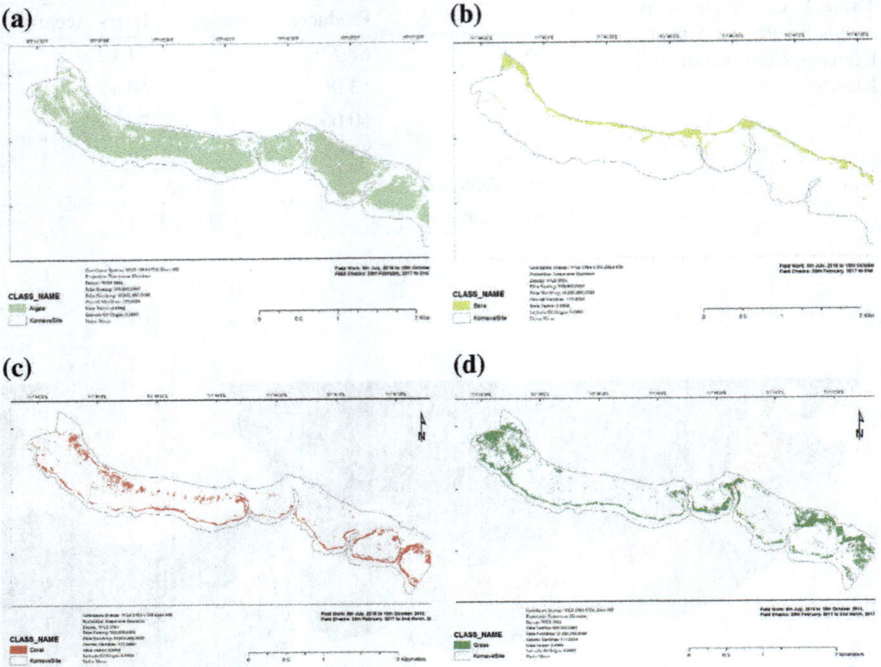

Fig. 3 Showing benthic cover in four classes Algae cover **(a)** , bare **(b)** , coral cover **(c)** , and grass cover **(d)**

Table 2 Percentage benthic cover on Cakaubalavu Reef, Cakau Lekaleka Reef, Vatumalawa Reef and Nalumu Reef

Benthic Cover (%)	Algae	Bare	Coral	Grass
Minimum	0.015	> 0.01	> 0.01	> 0.01
Maximum	35.13	6.39	1.46	2.07
Total	67.84	9.03	8.37	14.76

river channels and bays. Algal cover is particularly dense over the reef area. Seagrass are marine angiosperms that forms meadows in shallow areas (Roelfsema et al. 2014). They are sediment stabilizers as their rhizomes and roots bind sediments together, thus reducing soil erosion. Seagrasses provide important habitat for a diversity of fish and invertebrate species such as juvenile salmon (*Onchorhynchus sp.*) and Pacific herring (*Clupea harengus*) (Hemming and Duarte 2000; Borg et al. 2006; Chittaro et al. 2009; Robinson et al. 2011). The seagrass cover ranged from 0.01% to 2.07% (Table 2). Coral cover is patchy (Fig. 3c). Cover ranged from 0 to 1.46% of these four reefs investigated, with the highest levels of live coral cover seen in the reef area east of Komave. However, most of the section's reef area has less than 8.37% coral cover (Table 2).

Discussion

The majority of studies in the tropical regions have highlighted the ability of high-resolution satellite imagery to create maps of benthic marine habitat (Andrefouet et al. 2008). There is strong evidence from this study and others that optical remote sensing from high resolution multispectral sensors can provide crucial data for supporting conservation management in coastal tropical regions (Phinn et al. 2008; Roelfsema et al. 2009).

Fringing reefs extend along the Coral Coast of Fiji for approximately 63 km and between the coast and 1 km offshore. The reef is periodically broken by channels cut through the reef by influx of sediment and water from the nearby creeks and rivers. These channels provide suitable habitat for corals, other sessile forms, and their associated communities below the low tide (Fig. 2).

Seagrasses are one of the most important benthic habitats for this area. They provide ecological services including maintenance of marine biodiversity, regulation of the quality of coastal waters, protection of the coastline, and are indicators of the status of the coastal zone which can be used in coastal management strategies aiming at preserving and improving the environmental quality of the coastal zone (Beck et al. 2001; Green and Short 2003; Terrados and Borum 2004). Shallow water is preferred for seagrasses as they need optimum light penetration for photosynthetic activity. The discussion with the Komave community representative revealed that after Cyclone Winston the seagrass beds had deteriorated. Duarte (2002) states that vigorous disturbances caused by cyclones and hurricanes can lead to increased rates of seagrass losses. This is because strong winds generated during cyclones enhance the motion of the sand, damaging the grass beds and also leads to sea grasses being pulled out if the wind is forceful enough. Near the low tide end of the shore, most seagrasses were observed with few algal species as well. In the middle region of the intertidal zone it had mixed variation of flora and fauna species. Whereby, near the reef edge, mostly coral habitat was observed together with algal species.

Near the shore area is exposed to wave actions that will continuously disturb and destroy the stable growth of corals (Baker et al. 2004). Also, this area has the most abundant algae species, and algae can out-compete corals for space, and also hinder coral growth by decreasing the oxygen supply available for corals (Le Bec et al. 2000). The Komave fishing ground is heavily fished and over the last few decades, harvesting of corals has been a primary cause for the noticeable declines in fish stocks across the coral reef ecosystem (Clements et al. 2012). Wastewater pollution from villages, settlements, and nearby resorts contains pathogens, nutrients, and toxins that are not only a risk to human health, but also to environmental health. Given the importance of the Coral Coast for the local communities, the raised nutrient levels are of great concern (Mosley and Aalbersberg 2003). Additionally, bacterial levels in local creeks can reach extreme dangerous level due to the wastewater and pollution from piggeries. The distribution of coral and algae ultimately depends on providing of suitable substrate upon which to settle. There are many areas with high levels of exposed bedrock along the coast. The high levels of bedrock could be due to the environmental conditions in such areas that do not boost algal or coral growth. The distribution patterns suggest that algae are much better at colonizing suitable substrate than coral. Over-harvesting and destructive fishing practices in Coral Coast have depleted populations of marine resources and degraded the coral reef ecosystem (Pascal and Seidl 2013).

Almost 30% of the world's corals have disappeared in the last 20 years mainly because of bleaching, disease, starfish infestations, and pollution (Hoegh-Guldberg 2005). The decline suggests that coral reefs are losing their dominant coral populations by approximately 2% (Muscatine 1990). Recent findings suggest that global change will induce a lower rate of calcification as a result of change in ocean chemistry (Kleypas et al. 1999). These predictions indicate that the overall reef calcification ratio will increase. Combined with other events such as bleaching, and hurricanes, the reefs will slowly become less compact and more fragmented (Done 1999; Andréfouët et al. 2001; Bak et al. 2005).

Our results demonstrated that high resolution WorldView-2 satellite imagery for coastal ecosystem mapping provides an effective way for rapid identification of substrate types and the habitat cover in shallow water of coastal intertidal zone. The distribution and extent of each thematic category as well as the patterns of diversity can also be clearly recognized from the map. Therefore, high resolution imagery makes it possible to produce more accurate and detailed coastal ecosystem map. Our new map of coral reef resources will hopefully help the local communities understand this precious resource and implement management actions that minimize the trends seen locally and globally. This will be the significance of combining spatial tools, anthropological fieldwork, communal and natural science methods for the goal of aiding the design of marine protected areas.

Conclusion

One of the most significant needs to help manage the use of resources in a sustainable manner is to develop accurate, high spatial resolution base maps of coastal and shallow-water aquatic resources. Such maps are needed as a baseline against which to measure any future change, and to develop resource management plans and sustainable resource use strategies. It might be possible to create boundaries for Marine Protected Area's (MPA) without very detailed resource maps, however, in order to understand the baseline abundance and distribution of resources and how these resources change over time within any MPA or any other area, an accurate map is essential.

Our work highlights the importance of incorporating high-resolution WorldView-2 images in developing baselines for shallow water aquatic resources when delineating conservation programs. By incorporating gound-truthed field data and data from hyperspectral-like features, such the extra bands available in Digital Globe's WorldView-2, we have shown the benthic cover of Komave, Coral Coast.

While this study focused on a site within the South West Pacific, we believe that this work will be applicable to researchers working in other sites. Reef are threatened throughout the world (Burke et al. 2011) and there is a critical need for innovative conservation strategies to protect the intense beauty of these ecosystems.

The results from this study provides sufficient justification to manage Fijian reefs at a regional and non-regional scale. If possible, we would suggest a network of marine reserves within each of these regions that would operate in parallel, providing resilience and diversity within the coastal sites.

There are inherent limitations in mapping nearshore marine habitat using passive optical remote sensing technology; however, the combination of the most accurate outputs from both (e.g. WorldView-2 multispectral satellite imagery and field ground-truth) helped capture the spatial distribution of major habitats present at the study site. The ability to map benthic habitat using of WorldView-2 satellite imagery is also dependent on the depth of the habitat, the light attenuation characteristics of the overlying water column, and the reflectance contrast between the target habitat and the surrounding substrate (Green et al. 2000). These issues are particularly relevant for mapping nearshore habitats in tropical marine environments where benthic habitat occur outside of the resolution of passive optical sensors. WorldView-2 imagery is suitable for mapping fine-scale changes in habitat characteristics, however, acquiring imagery under ideal conditions can be challenging.

Given the challenges that corals and their symbionts encounter under a changing climate, the communities of dominant reef building corals are likely to become gradually rare, as is happening around the world (Bellwood et al. 2004; Trapon et al. 2011). A wise approach towards marine reserve management will help conserve Fiji's coral reefs (Jupiter and Egli 2011) and undergoing climate-related disasters can develop the rural poor's adaptation to climate change through community-based resource management (Takasaki 2015).

This paper presented an approach for tropical shallow water coastal ecosystem mapping using WorldView-2 satellite imagery. The produced coastal map provides the most up to date high spatial resolution information for Cakaubalavu Reef, Cakau Lekaleka Reef, Vatumalawa Reef and Nalumu Reef, Coral Coast, Fiji which can be used for inventory and monitoring activities. The map will be invaluable for coastal ecosystems assessment and management and as well as for future comparative study.

Application and Future Work

- The approach presented in this study used a combination of high spatial resolution (2×2 m pixel) WorldView-2 multispectral satellite imagery, detailed field data and object based image analysis to create habitat maps. For ecologists and managers, time series of these high spatial resolution maps can provide invaluable information about coral, algae and seagrass ecology.

- To allow assessment of the impacts of management actions on these habitats (e.g. bare, algae, coral, seagrass) the roles of various environmental drivers need to be determined through analysis of patterns in their abundance, distribution and composition, combined with corresponding environmental data (e.g. temperature, water quality).

Acknowledgements This study was supported by a grant from the USAID Pacific-American Climate Fund. Especial acknowledgement to the University of Fiji-PACAM project team members, Mr. Anish Maharaj and Ms. Michelle Kumar for their assistance during fieldwork and production of benthic maps and for their support during the tenure as the Project Manager (Principal Investigator) on the project namely "Developing base maps of tropical aquatic resources in the Pacific". The work was also supported in part by NASA grant NNX14AP62A to FMK. The manuscript is a contribution to the Marine Biodiversity Observation Network. The Government of Fiji granted permission to conduct research in Coral Coast and Komave Community were very cooperative and supportive during the study period.

References

Andréfouët, S., Muller-Karger, F., Hochberg, E., Hu, C., & Carder, K. (2001). Change detection in shallow coral reef environments using Landsat 7 ETM+ data. *Remote Sensing of Environment, 78*(1–2), 150–162.

Andrefouet, S., Costello, M., Rast, M., & Sathyendranath, S. (2008). Earth observations for marine and coastal biodiversity and ecosystems. *Remote Sensing of Environment, 112*(8), 3297–3299. https://doi.org/10.1016/j.rse.2008.04.006.

Bak, R., Nieuwland, G., & Meesters, E. (2005). Coral reef crisis in deep and shallow reefs: 30 years of constancy and change in reefs of Curacao and Bonaire. *Coral Reefs, 24*(3), 475–479.

Baker, A., Starger, C., McClanahan, T. & Glynn, P. (2004). Coral reefs: Corals' adaptive response to climate change. *Nature*, *430*(7001), 741.

Baker, A. C., Glynn, P. W., & Riegl, B. (2008). Climate change and coral reef bleaching: An ecological assessment of long-term impacts, recovery trends and future outlook. Estuarine, *Coastal and Shelf Science*, *80*(4), 435–471.

Beck, M., Heck, K., Able, K., Childers, D., Eggleston, D., Gillanders, B., et al. (2001). The Identification, Conservation, and Management of Estuarine and Marine Nurseries for Fish and Invertebrates. *BioScience, 51*(8), 633.

Bellwood, D., Hughes, T., Folke, C., & Nyström, M. (2004). Confronting the coral reef crisis. *Nature, 429*(6994), 827–833.

Biribo, N., & Woodroffe, C. D. (2013). Historical area and shoreline change of reef islands around Tarawa Atoll, Kiribati. *Sustainability Science*, *8*(3), 345–362.

Borg, J. A., Rowden, A. A., Atrill, M. J., Schembri, P. J., & Jones, M. B. (2006). Wanted dead or alive: high diversity of macroinvertebrates associated with living and dead Posidonia oceanica matte. *Marine Biology, 149*(3), 667–677.

Botha, E. J., Brando, V. E., Anstee, J. M., Dekker, A. G., & Sagar, S. (2013). Increased spectral resolution enhances coral detection under varying water conditions. *Remote Sensing of Environment, 131,* 247–261. https://doi.org/10.1016/j.rse.2012.12.021.

Burke, L., Reytar, K., Spadling, M., & Perry, A. (2011). *Reefs at risk revisited* (1st ed., p. 130). Washington, DC: World Resources Institute.

Cheng, P., & Chaapel, C. (2010). Pan-sharpening and Geometric Correction WorldView-2 Satellite. *GeoInformatics, 4,* 30–33.

Chittaro, P., Finley, R., & Levin, P. (2009). Spatial and temporal patterns in the contribution of fish from their nursery habitats. *Oecologia, 160,* 49–61.

Clements, C., Bonito, V., Grober-Dunsmore, R., & Sobey, M. (2012). Effects of small, Fijian community-based marine protected areas on exploited reef fishes. *Marine Ecology Progress Series, 449,* 233–243.

Collin, A., & Hench, J. L. (2012). Towards deeper measurements of tropical reefscape structure using the WorldView-2 spaceborne sensor. *Remote Sensing, 4*(12), 1425–1447. https://doi.org/10.3390/rs4051425.

Crowder, L., & Norse, E. (2008). Essential ecological insights for marine ecosystem-based management and marine spatial planning. *Marine Policy, 32*(5), 772–778.

Dawson, J. L., & Smithers, S. G. (2010). Shoreline and beach volume change between 1967 and 2007 at Raine Island, Great Barrier Reef, Australia. *Global and Planetary Change, 72*(3), 141–154.

DigitalGlobe. (2009). WorldView-2 Data Sheet. Retrieved May 05, 2012, from http://www.digitalglobe.com/downloads/WorldView2-DS-WV2-Web.pdf.

DigitalGlobe. (2014). Geolocation Accuracy of WorldView Products. Accessed September 16. http://global.digitalglobe.com/sites/default/files/WorldView_Geolocation_Accuracy.pdf.

Done, T. (1999). Coral Community Adaptability to Environmental Change at the Scales of Regions. *Reefs and Reef Zones. American Zoologist, 39*(1), 66–79.

Duarte, C. (2002). The future of seagrass meadows. *Environmental Conservation, 29*(02).

Eakin, C., Nim, C., Brainard, R., Elvidge, C., Gledhill, D., Muller-Karger, F., et al. (2010). Monitoring coral reefs from space. *Oceanography, 23*(4), 118–133.

Fiji Meteorological Service (2017). [online] Available at: http://www.met.gov.fj/. Accessed 12 July 2016.

Ford, M. R., & Kench, P. S. (2014). Formation and adjustment of typhoon-impacted reef islands interpreted from remote imagery: Nadikdik Atoll, Marshall Islands. *Geomorphology, 214,* 216–222.

Friedlander, A. M., Wagner, D., Gaymer, C. F., Wilhelm, T. A., Lewis, N. A., Brooke, S., et al. (2016). Co-operation between large-scale MPAs: successful experiences from the Pacific Ocean. *Aquatic Conservation: Marine and Freshwater Ecosystems, 26*(S2), 126–141.

Gardner, T., Côté, I., Gill, J., Grant, A., & Watkinson, A. (2005). Hurricanes and Caribbean coral reefs: impacts, recovery patterns, and role in long-term decline. *Ecology, 86*(1), 174–184.

Goodman, J.A., Samuel, J., Stuart, R. (2013). *Coral Reef Remote Sensing*. Springer Science.

Green, E., & Short, F. (2003). *World atlas of seagrasses* (1st ed.). Berkeley, Calif.: University of California Press.

Green, E. P., Mumby, P. J., Edwards, A. J., & Clark, C. D. (2000). *Remote sensing handbook for tropical coastal management*. Paris: UNESCO.

Guinotte, J., Buddemeier, R., & Kleypas, J. (2003). Future coral reef habitat marginality: temporal and spatial effects of climate change in the Pacific basin. *Coral Reefs, 22*(4), 551–558.

Hatziyanni, E. (2015). Implementing Regional Marine Policies at the Eastern Mediterranean, Through an Ecosystem Based Marine Spatial Planning. *Journal of Aquaculture & Marine Biology, 2*(1).

Hedley, J., Roelfsema, C., Chollett, I., Harborne, A., Heron, S., Weeks, S., et al. (2016). Remote Sensing of Coral Reefs for Monitoring and Management: A Review. *Remote Sensing, 8*(2), 118.

Hemming, M., & Duarte, C. (2000). *Seagrass ecology*. Cambridge University Press.

Heron, S. F., Johnston, L., Liu, G., Geiger, E. F., Maynard, J. A., De La Cour, J. L., ... & Iguel, J. (2016). Validation of reef-scale thermal stress satellite products for coral bleaching monitoring. *Remote Sensing, 8*(1), 59.

Hoang, T., Garcia, R., O'Leary, M., & Fotedar, R. (2016). Identification and Mapping of Marine Submerged Aquatic Vegetation in Shallow Coastal Waters with WorldView-2 Satellite Data. *Journal of Coastal Research, 75*(sp1), 1287–1291.

Hoegh-Guldberg, O. (2005). Low coral cover in a high-CO_2 world. *Journal of Geophysical Research, 110*(C9).

Hoegh-Guldberg, O., Mumby, P., Hooten, A., Steneck, R., Greenfield, P., Gomez, E., et al. (2007). Coral Reefs under Rapid Climate Change and Ocean Acidification. *Science, 318*(5857), 1737–1742.

Hughes, T. P., Baird, A. H., Bellwood, D. R., Card, M., Connolly, S. R., Folke, C., ... & Lough, J. M. (2003). Climate change, human impacts, and the resilience of coral reefs. *Science, 301*(5635), 929–933.

Hughes, T., Rodrigues, M., Bellwood, D., Ceccarelli, D., Hoegh-Guldberg, O., McCook, L., et al. (2007). Phase Shifts, Herbivory, and the Resilience of Coral Reefs to Climate Change. *Current Biology, 17*(4), 360–365.

Hughes, T. P., Kerry, J. T., Álvarez-Noriega, M., Álvarez-Romero, J. G., Anderson, K. D., Baird, A. H., ... & Bridge, T. C. (2017). Global warming and recurrent mass bleaching of corals. *Nature, 543*(7645), 373–377.

Hurrell, J., & Trenberth, K. (1994). Decadal atmosphere-ocean variations in the Pacific. *Climate Dynamics, 9*(6), 303–319.

Ichiro, N. (2009). World Review of fisheries and aquaculture (pp. 3–6). Rome: FAO Fisheries and Aquaculture Department. Retrieved from http://ftp://ftp.fao.org/docrep/fao/011/i0250e/i0250e.pdf.

Ismail, M. H., & Jusoff, K. (2008). Satellite data classification accuracy assessment based from reference dataset. *International Journal of Computer and Information Science and Engineering, 2*(2), 96–102.

Jennings, S. (1998). Artisanal fisheries of the Great Astrolabe Reef, Fiji: monitoring, assessment and management. *Coral Reefs, 17*(1), 82.

Jupiter, S., & Egli, D. (2011). Ecosystem-Based Management in Fiji: Successes and Challenges after Five Years of Implementation. *Journal of Marine Biology, 2011,* 1–14.

Kench, P. S., & Cowell, P. J. (2001). The morphological response of atoll islands to sea-level rise. Part 2: Application of the modified shoreface translation model (STM). *Journal of Coastal Research*, 645–656.

Kench, P. S., McLean, R. F., & Nichol, S. L. (2005). New model of reef-island evolution: Maldives, Indian Ocean. *Geology, 33*(2), 145–148.

Kench, P. S., Thompson, D., Ford, M. R., Ogawa, H., & McLean, R. F. (2015). Coral islands defy sea-level rise over the past century: Records from a central Pacific atoll. *Geology, 43*(6), 515–518.

Kleypas, J. (1999). Geochemical Consequences of Increased Atmospheric Carbon Dioxide on Coral Reefs. *Science, 284*(5411), 118–120.

Knutson, T., McBride, J., Chan, J., Emanuel, K., Holland, G., Landsea, C., et al. (2010). Tropical cyclones and climate change. *Nature Geoscience, 3,* 157–163.

Laffoley D, Gjerde K, Wood L.(2008). Progress with marine protected areas since Durban, and future directions. Parks.

Le Bec, N., Julliet-Leclerc, A., Corrège, T., Blamart, D., & Delcroix, T. (2000). A coral $\delta^{18}O$ record of ENSO driven sea surface salinity variability in Fiji (south -western tropical Pacific). *Geophysical Research Letters, 27*(23), 3897–3900.

Lovejoy, T., & Hannah, L. (2005). *Climate change and biodiversity* (1st ed.). New Haven, Conn.: Yale University Press.

Lyons, M. B., Phinn, S. R., & Roelfsema, C. M. (2012). Long term land cover and seagrass mapping using Landsat and object-based image analysis from 1972 to 2010 in the coastal environment of South East Queensland, Australia. *ISPRS Journal of Photogrammetry and Remote Sensing, 71,* 34–46.

Maglione, P. (2014). Coastline extraction using high resolution WorldView-2 satellite imagery. *European Journal of Remote Sensing*, 685–699.

Maritorena, S., Morel, A., & Gentili, B. (1994). Diffuse reflectance of oceanic shallow waters: influence of water depth and bottom albedo. *Limnology and Oceanography, 37,* 1689–1703.

Martin, J., Eugenio, F., Marcello, J., & Medina, A. (2016). Automatic sun glint removal of multispectral high-resolution WorldView-2 imagery for retrieving coastal shallow water parameters. *Remote Sensing, 8*(1), 37.

Mateo, M., Sanchez-Lizaso, J., & Romero, J. (2003). Posidonia oceania "banquettes": A preliminary assessment of the relevance for meadow carbon and nutrients budget. *Estuarine, Coastal and Shelf Science, 56,* 85–90.

McCarthy, M., & Halls, J. (2014). Habitat Mapping and Change Assessment of Coastal Environments: An Examination of WorldView-2, QuickBird, and IKONOS Satellite Imagery and Airborne LiDAR for Mapping Barrier Island Habitats. *ISPRS International Journal of Geo-Information, 3*(1), 297–325.

McCarty, J. J., Leary, O. F., Dokken, N. A., & DJ White, K. S. (2001). Climate change 2001: Impacts, adaptation, and vulnerability. *Contribution of working group II to the Third assessment report of the intergovernmental panel on climate change* (No. 574.5222 C639co 2001). Cambridge University Press.

McCarthy, M., Merton, E., & Muller-Karger, F. (2015). Improved coastal wetland mapping using very- high 2-meter spatial resolution imagery. *International Journal of Applied Earth Observation and Geoinformation, 40,* 11–18.

McLean, R. F., & Stoddart, D. R. (1978). Reef island sediments of the northern Great Barrier Reef. *Philosophical Transactions of the Royal Society, London, A, 291,* 101–117.

McLean, R. F., & Woodroffe, C. D. (1994). Coral atolls. In Carter, R. W. G., Woodroffe, C. D. (Eds.), *Coastal Ž. Evolution: Late Quaternary Shoreline Morphodynamics* (pp. 267–30). Cambridge: Cambridge University Press.

McLean, R., & Kench, P. (2015). Destruction or persistence of coral atoll islands in the face of 20th and 21st century sea-level rise? *Wiley Interdisciplinary Reviews: Climate Change, 6*(5), 445–463.

Miller, N. (2010). *Workshop report by the Research and Monitoring Expert Group*. Bonaire: DCNA.

Morgan, K. M., & Kench, P. S. (2016). Reef to island sediment connections on a Maldivian carbonate platform: Using benthic ecology and biosedimentary depositional facies to examine island-building potential. *Earth Surface Processes and Landforms, 41*(13), 1815–1825.

Mosley, L. M., & Aalbersberg, W. G. (2003). Nutrient levels in sea and river water along the 'Coral Coast'of Viti Levu, Fiji. *The South Pacific Journal of Natural and Applied Sciences, 21*(1), 35–40.

Mumby, P., & Harborne, A. (1999). Development of a systematic classification scheme of marine habitats to facilitate regional management and mapping of Caribbean coral reefs. *Biological Conservation, 88*(2), 155–163.

Mumby, P., Green, E., Edwards, A., & Clark, C. (1997). Coral reef habitat mapping: how much detail can remote sensing provide? *Marine Biology, 130*(2), 193–202.

Muscatine, L. (1990). The role of symbiotic algae in carbon and energy flux in reef corals. *Ecosystems of the world, 25,* 75–87.

NRC [National Research Council (USA)]. (2001). *Marine protected areas: tools for sustaining ocean ecosystems*. Washington, DC, USA: National Academy Press.

Nunn, P. D., & Mimura, N. (1997). Vulnerability of South Pacific island nations to sea-level rise. *Journal of Coastal Research*, 133–151.

Obura, D., & Mangubhai, S. (2001). Marine biological survey of the Kadava Ono Tinika, Fiji (Great and North Astrolabe Reefs). *Report for the World-Wide Fund for Nature—South Pacific Programme*. Suva, Fiji.

Pascal N. & Seidl A. (2013). *Economic benefits of marine protected areas: case studies in Vanuatu and Fiji*, South Pacific. Research report, AFD/IUCN. French Framework Agreement by CRIOBE (EPHE/CNRS), Mooréa, French Polynesia, 160 pages + annexes.

Perry, C. T., & Smithers, S. G. (2011). Cycles of coral reef 'turn-on', rapid growth and 'turn-off' over the past 8500 years: A context for understanding modern ecological states and trajectories. *Global Change Biology, 17*(1), 76–86.

Phinn, S., Roelfsema, C., Dekker, A., Brando, V., & Anstee, J. (2008). Mapping seagrass species, cover and biomass in shallow waters: An assessment of satellite multi-spectral and airborne hyper-spectral imaging systems in Moreton Bay (Australia). *Remote Sensing of Environment, 112*(8), 3413–3425.

Phinn, S., Roelfsema, C., & Mumby, P. (2011). Multi-scale, object-based image analysis for mapping geomorphic and ecological zones on coral reefs. *International Journal of Remote Sensing, 33*(12), 3768–3797.

PISCO. (2011). *The Science of Marine Reserves. 2nd Edition, Europe*. Corvallis: Partnership for Interdisciplinary Studies of Coastal Oceans.

Reshitnyk, L., Costa, M., Robinson, C., & Dearden, P. (2014). Evaluation of WorldView-2 and acoustic remote sensing for mapping benthic habitats in temperate coastal Pacific waters. *Remote Sensing of Environment, 153,* 7–23.

Roberts, C. M., Andelman, S., Branch, G., Bustamante, R. H., Castilla, J. C., Dugan, J. & McArdle, D. (2003). Ecological criteria for evaluating candidate sites for marine reserves. *Ecological applications,* S199-S214.

Robinson, C. L. K., Yakimishyn, J., & Dearden, P. (2011). Habitat heterogeneity in eelgrass fish assemblage diversity and turnover. *Aquatic Conservation: Marine and Freshwater Ecosystems, 21*(7), 625–635. https://doi.org/10.1002/aqc.1227.

Röckmann, Christine, Ulrich, Clara, Dreyer, Marion, Bell, Ewen, Borodzicz, Edward, Haapasaari, Päivi, et al. (2012). The Added Value of Participatory Modelling in Fisheries Management – What Has Been Learnt? *Marine Policy, 36*(5), 1072–1085. https://doi.org/10.1016/j.marpol.2012.02.027.

Roelfsema, C. M., Lyons, M., Kovacs, E. M., Maxwell, P., Saunders, M. I., Samper-Villarreal, J., et al. (2014). Multi-temporal mapping of seagrass cover, species and biomass: A semi-automated object based image analysis approach. *Remote Sensing of Environment, 150,* 172–187.

Roelfsema, C. M., Phinn, S. R., Udy, N., & Maxwell, P. (2009). An integrated field and remote sensing approach for mapping seagrass cover, Moreton Bay. *Australia. Journal of Spatial Science, 54*(1), 45–62.

Rowlands, G., Comley, J. & Raines, P. (2005). *The Coral Coast, Viti Levu, Fiji. A Marine Resource Assessment*.

Roy, P., & Connell, J. (1991). Climatic change and the future of atoll states. *Journal of Coastal Research,* 1057–1075.

Sherman, K., O'reilly, J., Belkin, I. M., Melrose, C., & Friedland, K. D. (2011). The application of satellite remote sensing for assessing productivity in relation to fisheries yields of the world's large marine ecosystems. *ICES Journal of Marine Science: Journal du Conseil, 68*(4), 667–676.

Spalding, M. D., Fox, H. E., Allen, G. R., Davidson, N., Ferdana, Z. A., Finlayson, M. A. X., et al. (2007). Marine ecoregions of the world: a bioregionalization of coastal and shelf areas. *BioScience, 57*(7), 573–583.

Spalding, M. D., & Brown, B. E. (2015). Warm-water coral reefs and climate change. *Science, 350* (6262), 769–771.

Takasaki, Y. (2015). Learning from disaster: community-based marine protected areas in Fiji. *Environment and Development Economics, 21*(01), 53–77.

Tawake, A. (2000). *Summary report on the 2000 Votua community marine resource management and monitoring workshops*. USP: IAS.

Terrados, J. & Borum, J. (2004). Why are seagrasses important?-Goods and services provided by seagrass meadows. *European seagrasses: an introduction to monitoring and management* (pp. 8–10).

Trapon, M., Pratchett, M., & Penin, L. (2011). Comparative Effects of Different Disturbances in Coral Reef Habitats in Moorea, French Polynesia. *Journal of Marine Biology, 2011,* 1–11.

Vahtmäe, E., Kutser, T., Kotta, J., Pärnoja, M., Möller, T., & Lennuk, L. (2012). Mapping Baltic Sea shallow water environments with airborne remote sensing. *Oceanology, 52*(6), 803.

Webb, A. P., & Kench, P. S. (2010). The dynamic response of reef islands to sea-level rise: Evidence from multi-decadal analysis of island change in the Central Pacific. *Global and Planetary Change, 72*(3), 234–246.

White, J., Botsford, L., Hastings, A., & Largier, J. (2010). Population persistence in marine reserve networks: incorporating spatial heterogeneities in larval dispersal. *Marine Ecology Progress Series, 398,* 49–67.

Woodroffe, C. D. (2008). Reef-island topography and the vulnerability of atolls to sea-level rise. *Global and Planetary Change, 62*(1–2), 77–96

Woodroffe, C. D., McLean, R. F., Smithers, S. G., & Lawson, E. (1999). Atoll reef-island formation and response to Ž. sea-level change: West Island. *Cocos Keeling Islands. Marine Geology, 160,* 85–104.

Woodroffe, C. D., & Morrison, R. J. (2001). Reef-island accretion and soil development on Makin, Kiribati, central Pacific. *CATENA, 44*(4), 245–261.

Yang, W., Yin, X., & Xia, G. S. (2015). Learning high-level features for satellite image classification with limited labeled samples. *IEEE Transactions on Geoscience and Remote Sensing, 53*(8), 4472–4482.

Exploring the Practicability and Applicability of Payment for Ecosystem Services in the Protection of Samoa's Selected Watershed Areas

Susana Taua'a

Abstract There is mounting evidence of studies conducted throughout the Developed and Developing world on the practicability and sustainability of implementing Payment for Ecosystem Services (PES) schemes particularly in relation to Watershed conservation and mangrove forest protection. There is a strong correlation between watershed conservation and mangroves protection, particularly in a small island state such as Samoa, where human activities in the watersheds can impact directly on the social–economic welfare of coastal communities and the marine resources they depend on. PES is a tool used to measure the economic value of ecosystem services provided by the natural environment. This study is a contribution to existing PES studies, focusing on the practicability of a PES program to protect and manage Samoa's watershed areas in the upland ridges extending to the coastal communities in the Apia urban area. The Vaisigano watershed is the study site selected to demonstrate what is happening in the watershed, justify the introduction of a PES scheme for water conservation, identify strengths and weaknesses of a PES scheme and concludes with a suite of suggested pathways for a PES model in Samoa. The Vaisigano watershed is a critical water source for the Apia coastal community.

Introduction

The natural environment provides a variety of ecosystem services critical to our social-physical-economic wellbeing. But, these services are often under-valued, or overlooked in decision-making at different levels from the individual-household unit up to national government bodies. Decision-making hastily made and lack people consultation and environmental considerations, and there are many thus far, for example, the hydropower scheme at Ta'elefaga, serve to illustrate how ill planned developments can return to plague communities. Nonetheless, evidence of

S. Taua'a (✉)
National University of Samoa, Le Papaigalagala Campus, P.O. Box 1622, Apia, Samoa
e-mail: s.tauaa@nus.edu.ws

W. Leal Filho (ed.), *Climate Change Impacts and Adaptation Strategies for Coastal Communities*, Climate Change Management,
https://doi.org/10.1007/978-3-319-70703-7_19

the value of the natural environment and benefits sourced from it include food, fiber and fuel, maintenance of water quality and supply, carbon storage, nutrient cycling, and healthy soils to name a few. But, these ecosystem services are seriously under threat from unabated human activities and poor decision making, hence the need to place ecosystem services at the center of local and global conservation dialogue and action.

Payments for Ecosystem Services (PES) is a recent instrument added to an existing range of market based instruments such as taxes, subsidies, user-charges and fines for non-compliance with environmental regulations. These instruments/ tools are designed to deal with the problems of landscape and environmental protection, biodiversity conservation and wetland restoration (Pirard 2012). PES essentially is a market based system in the sense that payments are made to farmers, landowners and land managers to adopt actions that enhance the quantity and quality of desired ecosystem services for the benefit of specific or general users who are dependent on these ecosystem services. In other cases, payments are made by beneficiaries of the environmental services, such as water users and hydropower companies, sometimes, national governments pay on behalf of their citizens, who are indirect beneficiaries (International Institute for Environment and Development 2015).

Rationale

The research study is a response to the need to collect and document evidence of adverse human activities within the vicinity of watershed areas to justify and support the implementation of a PES scheme. In the same vein, marine and coastal resources upon which many small island states rely on for their livelihoods will equally benefit from a holistic PES scheme considering the inter-connectivity between human activities in the upland areas and potential impacts they may have on the coastal lowlands. A Ridge to Reef approach as advocated in the United Nations 2014 Small Island Developing States Conference held in Apia, Samoa is an integrated approach to land, water, forest, biodiversity and coastal resource management that contribute to sustainable livelihoods and climate resilience (United Nations Department Economic Social Affairs 2014). Thus, the essence of this study is to begin an investigation of anthropological impacts beginning in the watershed as such activities do make an imprint on coastal ecosystems.

Study Limitations

Time and resource constraints hindered the extension of the research to include residents living on the Apia coastal villages which would have adequately satisfied the 'Ridge to Reef' integrated approach to discussing climate change vulnerability in coastal areas. Nonetheless, this study will initiate the conversation on mitigating factors to combat the impacts of climate change on Apia's coastal communities, particularly the marine coastal environment.

Study Objectives

The objectives of the study were as follows:

1. To conduct water sample testing from the streams- rivers of the watershed to determine the extent of water pollution to substantiate grounds for PES.
2. To seek the views of responsible stakeholders (land owners, community leaders, tourism operators, water providers, water users) in relation to establishing a PES scheme.

A spectrum of research methods were used to collect the information and data required to answer the research objectives.

Methodology

A spectrum of research methods was used to collect the info-data required to answer the broader research question to meet the objectives of the study:

1. The case study method is an empirical inquiry to explore a contemporary phenomenon within its real life context, when the boundaries between phenomenon and context are not clearly evident and in which multiple sources of evidence are used (Yin 1984, p. 23). The case study is suitable to address 'how' and 'why' type questions which are explanatory by nature and essentially the key questions underpinning this study. Case study researches are classified into three categories— exploratory, descriptive and explanatory (Schell 1992). This study adopts the positive elements of all three categories- preliminary research, surveys, histories, experimentation (water sampling), observation, and interviews, consultations and focus group discussions that are based on multiple sources of evidence (multiple triangulation) to provide a robust investigation into the feasibility and practicability of Payments for Ecosystem Services (PES) in Samoa. In relation to this study, the phenomenon explored is PES as applied to the Vaisigano watershed specifically and thereby derive a model that can be extended to other watersheds in the country.
2. Thematic analysis is the second method used. It is a method for identifying, analyzing, and reporting patterns (themes) within data (Braun and Clarke 2006). This method is suitable to answer questions related to people's experiences and views such as 'what do people think about paying a surcharge on their water bill for water resources conservation?' It suits question related to understanding and

representation such as, 'are water consumers willing to pay extra to help water policy makers and service providers to better manage our watersheds?' and 'how much are people willing to pay for ecosystem services?' From the formal and informal interviews with landowners and people living within the watershed areas, patterns of experiences can be listed from their direct quotes or common ideas that emerge from the interviews or conversations. Having identified the patterns, the next step is to classify these patterns into related themes (Aronson 1995). Themes are established from integrating parts of ideas and experiences of the informants that are meaningless when viewed alone but when combined they form a complete picture of their collective experience. To substantiate the established themes, the researcher needs to check back with the informants for feedback which can then be incorporated into the analysis. A comprehensive literature review should validate the selected themes in that it allows the researcher to make inferences that provide answer(s) to the research questions.

3. Focus group interviews of small groups between 4 to 7 people residing in the watershed area is also one method used to generate information on the collective views of the participants in the study. Gauging the participants' views is important to clarify, extend or challenge the data and information collected through other methods.
4. Other methods used in this study are structured questionnaires administered by the researcher and research assistants to selected residents living within the watershed area. Water quality testing of selected streams in the watershed is a quantitative tool used to collect supporting evidence to advocate for PES schemes. Key informant interviews are other critical methods of data collection to seek the views of water consumers from the community and big landowners in the watershed.

Results

A sample of 65 residents living within the villages of the Vaisigano watershed were entrusted with brief questionnaires primarily to gauge their views on the value they place on water and water resources, willingness to pay for water, and priority uses for water. Since the study targets the Vaisigano watershed, it is imperative that the respondents are drawn from a wide range of sites located within the specified study area that span upstream and downstream areas of the watershed. Table 1 shows the number of respondents and their gender. The gender of the respondent is statistically important because it impacts on the nature and type of farming activity the respondents engage in, for example, 41.6% [10/24] of males described themselves as farmers engaged in a range of farming activities such as taro, bananas, taamu, vegetable production and small scale cattle and pig farming. They admitted to using some kind of pesticide over time to manage the weed problem.

Table 1 Watershed villages and number of respondents by sex

Village	Total number of respondents	Male	Female
Tiapapata	5	3	2
Maagiagi	16	7	9
Vailima	9	3	6
Vaoala	8	2	6
Tanugamanono	5	1	4
Lelata	8	2	6
Motootua	4	3	1
Malifa	3	1	2
Maluafou	3	–	3
Leone	4	2	2
Total	65	24	41

A question in relation to the value placed by the respondents on water and water resources was asked to gage two important factors. First to confirm whether they value their water and second, how much value they placed on water relative to other basic needs. Ninety five percent of the respondents said 'yes' when asked whether they valued water. This response was expected, as was the case with a previous study on water demand and usage (Taua'a 2014). However, the follow up question, to quantify how much value they placed on water compared to other important and valuable things in their 'basket' of basic needs, a difference of views was quite evident (Table 2). While 80% highly value water in relation to items like electricity, sugar, shelter, and clothing (some of the items that were cited in the questionnaires as having less value compared to water), a significant 20% of the respondents valued water and/or placed an equal value of water to other items such as money, food, heath, and a place to stay.

On the question about their willingness to pay a fee, surcharge or a percentage of their water bill towards watershed protection activities, 44.6% of respondents were not forthcoming with the idea (Table 3). Several reasons were put forward to explain their position. First, some people were hesitant to pay for a "free" resource. They either do not see it or refuse to believe that water a precious and 'free'

Table 2 Value of water and how much value the respondents place on water

Question	Responses (number and percentage)		
Do you value water?	*Yes* 62 (95.4%)	*Blank/No Response* 3 (4.6%)	
How much value do you place on water?	*1 = Valuable* 5 (7.7%)	*2 = Equal Value to other Basic Needs* 8 (12.3%)	*3 = Highly Valuable compared to other Basic Needs* 52 (80.0%)

Table 3 Are you willing to pay a 'water conservation' fee or surcharge on top of your monthly water bill?

Question	Yes	No	Total
Are you willing to pay a water conservation fee or Surcharge on top of your monthly water bill?	36	29	65
Total (%)	55.4	44.6	100

resource is under threat from poorly managed human activities in the watersheds. The second most common reason for rejecting the idea of paying a water surcharge relates to income and 'lima vaivai' (state of being poor or economic hardship). They believe that further charges will add more financial burden to the economically vulnerable households. Again, this was one of the recurring themes that emerged from the focus group session. The third reason stemmed from the respondent's level and amount of water use in the home. According to 21% (6/29) of those who were unwilling to pay a water surcharge, their water use is minimal (as reflected in their monthly water bill) therefore it is not an issue for them. The surcharge should be imposed on those with excessive water consumption; it should not be a universal charge to all water consumers. Others felt that a water conservation surcharge is justified if the water supply was reliable, considering the almost constant water rationing, particularly during the dry spell (and more so during the wet season) it can be a futile exercise. And finally, some respondents felt that the current price of water is perceived as costly enough, for example, the cost of reconnection, so additional charges in the name of water conservation are unjustified and not welcome at all.

If it was mandatory to pay a percentage or surcharge towards water conservation would you pay? (Table 4). This was a hypothetical question put to the respondents to determine two things, first, the form of actions to consider when planning towards a National PES approach to upstream and downstream water resources management, and second to gauge the degree of sincerity to the question on their willingness to pay. Evidently, 66% of the respondents indicated that they would pay if it was mandatory. It is not surprising that people would pay if it was dictated by law, since the legal framework and enforcement are critical elements for a successful PES scheme anywhere in the world (Vonada, Herbert and Waage 2011, p. 24). Indeed a positive indication of how to initiate a national PES scheme.

Questions (6) and (7) (Table 5) were designed to gauge the respondents' views on what they consider as significant and priority uses of the rivers and their

Table 4 If it was mandatory by law would you pay a 'water conservation' fee or surcharge?

Question	Yes	No	No Response/ Not sure	Total
If it was mandatory by law, would you pay a water conservation fee or surcharge?	43 (66%)	15 (23%)	7 (11%)	65 (100%)

Table 5 Expressed views on how rivers and watershed resources ought to be used versus actions-behaviors on/towards rivers & watershed resources

Question	Do you support land clearance, and deforestation of watershed areas for Agriculture production?	Do you advocate/promote replanting and conserving watershed areas for water resources conservation?
(6) Do you prefer rivers/streams and watershed areas to be used for irrigation and agricultural cultivation?	29 (44.6%)	
(7) Do you prefer watershed areas to be conserved for water and aquatic ecosystem protection?		36 (55.4%)

catchment areas. The follow up question on associated 'actions' to match their 'preferred choice' was intended to reinforce and validate question 6. Interestingly, when people's livelihoods are pitched against resources conservation options, people's loyalties tend to waver slightly. It was obvious from the responses that for those people (44.6%) who depend on the watershed for their livelihood, they tend to prioritize land clearance and cultivation over replanting and conservation actions, an important issue to consider when designing a watershed based PES scheme.

Water Testing

Water quality is a significant attribute of hydrologic services. This is a measure of the chemicals, pathogens, nutrients, salts and sediments in surface and groundwater (Brauman et al. 2007).

Water samples from selected rivers of the Vaisigano watershed were abstracted to test for faecal contaminants, pathogens, biochemical oxygen demand pollutants, dissolved nitrogen, phosphorus on the 27th July, 2015. The purpose for this exercise was to provide quantitative evidence to support water quality monitoring and assessment of the Vaisigano watershed and to determine the type(s) of pollutants present in the rivers, to establish where these pollutants originate from and how they found their way into the rivers. The presence of contaminants in the rivers adds to the evidence collected in the past (Latu et al. 2012) to prove that human activities in and around the watershed are increasing with no sign of abating which strengthens the argument for PES schemes such as imposing a watershed conservation surcharge of 0.2% on total monthly water bill that would contribute to watershed protection activities such as replanting and reforestation, regular surface water testing and monitoring and safeguarding ecosystem services.

Imposing *surcharges* are done elsewhere in the world and go by various labels/names such as the sewer surcharge [stand-alone fixed amount paid by households per year in Ontario, Canada–$15.87], we have a fuel surcharge with our Electric Power Corporation. The argument is, if the Electric Power Corporation (one of the biggest user of water) is passing the cost of diesel import onto the consumers; why not introduce a similar scheme for water services, such as a flat rate per annum per household towards the maintenance of our watershed ecosystem services? A water surcharge is an equitable means of engaging people at all scales (industry, government ministries, private sector, households) in water conservation activities.

Table 6 presents the microbial results for water testing along the Vaisigano, Lelata and Maagiagi stream network on Monday 27th July. The results show the presence of Total Coliforms and Fecal coliforms, overgrown by heterotrophic bacteria in all sites sampled. Heterotrophic bacteria use organic nutrients for growth and are universally present in all types of water, food, soil, vegetation and air. This broad definition includes the primary and secondary bacterial pathogens such as coliforms.

Table 7 reports on the bacterial count from the Vaisigano River taken on the 28th July, 2015 to substantiate the results presented in Table 6. The microbiological method used to enumerate the genera of Total and Fecal Coliforms are as follows:

1 in 10 serial dilution to 1/10,000 then Vacuum Membrane Filtration 100ml using 0.45 micron Millipore.

Culture Media:

MEndo-Les agar for Total Coliforms.

MFC agar for Fecal Coliforms.

Incubation Conditions:

MEndo-Les agar incubated at 37 °C for 24 hours.

MFC agar incubated at 44.5 °C for 24 hours.

Distinct Colonies enumerated and recorded in terms of colony forming units per 100mls (cfu/100mls).

The results indicate a high level of contamination by Total Coliforms (up to 2,000 cfu/100mls) and Fecal Coliforms (up to 1000 cfu/100mls) indicating the presence of fecal matter of animal origin in the Vaisigano river system. This correlates with the high density of human and animal habitation along the river system. Old and leaking septic tanks could be a possible source of fecal contamination.

Tables 8, 9 and 10 present the sample analysis conducted for Dissolved Oxygen (DO) and Biochemical Oxygen Demand from Fale-o-le-Fe'e, Lelata, and Vaisigano river points. Table 11 provides a guide for the BOD data gathered. Naturally BOD levels are fairly high along the river points where much organic matter such as dead plants, leaves and manure correlate with densely settled areas. BOD levels from the three sample sites are above 6 mg/L which correlates with lots of organic material and many bacteria (Table 11). Likewise samples from Fale-o-le-Fe'e 2 and Vaisigano 2 are above 10 mg/L which indicate very poor, deteriorating water quality with large amounts of organic material in the water.

As stated earlier in the report, the purpose of water quality testing is to collect quantitative facts about the state of rivers/streams in the study area, as evidence to advocate for a watershed based PES scheme. Bacterial counts and BOD results as

Table 6 Microbiological results from the Vaisigano, Lelata, Maagiagi stream network

Sample Source	M Endo	MFC	Comments
Vaisigano 1. Sampling Time: 9:30 am	Heavy mixed growth of Coliform & heterotrophic bacteria. Single colonies not discernible	Heavy mixed growth with fecal coliforms present. Single colonies not discernible	Site description: 500 m from Vaisigano bridge. 500 mls of sample taken from north bank of Vaisigano river from approx. 10 cm depth. Water highly turbid with high detritus content. River slow moving
Vaisigano 2	Heavy mixed growth of coliform & heterotrophic bacteria. Single colonies not discernible.	Heavy mixed growth with fecal coliforms present. Single colonies not discernible	
Vaisigano 3	Heavy mixed growth of coliform and heterotrophic bacteria. Single colonies not discernible	Heavy mixed growth with fecal coliforms present. Single colonies not discernible	
Lelata, Sampling Time: 12:00 pm			Site description: approximately 15 m from north side of Lelata bridge. 500 mls of sample taken from approximately 10 cm depth. Water slightly turbid with moderate detritus content. River is fast moving. River center depth approx. 4–5 feet
Lelata 1	Heavy mixed growth of coliform and heterotrophic bacteria. Single colonies not discernible	Heavy mixed growth with fecal coliforms present. Single colonies not discernible	
Lelata 2	Heavy mixed growth of coliform and heterotrophic bacteria. Single colonies not discernible	Heavy mixed growth with fecal coliforms present. Single colonies not discernible.	
Lelata 3	Heavy mixed growth of coliform and heterotrophic bacteria. Single colonies not discernible	Heavy mixed growth with fecal coliforms present. Single colonies not discernible	

(continued)

Table 6 (continued)

Sample Source	M Endo	MFC	Comments
Maagiagi Uta Sampling Time: 2:00 pm			Site Description: Catchment area approx. 7–8 km from Maagiagi Cemetery. Access through Maagiagi village road in the southerly direction. Water turbidity low
Maagiagi Uta 1	Heavy mixed growth of coliform and heterotrophic bacteria. Single colonies not discernible	Heavy mixed growth with fecal coliforms present. Single colonies not discernible	
Maagiagi Uta 2	Heavy mixed growth of coliform and heterotrophic bacteria. Single colonies not discernible	Heavy mixed growth with fecal coliforms present. Single colonies not discernible	
Maagiagi Uta 3	Heavy mixed growth of coliform and heterotrophic bacteria. Single colonies not discernible	Heavy mixed growth with fecal coliforms present. Single colonies not discernible	

Table 7 Total and fecal coliform enumeration

Media	Undiluted	1/10	1/100	1/1,000	1/1,000
MEndo	1<	1<	1TC	2TC	2TC
MFC	1<	1<FC, 10HB	16FC	8FC, 22HB	1FC

Key FC-Fecal Coliforms, TC-Total Coliforms, HB- Heterotrophic Bacteria

Table 8 Dissolved oxygen sample Analysis on Monday 27 July, 2015

Sample	pH	Rep 1	Rep 2
Fale-o-le Fe'e [1]	6.4	3.6	4.6
Fale-o-le-Fe'e [2]	6.4	4.4	5.9
Lelata [1]	6.4	5.0	4.5
Lelata [2]	6.4	3.6	4.3
Vaisigano [1]	6.3	5.5	5.7
Vaisigano [2]	6.3	5.5	5.0

Table 9 Dissolved oxygen analysis on Friday 31 July, 2015

Sample	Rep 1	Rep 2	Rep 3
Fale-o-le-Fe'e [1]	4	5.6	4.5
Fale-o-le-Fe'e [2]	5	5.5	4.8
Lelata [1]	4	4.7	4.5
Lelata [2]	4	4.2	3.7
Vaisigano [1]	4.1	4.8	4.1
Vaisigano [2]	4.4	5	4.7

Table 10 Biochemical oxygen demand (BOD) results

Sample	BOD (mg/L)
Fale-o-le-Fe'e [1]	6
Fale-o-le-Fe'e [2]	11
Lelata [1]	10
Lelata [2]	6
Vaisigano [1]	16
Vaisigano [2]	11

Table 11 Interpretation of biochemical oxygen demand (BOD) levels

BOD Level (mg/L)	Status
1–2 mg/L	Clean water with little organic waste
3–5 mg/L	Moderately clean water with some organic waste
6–9 mg/L	Lots of organic material and many bacteria
>10 mg/L	Very Poor water quality. Large amounts of organic material in the water

Source *Vernier Software and Technology Website*

shown in the Tables below verify what we have suspected all along in relation to increased settlement and anthropogenic activities in and around the Vaisigano watershed.

Implications of Poor Surface Water Quality

Contaminated surface water can trigger negative/unwanted social and economic impacts. Nutrient pollution and algal blooms can destroy river aesthetics and fresh water ecosystems. Also threatened are water based recreational activities such as swimming and fishing. Given that the Lelata River empties into a section of the

Apia coastal water front, and most polluted freshwater ends up in the oceans (United Nations Environment Programme 2010, p. 7), the impact(s) on Samoa's Tourism industry and the general population residing along the coastal villages of Apia can be detrimental.

Increased pollution of freshwater ecosystems can hamper the livelihoods of people dependent on it for agricultural irrigation and drinking water for livestock. Families who have been residing along the Vaisigano and Lelata river for over 30 years reported a severe decline in fresh water shrimps (ula vai) and they can only reminiscent about the 'good old days' when the rivers flowed freely, and were very clean, cold and welcoming for a quick dip in the hot afternoon. Reference to the village cricket and rugby team named after a very important river resource (I'a o le Vai) is testament to the state of the river before it was ponded for electricity and water supply.

Poor water quality can also result in reduced property values. Clean and flowing rivers are a natural amenity that increases the economic value of land. For freehold properties drained by the rivers of the Vaisigano watershed, it is in the best interests of the owners to keep the rivers free of pollutants. This can be quite difficult if people residing along the different parts of the river do not play their role in keeping the river free from human induced pollution.

Key Informant Views

The views of key stakeholders such as small to medium sized business owners and large scale landowners were sought to determine their interests in supporting (or sponsoring) PES programs. The business sector plays a key role as buyers and sellers of ecosystem services. For some businesses PES is a new field they were only aware of from the interviews. Others (in the tourism sectors) reported activities in conservation as expenses relevant to their business operation, which by and large are the underlying principles of PES schemes, only they do not see it as such. Some business owners who were unaware of PES, but believe in the important role of ecosystem services on the profitability and longevity of their businesses supported the idea of PES in principle, which is a positive start to a PES partnership between government and private businesses. One tourism operator suggested that the business case for PES needs to be clarified, what are the ecosystems at stake and what are the best ways to pay for them?

The big landowners within the study area were quite enthusiastic about the idea of a PES scheme both as providers and beneficiaries of ecosystem services. Because many of them own property upstream and medium to large sized businesses downstream, it is in their best interest to ensure that hydrologic services produced from ecosystems in the watershed are sustainably managed to ensure constant water supply and potential flood risk to downstream properties are properly managed.

Other landowners and managers (Church and big land owners) were unaware of Payment for Ecosystem Services, but there was widespread recognition and acceptance of the critical role played by nature in providing important services for human existence. However, it is not enough to acknowledge and be aware of the importance of nature's role in providing for our ecosystem services, this awareness should be demonstrated with positive actions and commitment such as removing all forms of animal grazing and commercial agriculture from the vicinity of the watershed to eliminate possible leaching of organic and inorganic pollutants into the wider catchment network.

Extensive settlement in the uplands of Malololelei by the Catholic Church and private individuals will undoubtedly have an impact on the natural processes of water capture, storage, flowing, feeding, filtering, and evapotranspiration to name a few. In this respect, the government through the Planning, Urban Management Agency arm of the Ministry of Natural Resources and Environment need to strengthen their urban governance to mitigate further degradation of these critical landscapes.

Other Variables

There have been some other elements that came out in the conversations particularly with the key informants but cannot quite fit into the thematic analysis, which may be critical in the set up a PES scheme for Samoa. One of this relates to the types of opportunities presented for buyers and sellers of ecosystem services. An opportunities assessment therefore might be helpful in the initial planning stages of a PES scheme. Three types of opportunities assessment can be implemented. First, a beneficiary analysis to identify all those who may benefit from the scheme. This type of analysis allows for the identification of all possible sources of funding and to determine further avenues to expand the scope of PES. This presents a window of opportunity to extend PES to include climate change adaptation measures to address the vulnerability of Samoa's coastal areas to sea level rise, storm surge and flooding to name a few. Second, an extensive review of all relevant policies, programs and initiatives to pin point wider environmental and sustainability objectives that a PES scheme might be designed to address. These include programs concerning the integrated water resources management, flood risk assessment reports, river basin management plans, and state of the forests report to name a few. Funding analysis is the third type of opportunity assessment needed to identify potential sources of funding to prop up the scheme. This requires voluntary cooperation from all partners with a shared interest in the protection of ecosystem services who are prepared to pool resources to yield the best value from funding a PES scheme.

Conclusion

Hydrological watershed services are practical and ideal for PES markets because of easily identified direct and obvious users of water in a watershed, meaning there are clear delineations between beneficiaries' downstream and upstream providers (landowners who have a direct impact on land use). At the same time, a watershed based PES scheme is already embedded in the Integrated Water Resources Management (IWRM) framework; this is a good starting point to launch a universal water surcharge in conjunction with the Samoa Water Authority.

There is a lot of potential for PES schemes based on watershed conservation as demonstrated in the findings from the Vaisigano watershed study. Given the many examples of PES schemes in operation around the world, there is no need to reinvent the wheel. We only need to look at how other developed and developing countries are conducting their PES schemes, then learn from them, contextualize and adapt what they are doing to suit our needs. For example in Europe, the water utility companies pay the ecosystem providers in the upper watershed for conservation activities and pass these costs on to their clients. Given we have one water service provider (Samoa Water Authority) passing the cost of water conservation onto all water consumers can be in the form of a standard water surcharge. However, large businesses that demand excessive water usage such as the electric power corporation, beverage industries and water retailing businesses can be mandated to pay product taxes and fees towards the maintenance of ecosystem services. These taxes and fees can be determined by the service provider based on cost benefit analysis work, and rapid watershed scoping and appraisal activities to pin-point the extent of degradation of the watershed ecosystem and hence the wider ecosystem services derived from it.

The government also plays a critical role in ensuring the rules are in place to facilitate procuring some ecosystem services as public goods. In fact, government should be the main driver of PES scheme considering the scarcity of funding that can be sourced from the private sector even businesses that rely on water such as water retailers and beverage makers.

There is also the option of a nation- wide surcharge imposed on all water consumers connected to the Samoa Water Authority water service which is to be determined by the rate at which the country's water resources are being deteriorated. And one of the supporting evidence to justify a national water surcharge for water resources conservation is provided by scientific evidence of poor surface water quality as presented in this study.

Equally important, the study can be repeated in the coastal communities to detect the impact of polluted waters from the rivers of the watershed that make their way into the coastal waters of Apia to determine how and what their impacts are on the coastal marine ecosystem of the greater Apia coastal community. Evidence from a similar study will contribute to the existing pool of knowledge assembled over recent years to assist small island states such as Samoa address climate change adaptation strategies.

References

Aronson, J. (1995). A Pragmatic View of Thematic Analysis. *The Qualitative Report*, *2*(1), 1–3. Accessed July 4, 2015, http://nsuworks.nova.edu/cgi/viewcontent.cgi?article=2069&context=tqr.

Brauman, K. A, Daily, G. C; Duarte, T. K., & Mooney, H. A. (2007). The nature and value of ecosystem services: An overview highlighting hydrologic services. *Annual Review of Environment and Resources*, *32*, 67–98. Accessed September 28, 2015, http://environ.annualreviews.org.

Braun, V., & Clarke, V. (2006). Using thematic analysis in psychology. *Qualitative Research in Psychology*, *3*(2), 77–101. Accessed June 30, 2015, http://eprints.uwe.ac.uk/11735/2/thematic_analysis_revised.

International Institute for Environment and Development (2015). *Markets and Payments for Environmental Services*. Accessed June 26, 2015, http://www.iied.org/markets-payments-for-environmental-services.

Latu, F., Amosa, P., Imo, T., & Taufao, V. (2012). The microbiological survey of potential water borne pathogens in fresh water springs of the selected community located in the Upolu Island, Samoa. *International Journal of Environmental Science and Development, 3*(2), 89–93.

Pirard, R. (2012). Market-based instruments for biodiversity and ecosystem services: a lexicon. *Environ. Sci, 19*(20), 59–68.

Schell, C. (1992). *The Value of the Case Study as a Research Strategy*. Accessed June 30, 2015, http://www.finance-mba.com/Case%20Method.pdf.

Taua'a, S. (2014). Water attitudes, Consumption and Waste at the Household Level. *The Samoa Water and Sanitation Journal, 1*(1), 1–31.

United Nations Environment Programme. (2010). *Clearing the Waters: A Focus on water quality solutions*. Accessed August 16, 2015, http://www.unep.org/PDF/Clearing_the_Waters.pdf.

United Nations Department Economic Social Affairs. (2014). *Ridge to Reef. SIDS Action Platform.* UN Conference on Small Island Developing States, Apia Samoa, 1–4 September, Accessed March 11, 2017, http://www.sids2014.org/index.php?page=view&type=1006&nr=2373&menu=1507.

Vernier Software & Technology. (n.d). *Investigating Environmental Science Through Inquiry.* Accessed March 26, 2017, http://www2.vernier.com/sample_labs/ESI-16-bod.pdf.

Vonada, R., Herbert, T., & Waage, S. (2011). *Introduction to Payments for Ecosystem Services: A Reference Book for Uganda.* The Government of Uganda's National Environment Management Authority.

Yin, R. K. (1984). *Case study research: Design and Methods*. Newbury Park, CA: SAGE.

Predicting True Climate Change Risks and Opportunities in the Cook Islands: How Vulnerable Are Pacific Maritime Supply Chain Stakeholders?

Jack Dyer

Abstract With maritime resources dominating over 60% of Cook Islands, 83% Kiribati and 90% of Palau's economic activity, climate change uncertainty through increased impacts of storms, cyclones, floods, landslides, bushfires, heatwaves and tsunamis for future Pacific ports, shipping and maritime supply chains (MSC's) will significantly constrain coastal communities' capacity to survive and prosper. This paper identifies flaws in existing climate change risk management techniques; by proposing mechanisms to predict true risk events and associated impact costs for stakeholders to more reliably ascertain supply chain and coastal vulnerability beyond existing qualitative approaches. It establishes historic probabilities of climate change risk events across the Pacific as a basis for more accurately determining the probability of future extreme events. Providing a risk-perception survey for Pacific MSC' stakeholders, compares the extent of climate change risk awareness with reality. It presents a risk-vulnerability matrix of existing climate change across an entire Pacific MSC' using a Cook Islands case study. It proposes conditional probabilities of MSC' asset failure, projecting a risk analysis for vulnerable community stakeholders to prioritise risk. It identifies how these disruptive risks can be transformed into risk opportunities, through improving survey techniques and learning from existing Pacific, climateproofing, adaptation strategies.

Introduction, Climate Change and Pacific Maritime Supply Chain Overview

How resilient is our world to projected climate change? How significant are the risks and associated impact costs to coastal communities? Given adaptation costs and so many other significant challenges to development, should climate change really be prioritised? As coastal communities become increasingly vulnerable to existing and projected climate change impact consequences from an increased

J. Dyer (✉)
University of Tasmania, Maritime Way, Launceston, TAS 7250, Australia
e-mail: jack.dyer@utas.edu.au

W. Leal Filho (ed.), *Climate Change Impacts and Adaptation Strategies for Coastal Communities*, Climate Change Management,
https://doi.org/10.1007/978-3-319-70703-7_20

frequency, duration and intensity of temperatures, sea level rise, storms, cyclones, floods, landslides, bushfires, heatwaves and tsunamis (IPCC 2015), increasing research has concentrated on the Pacific as among the most significantly exposed and immediate locations, receiving significant global funding and attention. However existing research has concentrated on communities, agriculture, disaster risk management, health, education, overall economies, coastal ecosystems and key infrastructure but ignored the underlying ports, shipping and maritime supply chain/ ecosystem, which supports not just small, Pacific island nation states including the Cook Islands and Kiribati but Singapore, Hong Kong, Australia and the United Kingdom. Globally over 90% of the world's trade is seabourne, based on ports and shipping. However, coastal infrastructure and economies have yet to be sufficiently prioritised against emergent climate change risk and natural disasters.

Global sudden climate-change related events (defined as natural disasters) directly threaten lives, property, trade, economies, communities and ecosystems. These influence maritime supply chains directly from production to consumption with reduced resources, revenue, delayed cargo throughput, economic activity and increased risk exposure/economic impact costs. This threat is more evident for those nations/areas more vulnerable, especially the Pacific Islands region. Hurricane Katrina cost over $150 billion in direct economic impact costs to the USA. 2012 floods in the Solomon Islands (SPREP 2015), from demolished natural ecosystem, climate protection barriers including mangroves and sand dune erosion, cost over $55 million in taro crop production, residential property and physical infrastructure damage alone. A more recent, example occurs from Cyclone Pam's storm surge consequences on the small Pacific island, developing state of Vanuatu on 13th March 2015. Its 300 km/h winds caused damage to over 48,000 homes, 100,000 homeless and 24 deaths (Flannery and Steffen 2015), with as yet unascertained high economic and other opportunity costs. This is just among the latest in a history of increasingly severe natural disasters in one of the world's most highly geographically, economically and socially exposed regions. Therefore this paper seeks to ascertain just how vulnerable truly are Pacific maritime supply chain stakeholders to climate change, as a prototype method, not just for an individual risk event or stakeholders as for previous research but across an entire supply chain system.

Given existing supply chain risk management fails to identify, define, manage and adapt to climate change risks (section "How Existing Risk Management Methods Fail for Climate Change Risks on Maritime Supply Chains and Proposed Method") especially for low probability high impact events; this research seeks to address projected climate change uncertainty by ascertaining an integrated risk-vulnerability matrix method. This aims to more accurately determine which risks to prioritise, the probability of historic and future risks occurring, the conditional probability of a maritime supply chain (MSC') asset/system failure from a risk event. Sections "Climate Change Risk Identification and Perception Survey Results for a Pacific MSC'" and "Predicting Climate Change Risks for a Pacific Maritime Supply Chains: The Cook Islands Risk-Vulnerability Matrix" identify climate change risks for a Pacific maritime supply chain and stakeholders through a Ph.D. thesis, Cook Islands case study. This provides a Pacific example that has prioritised climate change risks but significantly ignored and underestimated risk from a maritime supply chain, stakeholder perspective. To address Matrix Stage II Climate Change Risk Identification, section "Climate Change Risk Identification and Perception Survey Results for a Pacific MSC'" identifies stakeholder profiles and survey results for Climate Change Risk Perceptions of Pacific Maritime Supply Chain Stakeholders (Stage III), to calculate the historic probability of a climate-change related risk occurring before projecting future risk events across a Pacific MSC'. This considers the extent to which climate change risk perceptions reflect empirical evidence, whether risks are over or underestimated in relation to potential impact costs and which risks to prioritise. Section "Predicting Climate Change Risks for a Pacific Maritime Supply Chains: The Cook Islands Risk-Vulnerability Matrix" applies this developed risk-vulnerability matrix results. It identifies this case study's background, demographics, maritime supply chain, environment/ecosystem and climate. It provides a Pacific MSC', Vulnerability-Risk Analysis (Matrix Stage IV), calculating existing/historic conditional probabilities of maritime supply chain asset/ecosystem/system failures based on climate change specific risk events and factors affecting risk probability of occurrence. Section "Cook Islands Climateproofing Adaptation Strategies" outlines Cook Islands climateproofing adaptation strategies that other Pacific and global MSC' stakeholders can implement to minimise vulnerability and enhance resilience. As summarised in section "Conclusions, Policy and Theoretical Significance for Stakeholders and Directions for Future Research". It provides a practical method application, further verifies the validity of this thesis devised, conceptual and analytical framework over existing method limitations. It enables Pacific maritime and global supply chain stakeholders including governments, researchers, businesses and individuals with limited resources and significant constraints in climate change risk prioritisation (Matrix Stage VI).

How Existing Risk Management Methods Fail for Climate Change Risks on Maritime Supply Chains and Proposed Method

Whilst more extensively developed in an IAME (2017) paper submission and undertaken University of Tasmania thesis, the lack of a consistent, existing risk management method to adjust to climate change risks, has significantly multiplied risk vulnerability and associated impact cost consequences for coastal communities, especially for dependent MSC' stakeholders. When existing risk management theory considers global supply chain disruption risks, these have primarily concentrated on landside not coastal stakeholders, resources and economic activities (Kern et al. 2012; Kazemia and Szmerekovsky 2015). However, this paper formally defines MSC's as a sequence of processes through which a commodity/ commodities is produced, transported, processed, distributed, sold and eventually

Fig. 1 A Pacific maritime supply chain

utilised or consumed, (occasionally recycled). These connect initial producer supply with final consumer demand, through beneficiation/manufacturing, ports, intermodal transport, customs, finance and the associated economic hinterland in Fig. 1. MSC's and dependent coastal communities, (including the Cook Islands) differ in being highly dependent on seaports and shipping. Maritime economies and livelihoods are completely dependent from production via fisheries to transport by sea and exports; whilst dominated by maritime ecosystem resources such as seafood and pearls in trade. These have virtually no land based economic hinterland or stakeholders, with significant maritime asset interdependency highly vulnerable to ocean and climate risks. Typhoons, tsunamis, storms, seal level rise, increased temperature, floods, droughts, changes ocean currents/thermohaline circulation and species migration are particularly maritime climate risk influenced. Shipping, seaports and coastal assets/ecosystems as key maritime stages, further emphasise how MSC's differ from conventional supply chains, though both remain vulnerable.

Previous climate change impact studies concentrated on aggregate economic sectors rather than specific stages, stakeholders and products/commodities. Only the most significant, relevant, qualitative impact studies and methods have recently concentrated on ports, shipping; intermodal transport and general supply chains as separate stages. Becker et al. (2011) proved among the first to utilise a survey to identify climate change disruption risks, impacts and adaptation solutions for seaports, consulting 208 International Association of Ports and Harbours stakeholders with 93 responses. 53% considered climate change would affect them but 88% agreed more research was necessary to understand its impacts. This survey method provided advantages of considering stakeholder perceptions towards climate change; the extent of their awareness and the degree to which they are prioritising supply chain adaptation solutions to incorporate into this paper's research method. UNCTAD (2011) similarly used a survey instrument among its 200 members to identify port vulnerability to climate change risks and associated stakeholder adaptation responses. However these primarily concentrates on port administrators. A number of climate change survey studies including UNCTAD (2011), and Kreie (2013 for global supply chains), also selectively ignore key supply chain stakeholders as economically peripheral. Or they involve too much time, resources and effort to incorporate a risk event's economic impact in study methods for stakeholders. However, this paper agrees with Becker and Caldwell (2015 for a, seaport case study in Gulfport USA), that climate change impact and adaptation strategies can no longer be constrained to just consulting direct port stakeholders involved. Ports alone and industry assessments of adapting supply chains to climate change (CSR 2015; BSR 2015) are insufficient and uninformed to resolve potential event disruptions on maritime supply chains. In contrast, this paper improves upon existing qualitative method studies by including as many supply chain stakeholders as possible for a single commodity. Stages include customs, the financial and insurance sector, value-added/beneficiation, consumers, subsistence fishermen, ecosystems and small entrepreneurs all ignored by past sources.

These and other existing risk management studies are primarily qualitative with few case studies utilising probability and empirical data. Qualitative over quantitative method advantages overcome common research issues of limited data specifically relevant to a study and in directly accessing the experience/knowledge of principal qualified stakeholders to obtain specific insight into key research objectives. Yet qualitative studies relying on stakeholder perceptions, combined with considered climate change uncertainty; significantly contributes to risk method flaws. These include including subjective understanding of risks, complexities in forming probabilities ignoring factors affecting risk; and providing unclear indications of how risk probabilities are calculated. Existing research emphasises climate change uncertainty for various supply chain stakeholders and coastal communities in risk identification, estimation, analysis and adaptation (McEvoy and Mullet 2013; Scott et al. 2013; Smith 2015).

This paper considers conventional risk management theory provides insufficient practical guidance to locating and assessing climate change risk impacts, true supply chain resilience and vulnerability to coastal communities and Pacific MSC' stakeholders. This especially applies for MSC' stakeholders and coastal communities with limited time, financial and other resources. The majority of conventional predictive models include existing event tree, cause and frequency analysis, Bayesian networks, HAZOP, What If? Markovian chain analysis, Monte Carlo simulations, VAR, real options approach, transactional costs and resource dependence theory and risk event/fault trees (Ellis et al. 2011; Kern et al. 2012; Ghadge and Kalawsky 2012). Such models assume risks remain static, time remains constant or based on

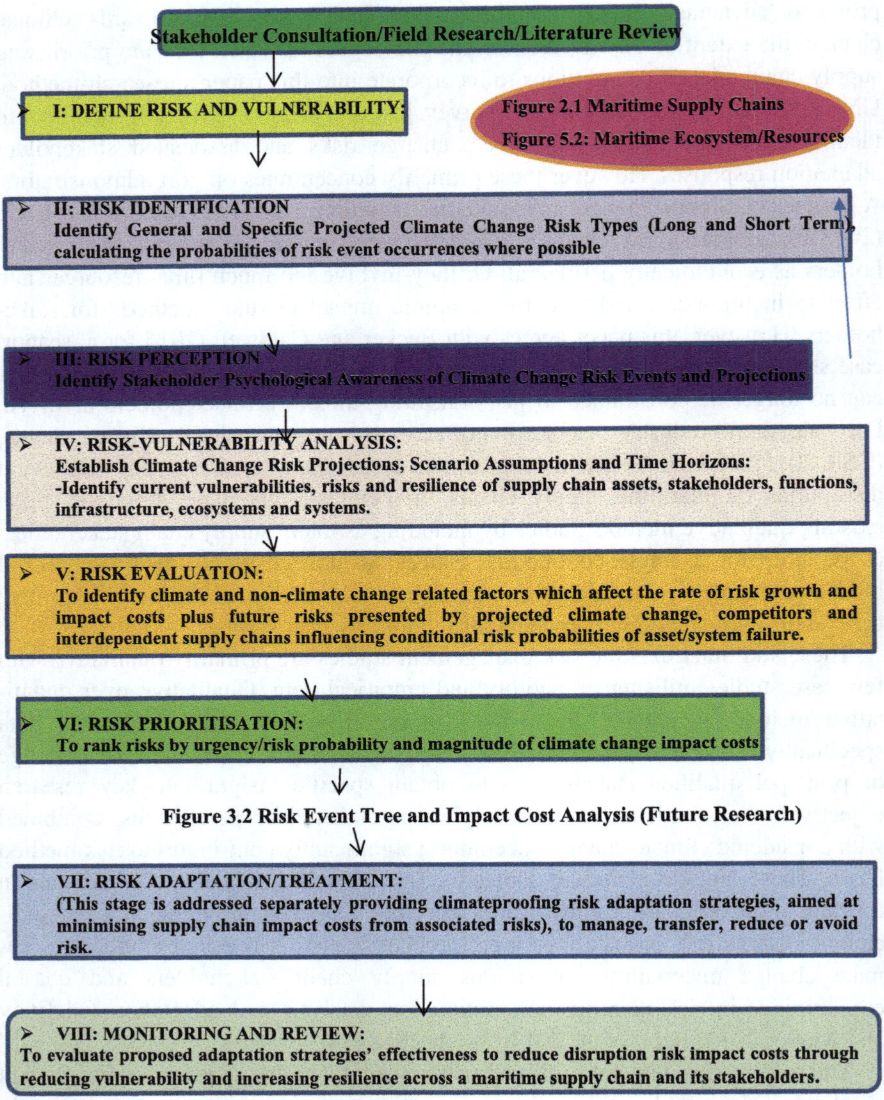

Fig. 2 Proposed vulnerability-risk analysis matrix for a maritime supply chain

historic time series data. These ignore the projected rate or increase in the probability of an event occurrence, its duration, frequency and intensity and how risks can be multiplied thorough vulnerability, resilience, adaptive capacity, constraints to adaptation and increased interdependence. Given potentially significant impact, maladaptation and opportunity costs, sections "Climate Change Risk Identification

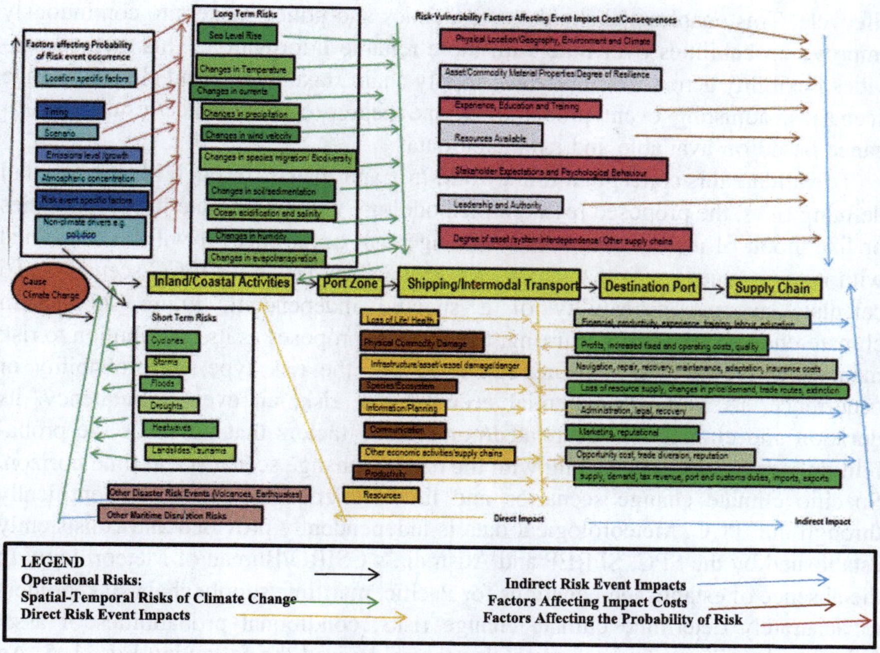

Fig. 3 Climate change risk and maritime supply chain impact event tree analytical Framework

and Perception Survey Results for a Pacific MSC'" and "Predicting Climate Change Risks for a Pacific Maritime Supply Chains: The Cook Islands Risk-Vulnerability Matrix" provide an empirical case study of the following integrated, risk-vulnerability matrix in Fig. 2 for the Cook Islands.

This paper summarises Stages I–VI. It considers climate change risk impacts for Pacific MSC's and coastal communities can only be accurately measured via a method, which links this matrix to a combined climate change risk and MSC' impact event tree. This method incorporates factors affecting the probability of a risk occurrence, factors which influence supply chain vulnerability/resilience to these risks, direct and indirect impact costs across the supply chain in Fig. 3. Unlike previous risk management probabilities assuming the status quo remains over an event or asset's lifetime, this framework considers climate change risk events as fundamentally dynamic rather than static. A static framework merely relies on historic time series data, ignoring given risk, uncertainty and climate change, Equations include increases in yearly, accumulative Pacific, climate change risk for Eq. 2. Accumulating risk considers taking existing risk projections. When estimated for a given percentage range increase over 25, 50, 100 years, this is converted to yearly percentage increases in alignment with stakeholders who consider risk preparation and management on an annual basis; instead of an asset's lifecycle. This enables stakeholders, academics and policy makers to continuously improve probabilities over time with more reliable information. This method provides flexibility across time horizons, supply chain stakeholders and climate change scenarios, adjusting event probabilities and degree of confidence/results significance based on available and simulated data.

To validate this conceptual framework, tree and matrix, through identifying and defining risks, the proposed research methodology will incorporate the probabilities or likelihood of a projected, climate change risk occurrence. It will be combined with its consequences (impact costs) across Pacific MSC's for IPCC scenarios. To calculate average probability of a specific independent, future short term climate-change risk event occurring, this method proposes as its contribution to risk management theory; a measure that integrates the risk type, its probability of occurrence, its past data/potential accumulative risk, an event's frequency, its duration and climate change related/non-climate factors that influence the probability of a risk occurrence along with the climate change scenario and time horizon. Specific climate change scenarios and time horizons are verified scientifically through the IPCC. Meteorological data is independently provided and consistently established by the SPC, SPREP and Australia's CSIRO/Bureau of Meteorology. In the absence of established equations for Pacific, maritime supply chain stakeholders to accurately determine climate change risks, conditional probabilities of asset failure and resilience/impact costs, this thesis devised the following Eqs. 1–5. An interaction or joint probability is necessary for calculating certain related events.—I. e. historical correlation between storms and flooding; earthquakes, tsunamis and landslides; precipitation and storms, sea surface temperature/wind velocity and cyclones; earthquakes and volcanoes simultaneously. Risk concentration increases asset vulnerability, increasing conditional individual/joint probability of an asset's failure.

To calculate the average probability of a specific independent, historic short term risk event occurring for Eqs. 1–5, the Poisson distribution is utilised. This calculates the probability of x occurrences per unit time for this paper's conceptual methodological contribution in adjusting existing probability risk management theory. For the Poisson distribution; probability, cumulative and joint probability can be calculated. The distribution allows for fewer observation values than the normal distribution with a normality and continuous distribution assumption of all possible values. To resolve problems of selective recall and limited information, the method proposes emphasising recent past events for which stakeholder data potentially exists. It estimates the expected, average number of risk events per year given historical actual events and projected increases in frequency/probability of occurrence. These are adjusted for increased accumulating risks per year to calculate future probabilities and impact costs. It equates to the expected value of damage in a future year plus the accumulated sum of impact costs. As a pioneering research method for climate change risk management for MSC's, the equations and method provide research advantages of being adaptable to divergent climate change risks, asset types, scenarios, stakeholders and stages. It can incorporate resilience, vulnerability, accumulated risk and factors affecting the probability of risk occurrence and adaptation costs.

Probability of a Historic Pacific, Climate Change Risk Event Occurring

$$= P(x) = \frac{e^{-\lambda}\lambda^{x}}{x!}\ for\ \frac{x\ = 0, 1, 2}{\lambda > 0} \tag{1}$$

where P(X) = The probability of X risk events for the given time period
λ = The expected/mean rate of climate change risk event per unit of time
e = Mathematical constant approximately 2.71828
x = Number of Climate Change Risk Events i.e. Storm (S), Flood (F), Bushfire (B), Cyclone (C), Drought (D), Gale (G), Heatwave (H), Landslide (L), Earthquake (E), Tsunami (T), Volcano (V).

Probability of a Future Climate Change Risk $= P(x)$

$$= \frac{e^{-\lambda}\lambda^{x}}{x!} + \sum P(x_1\Delta t - \Delta t_{-1}) + P(x_1)P(x_2) \quad for \frac{x = 0, 1, 2}{\lambda > 0} \tag{2}$$

where $P(x_1)P(x_2)$ = Joint Probability of an Event i.e. (C + F) or Cyclone combined with flood,
x_1 = Climate Change Event Risk 1 and x_2 = Climate Change Event Risk 2
$\sum P(x_1\Delta t - \Delta t_{-1})^{*}$ = Climate Change Accumulating Risk Probability based on scenario projections, engineering design standards, business forecasts and climateproofing adaptation measures.

Probability of a MSC′ Disruption $P(x_1 \leq x \leq x_n)$

$$= P(x_1) + P(x_1 + 1) + P(x_2 + 2) + P(x_n)$$
$$= e^{-\lambda}(\lambda^{x1} + \lambda^{x2} + \lambda^{x3} + \lambda^{xn}) \tag{3}$$

where n = Number of observations/stakeholders which can be individually calculated for each individual maritime supply chain stakeholder, stage, system, location and commodity.

Conditional Probability of a MSC' Asset Failure given a climate change risk event occurring i.e. P(Berth Damage| Flood)=

Assuming a joint density function and cumulative distribution function associated with this density $|y_x|yx = f_{x,y}(X,Y)$ for all x values such that $f_{(x)}x > 0$

$$\Phi(x,y) = \int_{-\infty}^{x} \int_{-\infty}^{y} \phi(x,y)dxdy = P(-\infty < X \le x, -\infty < Y \le y)$$

$$\boldsymbol{Pn = \frac{N!}{n!(N-n)!ft^n 1 - ft^{N-n}}}$$

$$\boldsymbol{\frac{dP}{dT} = (1 - p(t)f \text{ with initial condition } P(0) = 0) \text{ and } (1 - ft)^n = e^{nft}}$$

$$= e^{-nft}\frac{nft}{n!} + P(x_1 nft)P(x_2 nft)\frac{e^{-\lambda}\lambda^x}{x!} + \sum P(x_1\Delta t - \Delta t_{-1}) + P(x_1)P(x_2) \quad (4)$$
$$for \frac{x = 0, 1, 2}{\lambda > 0}$$

where the failure number during the interval is provided by the binomial distribution; f = failure and t = time period for asset failure. If P(ft) is the probability of failure at time t, $P(t) = 1 - e^{-ft}$ as cumulative probability distribution

$$\boldsymbol{p(t) = \frac{dP}{dT} = fe^{-ft}(\Delta t - \Delta t_{-1})} \quad (5)$$

Unlike previous studies which provide generalised, qualitative descriptions of overall macroeconomic climate change risk consequences and projected likelihood (Oswald 2011; Kiele et al. 2014), this paper will overcome existing literature gaps. It calculates and establish specific risk events across an entire Pacific MSC' case study, across multiple stakeholders and stages. This aims to provide specific guidance to stakeholders to further understand these consequences. Once a specific risk event probability is calculated from the above method; the conditional probability of an asset failure e.g. P(Flooding| Berth damage) can be calculated. This probability can be ascertained from evaluating historic and current time series data of risk event impacts from stakeholder consultation, field research and secondary source evaluation of asset properties and conditions as indicated in the survey. This method's research significance is that unlike previous literature (Furlow and Potter 2015; BSR 2015), which considers only either short or long term risks, this integrated methodology and figure combines both risk types and multiple interconnected climate change and supply chain dimensions. This assists stakeholders when analysing risks. It includes the flexibility to consider additional dimensions as information becomes increasingly available to reduce uncertainty over projected climate change risks and impact costs.

Climate Change Risk Identification and Perception Survey Results for a Pacific MSC'

To improve upon existing climate change uncertainty for Pacific MSC' stakeholders and coastal communities to understand impact costs and adapt, this paper proposes stakeholders need to identify specific local and global, climate change risks potentially affecting maritime supply chains. As existing Pacific data appears scattered across myriad sources, not previously accessible or locatable in a centralised source or system, acknowledged by the official stakeholders contacted, this paper derived from a thesis It provided long term, climate change projection risks and short term risk events based upon its centralised, historical, Pacific time series data from 1900 to 2015 and Table 1 Cook Islands example. Identifying risks enables supply chain stakeholders to anticipate the probability that various risk event types will affect them for a particular year/other time horizon or climate change scenario. It pinpoints which risks are worth marginalising/avoiding given finite resources. Utilising Eqs. 1–3, provided the expected probability of any number of climate change risk events for various risk types for the Cook Islands in Table 2 based on accumulated and expected risk. Equations 4 and 5 provide an estimation of future risk events to produce Table 3.

Given stakeholder adaptation constraints, established risk management theory often ignores which risks should be prioritised, how and why. It often fails to provide consistent objective criteria by which time series data can ascertain if stakeholders are becoming more or less vulnerable. Based on thesis section criteria in Fig. 4; this paper advocates prioritising risk utilising time series data, allocating resources towards the most frequent or probable risk event types by considering Fig. 5. This illustrates cyclones represent the highest proportion of existing recorded risks (33%), followed by storms (27%) unlike bushfires, earthquakes and volcanoes. For low probability, high impact risk events, considering projected increases in historic data initially, combined with climate change scenario, projected increased frequency, duration and intensity, counteracts stakeholder uncertainty. New risk types are considered highly unlikely by this paper to emerge, so stakeholders consciously ensure they know how risks threaten individual operations, underlying ecosystem and entire maritime/general supply chain system. Whilst the Cook Islands has more time to adapt related human, natural, technology, infrastructure, equipment, information, communication, system and other assets to long term, IPCC (2015) scenario risks; mitigation of emissions; need to be prioritised immediately to ensure supply chain and physical survival. This can avoid even greater risks.

The advantages of ascertaining historic and future risk through probabilities, time series data and pie charts emphasises the uniqueness of location specific risks affecting localised impact costs. Previous Pacific climate change research primarily treats risks as equally likely to increase for all areas as does conventional risk management for supply chains. E.g., where Cook Islands stakeholders could consider disruptions through heatwaves and cyclones as most significant and urgent,

Table 1 Cook Islands chronology of climate change-related risks 1900–2015

Year	No of risk events	Observed values 0	Year	No of risk events	Observed values 0
1900		0	1961		0
1901		0	1962		0
1902		0	1963	2S, 2C, 2(S + C)	6
1903		0	1964	H	0
1904	C	1	1965		0
1905	S, C	2	1966		0
1906		0	1967	2S, C	3
1907		0	1968		0
1908		0	1969	2C, H	3
1909		1	1970		0
1910		0	1971	H, G	2
1911		0	1972	C	1
1912	C	1	1973	H	1
1913		0	1974	H	1
1914	S, C, T	2	1975	H	1
1915		0	1976		0
1916		0	1977		0
1917		0	1978		0
1918		0	1979		0
1919	E	1	1980	2S, 1G	3
1920		0	1981	S	1
1921		0	1982	G	1
1922		0	1983		0
1923		0	1984		0
1924		0	1985		0
1925		0	1986		0
1926	S, C	3	1987	S, C, S + C	3
1927		0	1988	H	1
1928		0	1989	C	1
1929	C	1	1990	S, C, S + C	3
1930		0	1991	S	1
1931	S, C, S + C	3	1992	S	1
1932		0	1993		0
1933		0	1994	S	1
1934		0	1995		0
1935	S, C	2	1996	H, C, S, (S + C)	4
1936		0	1997	2H, 2C, S, G, S + C	7
1937		0	1998		0

(continued)

Table 1 (continued)

Year	No of risk events	Observed values 0	Year	No of risk events	Observed values 0
1938		0	1999	H	1
1939	D	1	2000	H	1
1940		0	2001	S, C, F	3
1941	S, C, S + C	3	2002	C	1
1942	S, C, S + C	3	2003		0
1943	2S, 2C, 2 (S + C)	6	2004		0
1944	S, C, (S + C)	3	2005	3C	3
1945		0	2006		0
1946	S, C, (S + C)	3	2007	H	1
1947		0	2008	H	1
1948	C	1	2009		0
1949		0	2010	T, E, C, S	4
1950	2C	2	2011		0
1951		0	2012		0
1952		0	2013		0
1953	H, S, T	2	2014		0
1954		0	2015	D	1
1955	H	1	2016		
1956		0	2017		
1957		0	2018		
1958		0	2019*		
1959	S, C, S + C	3	2020*		
1960	H, T	2			

Nauru's equatorial position and stable climate, favour only droughts and floods. This research utilises existing risk related disasters as a basis for ascertaining future expected probability of risk event occurring depending on the climate change risk type and MSC' stage/geographical location. This contrasts with previous climate change impact studies that merely identify a projected increase in severity, intensity and frequency for all types.

This paper and thesis's analytical contribution also establishes a survey. This considers the extent to which Cook Islands stakeholders perceived climate change risk accurately, their ability to learn and experience events. This validates its theory proposed, risk perception management stage, unestablished in conventional risk management theory. This aimed to overcome subjective stakeholder perceptions estimating risk as Low, Medium or High or a Likert scale from 0–5, with no standard risk estimation criteria to evaluate their psychological capacity to accurate determine, estimate, value and understand risks sufficiently. Given subjectivity of stakeholder risk perceptions, this paper proposes this stage to consider the extent to

Table 2 Historic climate change risk probabilities for Cook Islands maritime supply chains

Expected probability of a Cook Islands climate change related risk event 1900–2015					
Total average no of events = 107	λ = 0.9304	Landslides = 0	λ = 0	Drought = 2	λ = 0.0435
P(X = 0)	0.3944	P(X = 0)	0	P(X = 0)	0.9828
P(X = 1)	0.3669	P(X = 1)	0	P(X = 1)	0.0171
P(X = 2)	0.1707	P(X = 2)	0	P(X = 2)	0.0001488
P(X = 3)	0.0529	P(X = 3)	0	P(X = 3)	1.854E−07
P(X = 4)	0.0123	P(X = 4)	0	P(X = 4)	3.75E−08
P(X = 5)	0.002291	P(X = 5)	0	P(X = 5)	1.3061E−11
Bushfire = 0	λ = 0	Tsunami = 2	λ = 0.0435	Earthquake = 0	λ = 0
P(X = 0)	0	P(X = 0)	0.9828	P(X = 0)	0
P(X = 1)	0	P(X = 1)	0.0171	P(X = 1)	0
P(X = 2)	0	P(X = 2)	0.0001488	P(X = 2)	0
P(X = 3)	0	P(X = 3)	1.854E−07	P(X = 3)	0
P(X = 4)	0	P(X = 4)	3.75E−08	P(X = 4)	0
P(X = 5)	0	P(X = 5)	1.3061E−11	P(X = 5)	0

Volcano = 0	λ = 0	Flood = 1	λ = 0.0867	Storms = 29	λ = 0.2522	Cyclone = 36	λ = 0.313
P(X = 0)	0	P(X = 0)	0.9913	P(X = 0)	0.7771	P(X = 0)	0.7312
P(X = 1)	0	P(X = 1)	0.0088	P(X = 1)	0.196	P(X = 1)	0.2289
P(X = 2)	0	P(X = 2)	0.0000375	P(X = 2)	0.0247	P(X = 2)	0.0358
P(X = 3)	0	P(X = 3)	0.000001088	P(X = 3)	0.002078	P(X = 3)	0.003737
P(X = 4)	0	P(X = 4)	2.3664E−10	P(X = 4)	0.000131	P(X = 4)	0.0002924
P(X = 5)	0	P(X = 5)	4.1775E−13	P(X = 5)	0.00006607	P(X = 5)	0.00001831

(continued)

Table 2 (continued)

Volcano = 0	λ = 0	Flood = 1	λ = 0.0867	Storms = 29	λ = 0.2522	Cyclone = 36	λ = 0.313
Gale = 5	λ = 0.0435	Heatwave = 17	λ = 0.1478	S + C = 14	λ = 0.1217		
P(X = 0)	0.9524	P(X = 0)	0.8626	P(X = 0)	0.8854		
P(X = 1)	0.0416	P(X = 1)	0.1274	P(X = 1)	0.1078		
P(X = 2)	0.0009059	P(X = 2)	0.009422	P(X = 2)	0.006557		
P(X = 3)	0.00001313	P(X = 3)	0.0004642	P(X = 3)	0.000266		
P(X = 4)	1.428E−07	P(X = 4)	0.00001715	P(X = 4)	0.000000805		
P(X = 5)	1.24E−09	P(X = 5)	0.00000507	P(X = 5)	0.000000197		

Table 3 Predicting current/future climate change risk event probabilities for Cook Islands MSC's

	Current	2017	2018	2019	2020		Current	2017	2018	2019	2020
Flood = 1	λ = 0.0867					**Storms = 29**	λ = 0.2522				
P(X = 0)	0.9913	0.9833	0.9753	0.9673	0.9593	P(X = 0)	0.7771	0.7691	0.7611	0.7531	0.7451
P(X = 1)	0.0088	0.0168	0.02488	0.03288	0.04008	P(X = 1)	0.196	0.204	0.212	0.22	0.228
P(X = 2)	0.0000375	0.0000375	0.0000375	0.0000375	0.0000375	P(X = 2)	0.0247	0.0247	0.0247	0.0247	0.0247
P(X = 3)	0.000001088	0.000001088	0.000001088	0.000001088	0.000001088	P(X = 3)	0.002078	0.002078	0.002078	0.002078	0.002078
P(X = 4)	2.366E−10	2.366E−10	2.3664E−10	2.3664E−10	2.366E−10	P(X = 4)	0.000131	0.000131	0.000131	0.000131	0.000131
P(X = 5)	4.178E−13	4.178E−13	4.1775E−13	4.1775E−13	4.178E−13	P(X = 5)	0.00006607	0.00006607	0.00006607	0.00006607	0.00006607
Future	**Current**	**2017**	**2018**	**2019**	**2020**		**Current**	**2017**	**2018**	**2019**	**2020**
Bushfire	λ = 0					**Tsunami = 2**	λ = 0.0435				
P(X = 0)	0	0	0	0	0	P(X = 0)	0.9828	0.9748	0.9668	0.9588	0.9508
P(X = 1)	0	0	0	0	0	P(X = 1)	0.0171	0.0251	0.0331	0.0411	0.0491
P(X = 2)	0	0	0	0	0	P(X = 2)	0.0001488	0.0001428	0.0001428	0.0001428	0.0001428
P(X = 3)	0	0	0	0	0	P(X = 3)	1.854E−07	1.854E−07	1.854E−07	1.854E−07	1.854E−07
P(X = 4)	0	0	0	0	0	P(X = 4)	3.75E−08	3.75E−08	3.75E−08	3.75E−08	3.75E−08
P(X = 5)	0	0	0	0	0	P(X = 5)	1.3061E −11	1.306E−11	1.3061E −11	1.3061E −11	1.3061E −11
Future	**Current**	**2017**	**2018**	**2019**	**2020**		**Current**	**2017**	**2018**	**2019**	**2020**
Heatwave = 17	λ = 0.1478					**T + E**	λ = 0				
P(X = 0)	0.8626	0.8546	0.8466	0.8386	0.8306	P(X = 0)	0	0	0	0	0
P(X = 1)	0.1274	0.1354	0.1434	0.1514	0.1594	P(X = 1)	0	0	0	0	0
P(X = 2)	0.009422	0.009422	0.009422	0.009422	0.009422	P(X = 2)	0	0	0	0	0
P(X = 3)	0.0004642	0.0004642	0.0004642	0.0004642	0.0004642	P(X = 3)	0	0	0	0	0
P(X = 4)	1.715E−05	1.715E−05	0.00001715	0.00001715	1.715E−05	P(X = 4)	0	0	0	0	0
P(X = 5)	5.07E−06	5.07E−06	0.00000507	0.00000507	5.07E−06	P(X = 5)	0	0	0	0	0
	Current	**2017**	**2018**	**2019**	**2020**						
Cyclone = 36	λ = 0.313										
P(X = 0)	0.7312	0.7232	0.7152	0.7072	0.6992						
P(X = 1)	0.2289	0.2369	0.2449	0.2529	0.2609						
P(X = 2)	0.0358	0.0358	0.0358	0.0358	0.0358						

(continued)

Table 3 (continued)

	Current	2017	2018	2019	2020						
P(X = 3)	0.003737	0.003737	0.003737	0.003737	0.003737						
P(X = 4)	0.0002924	0.0002924	0.0002924	0.0002924	0.0002924						
P(X = 5)	1.831E−05	0.00001831	0.00001831	0.00001831	0.00001831						
	Current	**2017**	**2018**	**2019**	**2020**		**Current**	**2017**	**2018**	**2019**	**2020**
Earthquake = 0	λ = 0					**Gale = 5**	λ = 0.0435				
P(X = 0)	0	0	0	0	0	P(X = 0)	0.9524	0.9444	0.9364	0.9284	0.9204
P(X = 1)	0	0	0	0	0	P(X = 1)	0.0416	0.0496	0.0576	0.0656	0.0736
P(X = 2)	0	0	0	0	0	P(X = 2)	0.0009059	0.0009059	0.0009059	0.0009059	0.0009059
P(X = 3)	0	0	0	0	0	P(X = 3)	1.313E−05	1.313E−05	1.313E−05	1.313E−05	1.313E−05
P(X = 4)	0	0	0	0	0	P(X = 4)	1.428E−07	1.428E−07	1.428E−07	1.428E−07	1.428E−07
P(X = 5)	0	0	0	0	0	P(X = 5)	1.24E−09	1.24E−09	1.24E−09	1.24E−09	1.24E−09
	Current	**2017**	**2018**	**2019**	**2020**						
S + C = 14	λ = 0.1217										
P(X = 0)	0.8854	0.8774	0.8694	0.8614	0.8534						
P(X = 1)	0.1078	0.1158	0.1238	0.1318	0.1398						
P(X = 2)	0.006557	0.006557	0.006557	0.006557	0.006557						
P(X = 3)	0.000266	0.000266	0.000266	0.000266	0.000266						
P(X = 4)	8.05E−07	0.000000805	0.000000805	0.000000805	0.000000805						
P(X = 5)	1.97E−07	0.000000197	0.000000197	0.000000197	0.000000197						

- Probability of Event then Asset Failure/ Urgency
- Frequency
- Rarity –Ecological Sustainability
- Impact Cost Magnitude/Duration
- Vulnerability/ Resilience
- Revenue Earning Capacity/Functional Significance and Asset Interdependency (utilities, roads, port, transport, bridges, communication/information/disaster risk management/emergency services
- Time/Recovery Time
- Constraints to Adaptation
- Adaptive Capacity
- Resources Available

Fig. 4 Criteria to prioritise climate change risks

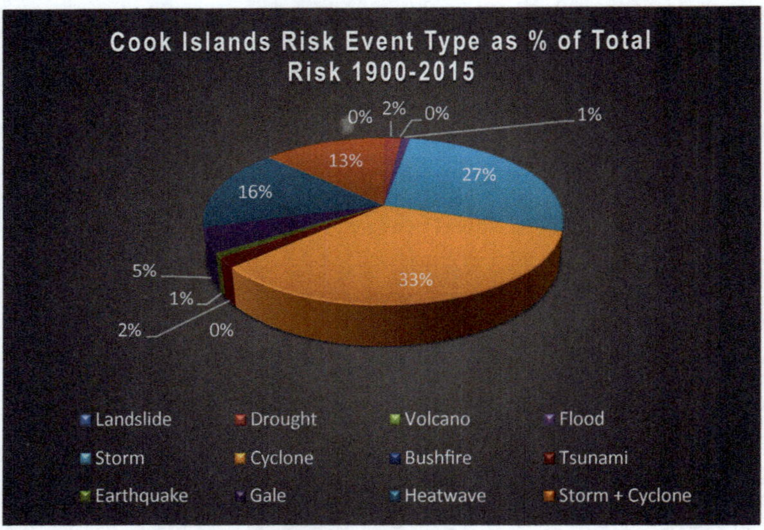

Fig. 5 Prioritising Cook Islands climate change risks for a Pacific MSC'

which stakeholder risk awareness is actually measured accurately. It seeks to minimise risk omission, under and overestimation and ascertain the validity of assigned probabilities. It advocates considering stakeholder identification of past risks—event frequency, duration and intensity/impact costs along with asset failure against existing risk events. It presents objective, risk identification criteria for future research.

To ascertain how the Cook Islands are vulnerable to climate change risks, this thesis derived paper contacted 350 MSC' stakeholders of which 147 responded, providing a response rate of 42%. Of those 147, 119 were familiar with climate change risk generally for national and international climate change legislation and disaster risk management policy strategies, of potential adaptation funding sources and responses by international professional associations; research and competitors. Curiously it was primarily the 3 international foreign banks, 2 insurance and 6 shipping companies and 8 aquaculture/fishing companies based abroad that remained both unaware and unconcerned about how climate change risks personally affected their operations and supply chains, unaware of local climate change efforts. This indicated that foreign businesses in particular especially underestimated the vulnerability of their local Cook Islands enterprises. Only the 22 government and 8 NGO's who responded, actually could provide an accurate indication of climate change risk perceptions rating it in alignment with the data; since key public assets have significantly benefited from existing efforts to emphasise climate change awareness for these sectors. In particular smaller businesses, organisations and individuals including 37 subsistence fishermen, 5 local aquaculture/2 value adding, 12 retail; 12 logistics and distribution, 5 recreational, 22 consumers, 2 marketing, and the Customs Department. Even larger insurance companies were unable to determine which were the most significant historic, current and future risks, citing limited resources and no centralised information available. Whilst virtually all interviewed demonstrated awareness of climate change given significant aid agency and local government efforts, stakeholders considerably underestimated long term and short term risk event frequency, beyond the last 15–20 years, reflecting age, experience and lack of data/resources/priorities. All agreed time series information, accurate charts involving risks and more specific field research rather than greater general information about climate change awareness would benefit them.

Stakeholders on average differed from 7–27 years of experience, seldom exceeding 30 years yet were practically experienced with climate change indicating considerable past experience of climate change events in contrast to developed nations encountering a greater frequency of low probability, high impact events. Education levels ranged from primary level for subsistence to postgraduate degrees for international companies and government. However the survey results considered that stakeholders relying on memories and limited records underestimated the number of disasters by 70–80 on average, indicating problems of risk underestimation in depending on memory. Survey results indicated the most significant, long term risks were sea level rise, changes in species migration and precipitation. The least significant were changes in temperature and current, although stakeholders had not considered currents, wind velocity, soil composition and wave energy. The most significant, real short term risk events of cyclones, storms and heatwaves were considered to be among the least significant, stakeholders considering floods, tsunamis and landslides as more likely, which does not reflect Tables 1 and 2 probabilities and historic time series data. This further validates the need for basic information over climate change risks and vulnerabilities that affect individual stakeholders and entire MSC' system. This would improve existing climate change risk identification and management to minimise vulnerability in the Cook Islands, Pacific and globally.

Predicting Climate Change Risks for a Pacific Maritime Supply Chains: The Cook Islands Risk-Vulnerability Matrix

Globally, tropical Pacific Ocean MSC' boundaries expand over 155,557,000 km^2 of ocean territory (8,497,017 sovereign nation land area) with 38,039,400 people (UNESCAP 2016). Yet world and localised climate change risk consequences extend beyond individual stakeholders and supply chains within dependent island territories and 16 sovereign states. Whilst existing climate change and supply chain risk management impact studies are land centred and separated; this paper contrasts in emphasising the maritime sector as more significant and vulnerable for Pacific and global trade, resources, climate change risks, economies, ecosystems, supply chains and stakeholders. With limited land area and natural resources, most Pacific MSC's and island nations are vulnerable to other nations and ocean-spanning, seaborne trade route connections. This paper's thesis derived case study, the Cook Islands, capital Avarua on Rarotonga Island a self-governed constitutional monarchy since 1965 consist of 20,700 people over 15 coral atoll islands and 2 submerged reefs (Fig. 6). It includes just 236 km^2 land area but 2.200,000 km^2 of Pacific Ocean (SPC 2015). Its climate and climate change projections are summarised below. Its economy is dominated by tourism and offshore sector banking services but as Table 4 summarises, it essentially depends upon the Pacific Ocean for MSC' centred products including pearls and fish exports as 87.7%, with refined petroleum imports of 25%). Thehe second highest percentage occurs from vessels/vehicles

Fig. 6 The Cook Islands. Author unknown (2016), http://www.vidiani.com/map-of-cook-islands

Table 4 Cook Islands trade overview 2015

GDP $ million	Exports (type)	Value $	% Exports	Imports (type)	Value $	% Imports
309.23	Total	34,000,000	100	Total	143,000,000	100.0
	Fish	20,040,000	60	Petroleum	35,900,000	25.1
	Pearls	9,418,000	27.7	Vessels/ vehicles	7,860,000	5.5%
	Citrus fruit	476,000	1.4	Iron structures	2,950,000	2.1%

(4.7%); the third highest includes iron structures (2.1%) to replace frequent coastal environment and adverse risk event exposure. Top 5 export trade partners include Japan, Turkey, Thailand, China and South Africa. Top import partners include New Zealand, China, Fiji, Australia and Turkey.

A previous published thesis book chapter involved 'Adapting Climate Change Projections To Pacific Maritime Supply Chains,' (Dyer 2017). This aimed to reduce stakeholder uncertainty in downscaling global and regional projections. The Cook Islands and other Pacific nations experience a near equatorial tropical climate with similar characteristics in air/sea surface temperatures, seasonal variations, humidity, winds and other climate related factors (Mori et al. 2013; New Zealand Office of Chief Science Advisor 2013; McCubbin et al. 2015). Precipitation and cyclone seasons frequently occur between November–April, extending between October–May during El Nino conditions, the dry season between May–October. The Cook Islands, consists of 2 archipelagos. North and South Group. Climate varies from a minimum temperature range of 19–24 °C in July (Fig. 7) to a maximum of 27–30 °C in December. Based on local data and Pacific Climate Future, model projections, (Fig. 8), mean climate change temperature is projected to increase from a 0 °C baseline in 2000 for all three scenarios to a 0.5° increase under a B1, 1.1° (A1B) and 1.5° (A2) scenarios for the short term adaptation, (2030) time horizon. By 2055 for a medium term adaptation time horizon; mean projected temperature is expected to increase by 1° under a B1, 1.5° (A1B) and 3° (A2) scenarios. This accelerates to 2° under a B1, 3° (A1B) and 5°C (A2) scenarios under a long term, adaptation time horizon (2090–2100). Average wind speed is projected to increase between 5–15%. Ocean PH will decrease of 0.1–0.3, increasing acidification by 2050, damaging ecosystems and increasing infrastructure corrosion costs. Current southeast trade winds shifting latitude and direction and a decrease in the South Pacific gyre; weakening currents are anticipated to further complicate navigation and reduce marine ecosystem nutrient flows.

This section proposes a more accurate means to ascertain the vulnerability of Pacific maritime economies, supply chain systems, operations and stakeholders to the above projected risks. It is based on the risk-vulnerability matrix section "How Existing Risk Management Methods Fail for Climate Change Risks on Maritime Supply Chains and Proposed Method". This is proposed, in a summarised form for the Cook Islands as a prototype for future research. As Fig. 1 illuminated, a Cook

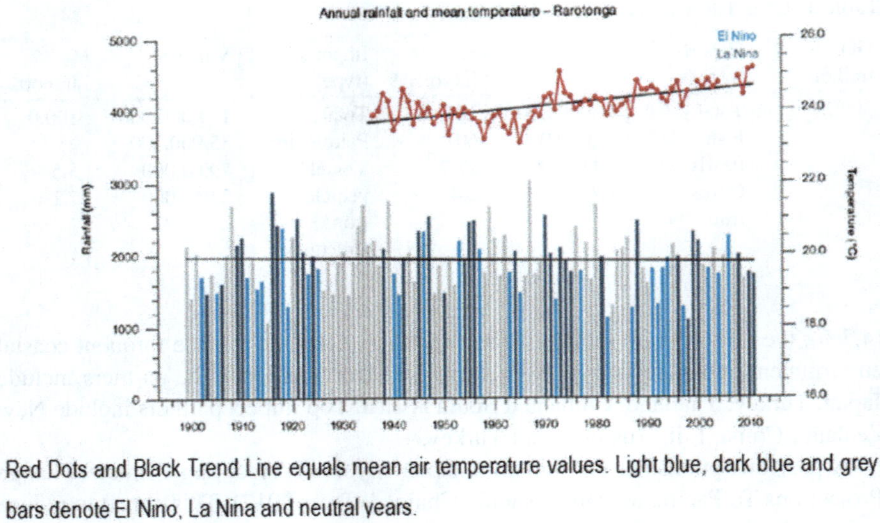

Fig. 7 Cook Islands climate: Mean annual precipitation and temperature *Source* Australian Bureau of Meteorology and CSIRO 2014, p. 32

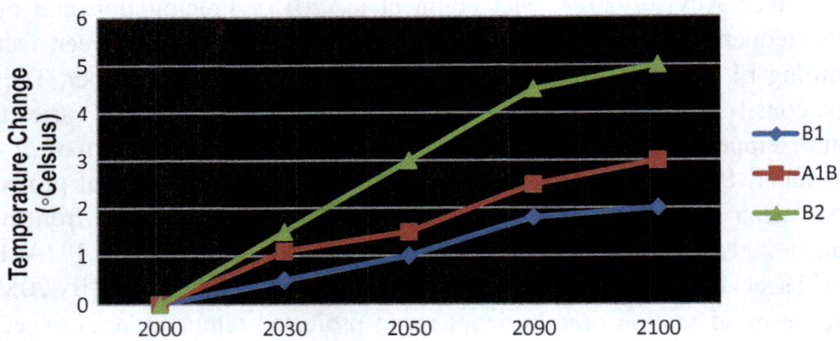

Fig. 8 Cook Islands climate change, projected mean temperature rise

Islands MSC' consists of resource extractors/producers including subsistence, recreational, domestic and international commercial fishing/aquaculture; industry, value adding and beneficiation including pearls and pharmaceuticals. It extends to the port, customs authority and government, importers/exporters and freight forwarders; logistics transport and distribution. It adds wholesale/retail; marketing/publicity, consumers and the financial banking/insurance sector directly along with other supply chains/economic/state sectors indirectly. However, a significant flaw of established risk management theory is that it completely ignores not only climate change risk uncertainty but also the status and sustainability of underlying environments and ecosystems that provide supply chain resources. This thesis derived paper's further original contribution proposes before any supply chain, risk

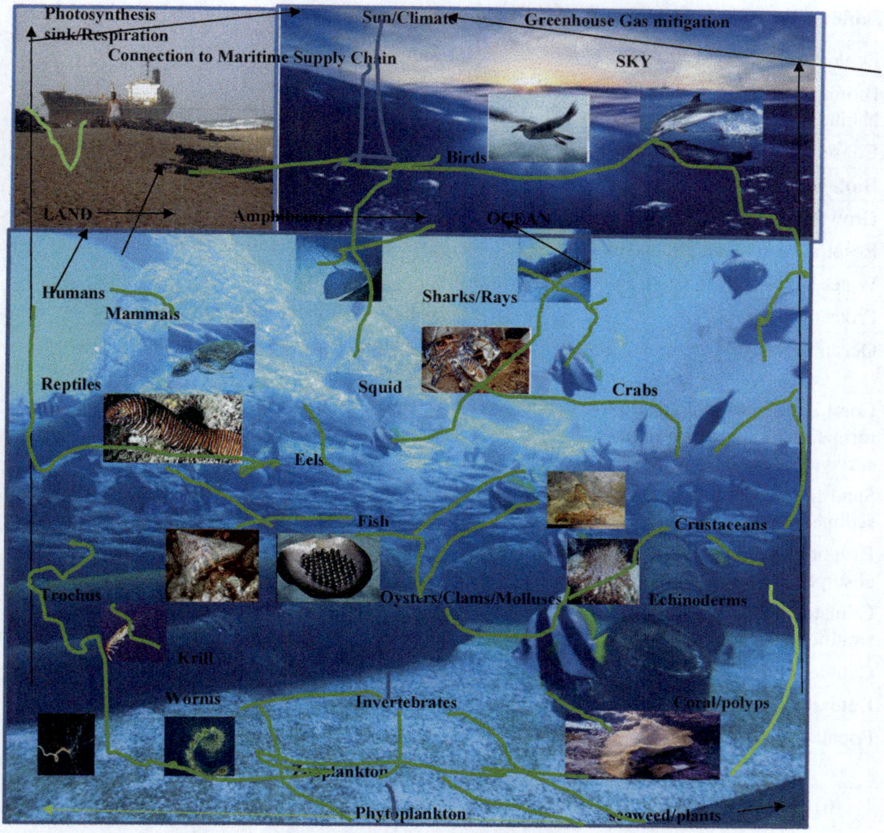

Fig. 9 A Pacific maritime ecosystem/maritime supply chain resources

assessment can be undertaken, it is necessary to consider existing environmental conditions prior to evaluating climate change risks to future resources. This ensures future prosperity, sustainability, performance and ultimate risk survival. Figure 9 illustrates a Pacific maritime ecosystem as the basis of Fig. 1 Cook Islands and other MSC's. Whilst ecosystem purposes are summarised in Table 5, existing risk management theory fails to protect maritime supply chain stakeholders by ignoring how assets, stakeholders, operations and systems will be adversely affected from risks and impact cost consequences. Consequences include a loss of coastal protection, natural resources, oxygen, water supply and other essential Pacific maritime ecosystem functions, reducing natural resilience/enhancing impact cost consequences. This paper therefore proposes active risk monitoring, identification, analysis and adaptation extends to ensuring both marine and terrestrial ecosystems remain functional and thrive.

The Cook Islands environment includes extinct volcanoes, coral reefs and atolls, seagrasses but no mangroves. As with other Pacific and global MSC's, its ecosystem contributes to economic activity through sand formation for beaches/

Table 5 Ecosystem Functions for MSC'/Economy Stakeholders

Ecological	Economic
Biomass/biodiversity life formation and habitat	Life, food, material
Conservation	Supply of natural resources, reduced imports
Biological/physical/chemical	Redundancy against uncertainty
Growth, reproduction	Trade, production, consumption, income/profit
Respiration/oxygen/photosynthesis	Greenhouse gas mitigation source sink
Water supply/purification. Food security/nutrition	
Protection	Protection—Vulnerability and Resilience
Ocean Chemistry, currents, salinity	Risk Identification, monitoring, prioritisation, adaptation
Coral atolls—geographical physical formation, continued growth and survival	Risk enhancement if ignored—legal, reputational, insurance, security, operational, impact costs
Sand formation, nourishment and sediment	Opportunity
Evaporation, condensation and absorption	Insurance against maladaptation
Climate regulation—calcification, stratification	Future sustainability and survival
Counter eutrophication	Knowledge—existing and potential/spiritual
Detoxification	Stability/security/increased adaptive capacity
Population equilibrium	Aesthetic/cultural/social
	Tourism

coral reef formation for coastal tourism to geographical physical island formation to wave energy dispersal barriers. These serve as natural resilience against sudden disruption risks and influences the quality/quantity/habitats and survival of interconnected Fig. 9 maritime and other terrestrial resources for fisheries, forestry, mining, jewellery and other economic activity. Although risk management prioritises business operations to non-natural assets, the Cook Islands Bishop Museum identifies 1 trochus species, 66 seaweed, 11 oysters, 83 seabirds, 2 squid, 5 ray, 19 shark, 20 reptiles, 22 maritime mammals, over 664 fish, 232 crustaceans and 109 echinoderms. It recognises 539 molluscs, 70 clam/6 giant clam, 11 oysters and 34 eel species aside from plants, coral reefs and land based ecosystems, which this paper considers directly or indirectly influence maritime supply chain and tourism activity. Therefore this risk-vulnerability matrix provides among the first to assess climate change risks to resources as a rudimentary stage to assessing risks to producers and whether disruptions remain temporary or become permanent.

Only 0.2% of world possesses coral reefs, yet home over 33% of all planetary global marine biodiversity and ecological capital. The Cook Islands provides a significant Pacific maritime supply chain case study its capital island Rarotonga encircled by reef. With Kiribati it possesses among the planet's largest Exclusive

Economic Zones of maritime terrain and potential maritime resources, yet limited land areas. However, existing and projected risks including increased ocean acidification and salinity, coral bleaching and disease, increased sea level rise, decreased salt crystals and increased salt spray pressure along with higher sea surface, land and air temperature rises (Rongo and Dyer 2015) have contributed to a significant increase in reducing coral reef areas. Areas declined 30% from 1990s to 2001 and 6% further by 2015. In the 1970s coral took 10 years to recover. In 1990s 21 years to recover but increased sea surface temperature threatens this as with a projected increase in coral bleaching events. Repeated reef bleaching and disease risk events from 1991–2014 from cyclones, sea surface temperatures, and invasive species (1991, 1994, 1995–1998, 2006, 2010, 2014) emphasise ecosystem vulnerability. Rongo and Dyer (2015) consider even with existing climate change and other pressures, the following Table 6 Cook Islands species have declined rapidly from 1970–2014 in a Rarotonga coastal reef survey with the exception of predator porcupine fish, invasive Indian mynahs and oceanic sharks from a decline in deeper water from trawling, combined with the Cook Islands, 2012 declaration of a Shark sanctuary. Therefore this paper considers the high vulnerability of Cook Islands maritime resources; specifically affected by historic, Pacific, climate change risks, decreasing species numbers and habitats. This threatens MSC' commercial resource security for production, value adding/aquaculture and remaining stages. Yet each of these species is worth investing in as ecological capital, given existing commercial demand and future potential research, whilst 20 species (UNFAO 2010) possess value being favoured ornamental/aquarium fish.

Given IPCC (2015) climate change projections range from 0.5–1.5 metre sea level rise, 1–5 °C (2030–2100), 0.1–0.3 ocean PH increase from 8.08 (1980–1999) and 1–8% precipitation increase, the above species and interconnected Fig. 5.2 ecosystem are further expected to decrease rapidly in numbers, sizes and quality. From 1980–1999, average Cook Islands sea surface temperature was 26.5 °C. Coral with maximum temperatures of 25–29° as with other maritime ecosystems have resilience limits, which collapse. PCCASP, SPC and SPREP (2015) project further coral area eradication of 25–65% by 2030, 50–75 (2055) and 90–100% by 2100 unless reefs are restored. Increased sea level rise, wind velocity, wave energy, reduced surface runoff during El Nino/droughts for the Islands, risks to river flow, changes to oceanic currents, freshwater, lagoon/estuaries and soil sedimentation along with changes in species migration and reduced biodiversity provide future climate change risks. ENSO has already affected skipjack tuna migration as a key MSC' staple. Its catch rate, quality and size have significantly decreased. These are projected to initiate a 20% reduction in nutrient supply, producing fewer zooplankton, algae and primary biomass, and lower coral cover/slower recovery rates which affects all other species. The lower the biodiversity, the more significant the existing species in preserving the functionality and value of local ecosystems and environments.

Without recovery (further pressurised by invasive species including crown of thorn starfish) and prioritising future ecological capital, maritime ecosystem resources decline. This increases extinction rates as catches unsustainably increase,

Table 6 Cook Islands risks and maritime ecosystem species decline 1970–2015

Species terrestrial/ freshwater	Cook Islands location	Climate change/variability risk impacts
Whitebait	Rarotonga	Extinct from droughts–formerly favoured floods
Dusky sleeper	Mangaia, Atiu, Mauke	Droughts
Western gambusia	Rarotonga	Drought/drained swamp
Tilapia[a]	Mangaia, Atiu, Mauke	Increased sediment/siltation/drought
Freshwater eel	Rarotonga, Avatiu	Drought, increased salinity
Marine species		
Hatchet/giant seahares	Mangaia, Rarotonga	Drought/changes in species migration/loss
Lined seahare	Mangaia, Rarotonga	Droughts/change in biodiversity
Serrated swimming crab	Aitutaki, Rarotonga	coastal erosion
Coconut Crab	Mangaia, Rarotonga	Preferred delicacy, coastal erosion, soil sedimentation
Sargassum	Mitiaro, Atiu, Mangaia	Cyclones, storms, tsunamis
Tangled hair seaweed	Mangaia, Rarotonga	Cyclones, storms, tsunamis
Sea grapes	Atiu, Aitutaki, Mangaia	SST
Sponge seaweed	Mangaia, Rarotonga	SST
Seagrass parrotfish	Mauke, Mitiaro, Atiu	Change in species migration and biodiversity SST, Air, SLR, Flooding Increased ocean acidification
Scribble/silver rabbitfish	Mauke, Mitiaro, Rarotonga	
Forktail rabbitfish	Mauke, Mitiaro, Rarotonga	
Rudderfish	Mangaia, Mitiaro, Mauke	**Marine reef Fish affected include**
Fringelip/warty lip mullet	Atiu, Aitutaki, Rarotonga	
Marbled/brown cod	Mangaia, Rarotonga	Marbled grouper
lunar tail/peacock cod	Mangaia, Rarotonga	Yellowfin goatfish
Brown moray	Mangaia, Rarotonga	Barracuda
Red snapper	Rarotonga, Atiu	Convict/Black/yellowfin surgeonfish
Rose mouthed turban	Mangaia, Rarotonga	Mackerel scad
Branching coral	Rakahanga, Manihiki	Bull's eye
Winged mussel	Rakahanga, Manihiki	Green triggerfish
Black lipped pearl oyster	Rakahanga, Manihiki	Orange spotted emperor
Trochus, giant clam	Rarotonga	Napoleon wrasse
Brown pencil urchin	Mangaia, Mitiaro, Mauke	Big eye bream

(continued)

Table 6 (continued)

Species terrestrial/ freshwater	Cook Islands location	Climate change/variability risk impacts
Star shaped limpet	Mangaia, Mitiaro, Mauke	Topsail drummer
Pelagic		Unicorn fish
Reef sharks	Aitutaki, Rarotonga	5 Banded parrotfish
Skipjack/yellowfin tuna	Mangaia, Mitiaro, Nassau	Squirrelfish

Rongo et al. (2015). [a]Introduced

multiplying ecosystem pressure further. Lower tides increase exposure for subsistence fishermen who need motorised boats to increasingly access catches, affecting species. Increased ocean currents and turbulence, wind velocity combine to provide rougher ocean conditions, exposing existing delicate species such as algae and crustaceans to further pressure. Terrestrial ecosystems and supply chains are similarly affected. Changes include mangos blooming earlier in July whilst apples and other, less temperature resistant fruit have stopped germinating. This research concentrated on MSC's specifically. Unlike other global regions including North America, South America, Europe, Russia, China, India and Japan, significant global and Pacific climate change research exists for ecosystems and biodiversity, (UNFAO 2010). This is even provided in a specialised Cook Islands Directory of Sources. This paper proposes affected stakeholders utilise sources when checking risks to maritime ecosystems and species, considering which are rarest, most vulnerable and most crucial for the maritime supply chain to prioritise conservation/ resources/reserves as most saving with eco agencies, local community and government stakeholders.

The future of Pacific maritime resources for MSC's are further threatened by existing environmental pressures across supply chain locations and constraints as small developing island states identified in this survey. For the Cook Islands, these specifically include pollution, over-fishing over-development, sensitive habitats, subsistence lifestyles, overpopulation, shore vegetation removal, competing land use, waste dumping, coastal erosion and physical topography. It includes beach sand mining and limited freshwater supply for atolls/Mangaia, Rarotonga and Aitutaki lagoon contamination, enhancing climate change risks further. Volcanic islands with limited land areas, soil fertility, low crop yield/productivity, and soil for agriculture, increase coastal fisheries dependence. Draining wetlands for taro production lowers biodiversity and flood protection. Lacking environmental reserves and sustainability, Cook Islands ecosystems will be significantly constrained in satisfying future Table 5 MSC' performance, revenue, resilience and production requirements increasing potential impact costs as in future research. It multiplies risks to subsequent MSC' stages of production/value adding and beneficiation/manufacturing. Without managing fisheries and agriculture ecosystem risks, stakeholders will lack the resources to ensure future production, income and recovery from long and short term risks, given that without renewing resources, the Cook Islands and other Pacific nations will increasingly be unable to afford adaptive capacity or participate in global MSC' activities, lacking minerals, fuel and other products. This may necessitate even more aid, long term economically unsustainable. Developed nations will face increased security and other costs from immigration pressures, as more coastal exposed developing nations face collapsing ecosystems, economies and submerged land areas from higher instability, unless risk is pre-empted by increasing natural risk mitigation systems. This paper further proposes future research concentrates on specific climate change risk implications for Pacific MSC' stakeholders and ecosystems. This is significant as specific location, ecosystem and supply chain risk type consequences differ yet have been ignored. Increased landslides multiply soil sedimentation, coastal erosion and species habitat loss. They present fewer maritime risks then tsunamis, storms, increased wave energy, floods, cyclones and droughts to coral reefs and less mobile/ more fragile maritime species assets such as algae, trochus and black lipped oysters. Whilst seabirds, crustaceans, reptiles, invertebrates, fish and mammals including humans possess limited adaptive capacity with sufficient warning to prepare for sudden risks temporarily, certain physical, economic and environmental constraints prohibit permanent adaptation to long term risks without sufficient prioritisation and investment in maritime ecosystem functions as climate proofing risks.

For this thesis and derived paper; section "Climate Change Risk Identification and Perception Survey Results for a Pacific MSC's" risk perception survey indicated that whilst 107/147 stakeholders indicated certain climate change awareness of ecosystems, none had considered these functions or need to integrate ecosystems as effective risk management and resource security for MSC's. Given low market prices for consumers and existing environmental pressures, the maritime ecosystem remains undervalued relative to potential risk and impact cost aversion. In particular international financial services/insurance ignored the historically successful role of coastal asset protection, where ecosystems were valued. Yet international companies were particularly ignorant, lacking concern about potential risks to Pacific maritime/land ecosystems and their future resources even those with resources to adapt. More attitudes focused on prioritising short term risk events (89.2% of stakeholders) without considering long term risk pressures which undermine maritime ecosystem resource survival for production, aquaculture, own or other supply chain operations.

This paper proposes utilising localised existing sources including geophysical topographic maps, statistics, stakeholder consultation and physical field research observation can more effectively determine localised climate change risk and vulnerable stakeholders. These and Fig. 9 confirmed for Rarotonga that of 237 registered businesses and 7108 potential stakeholder inhabitants—as producers, employees or consumers; over 81% were concentrated around the coast and within 0.5–1.5 m of the ocean/a potential river flood source. These were less than 100 m above sea level. It similarly indicated the vulnerability of the airport, seaport, fuel supply tanks, and key coastal roads with limited accessibility into the mountainous hinterland with only 2 main roads circulating the island. High precipitation levels

Fig. 10 Avarua, Rarotonga. Cook Islands Government (2016)

and soil composition indicate potential for landside and soil erosion damage. Whilst the coral reef provides partial coastal protection, Aviatiu's port expansion has partially eliminated this. Field research identified key infrastructure in electricity, telecommunications, the Cook Islands Trading Company, 2 local shipping companies, the airport, Port Authority, government including National Environmental Service; Triad Petroleum and Pacific Energy were all prioritising climateproofing for risk adaptation and had minimised supply chain delays/vulnerabilities. Other parts of the MSC remain neglected, uninformed and risk exposed (Fig. 10).

This paper identified that only 25 of the key Pacific MSC' stakeholders that responded were partially prepared psychologically/commercially. Yet all 350 remained physically exposed to the above projected climate change risk as in Fig. 9. Existing fishing operations, aquaculture, pearl, seaweed and recreational lack coral reef and natural coastal protection to infrastructure, people, transport and equipment along with environmental pressures threatening the future of maritime resources; especially ocean acidification, salinity, temperature, currents, wave energy, wind velocity, precipitation, cyclones and storms. This subsequently affects e vulnerability to the resource inputs, outputs, quality, capacity and performance of existing manufacturing, industry and beneficiation, similarly dependent on resources, exposed physically and remain underprepared for existing and projected risk events. Although the sea and airports, with key infrastructure are climate-resilient to an extent despite being physically exposed; none of the importer/exporter or 12 logistics/transport/private sector storage companies analysed nor the wholesale/retail upon which they rely. Nor were the 2 marketing/publicity companies concerned and aware that affect reputation, nor the 3 major international

Fig. 11 Cook Islands, Rarotonga topographical map of climate change vulnerabilities. Cook Islands Government (2016)

banks and several local insurance firms conscious of projected risks; impact costs and adaptation strategies (Fig. 11).

This paper forecasts in a major risk event occurrence i.e. a tsunami, cyclone, flood or storm; the majority of producers, key infrastructure and consumers would experience significant impact cost damage if not collapse. Even those few stakeholders who were aware of climate change risk, prioritising emissions mitigation and adaptation including the port; would be overwhelmed by significant congestion, of the subsistence, recreational and private sector, MSC' stakeholders and businesses who remain completely vulnerable and unprepared for existing risks; regardless of projected IPCC (2015) increases in risk event frequency, duration and intensity. Insufficient stakeholder cooperation, research prioritisation and information sharing has historically and remains projected to loss of life and ecosystems, physical commodity/resources damage—to stockpiles, crops, fishery yields/catches. Supply chains experience other costs include port, cargo and supply chain infrastructure, clean-up, damage costs, repair costs, asset replacement costs, port recovery costs, opportunity costs of resources. Examples include port approaches, port limits, breakwaters, turning basins, fairways, berths, docks, channels, pavements, container stacking areas, quay walls, port authority, customs, damage to terminals, cargo warehouses, offices, hazardous cargo storage zones, commercial businesses, water, electricity, sewerage and bunkerage. These include equipment damage, repair and maintenance costs. Examples include operator vehicles, synchrolifts, stacking equipment, cranes, container scanning facilities, reach stackers and container reefer points.

Other projected port impact costs, yet to be estimated, analysed or calculated for the Cook Islands or elsewhere in research prior to this paper's thesis include port and supply chain service costs. Examples include pilotage, mooring, tugs and towing, salvaging, dredging, customs, stevedoring, drydocks/repairs, waste disposal, navigation, vessel tracking, communication, information, security, hazard warning systems, inspection, freight forwarding. These include technology damage—security, navigation, customs and administration and physical damage to road, rail, shipping and air as intermodal transport types. Stakeholders can experience increased cleaning/storage costs, cargo loading/unloading costs and other port/cargo dues from delays. Shipping cost losses include reduced economies of scale, increased fuel consumption costs, increased daily fixed shipping costs per TEU. Increased shipping variable costs i.e. crew wages, stores, port dues plus time, voyage and spot charter costs, brokerage costs, contract of affreightment and carriage penalty charges. Stakeholders throughout the supply chain experience increased evacuation/relocation costs for businesses along with production/disruption/sales revenue/input costs; increased information, communication, administration, marketing and recovery costs to minimise reputational costs and recover to business as usual. These subsequently reduce port throughput volumes, values and revenue. It reduces potential customs/government revenue; imposes higher consumption/consumer costs and financial sector loss including loan/investment costs/loss of profit, capital potential, bankruptcy, insolvency and insurance throughout the entire supply chain.

The vulnerability of individual Cook Islands and MSC' system stakeholders to risks provides higher impact costs beyond; influencing and influenced by risks and consequences to domestic land supply chains, and increasingly globalised systems, Costs include reputational costs, higher imports, reduced exports, increased costs to foreign exchange and balance of payments with higher unemployment. Increased maintenance costs include routine, periodic, rehabilitation, deferred and unexpected/unpredictable. Other costs include trade diversion and contingency rerouting costs, increased opportunity costs to business/supply chains, higher legal, technical and regulatory compliance costs and costs to tax revenue which affects financing potential of supply chain infrastructure/functions. Evaluating the extent of costs emphasises the extent to which not just local but global stakeholders have continued to underestimate the true vulnerability of international trade and global maritime supply chains to existing and projected climate change risks. This excludes existing risk management and climate change, impact studies existing failure to currently ascertain increased ecosystem damage/coastal erosion costs to resources and natural/coastal protection/biodiversity cost. This affects current, past and future vulnerability and risk exposure. It excludes risks to/by other supply chain partners and subsequent climate change mitigation, adaptation asset replacement, elevation, relocation, recovery, retreat/surrender, migration and ecological rehabilitation response strategy costs. In ignoring the extent of true risks and impact costs, these factors not only amplify true climate change vulnerability for the Cook Islands and others but the capacity to survive by financing existing climateproofing adaptation strategies summarised in section "Cook Islands Climateproofing Adaptation Strategies".

Cook Islands Climateproofing Adaptation Strategies

The Cook Islands have sought to reduce climate change risk event vulnerability through the following adaptation strategies.

- 1987: Relocation of coastal breakwater from cyclone.
- 1992: Ratified and implemented Kyoto.
- 2005: Airport protected its radar and other electronics and secured 3 months fuel supply. Pacific Energy and Triad Petroleum were persuaded to improve stockpiles up to 6 weeks, training and invest in underground cables/pipelines and more elevated tanks.
- 2007–2014: $27,600,000 Climateproofing of Avatiu Harbour with dredging and expanding port areas, strengthening and constructing seawalls and offshore breakwaters, elevating container stacking areas, improving lighting and relocating vulnerable sheds. This aims to withstand up to 500 cm in sea level rise, higher wind speeds up to 150 km/h, increased precipitation wave height and energy (Blacka et al. 2013; Cook Islands Port Authority 2015; Asian Development Bank 2015). It invested in mobile forklifts, cranes and a 5 ton tug that can be easily removed and stored along with updated crew training. It committed $4,000,000 to Penrhyrn, $3,200,000 to Aitutaki and several million to improving outer island ports. Aware of the need to secure other vulnerable targets, it extended adaptation to other maritime supply chain stages and economic hinterland connections including $2,900,000 in sealing Rarotonga roads and $2,400,000 in outer Island roads.
- 2010: Cyclone Pat Recovery and Reconstruction Plan for Aitutaki established $530,416 committed to improving disaster risk reduction in foreign aid with $200,000 to improving waste management, $24,000 on a climate risk warning system and $200,000 on improving general infrastructure resilience. It also allocated $1,356,870 to local economic recovery, aid being conditional on stakeholders considering climate change awareness to enhance future resilience, which developed countries have yet to consider. $1,000,000 of this was allocated to small entrepreneurs, $194,870 to restore livestock and $55,000 for more climate resistant taro production.
- 2012: The National Infrastructure and Preventive Infrastructure Investment Plan require considering climate change for any potential maintenance/upgrades to preserve and enhance performance, flexibility and productivity.
- 2012–2016: extend maritime sanctuaries, impose fishing restrictions in days and quotas, increase community based enforcement, ecotourism and prioritise aquaculture to ensure future maritime food and resource security in its 2012–2016 National Aquaculture Development Plan.
- 2013: Nationally, $512,343 was invested in improving disaster risk event response training to Emergency Management Cook Islands.
- 2016: Ratified COP21 Paris.

However; this research identifies the continued lack of international and domestic private business, community and individual stakeholder, climateproofing risk identification and adaptation efforts which it proposes are prioritised in future research. It proposes evaluating the extent to which existing climateproofing strategies have enhanced resilience, reduced vulnerability; preserved stakeholder requirements, ecological capacity sustainably and minimised disruption cost/

accelerated recovery times/rates. These could exist in a risk event monitoring and review stage, against projected existing and accumulated risk. It suggests providing standardised criteria to ascertain each asset, system, operation and ecosystem's potential vulnerability to climate change including provide and identify the location/geography, potential climate change and non climate change risks, history of risk exposure, asset characteristics, ecosystem/resilience/vulnerability, number of potentially affected stakeholders, risk type/probability of risk plus conditional probability of an asset failure, degree of asset/supply chain interdependence, performance and stakeholder requirements are affected.

Conclusions, Policy and Theoretical Significance for Stakeholders and Directions for Future Research

In conclusion, this paper aimed to address how aware, resilient and vulnerable Pacific MSC systems, stages and stakeholders are to current and future climate change risk events through a Cook Islands case study. Given these risks project an uncertain future for Pacific and global ports, shipping and MSC's; it considers the failure of conventional risk management methods to adequately predict true stakeholder vulnerability. Managing risk,' infers readjusting to normal conditions and a stable long run equilibrium or 'business as usual. However, IPCC (2015) climate change projections infer multiple extreme events increasing in risk frequency, duration and intensity as the new normal. However, conditions may not be normalised or stabilised, especially if emissions are not substantially reduced and the underlying maritime ecosystem/resources restored. Risk Management often ignores the role of stakeholder consultation and how to overcome existing challenges of stakeholders in sufficient and accurate risk determination, the role of psychological expectations influencing risk probability and impact cost consequences. Risk management merely considers existing scenarios as the extent of maximum possible risk, underestimating worst case examples with dependence on stakeholder risk perceptions rather than measuring historic risk through time series data. Long run equilibrium assumes constant growth rates, a supporting population, functioning economy and resources able to effectively monitor, respond, identify and adapt to risk events over time, ignoring climate change which infers risk adaptation and resources are sustainably secured whilst still possible. It assumes risk as static rather than fundamentally dynamic. It ignores how globalisation has increased supply chain asset/system interdependency and risk vulnerability to single catalyst events. This not only presents theoretical limitations but restricts key supply chain stakeholders from a managerial/policy perspective in predicting true climate change risks and opportunities.

To validate its conceptual framework of empirical probabilities rather than qualitative stakeholder perceptions; this paper proposed an improved risk-vulnerability matrix for risk management stakeholders for more accurate risk identification, given scarce resources and other constraints to adaptation. It established equations and a risk-event tree from a thesis to enable policy makers and researchers to more effectively determine projected climate change risks, impact costs and adaptation strategies. For the Cook Islands, it produced the first centralised Pacific climate change risk time series data for all risk events from 1900 to 2015 with 107 events. Cook Islands stakeholders can therefore prioritise cyclones (33% of all events) followed by storms (27%) based on thesis and paper derived criteria including resources available, stakeholder adaptive capacity, resilience, vulnerability and frequency, ignored by previous risk management research. The least significant were volcanoes, earthquakes and bushfires Field research established that 119/142 Pacific maritime supply chain stakeholder respondents were generally aware of projected climate change risks but all stakeholders generally underestimated the total historic number by 70–80; illustrating the problems of relying on stakeholder perceptions and consultation rather than laboriously determining all risks. The survey sample indicated the need to distinguish between physical adaptive capacity/risk and psychological.

This paper further proposes utilising downscaled Pacific climate and climate change projections to predict true risks and opportunities, indicating the need for stakeholders to adapt to 2–5° higher average temperatures and a 30% reduction in coral reefs 1990–2001 and a further 6% by 2015. It projects a 25–65% loss of reefs and related maritime ecosystem resources of 25–65% by 2030. It identified government has invested in over $27,600,000 climateproofing for Aviatu port, $4,300,000 in roads and $530,416 in disaster risk management training; utilising topographical maps, stakeholder consultation and physical field research observation. However, both private and government, maritime supply chain stakeholders remain significantly physically vulnerable. The community/subsistence and commercial/private sector remain vulnerable from a resources, unadapted operational and human psychological risk underestimation, perspective. With over 237 businesses and 7108 potential producers, employees and consumers vulnerable for the Cook Islands; it considers stakeholders face significant direct and indirect projected costs to lives, properties, ecosystems, resources, systems, for relocation, recovery and adaptation, despite over 3 decades of climate change adaptation efforts. Future research proposes determining the effectiveness of adaptation solutions. Though vulnerable, certain Pacific case studies appear more psychologically prepared for climate change risks based on existing natural disaster events, given limited physical capacity. Existing funding/research remains Pacific orientated. This is in overwhelming contrast to developed world ports, supply chains and stakeholders, which fail to prioritise climate change. They are considered even less prepared as a direction for future research. Previous research illuminated limited climate change awareness, effective methods, case studies and action priorities among global supply chain researchers and stakeholders. These fail to recognise it as a key problem and could learn from the Pacific and existing risk to consider an uncertain future; as these stakeholders learn from physical experience not to underestimate risk but to prioritise climate change.

References

Asian Development Bank. (2015). *Cook Islands: Aviatu Port development project*. Asian Development Bank project completion report, Manila.

Author Unknown. (2016). Figure VI: The Cook Islands. http://www.vidiani.com/map-of-cook-islands.

Becker, A., Acciaro, M., Asariotis, R., Cabrera, E., Cretegney, L., Crist, P., et al. (2011). A note on climate change adaptation for seaports: A challenge for global ports, A challenge for global society. *Climatic Change, 110*(1), 5–29. Accessed 19 January 2015. http://dx.doi.org/10.1007.

Becker, A., & Caldwell, M. (2015). Stakeholder perceptions of seaport resilience strategies: A case study of Gulfport (Mississippi) and Providence (Rhode Island). *Coastal Management, 43*(1), 1–34.

Blacka, M., Flocard, F., & Parakoti, B. (2013). *Coastal adaptation needs for extreme events and climate change, Avarua, Rarotonga, Cook Islands*. Water Research Laboratory University of New South Wales report, Sydney.

Business for Social Responsibility (BSR). (2015). *Business action for climate-resilient supply chains: a practical framework from identifying priorities to evaluating impact*. Business for Social Responsibility report, Cambridge.

Climate Change Cook Islands Office of the Prime Minister. (2012). *An assessment of national climate change priorities alignment to the regional framework*. Cook Islands Government report, Avarua.

Cook Islands Government. (2016). Refers to sources acquired during field research.

Cook Islands Ministry of Marine Resources. (2012). *Cook Islands aquaculture development plan 2012–2016*. Cook Islands Government report, Avarua.

Cook Islands Port Authorit. (2015). Annual financial report for the year ending in June 2014. Cook Islands Port Authority Report, viewed 17 May 2015, http://www.ports.co.ck/doc.

Corporate Social Responsibility (CSR). (2015). Climate change adaptation: Engaging business in Asia. Corporate social responsibility Asia report, viewed 6 April 2015, http://www.csr-asia.com/report.

Dyer, J. (2017). 'Adapting climate change projections to Pacific maritime supply chains', *Climate change adaptation in Pacific countries, fostering resilience and upkeeping life quality*. London: Springer Press.

Ellis, S., Shockley, J., & Henry, L. (2011). Making sense of supply disruption risk research: A conceptual framework grounded in enactment theory. *Supply Chain Management: An International Journal, 47*(2), 65–96.

Flannery., & Steffen. (2015). Climate council briefing statement: Damage from cyclone pam was exacerbated by climate change, Media release, 18 March, Australia Climate Council. Accessed 30 April 2015, https://www.climatecouncil.org.au.

Furlow, J., & Potter, J. (2015). Successful strategies to build resilience in developing countries. Paper presented to *First international conference on surface transportation system resilience to climate change and extreme weather events September 16, 2015, Washington*.

Ghadge, A., & Kalawsky, S. (2012). Supply chain risk management: Present and future scope. *The International Journal of Logistics Management, 23*(3), 313–339.

Intergovernmental Panel on Climate Change, (IPCC). (2015). Fifth assessment report of the Intergovernmental Panel on Climate Change. In Edenhofer, O., Pichs-Madruga, R., Sokona, Y., Farahani, E., Kadner, S., Seyboth, K., et al. Cambridge: Cambridge University Press.

Kazemia, Y., & Szmerekowsy, J. (2015). Modelling downstream petroleum supply chain: The importance of multi-mode transportation to strategic planning. *Transportation Research Part E, 83,* 111–125.

Kern, D., Moser, R., Hartmann, E., & Moder, M. (2012). Supply risk management: Model development and empirical analysis. *International Journal of Physical Distribution and Logistics Management, 42*(1), 60–82.

Kiele, R., Mageto, A., Numbasa, G., & Miria, G. (2014). Port Moresby Papua New Guinea Climate Change Vulnerability Assessment. UNHABITAT Cities and Climate Change Initiative Report, Port Moresby.

Kreie, A. (2013). *The Adaptation of Supply Chains to Climate Change*. Ph.D. Research Thesis, Heriot Watt University, Edinburgh.

Mccubbin, S., Smit, B., & Pearce, T. (2015). Where does climate fit? Vulnerability to climate change in the context of multiple stressors in Funafuti, Tuvalu. *Global Environmental Change, 30*, 43–55.

Mcevoy, D., & Mullet, J. (2013). Enhancing the resilience of seaports to a changing climate: Research synthesis and implications for policy and practice. Accessed 15 May 2015, http://www.nccarf.edu.au/sites/default/files.

Mori, N., Shimura, T., Yasuda, T., & Mase, H. (2013). Multi-model climate projections of ocean surface variables under different climate scenarios. Future change of waves, sea level and wind. *Ocean Engineering, 71*, 122–129.

New Zealand Office of the Chief Science Advisor. (2013). *New Zealand's changing climate and oceans. The impact of human activity and implications for the future*. Office of the Prime Minister's Chief Science Advisor Committee report, Wellington.

Oswald, M. (2011). Development of a decision support tool for transportation adaptation practises in response to climate change. PHD thesis, University of Delaware, Newark.

Rongo, T., & Dyer., C. (2015). *Using local knowledge to understand climate variability in the Cook Islands*. Office of the Prime Minister Cook Islands report, Rarotonga.

Rongo, T., Tautu, B., Mcdonald, G., & Hanchard, B. (2015). *Rarotonga fore reef community survey for 2014*. Government of the Cook Islands, report, Rarotonga.

Scott, H., Mcevoy, D., & Chhetri, P. (2013). *Climate change adaptation guidelines for ports, enhancing the resilience of seaports to a changing climate report series*. National Climate Change Adaptation Research facility RMIT University Report, Southport.

Secretariat of the Pacific Community. (2014). *Sustainable maritime transport in the Pacific Islands region*. Secretariat of the Pacific Community report, Suva.

Secretariat of the Pacific Community. (2015). *Cook Islands country programme*. SPC Cook Islands 2014 Programme report, Suva.

Secretariat of the Pacific Regional Environmental Programme. (2015). *PACC demonstration guide: 'Climate Proofing' coastal development on Mangaia Island, Cook Islands*. Pacific adaptation to climate change project report, Suva.

Smith, R. (2015). *The good, the bad, and the robust: Climate change adaptation choices for the port of Rotterdam, Port of San Diego, and Naval Base Kitsap - Bremerton*. Master's Thesis, University of Washington, Seattle.

United Nations Food and Agricultural Organisation. (2010). *Cook Islands national fisheries sector overview*. United Nations Food and Agricultural Organisation report, viewed 22 February 2016, http://www.fao.org/home/en.

United Nations Conference on Trade and Development (2011). Ad Hoc expert meeting summary on climate change impacts and adaptation: A challenge for global ports. UNCTAD Report, Geneva.

United Nations Economic and Social Commission for Asia and the Pacific (UNESCAP). (2016). Overview of natural disasters in Asia and the Pacific 1970–2014. UNESCAP Databank, viewed 12 February 2017. http://data.unescap.org/escap_stat.

Climate Change and Integrated Coastal Management: Risk Perception and Vulnerability in the Luanda Municipality (Angola)

Bernardo Castro, Walter Leal Filho, Fernando J. P. Caetano and Ulisses Miranda Azeiteiro

Abstract Climate change has imposed significant changes in the structure and natural dynamics of coastal ecosystems, compromising their resilience capabilities. The socio-environmental vulnerability in many coastal areas is aggravated by the inadequate articulation of an integrated territorial management and policies and plans to adapt to climate change. This work was carried out in the municipality of Luanda where erosion, floods and mass movements are an increasingly worrying reality. The research methodology for this study was based on questionnaire surveys and semi-structured interviews, with a focus on qualitative aspects, applied to policy makers, population and NGOs. The results indicate a lack of active participation of citizens, training and taking initiatives in environmental policies and integrated management of the coastal zone, and a lack of environmental information and integrated coordination of institutions with environmental responsibility.

B. Castro · U. M. Azeiteiro (✉)
Department of Biology, University of Aveiro, Aveiro, Portugal
e-mail: ulisses@ua.pt

B. Castro
e-mail: bernardocastro9@gmail.com

W. L. Filho
Manchester Metropolitan University, Manchester, UK
e-mail: walter.leal2@haw-hamburg.de

B. Castro · F. J. P. Caetano
Department of Sciences and Technology, Universidade Aberta, Lisboa, Portugal
e-mail: fernando.caetano@uab.pt

F. J. P. Caetano
Instituto Superior Técnico, Centro de Química Estrutural, Lisboa, Portugal

W. Leal Filho (ed.), *Climate Change Impacts and Adaptation Strategies for Coastal Communities*, Climate Change Management,
https://doi.org/10.1007/978-3-319-70703-7_21

Introduction

Strategies for adaptation to climate change are today considered as a general instrument intrinsic to the development process. Angola is a country whose economy is largely dependent on climate and with more than 30% of its territory under climatic risk. Excessive anthropogenic pressures in coastal zones (CZ) contrast with the strategies of integrated management and use per the criteria of the sustainability ethic. Thereby constituting spaces with greater potential for risks and conflicts. Coastal erosion, uncontrolled occupation, and pollution are a reality whose impacts pervade life, local economies and different coastal and marine ecosystems. The absence of a culture based on the precautionary principle, and the inability to take the socio-environmental risk posed by climate change as a decision-making tool and vigilance in public policy, has given rise to the growing threat to the integrity and ecological and environmental balance in many CZ.

Despite the antiquity of human presence in some coastal areas (CA), dynamics and the complexity of their ecosystems continue to surprise (namely resilience). Disorganized constructions on drainage lines or on steeply sloped areas have imposed significant transformations on the biological or chemical properties and the natural landscape of ecosystems with consequences for the life and patrimony of Humanity. Conflicts and struggles for spaces are a reality, with greater expression, in CA whose ecosystems today have high anthropogenic pressures. Coastal areas mischaracterization processes increasingly disturb the dynamics and natural evolution of their ecosystems.

The scarcity of certain coastal resources is increasingly worrying, having imposed not only some limitations or vulnerabilities, but also the retreat of the regeneration capacity of these spaces. The destruction of the vegetation cover of many coastal space clippings is followed by an illiteracy that is somewhat generalized about the amount of water entering and being lost through evapotranspiration and evaporation in the country, which is essential to know such as the Water balance or the yield of the basins in the coastal areas.

This work intends to deepen the knowledge about the coastal zones in the light of the socio-environmental risks and vulnerabilities arising from Climate change (CC) in the municipality of Luanda (Angola) with a focus on local adaptation based on the identification of local risk conditions and vulnerabilities.

Study Area

Location of the Municipality of Luanda

Luanda is one of the municipalities of the capital of the Republic of Angola in Southern Africa. The municipality is bordered by the Atlantic Ocean and has an area of 113 km^2 according to Law no. 29/11 of 1 September and of Decrees no.

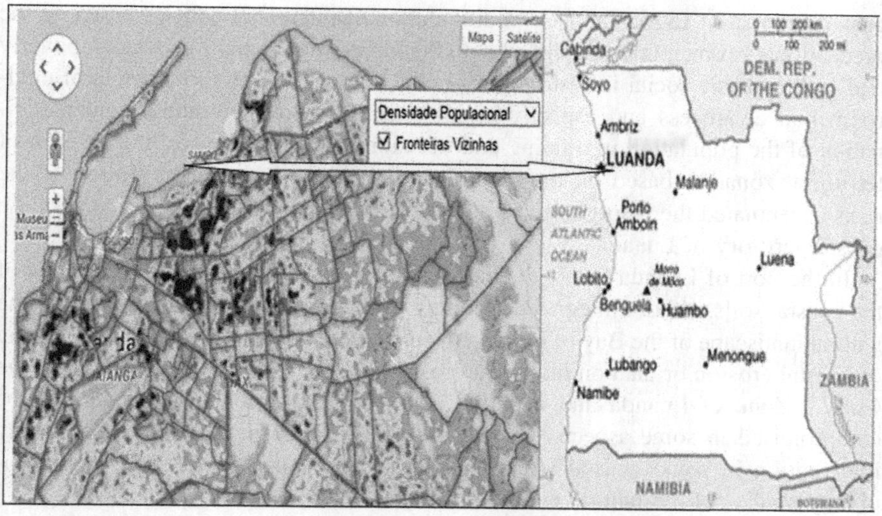

Fig. 1 Luanda Municipality

277/11 and 47/12 which defines the New Political Division and Administrative Organization of Luanda. It lies in the region corresponding to the parallels 8°49′13″ of South Latitude and 13°13′09″ of Longitude East. The figure shows the geographic context of the municipality Fig. 1.

Climate

The climate of Luanda is characterized by two seasons: one hotter and wetter and one colder and drier. The climate is tropical humid, but dry due to the cold current of Benguela marked by a weak rainfall with average of 323 mm annual, irregularly, distributed and, with a coefficient of variation, between the highest of the world (about 40%). The trade winds are constant with average speed between the West and Southwest directions (Trindade 2000). Regarding precipitation according to Lotz-Sisitka and Urquhar (2014) and Urquhart et al. (2014), the information in Angola is not entirely reliable considering that only 20 rainfall stations are in operation. Between 1970 and 2004 there was an increase in surface temperature in Angola between 0.2 °C and 1.0 °C in the coastal zones and North. The rate of air temperature increase in Luanda is 0.2 °C per decade, accumulating between 1.911 and 2005, a total of 1.9 °C (Lotz-Sisitka and Urquhar 2014). From a historical point of view, the constructions in Luanda were made in the margin of the foundations and norms of planning and territorial planning. According to Correia (2012), the colonial installation in Luanda presided high for strategic military reasons with the first plant in Luanda dating to 1621, almost half a century after the founding of the city of Luanda in 1576. The lack of articulation of integrated policies based on the precautionary principle is associated with the increasing population density; poverty and lack of basic social infrastructure; spatial segregation and low levels of environmental awareness and capacity building; the lack of information and participation of the population in training and decision-making processes; the absence of territorial zonation based on the identification, monitoring and prevention of the risks accentuated the disorganized settlements imposing physical alterations in the coastal territory of Luanda.

In the port of Luanda heavy engineering works and other inappropriate uses of the coastal soils without integrated planning of the territory have altered deeply the natural landscape of the Bay of Luanda in recent years. Since there is no monitoring of coastal erosion or data on historical erosion rates in Luanda, the Planning of the Coastal Zone of Luanda and eventual adaptation scenarios for erosion may be compromised in some aspects. One of the observable aspects has been the early intervention in some coastal stretches without being known through multidisciplinary studies of the natural evolution of the coastline.

Flood Characterization in Luanda Floods are a hydrological phenomenon resulting from a set of natural or entropic factors that are defined as the partial or total submersion of an area per rule immersed by trans-shipment or accumulation of water (Garcia 2013). In Luanda, floods have the following characteristics:

(a) rapid floods with short rains due to the geological structure of reduced permeability in some areas that makes it difficult to infiltrate water by increasing the flow of surface runoff
(b) marine floods generated by the combined action of a meteorological storm and tide, locally called Calemas or "higher waves" that have caused coastline gouges in some areas
(c) flooding by rising of the water table after rainy days in some areas.

The main entropic factors observed that introduced changes in the hydrological cycle are

(a) destruction or replacement of the vegetation cover that increases the surface runoff as it reduces the evapotranspiration
(b) construction on or at the drainage lines
(c) lack of cleaning of the river channels; inadequate drainage and disposal of sewage and sanitation infrastructure.

According to Cain (2014), the floods caused by the sub-basins of the Cambamba, Mulenvos/Seco and Cambolombo rivers pose the greatest risk.

Methodology

From the methodological point of view, three questionnaires were used for application to three different sample components: (a) population; (b) NGOs and (c) policy makers. The overall objective of this survey of the three sample components was to understand risk perception and knowledge about the management of the coastal zone of Luanda by surveying perceptions of the population, identifying and describing socio-environmental capacities, risks and vulnerabilities. CC proposing local adaptation and resilience scenarios encompassing the key actors. In the sample component 'Population', the questions relating to the profile of the respondent, perceptions about Climate Changes, local knowledge, attitudes and capacities were essentially established. This component had a sample of 200 individuals for each 6 urban districts of Luanda (a total of 1200 individuals). The ages were distributed as follows: from 18 to 25; 26–35; 36–45; 46–55, and over 56 years without class or gender discrimination.

The survey was launched in the last fortnight of November 2012 with variations for the different urban districts of Luanda according to the authorizations to the petitions made. For this component, the interviews occurred in places of greater population concentration from the informal markets in the outlying districts of the city of Luanda to the formal commercial centers in the city, schools and residences through the interception of passers-by and, in other cases, by prior notification with the Help of the heads of the Residents' Commission. Considering the high illiteracy rate in Angola and the novelty of the subjects on environmental issues, the interviewers sought, in some cases, to translate certain concepts into the language closest to local experiences and knowledge.

For the 'Political Decisions' component of the sample, the key issues were related to institutional capacities, political, legal and scientific tools produced in the light of local sustainable development, issues of land management processes and citizen participation In training and decision-making, and finally, questions about the state of health infrastructure and environmental and public health. The general objective of the survey was to understand the extent to which local public policies influence the emergence of socio-environmental risks and vulnerabilities in Luanda.

Interviews were prearranged and conducted at the interviewee's service sites. In some cases, the interviews were done by email, skype and telephone. Fifteen public institutions were surveyed, of which five are ministries and other national, municipal and communal and urban district directions. For the evaluation of the public policies the following indicators were considered: the pertinence, the opportunity, the effectiveness/efficiency and the impacts. For the 'NGO' component of the sample, it was intended to understand the degree of involvement of civil society in environmental issues, dialogue and closeness between citizens and state institutions and to understand how local policies are assessed by civil society. This component of the survey had four NGOs. These interviews were arranged and took place in service places. The research was developed in accordance with the qualitative foundations; the techniques used were the documentary research through bibliographical consultations and the direct observation that was translated in the contact with the key informants. The participants included members of the Committees of Residents for the six districts of the Municipality of Luanda, Political decision makers such as the Municipal Administrations and directors of various public and private institutions. Ethical aspects were observed such as voluntary participation, consent and confidentiality throughout the process for both the interviews and the questionnaire.

Finally, after collection, the data and all information were subjected to a classification, analysis, treatment and discussion of results.

Results and Discussion

According to DOING BUSINESS (2012) Angola occupies the lowest place in Sub-Saharan Africa with position 178° in the item referring to "employed worker". This is despite having implemented the Strategy First Job law and public policies to promote employment, social security and having ratified the ILO's eight fundamental Conventions.

Knowledge of diseases caused by water contamination are represented in Fig. 2, and it was verified that 67% of 200 interviewees in areas of higher risk are aware of the viral and bacterial origin of diseases caused by contamination of water bodies, while 33% do not know.

Despite the high level of knowledge of diseases caused by contamination of water bodies, CEI/UC (2013) states that the epidemiological picture of Angola in 2013 continued unchanged with the leadership of malaria as the most reported disease with 56%; Acute respiratory diseases (21%), acute diarrhoea with 9%, typhoid 5%, dysentery 4%. According to OPSA and ADRA (2014), health expenditure decreased by 1.3% points from 5.6% of total expenditure in 2013 to only 4.3% in 2014. For the first-time Dengue cases numbered 1241, with the registration of 11 deaths MINSA (2013).

The Ministry of Planning cited by Oliveira (2012) states that malaria is the leading cause of mortality and morbidity in Angola accounting for 35% of the

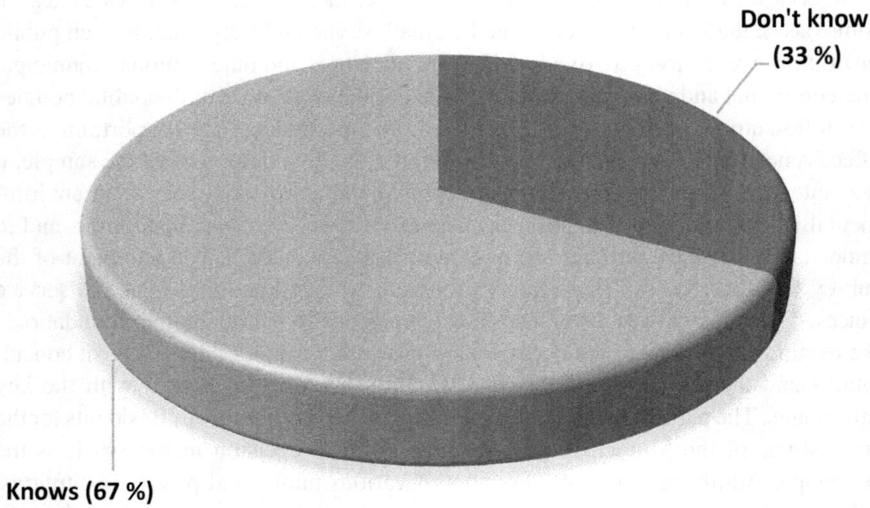

Fig. 2 Knowledge of waterborne/waterborne diseases (n = 1200)

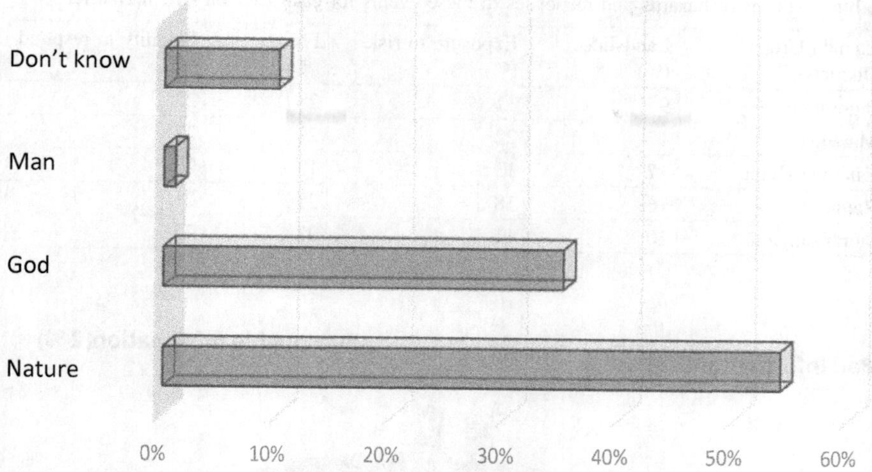

Fig. 3 Events of climate change (n = 1200)

Table 1 Estimated level of ignorance on the effects of the climate changes (values in percentage)

Urban Districts of Luanda	
Maianga	49%
Rangel	39%
Kilamba	36%
Samba	34%
Sambizanga and Ingombota	33%

demand for curative care, 20% for hospital admissions, 40% for prenatal deaths and 25% for postnatal cases. The responses from a sample of 1200 respondents are summarized in Fig. 3.

The debate about Climate Changes in Luanda and in the country in general, is still not a social reality. In public or private information spaces, there are scientific productions and publications on CC. According to Angola's Initial Communication to the United Nations Framework Convention on Climate Change, Angola has a well-established record of warming. Surface temperatures increased between 0.2 °C and 1.0 °C in the 1970 and 2004 at the coastal zones and Northern regions with a range between 1.0 °C and 2.0 °C in the Center and East. The air temperature in Luanda shows a growth rate of 0.2 °C per decade, resulting in a cumulative total of 1.9 °C between 1911 and 2005, and higher increases in the cold season (Lotz-Sisitka and Urquhar 2014).

The high level of ignorance on the effects of the Climate Changes, valued at around 72%, reflects the degree of peripherization of environmental issues in everyday considerations and achievements. In the joint analysis of variables, (a) inability to respond, and (b) exposure to risks and climatic events, the results are distributed in order of vulnerability, high, medium or low according to the following data for the various districts of Luanda (Table 1).

Table 2 Types of hazards and responses to these events for each Luanda Urban District

Luanda Urban District	Landslides (%)	Exposure to risk (%)	Floods (%)	Inability to respond (%)
Ingombota	9	11	28	22
Maianga	17	22	21	27
Kilamba-Kiaxi	17	13	13	23
Rangel	16	18	31	21
Sambizanga	10	11	28	22

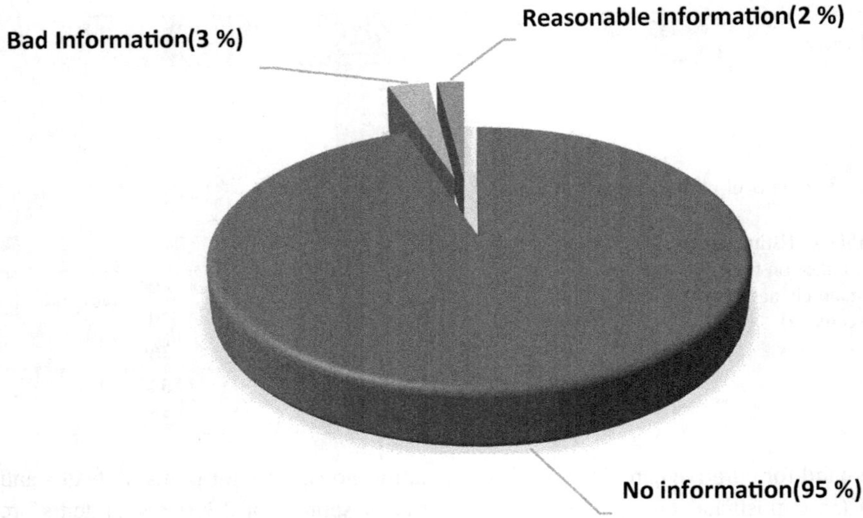

Fig. 4 Adaptation to climate change perceptions (n = 1200)

The different types of hazards and responses to these events by each Luanda Urban District in a sample of 1200 respondents are presented in Table 2.

Figure 4 shows the degree of information on Adaptation to climate change with 95% of 1200 respondents indicating they did not have information, while 3% pointed to very poor information. Only 2% indicated reasonable information about CC Adaptation.

The context of adaptation in Angola is marked by a high degree of scarcity of data and historical or scientific information (where and when). Uncertainties about the projections and the lack of real identification of local capacities or vulnerability are a reality, especially in the agricultural sector.

Adaptation measures essentially require two approaches. Firstly, analysis of risks and vulnerabilities conducted by government offices and secondly, it should be taken into consideration the soil-climatic conditions and practices in land use and occupation, especially those of a traditional nature which can also increase the risks of climate change. The following elements should be included in the structure.

(a) The principle "No Regrets", especially in situations of need and uncertainty, it is a vigilant measure that must embody and influence the whole process of policy planning. The No Regrets principle is regarded as a risk-benefit measure irrespective of whether or not projected extreme weather events occur. The institutional framework for natural disaster risk management in the country is largely monopolized and focused on public accountability. It is based, above all, on reactive adaptation and the margin of integrated readings and systemic understanding of a given territory. This prevents the measurement of the different impacts at national level and the optimization of resources. Therefore, insertion into current adaptation policies of "No Regrets" measures can help to minimize efforts that in a reactive adaptation situation would be burdensome and costly;
(b) The second policy option in adaptation to climate change for the agriculture sector in Angola is the iterative management of risk. It is not enough to plan or integrate the climate focus into policy or policy instruments. As risk and sensitivity or vulnerability factors are associated with the different trajectories and development contexts of the most diverse systems in the country, it is important to adopt permanent evaluation tools. Interactive management is an important measure in adaptation, because it accepts planning not just as a single moment and secondly because it evaluates, permanently, the indicators of verification of ab initio results. Iterative risk management offers methods to deal with uncertainties through a continuous process of evaluation, action, re-evaluation and response. (OCDE 2011) presents the network of interactions, the core of which comprises five stages of climate change adaptation planning: the design and identification of baseline planning elements; assessment of current risks and vulnerabilities; assessment of future risks and vulnerabilities; formulation of the adaptation strategy and continuous adaptation process;
(c) The third axis is the dialogue between modern and traditional epistemologies. In adaptation, it is important to articulate different experiences, practices and knowledge, especially with epistemological dialogues that structure and reinforce each other for resilience. The different policy and regulatory instruments relating to the agriculture and environment sector consulted show a high deficit of dialogue and articulation with traditional knowledge. Adaptation to climate change has its framework for building effectiveness, primarily from the location informed by specific behaviors and knowledge. The lack of local historical capital, could compromise strategies to deal with historical climate variability, in a country without territorial coverage of meteorological information instruments or historical databases, to the ways of conserving seeds or to read seasonal calendars and knowledge of consumption habits or crops and their tolerances may lead to bad adaptation;
(d) The fourth policy option in adaptation is consistency. For OCDE (2011) the articulation and interconnectivity of both political and normative instruments in adaptation is fundamental. It is important that, from the point of view of horizontal coherence, it is ensured that the individual and policy objectives of different local actors are mutually reinforcing. This requires not only the

construction of objectives and the definition of goals to be achieved, but also the coherence of practices at different levels (vertical coherence) that must be complicity with the general commitments of the country. From a temporal point of view, short-term instruments or measures do not contradict long-term policies or commitments;

(e) The fifth principle to be taken into account in adaptation policy scenario is security. The concept of security is welcomed here as a relative of trust. It is important to understand that adaptation is a system of adjustments that provides a certain system of instruments, including behavioural or technical tools that enable it and contribute to its resilience. This is not possible in a troubled socio-land tenure. In Angola, there is no specific land management policy or criteria for the spatial location of certain investments, taking into account the precautionary principle and sustainability.

Possible indices of different types of soil pollution in the country are not known. Land conflicts have imposed a climate of uncertainty and insecurity on farmers and/or peasants. The current statute that imposes the obligation of plans or risk maps in each public administration is an important step to promote caution or minimize losses due to natural disasters in the country. This principle appeals to the community spirit, responsibility and local complicity in adaptive capacity building policies and resilience. Luanda must evolve structural and non-structural measures such as the empowerment of local communities from the promotion of quality education, integrated policies on health to the active participation of the citizens in training processes and public decision-making. Acquisition of technical and monitoring instruments is important. It is advisable to encourage research and development, to establish public-private partnerships in coastal management, institutional strengthening and dialogue between different powers and epistemologies as well as constructions adapted to increasingly uncertain climatic events.

Examples of boundaries and settlement rules along the North Carolina coastline (Freire 2011) or others may promote a healthy environment and sustainability of coastal resources. Construction of light or heavy works should take into account, in addition to other aspects, the historical rate of erosion. As for the knowledge about the socio-environmental risks arising from climate changes, the results are as follows: 949 indicated a value of 0, without information, corresponding to 79% of the sample while 16% is with very poor information, therefore value 1. Only 5% considered to have good information as can be seen in Fig. 5.

Political Decisions Makers

The perceptions about the causes of disordered settlements in Luanda are in general agreement with Bettencourt (2011) in pointing to the lack of political will. Figure 6 helps to perceive the obtained results.

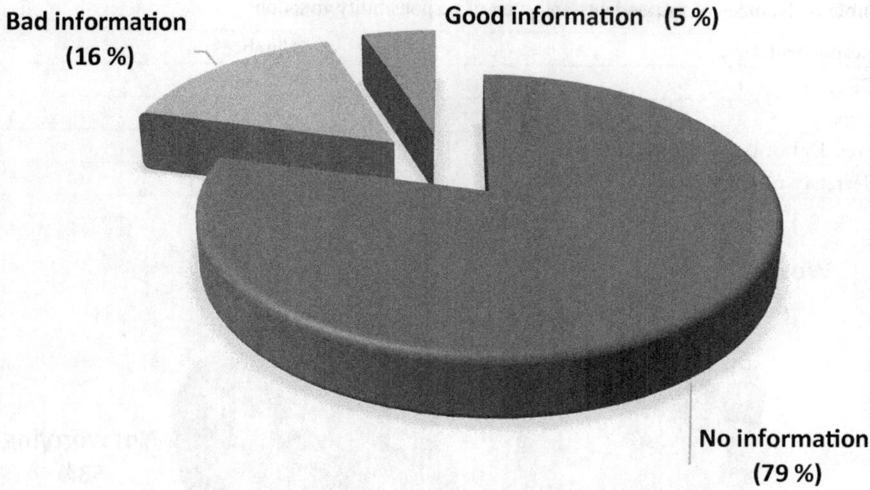

Fig. 5 Socio-environmental risk perceptions (n = 1200)

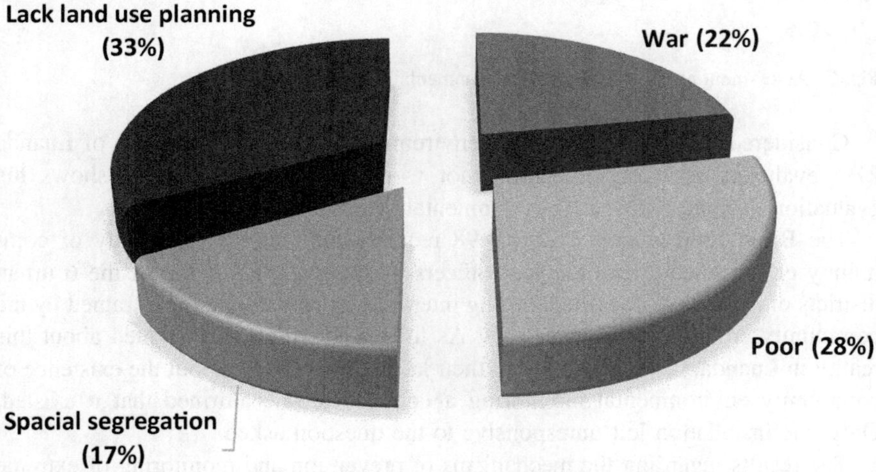

Fig. 6 Perceptions about the cause of disordered settlements (n = 1200)

For Luanda, Bettencourt (2011) points to the lack of political will and bases the statement by listing fifteen instruments or programs that prove the ineffectiveness of policies and instruments of urban management in Luanda since 1943 with the Urbanization Plan approved by the Luanda City Council. As for the other environmental policies, available to the institution, the results point to only four institutions with emergency plans, mainly in cases of occupational accidents.

These four institutions denied that workers had participated in the preparation of such policies. The 15 respondents interviewed are presented in Table 3.

Table 3 Number of respondents and type of responsibility function

Respondents	Number
Director	8
Deputy	4
President of the Board of Directors	1
Secretary of State	2

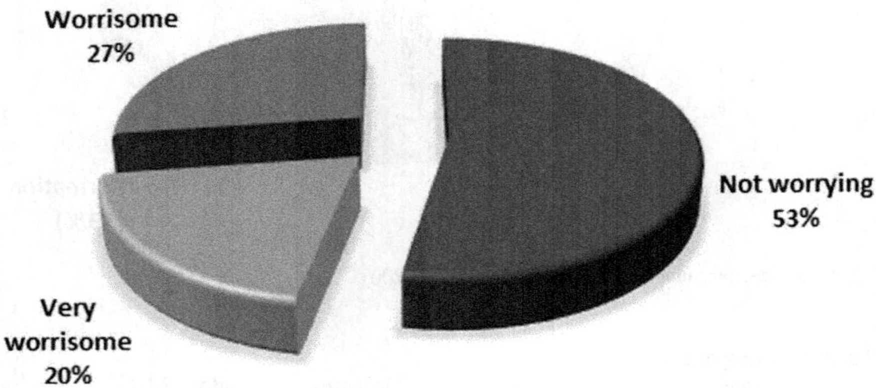

Fig. 7 Assessment attributed to socio-environmental impacts

Considered very highly the socio-environmental impacts in the city of Luanda 27% evaluates as worrying. Those not worried are 53%. Figure 7 shows the evaluation attributed to socio-environmental impacts in Luanda.

The Basic Environmental Law 5/98 requires the creation of a body of community environmental enforcement officers. However, the volume of the 6 urban districts of Luanda was verified, during interviews, that they were not trained by the community environmental inspectors. As fifteen institutions questioned about this reality in Luanda, they demonstrated their lack of knowledge about the existence of community environmental monitoring agents, while one affirmed that it existed. Only one installation left unresponsive to the question asked.

The results regarding the mechanisms of prevention and monitoring of extreme climatic events pointed out that eleven institutions that are unaware of the existence of mechanisms of prevention and environmental monitoring at the level of the State. The respondents stated that they do not exist in the institution.

Seven of the eleven institutions stated that they had a means of warning in cases of fire.

Three said they did not have them in the institutions.

Only one institution claims the existence of instruments for prevention and monitoring of extreme events in Luanda. Of the fifteen institutions, including the Luanda Municipal and Districts Administrations, schools, Civil Protection and Fire Services, Ministry of Construction and others, thirteen are unaware whether or not there is any study on socio-environmental risks and vulnerabilities. To the question of whether there is a risk map in the coastal zone, twelve institutions are unaware of the fact that three have claimed it does not exist.

Regarding the existence of ecological reserve units, eleven institutions said they were unaware of their existence while three said they were not sure. Only one claimed it existed. Regarding the way land is affected along the coastal zone of Luanda, five institutions claim to have disorder in the form of affectation and access to land, three chose the words anarchy and lack of transparency, four affirmed that the process of affectation occurs normally while three have alleged weaknesses in supervisory bodies.

NGOs

Public approaches to environmental issues are in practice non-existent or irrelevant in terms of scope. Table 4 summarizes how environmental issues are addressed in the community.

Regarding the topics most addressed in the urban districts of Luanda, the results are shown in Fig. 8.

It is clear that issues such as environmental monitoring, mitigation and adaptation to CC are not familiar to the communities of Luanda and have no communication regarding them.

As for the forms of participation of the citizens, it is concluded that they are little involved in the implementation of environmental projects. Figure 9 summarizes how communities participate in environmental projects.

Figure 10 shows the degree of knowledge about environmental legislation, and it was verified that in a sample of 1200, 1007 there is no legal instrument on the environment, which in percentage terms represents 83%, whereas 17% claimed to know of some legislation by hearsay. Of the 17% with knowledge about environmental legislation only 2% cited some laws such as the Environmental Law and Environmental License Act. Figure 11 summarizes the results regarding knowledge of environmental legislation in Luanda.

Angola has a total of 54 legal or legislations produced or received concerning, directly or indirectly, environmental issues. It is therefore a considerable piece of legislation, although the following resolutions have not been published: (a) Convention on the Prohibition of the Import of Hazardous Wastes and the

Table 4 Addressing the environmental issues in the community

Dialogue Estate/NGOs	Yes	38%
	No	62%
Dialogue ONGs/Civil Sociaty	Yes	76%
	No	24%
Dialogue ONGs/ONGs	Yes	83%
	No	17%

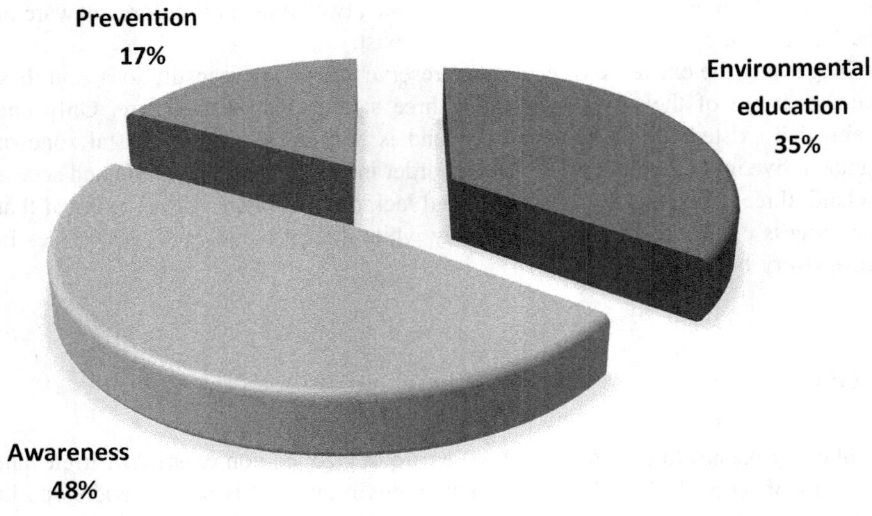

Fig. 8 Thematics that are more addressed by NGOs

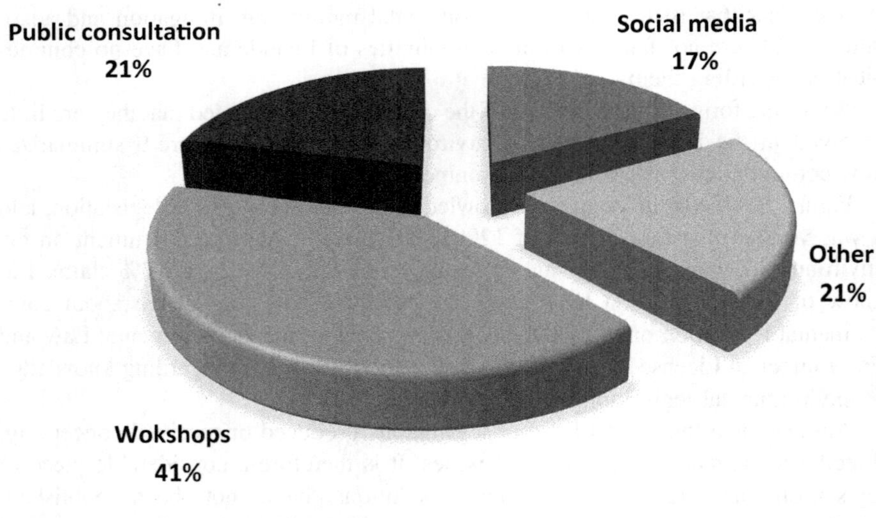

Fig. 9 Forms of community participation

Control of the Transboundary Movement of such Wastes in Africa; (b) the Convention on the Conservation of Nature and Natural Resources in Africa; (c) Convention on the Control of Transboundary Movements of Hazardous Wastes and their Disposal; (d) Convention on Wetlands of International Importance and (e) the main International Conventions on the Prevention of Oil Pollution by Sea. To a large extent, legislation is produced outside the active participation of citizens and socio-cultural foundations, which jeopardises its ability for effectiveness.

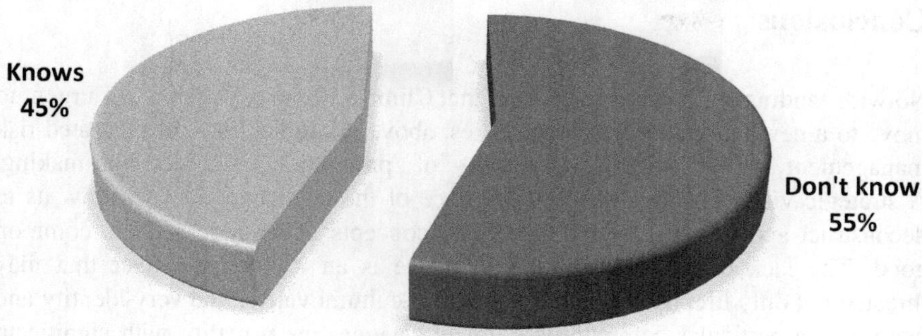

Fig. 10 Knowledge of environmental legislation

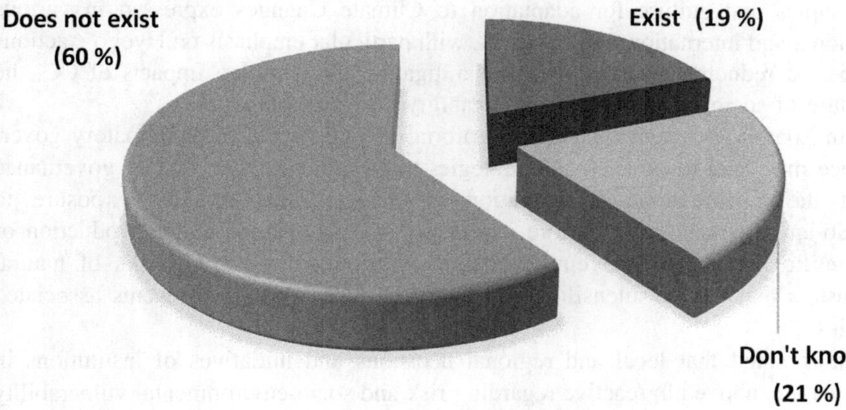

Fig. 11 Information about a programming in the community

National NGOs emerge within a recent historical-political context of Angola marked by democratic openness. The institutionalization of the environment in Angola since 1992 parallels the period of establishment of NGOs. According to the CEI/UC (2013), the initiative of NGOs and civil society is scarce in order to follow the management of public policies and Control of government action. The 2012 Social Report of the CEI/UC (2013) also alleges the lack of political will in the exercise of civic participation.

The question of whether or not there is an ongoing program in the community on environmental issues, the table is as follows: 64% in a sample of 1200 claims there is no on-going program in the community while 23% have no knowledge or information on the implementation of a program with the theme in question. The 13% claim to have some program, but not identified. Figure 11 shows the indexes of information about the existence of a program on environmental issues in the community.

Conclusions

Notwithstanding the complex scenarios that Climate Changes require, it is urgent to move to a new rationality that emphasizes, above all, the culture of integrated risk management as a process instrument of production and decision-making. A strategic vision of governance in the face of the challenges of CC allow us to deconstruct and reformulate the traditional concepts of security and the common good. The lack of environmental governance is an imminent danger that may threaten not only life, natural heritage but also cultural values, the very identity and history of a particular community. Climate Changes are a reality with significant impacts on the natural resources and agriculture of the Angolan people, which is essentially pluvial in a context where droughts have greater expression in Africa. As long as African governments do not translate into social discourse and life the principles and culture for adaptation to Climate Changes expressed in various regional and international instruments, with particular emphasis on Hyogo's actions aimed at reducing disaster risks and mitigating the resulting impacts of CC, the picture of socio-environmental vulnerability may be even worse.

In Luanda, deficient integrated territorial management and participatory governance may lead to policies and strategies that promote better climate governance and the construction of adaptation mechanisms that mitigate exposure to socio-environmental risks. More investment will be needed in the production of scientific knowledge and empowerment of populations on the risks of natural disasters that can be intensified with the occurrence of climate events associated with CC.

It is noted that local and regional decisions and initiatives of institutions in Africa are, above all, reactive regarding risk and socio-environmental vulnerability arising from Climate Changes. Therefore, the difficulties in many sectors of decision making to associate the environmental risk with the socio-political vulnerability that in Africa is very visible. The approach to environmental risk arising from CC is minimised by linking only the environmental risk to the instability of natural biophysical systems. While the influence of physical, biological, or chemical factors on economic systems in man or other living beings, is a structured and complex process that cannot always be monitored by human perception. Little attention is paid to policies considered as vigilant in the management of resources in the framework of sustainability. The low ecological awareness in sectors of Angola's political and social life is well-known.

In Luanda, exposure to risk is associated with poverty, understood as exclusion deprivation and new dynamics in space management. In a situation of poverty, obviously, the alternatives and the options of access to goods and services or structures of power are obstructed, which increases the propensity for vulnerability reducing the capacity of response to the needs of the populations in their daily life. The peripheries of the great metropolis are the picture of this reality. Land occupations are generally not governed by territorial plans and the criteria for locating housing areas are not known due to lack of soil qualification due to their soil and climatic conditions or aptitudes.

The changes mean little or nothing. The big question is the impacts they trigger by undermining the life and heritage of humanity. Policy makers need to understand Climate Changes from local response capacities and the ways in which human societies position themselves in the face of environmental risk. The magnitude of Climate Changes impacts depends on the degree of institutional, socioeconomic and cultural vulnerability of those that are primarily responsible. In some sectors, geopolitical and economic interests continue to disrupt the understanding of Climate Changes by overriding personal and group benefits to human losses and scientific evidence. In African countries, the climate change approach is still timid, although there is recognition of the emergence of partnerships among some universities in the region on CC.

The budgets for adaptation in Climate Changes are still small. The technological risks associated with the methods of production and consumption in Luanda raise some concern. Some of the objectives and priorities of the NAPA, in addition to being disjointed with the general environmental policy orientation of the country, are at great risk of efficiency because of the lack of resources to make it sustainable.

References

Bettencourt, A. C. de. A. (2011). *Qualificação e Reabilitação de Áreas Urbanas Críticas. Os Muceques de Luanda*. 97p. Tese de Mestrado, Universidade Técnica de Lisboa, Faculdade de Arquitectura, Lisboa, Portugal. http://hdl.handle.net/10400.5/3654.

Cain, A. (2014). *Adaptação às Alterações Climáticas e Gestão de Recursos Hídricos nos Assentamentos Costeiros em Angola*. 40p. Development Workshop, Luanda, Angola. Available at: https://pt.slideshare.net/DevelopmentWorkshopAngola/allan-cain-alteracoes-climaticas-e-assentamentos-costeiros-dw-debate-18072014.

Centro de Estudos e Investigação da Universidade Católica de Angola—CEI/UC. (2013). *Relatório Social de Angola 2012*. Luanda, Angola. Accessed in September, 2014, available at: http://www.ceic-ucan.org/wp-content/uploads/2014/04/RELATÓRIO-SOCIAL-DE-ANGOLA-2012.

Correia, M. A. V. A. M. (2012). *O Património do Movimento Moderno em Luanda (1950—1975)*. 226p. Tese de Mestrado, Universidade de São Paulo, Brasil. Não publicado.

Doing Business. (2012). *Fazer Negócios em um Mundo mais Transparente*. 36p. Banco Internacional de Reconstrução e Desenvolvimento e Banco Mundial (2011), Washington, USA. Accessed in November 2014, available at: http://www.doingbusiness.org/Reports.

Freire, L. F. (2011). *Erosão Costeira: Alternativas de Protecção e Proposta de Solução*. 82p. Dissertação de Mestrado, Instituto de Tecnologia para o Desenvolvimento—LACTEC, Instituto de Engenharia do Paraná. Programa de Pós-Graduação em Desenvolvimento de Tecnologia—PRODETEC, Curitiba, Brasil. Available at: http://sistemas.institutoslactec.org.br/mestrado/dissertacoes/arquivos/LuizFreire.pdf.

Garcia, L. J. S. (2013). 2º *Ciclo de Estudos em Sistemas de Informação Geográfica e Ordenamento do Território. Modificação Recentes no Uso do Solo e na Dinâmica Fluvial em Luanda: a bacia hidrográfica da ribeira da Samba*. 69p. Dissertação de Mestrado, Universidade do Porto, Portugal. Available at: https://repositorio-aberto.up.pt/bitstream/10216/68692/2/28096.pdf.

Lotz-Sisitka, H. e Urquhar, P. (2014). *Enquadramento de Co-Produção de Conhecimentos*. 398p. Estudo de Planificação da Sarua sobre Ocorrências de Alterações Climáticas, África do Sul. Available at: http://www.sarua.org/files/SARUA%20Climate%20Change%20Vol1No1%20KCPF%20PORTUGU%C3%8AS.pdf.

Ministério da Saúde—MINSA. (2013). *Relatório Anual das Actividades Desenvolvidas*, Angola. Accessed in October 2014 http://www.minsa.gov.ao/2014/02/MINSA-RELATÓRRIO-ANUAL-2013.pdf.

OECD (2011). Integração da Adaptação às Alterações Climáticas na Cooperação para o Desenvolvimento: Guia para o Desenvolvimento de Políticas, OECD Publishing. Available at: http://dx.doi.org/10.1787/9789264110618-pt.

Oliveira, S. (2012). Olhar a Pobreza em Angola: Causas, Consequências e estratégias para a sua Erradicação. *Ciências Sociais Unisinos* 48(1), 29–40, https://doi.org/10.4013/csu.2012.48.1.04.

OPSA e ADRA. (2014). *Posição do OPSA e da ADRA sobre o OGE 2014*. Luanda, Angola. Accessed in March, 2015, available at http://www.africa21online.com/artigo.php?a=7596&e=Estudos.

Trindade, A. J. P. (2000). *O Fenómeno Urbano na África Subsahariana: o caso de Luanda*. 488p. Tese de Mestrado, Universidade Técnica de Lisboa, Instituto Superior de Ciências Sociais e Políticas, Lisboa, Portugal. (Not published).

Legislation

Diário da República, 1ª Série – Nº 168 –1st September 2011. Available at: http://admbelas.org/attachments/article/1683/lei_2911_de_1_de_setembro.pdf.

Diário da República, 1ª Série – Nº 210 – 31st October. Available at: http://www.scm.gov.ao/diploma_texto.php?diplomaID=110255.

Diário da Repúblia, 1ª Série – Nº 56 – 22nd March. Available at: http://www.lexlink.eu/conteudo/geral/ia-serie/76730/decreto-presidencial-no4712/14793/por-tipo-de-documentolegal.

Diário da República, 1ª Série – Nº 5 – 2nd February 2001. Available at: http://faolex.fao.org/docs/pdf/ang118157.pdf.

Diário da República, 1ª Série – Nº 27 – 19th June 1998. Available at: http://www.vda.pt/xms/files/Newsletters/Flash_VdAtlas_Angola___Novo_Regulamento_sobre_a_Gestao_de_Residuos_em_Angola_-DM_2312849_1-.PDF.

Diário da República, 1ª Série – Nº 114 – 21st June 2010. Available at: http://bibliotecaterra.angonet.org/sites/default/files/ao_desp_29-10_2106_comissao_tecnica_ordenamento_territorial_e_desenvolvimento_economico_e_social_de_luanda.pdf.

A Psychological Model of Climate Change Adaptation: Influence of Resource Loss, Posttraumatic Growth, Norms, and Risk Perception Following Cyclone Winston in Fiji

David N. Sattler, Albert Whippy, James M. Graham and James Johnson

Abstract This chapter examines behavioral intentions to prevent climate change and climate change risk perceptions among people living in coastal communities in Fiji following Cyclone Winston, a natural disaster whose strength was likely increased by climate change. Cyclone Winston was one of the strongest cyclones ever recorded in the Southern Hemisphere and the strongest to make landfall in Fiji. The study extends our psychological climate change risk perception model to examine how posttraumatic growth following Cyclone Winston influenced behavioral intentions to prevent climate change (Sattler and Graham 2017). Posttraumatic growth can occur in response to experiencing a traumatic stressor and involves reflecting on life priorities and what gives live meaning (Calhoun and Tedeschi 2001). We also tested van der Linden's (2015) climate change risk perception model by considering how individual and socio-cultural variables influence climate change risk perception, and extended it to predict behavioral intentions to prevent climate change. For this study, we used behavioral intentions as a proxy for behavioral adaption. The participants, 274 persons (160 men, 114 women) in coastal communities in Fiji (age: $M = 39$, $SD = 14$), completed measures assessing climate change risk perception; knowledge, affect, and social norms concerning climate change; behavioral intention to prevent climate change, and demographics.

D. N. Sattler (✉) · J. M. Graham
Department of Psychology, Western Washington University, Bellingham
WA 98225-9172, USA
e-mail: David.Sattler@wwu.edu

J. M. Graham
e-mail: jim.graham@wwu.edu

A. Whippy
Institute of Applied Science, University of the South Pacific, Suva, Fiji
e-mail: bertwhippy@gmail.com

J. Johnson
School of Social Sciences, University of the South Pacific, Suva, Fiji
e-mail: james.johnson@usp.ac.fj

W. Leal Filho (ed.), *Climate Change Impacts and Adaptation Strategies for Coastal Communities*, Climate Change Management,
https://doi.org/10.1007/978-3-319-70703-7_22

The findings show three pathways to climate change adaptation/behavioral intention to prevent climate change: (1) posttraumatic growth mediates the relationship between resource loss due to the cyclone and education with behavioral intentions to prevent climate change, (2) resource loss due to the cyclone activates social norms concerning climate change action, which in turn leads to behavioral intention to prevent climate change, and (3) climate change risk perception mediates the influence of social norms, knowledge, and affect on behavioral intentions to prevent climate change. The findings support and extend our psychological model and van der Linden's model. Implications of the findings for climate change adaptation and education are discussed.

Introduction

This chapter examines behavioral intentions to prevent climate change and climate change risk perceptions in the wake of Cyclone Winston, one of the strongest storms ever recorded in the Southern Hemisphere and the strongest to make landfall on Fiji. On February 20, 2016, the category 5 storm devastated communities across Fiji (Earth Observatory 2016). Climate change was implicated as a contributor to the storm's strength as a result of increasing ocean temperature. Heat energy in the ocean is a source of fuel for cyclones (also referred to as hurricanes and typhoons) and warmer waters can result in stronger storms with greater destructive potential (National Oceanic and Atmospheric Administration 2013).

Cyclone Winston impacted more than one-third (40%) of Fiji's population (United Nations Development Programme 2016) and damaged or destroyed more than 28,000 houses. Nearly all of the buildings were leveled in several communities. Property damage was approximately US $1.4 billion or approximately one-third (30%) of Fiji's gross domestic product (Asian Development Bank 2016). The cyclone destroyed significant portions of the ecosystem, including crops and coconut trees.

The gravity of the threats imposed by climate change on South Pacific island nations and coastal communities around the world was underscored by Fiji President Jioji Konousi Konrote at the COP23 United Nations Climate Change Conference. President Konrote stated, "…Fiji intends to do whatever is in its power to persuade the global community about the root cause of extreme weather events such as Winston that are causing pain for our people and other vulnerable nations around the world… We must get the community of nations to continue reducing carbon emissions. This is a fight we must win. Our entire way of life is at stake" (Armbruster 2017).

Climate Change Risk Perception Models

Few studies have examined in detail how psychological reactions to surviving a natural disaster whose strength is associated with climate change influence climate change adaptation, or how individual and social-cultural variables influence climate change adaptation following disaster exposure. These are significant limitations. Natural disasters threaten lives and property and have a profound influence on quality of life, health and safety, and individual and societal functioning. As a result of surviving a disaster, people may gain new insight into the vulnerability of their community and the immense challenges to recovery (Sattler et al. 2000). Research is needed to examine how such experiences influence climate change perceptions and adaptation. The present study is among the first to examine how coping and mental health responses as well as social-cultural variables influence climate change adaptation and risk perceptions following exposure to a catastrophic natural disaster whose strength was associated with climate change.

We examined behavioral intentions to prevent climate change and climate change risk perceptions in coastal communities that experienced significant damage as a result of Cyclone Winston. The project was guided by our climate change risk perception model that details how psychological reactions to a catastrophic cyclone are associated with climate change risk perceptions (Sattler and Graham 2017; Sattler et al. 2016), and van der Linden's (2015) comprehensive climate change risk model, which considers individual and social-cultural influences.

A psychological climate change risk perception model. We developed a psychological climate change risk perception model based on our studies examining the associations between personal experience with a cyclone, mental health, and climate change risk perceptions in the Philippines and Fiji (Sattler and Graham 2017; Sattler et al. 2016). These studies were among the first to examine in detail the relationships between reactions to a traumatic event and climate change risk perceptions. Survivors completed measures assessing resource loss due to the cyclone, posttraumatic stress symptoms, coping, social support, posttraumatic growth, and climate change risk perceptions. The resource loss measure assessed loss of object resources (e.g., home, possessions), condition resources (e.g., employment), energy resources (e.g., time for adequate sleep), and personal characteristics (e.g., sense of optimism; Hobfoll 2012). The posttraumatic stress measure assessed symptoms associated with posttraumatic stress, which may occur in response to a life threatening traumatic event wherein the individual experiences flashbacks about the event, nightmares and severe anxiety, and intrusive thoughts about the event (National Center for PTSD 2017). Posttraumatic growth can occur in response to experiencing a traumatic stressor and involves the individual reflecting on life priorities and what gives life meaning (Calhoun and Tedeschi 2001).

Our model was developed with data we collected after Super Typhoon Haiyan in the Philippines and confirmed with data we collected after Cyclone Winston in Fiji.

The model shows that resource loss as a result of a cyclone contributes to posttraumatic stress symptoms, which in turn activates coping, which then activates posttraumatic growth, and which then influences climate change risk perceptions. Posttraumatic growth mediated the influence of these variables on climate change risk perception.

The model suggests that climate change risk perceptions are greatest when people experience loss due to an environmental catastrophe associated with climate change, and then reflect on the meaning of life and their values. This finding underscores the importance of resource loss, mental health functioning, and self-reflection in influencing climate change risk perceptions.

A comprehensive climate change risk perception model. van der Linden's (2015) comprehensive climate change risk perception model considers how knowledge, experiences, and socio-cultural factors influence climate change risk perception. *The cognitive component* includes knowledge about the causes of climate change, the impact of climate change on the environment and humans, and actions that can reduce climate change. More knowledge is associated with higher risk perceptions. *The experiential processes component* includes affective or emotional responses to and personal experiences with climate change, such as personal loss and distress. More negative emotional responses and higher levels of loss and distress are associated with higher risk perceptions. *The socio-cultural component* includes broad value orientations and social norms; both reflect cultural influences. Three broad value orientations are identified: biospheric values concern being sensitive to and attending to the environment, egoistic values focus on increasing outcomes for the individual, and socio-altruistic values focus on being sensitive to and considering other people. These values are considered primary guiding principles that influence behavior (Stern 2000; van der Linden 2015), with biospheric and socio-altruistic values being positively associated with climate change risk perception. Descriptive and prescriptive social norms influence risk perceptions. Descriptive norms describe other people's actions and prescriptive norms suggest behavioral expectations that an individual should perform. Social norms to take action against climate change are associated with higher risk perceptions. Finally, the model identifies a *socio-demographic component* which includes gender, age, and education. The model considers these as control variables rather than as directly associated with risk perception.

van der Linden (2015) found support for the model with a sample of residents in the United Kingdom. Gender, political party, knowledge, social norms, value orientation, affect, and personal experience predicted climate change risk perceptions in the expected directions. The experiential and socio-cultural components accounted for more risk perception variance than the cognitive and socio-demographic components. van der Linden suggests that an implication of these findings is that risk communication may be more effective if it includes information addressing the three main components in the model: knowledge (causes, impact, actions to reduce climate change), experiential processes, and socio-cultural values and norms.

Present Study: Climate Change Adaptation/Behavioral Intentions to Prevent Climate Change

In the present study, we applied our psychological climate change risk perception model and van der Linden's climate change risk perception model to examine behavioral intentions to prevent climate change, and to address questions raised in prior studies. For this study, we consider behavioral intentions to be a proxy for behavioral adaption.

We addressed several questions raised by van der Linden (2015). Because participants in that study had limited experience with large scale weather events or disasters associated with climate change, the first question was "Will the climate change risk perception model hold for people who experienced a large scale catastrophic disaster associated with climate change?" The second question was "Will the model generalize to a location and culture other than the United Kingdom?" The third question was "Would the factors in the model predict both climate change risk perceptions as well as behavioral intentions to prevent climate change?"

We also considered our psychological climate change risk perception model which shows posttraumatic growth mediates the relationships between resource loss, posttraumatic stress symptoms, coping, and social support and climate change risk perceptions for people with living in communities severely impacted by a cyclone. Because we previously found that posttraumatic growth mediate these relationships, we only included the resource loss and posttraumatic growth components of our model in the present study. This allowed us to integrate our model with van der Linden's model more simply. We included resource loss because it is a measure of the degree to which the disaster impacted individuals, and as such, may influence responses and perceptions.

In order to extend these models, we developed a behavioral intentions scale to assess actions relevant to people living in small, rural coastal villages in Fiji. We developed the scale to assess behaviors that have the potential to reduce or prevent climate change (based in part on Stern 2010). These behaviors include changing household consumer behaviors; supporting sustainable practices, goals, and policies; and engaging in environmental activism.

Because the project was exploratory, we did not have specific hypotheses.

Method

Participants

Fifteen months after Cyclone Winston made landfall, 274 persons (160 men, 114 women) in coastal communities in Fiji completed the survey (age: $M = 39$, $SD = 14$, range: 18–79 years). Nearly half (46%) had less than a secondary school education, about one-quarter completed secondary school (22%), and about one-third (32%) had some college or a college degree. They lived within 2 km of the coast. Most individuals agreed to participate; the response rate was 92%.

Assessment Instruments

A cover letter introduced the study, presented informed consent information, and indicated responses were anonymous. Table 1 shows the means, standard deviations, and Cronbach alpha reliabilities for the instruments. Participants completed the instruments in the following order.

Climate change risk perceptions and affect. Five items assessed the degree to which participants believed they would experience negative consequences as a result of climate change and their level of concern. The items were adapted by van der Linden (2015). An example is "In your lifetime and in your judgment, how likely are you to experience serious threats to your health or overall well-being as a result of climate change?" We summed the items to create a total score, and higher scores indicate stronger perceived risk as a result of climate change.

Two items assessed affect concerning climate change, based on van der Linden (2015). An example is "Do you think climate change is something that is unpleasant?" Higher scores indicate more negative affect toward climate change. Participants used a 5-point scale (from 1 = not at all to 5 = very much) to indicate their answers.

Table 1 Correlations and descriptive statistics (N = 274)

Variable	1	2	3	4	5	6	7	8	9		
(1) Education	–										
(2) Personal Loss	0.02										
(3) Post-Traumatic Growth	0.31	0.19									
(4) Values Orientation	0.17	0.18	0.42								
(5) Social Norms	0.05	0.20	0.12	0.28							
(6) Knowledge	0.36	0.09	0.50	0.35	0.029						
(7) Holistic Affect	−0.10	0.09	−0.13	0.16	0.32	0.18					
(8) Risk Perception	0.09	0.28	0.17	0.32	0.37	0.31	0.51			–	
(9) Behavioral Intent	0.20	0.26	0.22	0.25	0.42	0.31	0.20	0.39			
Mean	2.85	4.59	3.86	5.11	3.70	0.64	4.17	4.25	4.16		
Standard Deviation	1.10	0.61	0.73	1.02	0.80	0.16	0.93	0.72	0.68		
Cronbach's Alpha	–	−0.68	0.91	0.90	0.77	0.82	0.64	0.73	0.86		

Note Correlations greater than 0.21 are statistically significant at the $p < 0.001$ level, greater than 0.16 at the $p < 0.01$ level, and greater than 0.12 at the $p < 0.05$ level

Personal experience with extreme weather (Cyclone Winston). Four items developed by the first author assessed hardship and loss created by Cyclone Winston. An example is "How much hardship did Cyclone Winston create for you." Participants used a 5-point scale (from 1 = not at all to 5 = very much) to indicate their answers. Higher scores indicate greater experiences of hardship and loss.

Knowledge about climate change. We used van der Linden's items to assess knowledge and followed the procedure to score each item as correct or incorrect and sum the items. Ten items assessed *knowledge concerning the causes of climate change*. Participants used a 4-point scale (major cause, minor cause, does not contribute, and do not know) to indicate their answers. Ten items assessed *knowledge concerning the consequences and impacts of climate change*. Participants used a 4-point scale (likely to decrease, no change, likely to increase, do not know). Ten items assessed *knowledge concerning actions to reduce climate change*. Participants used a 4-point scale (reduce it a lot, reduce it a little, not going to reduce it, do not know).

Social norms concerning climate change. Seven items developed by van der Linden (2015) assessed social norms. An example is "Most people who are important to me are doing something to reduce the risk of climate change." Participants used a 5-point scale (from 1 = not at all to 5 = very much) to indicate their answers. Higher scores indicate social norms more supportive of climate change action.

Broad value orientations. Eight items developed by van der Linden (2015) assessed biospheric and socio-altruistic values. Participants used a 6-point scale (1 = very much opposed to 6 = very important) to indicate how well each item matched their values. An example is "Respecting the Earth and living in harmony with other species." Higher scores indicate more biospheric and socio-altruistic values.

Posttraumatic growth. The 21-item Posttraumatic Growth Inventory assessed degree of growth (Calhoun and Tedeschi 2001). An example is "Priorities about what is important in my life." Participants used a 7-point scale (1 = *great decrease* to 4 = *no change* to 7 = *great increase*). Higher scores indicate a greater degree of posttraumatic growth.

Behavioral intentions to prevent climate change. Twelve items written by the first and fourth authors assessed the degree to which participants would be willing to take action regarding climate change. Examples include "Participate in a march to raise awareness about climate change" and "Use less energy at home to help stop climate change." Participants used a 5-point scale (from 1 = not at all to 5 = very much). Higher scores indicate greater willingness to take action to prevent climate change.

Demographics. Three items asked for gender, age, and education. Participants checked their choices or wrote in a number to indicate their answers.

Procedure

The study was approved by the Human Participants Research Committee at Western Washington University and the Institutional Review Board at the University of the South Pacific, and followed the American Psychological Association ethical guidelines. Prior to administering the survey, the first author trained 10 advanced university students in questionnaire administration. We administered the questionnaire in five villages that experienced extensive damage as a result of Cyclone Winston. We approached people in their homes. Participation was voluntary and no inducements were offered. It took about 20 min to complete the survey.

Results

Model Building Data Analytic Plan and Procedures

We used a model building approach to create and test a path model describing predictors of behavioral intentions to prevent climate change and climate change risk perception. During model development, determining which paths should be removed based on a single sample may cause it to be unduly influenced by sampling error. To address this problem, we randomly divided the data and used one sample to create the model (the exploratory sample, $N = 146$) and the second sample to confirm the model and test its generalizability (the confirmatory sample, $N = 128$). We used Amos version 24 (Arbuckle 2016) to conduct the path analyses with maximum likelihood estimates.

Several fit indices assessed model fit: chi square test, Standardized Root Mean Residual (SRMR), Comparative Fit Index (CFI), and Root Mean Square Error of Approximation (RMSEA). The chi square test examines the null hypothesis that the data did not fit the hypothesized model; non-significant chi square values indicate acceptable fit. SRMR is a measure of absolute fit, with values less than 0.08 indicating a good fit (Hu and Bentler 1999). We used CFI as a measure of incremental fit; it generally has a maximum value of 1, with larger numbers indicating better fit. A CFI of 0.95 or greater is considered a good fit (Hu and Bentler 1999). RMSEA is a measure of fit that rewards parsimonious models. It has a minimum value of 0, with lower numbers indicating better fit; cut off points values of 0.01, 0.05, and 0.08 indicate excellent, good, and poor fit (MacCallum et al. 1996).

We separated the variables into four levels, based on the theoretically-anticipated causal direction of the variables:

Level 1: Education and personal experiences of loss
Level 2: Posttraumatic growth, value orientation, social norms, knowledge, and holistic affect

Level 3: Climate change risk perception
Level 4: Behavioral intentions to take action to reduce climate change

We first created a fully saturated path model wherein every variable was allowed to relate to every other variable. Variables were allowed to correlate to variables at the same level and predicted variables at subsequent levels.

The fully saturated model resulted in a "just identified" model that perfectly replicated the sample. We ran this model, examined the correlations and regression paths, and removed those with the lowest critical ratio (those that were least statistically significant). We re-ran the new, simplified model, examined the correlations and regression paths, and again removed those with the lowest critical ratios. We continued this process until all of the remaining paths were significant at the $p < 0.05$ level, and paid careful attention to the fit indices. Removing paths results in a worsening of fit compared to the original fully saturated model; our aim was to create as simple a model as possible while maintaining acceptable fit.

Early in this process, we removed the paths from value orientation to risk perception and from value orientation to behavioral intent because they were not statistically significant. Therefore, we removed value orientation from the model because did not predict any primary variables of interest. Removing value orientation does not suggest that it is not a useful predictor. In fact, the correlations between value orientation and risk perception ($r = 0.32$) and behavioral intention ($r = 0.25$) suggest that values do predict outcomes. The removal of value orientation here is the result that, in the context of the other predictors, it does not result in additional explanatory power. The same process with a different sample might result in the retention of value orientation.

Predicting Climate Change Risk Perceptions and Behavioral Intentions: A Path Model

Figure 1 shows the final path model. The fit of this model in the exploratory sample was outstanding, as shown by the statistically non-significant chi square, X^2 (12) = 13.66, $p = 0.323$. The other fit indices also show good overall fit of the model: SRMR = 0.051, CFI = 0.991, RMSEA = 0.031.

To account for the possibility that sampling error might influence the model, we tested the fit of the final model with the confirmatory sample. Predictably, although the fit of the model was not as strong in the confirmatory sample as in the exploratory sample, the fit of the confirmatory sample was satisfactory. While the chi square was significant, X^2 (12) = 21.63, $p = 0.05$, it was only just so, suggesting marginal fit. All other fit indices were within the acceptable ranges: SRMR = 0.075, CFI = 0.961, RMSEA = 0.078. This suggests that the model generalizes reasonably well to the other sample.

Predicting behavioral intentions to prevent climate change. Three main predictors explained 18% and 32% of the variance in behavioral intention to take

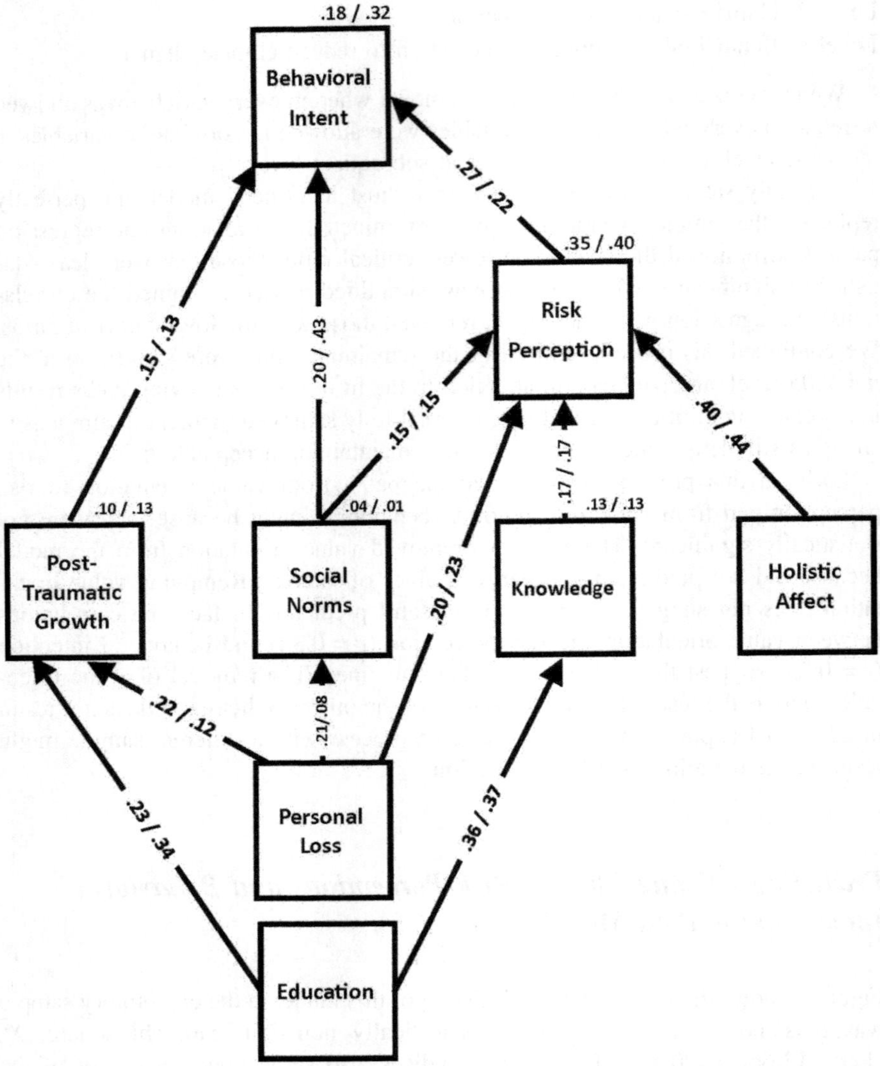

Fig. 1 Standardized path coefficients for the final path model. Coefficients before the "/" are for the exploratory sample (N = 146). Coefficients following the "/" are from the confirmatory sample (N = 128). The offset values on endogenous variables are R^2 effect-sizes. While not shown here, social norms, knowledge, and holistic affect correlated with one another, and posttraumatic growth correlated with knowledge and holistic affect

action to prevent climate change for the exploratory and confirmatory samples, respectively. Social norms was a main predictor. Figure 1 shows that participants who reported being surrounded by others who are committed to taking action to prevent climate change were more likely to express intent to take action themselves. This finding may reflect a potential "snowball effect" for climate change action. As more individuals commit to taking action against climate change, the more normative such action becomes. Likewise, as normative action increases, the more individuals commit to action.

Figure 1 also shows that climate change risk perceptions predicted behavioral intent to prevent climate change. Higher levels of climate change risk perceptions were associated with greater behavioral intention to reduce climate change. This finding suggests a possible extension of van der Linden's (2015) climate change risk perception model. Risk perception appears to mediate the effect of most climate change risk perception predictors on behavioral intention to reduce climate change. Future research might further extend van der Linden's (2015) climate change risk perception model by using risk perception to predict behavioral intent and, ultimately, behavioral action.

Finally, Fig. 1 shows that posttraumatic growth played a role in predicting behavioral intent. In line with our previous study (Sattler et al. 2016), loss due to Cyclone Winston was associated with higher levels of posttraumatic growth. However, unlike our previous study, posttraumatic growth did not predict climate change risk perceptions in the context of the other predictors. Here, posttraumatic growth predicted behavioral intentions to prevent climate change. As such, the reprioritization of values associated with posttraumatic growth mediated the effect of loss on behavioral intentions. The extent to which Cyclone Winston survivors were able to use their traumatic experiences due to the storm and re-evaluate their priorities predicted their intention to take action concerning climate change. Finally, level of education predicted posttraumatic growth, with higher levels of education being associated with more posttraumatic growth.

Additional analysis: Posttraumatic growth, social norms, and behavioral intentions to reduce climate change. To further examine the role of posttraumatic growth in predicting behavioral intent, we ran a separate analysis on the entire sample using Hayes's (2012) PROCESS macro for SPSS, predicting behavioral intent with social norms, posttraumatic growth, and the interaction between the two. The predictors explained 25% of the variance in behavioral intention, $F(3, 270) = 30.63$, $p < 0.001$. All predictors were significant in the expected directions. Social norms and posttraumatic growth were positively associated with high levels of behavioral intention. The interaction suggested that higher levels of social norms weaken the relation between posttraumatic growth and behavioral intent, $\beta = -0.19$, $p < 0.001$. For individuals whose social norms were less supportive of climate change action, posttraumatic growth was a stronger predictor of behavioral intent. For individuals with stronger social norms, posttraumatic growth was relatively unrelated to behavioral intent. These findings suggest that posttraumatic growth is a particularly important pathway to behavioral intentions for individuals whose social norms do not already support climate change action. For individuals who have a strong social context supporting action, posttraumatic growth is less important, and social norms might better determine behavioral intentions.

Predicting climate change risk perceptions. In general, the path model for our data supports van der Linden's (2015) climate change risk perception model. Experiencing loss due to Cyclone Winston, knowledge about climate change, affect concerning climate change, and social norms concerning climate change actions explained 35% and 40% of the variance in climate change risk perceptions in the exploratory and confirmatory samples, respectively. Figure 1 shows that individuals who experienced greater loss due to Cyclone Winston reported higher levels of climate change risk perceptions. While level of education did not directly predict climate change risk perception, it was associated with greater knowledge of the causes of climate change that, in turn, were associated with higher levels of climate change risk perception. Thus, knowledge about climate change mediated the effect of education on climate change risk perception. The findings also show that individuals with negative affect about climate change and who reported social norms consistent with behavioral intention to prevent climate change reported higher levels of climate change risk perceptions.

Discussion

The findings are the first we know of to illustrate the relationships between posttraumatic growth, climate change risk perceptions, and behavioral intentions to prevent climate change following exposure to a catastrophic cyclone in the South Pacific. The findings show three pathways to behavioral intention to prevent climate change (see Fig. 1):

Pathway 1: Posttraumatic growth mediates the relationship between both resource loss due to the cyclone and education with behavioral intentions to prevent climate change. Higher levels of resource loss and education each independently lead to posttraumatic growth, which in turn leads to behavioral intention to prevent climate change.

Pathway 2: Resource loss due to the cyclone activates social norms concerning climate change action, which in turn leads to behavioral intention to prevent climate change.

Pathway 3: Climate change risk perception mediates the influence of social norms, knowledge, and affect on behavioral intentions to prevent climate change. These findings support and extend van der Linden's (2015) model by showing they predict both climate change risk perceptions as well as behavioral intentions to prevent climate change.

The finding that posttraumatic growth influences behavioral intention to prevent climate change extends our psychological climate change risk perception model to behavioral intentions. Why might posttraumatic growth play a key role in behavioral adaptation and risk perception? According to theory and research, coping with a traumatic event can lead people to experience posttraumatic growth, wherein they reflect on their values, life meaning, and life priorities. This process of reflection may lead people to experience positive life change and gain new or heightened awareness and appreciation of life, of relationships with others, and an enhanced sense of self-efficacy (Groleau et al. 2012; Linley and Joseph 2004; Sattler et al. 2014). Posttraumatic growth has been documented among survivors of large scale natural disasters (Sattler et al. 2006, 2014). The heightened sense of self-efficacy may increase behavioral intentions to take action to mitigate a threat, such as one posed by a cyclone whose strength can be magnified by climate change. Engagement, a positive affective motivational state that can develop in response to experiencing a traumatic event or stressful event, may encourage self-efficacy, self-esteem, and feelings of control (cf. Hobfoll 2012). These processes may explain, in part, the relationship of posttraumatic growth and behavioral intention to prevent climate change. To examine this possibility, future research could examine in more detail self-efficacy and feelings of engagement, the activities that may have promoted them, and their relationship to behavioral intentions to prevent climate change. Research also might compare and contrast the relationship among these variables with those who have little or no experience with climate change related disasters and those who have had significant exposure.

Why might social norms (1) mediate the relationship between loss of resources and behavioral intention to prevent climate change, and (2) directly influence climate change risk perceptions? We suspect this finding may reflect the effectiveness of the informational campaigns to raise awareness about climate change and the behaviors that contribute to it. In the wake of Cyclone Winston, campaigns in Fiji presented information about the threats imposed by climate change and the behaviors that contribute to it. These campaigns, as well as other experiences following the cyclone such as discussions with friends and neighbors, may have promoted the development of or reinforced existing social norms—expectations about appropriate behavior—concerning actions relevant to climate change. For example, the 350pacific.org organization promoted the Pacific Climate Warrior campaign to educate Fijians and people in other countries about how their own decisions concerning carbon dioxide emission contributes to climate change and thereby adversely affects countries in the South Pacific and around the world. Newspapers and television programs regularly published climate change stories to highlight each citizen's role in preventing it (e.g., Delaiba Tiki 2017; Panapasa 2017). Schools incorporated climate change education at all grade levels (Ministry of Foreign Affairs Ministry of Foreign Affairs, Fiji 2017). Social norms were likely salient in the communities sampled in this study, given that they were small, rural, and located along the coast in areas greatly impacted by the storm. Social norms that prescribe expected behavior, rather than those that describe behavior, may be more effective in promoting desired behavior (e.g., Cialdini et al. 2006).

The findings extend van der Linden's (2015) model by showing that the cognitive, socio-cultural, experiential, and affective components influence behavioral intentions to prevent climate change. However, whereas van der Linden found that that the experiential and socio-cultural components were most influential in climate change risk perception, we found that the experiential (viz., resource loss due to the cyclone) and affect components were most influential for behavioral intentions. It is likely that this difference reflects degree of disaster experience between the two samples and outcome. The participants in our study were living in communities severely damaged by the cyclone, had to rebuild homes and lives, and were exposed daily to reminders of the devastation in their community. Participants in van der Linden (2015) experienced a lesser degree of damage. These findings also support, in part, a model of antecedents of climate change behavior developed by Reser et al. (2012) in Australia. The model generally states: (1) beliefs about climate change influence concern for climate change and climate change risk perceptions, (2) concern influences feelings of self-efficacy, which then influences distress about climate change, (3) distress is influenced by beliefs about climate change, concern, and risk perceptions, and (4) distress leads to feelings of responsibility, which leads to adaption and adaptive behaviors. We also note that the finding that personal experience influences judgement and behavioral intentions is consistent with the availability heuristic, which predicts that judgments are influenced, in part, by the degree to which thoughts come to mind. Risk perceptions and behavioral intentions may be stronger among people who have personal and tangible negative experience with climate change related weather events (Viscusi and Zeckhauser 2015).

We note a few limitations to the study. Because we were unable to use a pure random sampling strategy, the findings may not generalize to all persons affected by the storm or in other regions. We do not know the interrater reliability across those who administered the survey to participants.

Conclusion

This study is among the first to examine in detail how resource loss and psychological reactions such as posttraumatic growth influence climate change behavioral intentions in coastal communities in the wake of a catastrophic natural disaster whose strength was associated with climate change. The findings show that posttraumatic growth—engaging in a process of self-reflection concerning life priorities and what gives life meaning—led to behavioral intentions to prevent climate change, or adaptation. Higher levels of resource loss as a result of the cyclone and level of education each independently led to posttraumatic growth. One implication of this finding is that focusing people on what is important in their lives and enhancing a sense of self-efficacy that they can may a difference by taking action to reduce threats may motivate behavior change. The findings also show that losing valued resources activates social norms concerning climate change action, which in turn led to behavioral intention to prevent climate change. These findings support and extend our psychological model of climate change risk perceptions. The findings also extend van der Linden's (2015) model by showing climate change risk perception mediates the influence of social norms, knowledge, and affect on behavioral intentions to prevent climate change.

Taken together, the findings provide direction to increase the effectiveness of educational efforts designed to promote behaviors that mitigate climate change.

Educational efforts could underscore the connection between disasters such as hurricanes with climate change and the threats they create to life goals, values, and property, and present clear and tangible actions people can take to reduce and minimize behaviors that contribute carbon dioxide to the atmosphere (viz., minimize their carbon footprint). When people have not experienced the consequences of climate change (e.g., life threat, loss, or engaged in the process of posttraumatic growth) and/or are not aware of how the consequences affect their lives, then educational campaigns could present information in vivid ways about how climate change impacts people. Vivid information influences behavior change to a greater degree than abstract or statistical information (cf. Sattler et al. 1995). One approach would be to share the experiences of survivors of hurricanes associated with climate change and of others whose lives were negatively impacted by climate change (cf. Blennow et al. 2012; Taylor et al. 2014). Research is needed to develop such educational campaigns and assess their effectiveness.

Acknowledgements This project was funded by a grant to the first author from the Office of Research and Sponsored Programs and funding from the Department of Psychology, Western Washington University. We thank Larry Symons, Mehnaaz Sattler, Epi Druavesi, Torika Korocowiri, Semisi Leweni, Seremaia Qereqeretabua, Sekove Ravualala, Victor Salele, Timoci Tuibua, Aisake Vatiliai, and Laraini Waucu for their assistance. We are especially grateful to the participants and communities in Fiji for participating in this project.

References

Arbuckle, J. L. (2016). *Amos (Version 24.0)* [Computer Program]. Chicago: IBM SPSS.

Armbruster, S. (2017, February 20). A year after Cyclone Winston, Fiji calls for global action on climate change. *Special Broadcasting Corporation*. Retrieved from http://www.sbs.com.au/news/article/2017/02/20/year-after-cyclone-winston-fiji-calls-global-action-climate-change.

Asian Development Bank. (2016). ADB provides $50 million for Fiji cyclone relief. Retrieved from https://www.adb.org/news/adb-provides-50-million-fiji-cyclone-relief.

Blennow, K., Persson, J., Tome, M., & Hanewinkel, M. (2012). Climate change: Believing and seeing implies adapting. *PLoS ONE, 7*(11), e50182. doi:10.1371/journal.pone.0050182.

Calhoun, L. G., & Tedeschi, R. G. (2001). Posttraumatic growth: The positive lessons of loss. In R. A. Neimeyer (Ed.), *Meaning reconstruction and the experience of loss* (pp. 157–172). Washington, DC: American Psychological Association. doi:10.1037/10397-008.

Cialdini, R., B., Demaine, L. J., Sagarin, B. J., Barrett, D. W., Rhoads, K., & Winter, P. L. (2006). Managing social norms for persuasive impact. *Social Influence, 1*(3-15). doi:10.1080/15534510500181459.

Delaiba Tiki, N. (2017, May 12). Let us back climate change, ocean campaign. *Fiji Sun Online*. Retrieved from http://fijisun.com.fj/2017/05/12/let-us-back-climate-change-ocean-campaign.

Earth Observatory. (2016). Tropical Cyclone Winston slams Fiji. Retrieved from https://earthobservatory.nasa.gov/NaturalHazards/view.php?id=87562.

Groleau, J. M., Calhoun, L. G., Cann, A., & Tedeschi, R. G. (2012). The role of centrality of events in posttraumatic distress and posttraumatic growth. *Psychological Trauma: Theory, Research, and Policy, 5*(5), 477–483. doi:10.1037/a0028809.

Hayes, A. F. (2012). *PROCESS: A versatile computational tool for observed variable mediation, moderation, and conditional process modeling* [White paper]. Retrieved from http://www.afhayes.com/public/process2012.pdf.

Hobfoll, S. E. (2012). Conservation of resources and disaster in cultural context: The caravans and passageways for resources. *Psychiatry, 73*, 227–233. doi:10.1521/psyc.2012.75.3.227.

Hu, L., & Bentler, P. M. (1999). Cutoff criteria for fit indexes in covariance structure analysis: Conventional criteria versus new alternatives. *Structural Equation Modeling, 6,* 1–55. doi:10.1080/10705519909540118.

Linley, P. A., & Joseph, S. (2004). Positive change following trauma and adversity: A review. *Journal of Traumatic Stress, 17,* 11–21. doi:10.1023/B:JOTS.0000014671.27856.7e.

MacCallum, R. C., Browne, M. W., & Sugawara, H. M. (1996). Power analysis and determination of sample size for covariance structure modeling. *Psychological Methods, 1,* 130–149. doi:10.1037//1082-989X.1.2.130.

Ministry of Foreign Affairs, Fiji. (2017). *Media release*. Retrieved from http://www.foreignaffairs.gov.fj/media-resources/media-release/77-climate-change-in-school-curriculum.

National Center for PTSD. (2017). *What is PTSD?* Retrieved from https://www.ptsd.va.gov/public/ptsd-overview/basics/what-is-ptsd.asp.

National Oceanic and Atmospheric Administration. (2013). Hurricanes form over tropical oceans, where warm water and air interact to create these storms. Retrieved from http://oceanexplorer.noaa.gov/facts/hurricanes.html.

Panapasa, G. (2017, March 26). Shine a light for climate change. *Fiji Times Online*. Retrieved from http://fijitimes.com/story.aspx?ref=archive&id=394253.

Reser, J. P., Bradley, G. L., Glendon, A. I., Ellul, M. C., & Callaghan, R. (2012). *Public risk perceptions, understandings, and responses to climate change and natural disasters in Australia, 2011 and 2011*. Gold Coast, Australia: National Climate Change Adaptation Research Facility.

Sattler, D. N. (2006). Family resources, family strains, and stress following the Northridge earthquake. *Stress, Crisis, and Trauma, 3–4,* 187–202. doi:10.1080/15434610600854038.

Sattler, D. N., Adams, M. G., & Watts, B. (1995). Effects of personal experience on judgments about natural disasters. *Journal of Social Behavior and Personality, 10,* 891–898.

Sattler, D. N., Assanangkornchai, S., Moller, A., Kesavatana-Dohrs, W., & Graham, J. M. (2014). Indian Ocean Tsunami: Relationships among posttraumatic stress, posttraumatic growth, resource loss, and social support at three and fifteen months. *Journal of Trauma and Dissociation: Systemic Trauma, Special Issue, 15*, 219–239. doi:10.1080/15299732.2014.869144.

Sattler, D. N., & Graham, J. (2017, August). Climate change risk perceptions among hurricane survivors: Role of resource loss, posttraumatic stress, coping, and resilience. Paper presented at the International Conference on Environmental Psychology. A Coruna, Spain.

Sattler, D. N., Kaiser, C. F., & Hittner, J. (2000). Disaster preparedness: Relationship between prior experience, personal characteristics, and psychological distress. *Journal of Applied Social Psychology, 30*, 1398–1420. doi:10.1111/j.1559-1816.2000.tb02527.

Sattler, D. N., Whippy, A., Johnson, J., Vucago, N., & Graham, J. (2016, July). Cyclone Winston in Fiji: Climate change risk perceptions, posttraumatic stress, coping, and resilience among survivors. Paper presented at the Symposium on Climate Change Adaptation in the Pacific Region. Lautoka, Fiji.

Stern, P. C. (2000). New environmental theories: Toward a coherent theory of environmentally significant behavior. *Journal of Social Issues, 56,* 407–424. doi:10.1111/0022-4537.00175.

Stern, P. C. (2010, October). Environmentally significant behaviour and how to change it. Keynote Address to the National Congress on Behaviour Change for Sustainability. Sydney, Australia.

Taylor, A., Bruin, W., & Dessai, S. (2014). Climate change beliefs and perceptions of weather-related changes in the United Kingdom. *Risk Analysis, 34*, 1995–2004. doi:10.1111/risa.12234.

United Nations Development Programme. Response to Cyclone Winston Update. Retrieved from http://www.pacific.undp.org/content/pacific/en/home/presscenter/articles/2016/03/24/responding-to-cyclone-winston-undp-update-as-of-22-march-2016-.html.
van der Linden, S. (2015). The social-psychological determinants of climate change risk perceptions: Towards a comprehensive model. *Journal of Environmental Psychology, 41*, 112–124. doi:10.1016/j.jenvp.2014.11.012.
Viscusi, W. K., & Zeckhauser, R. J. (2015). The relative weights of direct and indirect experiences in the formation of environmental risk beliefs.Risk Analysis, 35, 318–331. doi:10.1111/risa.12271.

Author Biographies

David N. Sattler is a social psychologist and Professor of Psychology at Western Washington University. He conducts international projects in the wake of disasters, has developed interventions to educate citizens about disaster threats and climate change, established in the International Tsunami Museum in Thailand following the Indian Ocean Tsunami, and produced educational videos to promote tsunami awareness for children and emergency response in schools. He is a Behavioral and Social Science Editor for Natural Hazards Review journal.

Albert Whippy is Master's degree candidate at the Institute of Applied Science, University of the South Pacific, Suva, Fiji. His thesis is based on Kappaphycus alvarezii and focuses on farming and drying methods. The project considers environmental and social factors that contribute to the growth of the seaweed.

James M. Graham is a Professor of Psychology at Western Washington University. He received his Ph.D. in Counseling Psychology from Texas A&M University. His substantive research focuses primarily on adaptive processes in romantic relationships, while his methodological research focuses on score reliability and best practices in quantitative modeling.

James Johnson is a social psychologist and Head of the School of Social Sciences at the University of the South Pacific. His research interests are in interpersonal conflict and all forms of intergroup and intragroup stereotypes. He has secured over 4 million dollars in grant funding for his work, which has been featured in international news outlets such as CNN, Associated Press, and the Washington Post.

'Yakasisi' in Planning for a More Sustainable Future of Coastal Communities Impacted by Climate Change, Milne Bay, Papua New Guinea

David K. Mitchell and George Aigoma

"I want the same type of respect that was used in the past to be taught to our children today" Women's Leader, Ware Island Women's Leader (2016).

Abstract Subsistence agriculture/fisher reliant island communities of the Louisiades, Milne Bay Province, PNG are seeing changes in their environment and in their ability to support themselves from its natural resources and services. Change is seen as the cumulative impacts upon these resources rather than disaggregating the individual causes. The lack of productivity and depletion of resources is seen as resulting from a combination of over-utilisation/over-harvesting by a growing population for food or cash, from rising sea level, catastrophic cyclones and droughts. Another important factor however emerged, the loss of 'Yakasisi'. In Bwanabwana language 'Yakasisi' is respect; respect for elders' guidance and decisions based on their understanding of changes they have coped with over their lifetimes; respect for their customary ecological knowledge and its application in the changing present and future management of clan resources; respect for the environment. 'Yakasisi' therefore has social and environmental dimensions, of people and place. The realisation of the value of 'Yakasisi' became apparent, along with its current erosion and disrespect through communities watching self-made videos of their people story-telling about the state of their natural resources and the challenges they face in coping with this. This has led to discussion in recognising, reinforcing and supporting 'Yakasisi', combined with current knowledge, to plan for a more sustainable future as impacts such as from climate change increase.

D. K. Mitchell (✉) · G. Aigoma
Eco Custodian Advocates, Milne Bay Province, Papua New Guinea
e-mail: dmitchell.eca@gmail.com

W. Leal Filho (ed.), *Climate Change Impacts and Adaptation Strategies for Coastal Communities*, Climate Change Management,
https://doi.org/10.1007/978-3-319-70703-7_23

Background

Communities from four small islands of the Western Louisiade Archipelago of the Milne Bay Province; Kwaraiwa (10°37'22"S 151° 17'23"E), Tubetube (10°35'3"S 151° 11'15"E), the isolated island of Ware (10°57'46"S 151° 2'46"E) and later Koyagaugau (10°23'15"S 151° 24'50"E) took part in the *Spreading the Reach of Community-Based Marine Management in Milne Bay* project. This was an initiative by Conservation International (CI) in Papua New Guinea [which transitioned to the local NGO, Eco Custodian Advocates (ECA)] that was selected for implementation as part of the first cycle of USAID PACAM (Pacific-American Climate Fund) funded projects across the Pacific with the aim of reducing long term vulnerabilities associated with climate change (USAID PACAM 2015).

Within this project one of the primary activities was the creation of video products derived from traditional and contemporary community storytelling revolving around changes in the environment from the past to today. Through this activity, we expected that communities would become empowered thorough the understanding of their resource history and potential futures. CI PNG found that the process of people collecting their own stories and evaluating their own situation in terms of resource management had a more powerful impact on communities than general awareness messages generated by the NGO or government. The sharing of these community-derived stories on the changes in the environment, both within the community and with neighbouring communities, enhanced the adaptive capacity of the communities and influenced improved decision-making and ecosystem management processes to reduce the communities' vulnerability from resource declines and change climate impacts (USAID PACAM 2015).

Methodology

The initial concept in facilitating community-driven video stories as an approach to identifying climate change impacts in coastal Milne Bay, was to place the right accessible resources, including photographic and computer equipment in three communities. This was supported by the provision of a user-friendly manual on camera use and targeted skills development in picture and video taking.

A small waterproof/shockproof digital camera (Panasonic Lumix DMC-TS5) and a robust laptop (Apple MacBook Pro) with a flexible solar panel was given to community schools on Kwaraiwa, Tubetube and Ware. The project provided a small camera because it was seen as less intrusive than larger camera/video recorders, whilst still capable of high definition images, with reasonable sound recording capability up to 4 meters (Mitchell and Aigoma 2017). Also, a laptop that could withstand tropical temperatures and humidity was chosen to download video files onto and potentially to process into short video clips. As the islands do not have electricity, a foldable solar panel was also supplied to charge the laptop and camera. The placement of these in the school was because it was viewed by many as a community institution with a secure office where these could be stored and borrowed from a central location. Many teachers in these schools have some computer skills with which to utilise and manage the equipment. Being educators, they also can offer guidance to others in the taking of pictures and video (Mitchell and Aigoma 2017).

Gender Evaluation

Prior to the distribution of audio-visual equipment, a survey was undertaken with women on two of the islands Kwaraiwa and Koyagaugau (Ketaloya 2015) to determine the potential effectiveness of the proposed project methodology. The survey assessed the women's acceptance of different extension techniques and ways to express their views. In the latter series of questions, women were asked if they felt comfortable to record their views by taking pictures or video. Across the different leadership structures of the society ('ordinary' or village women; public servants who were teachers or health workers; church leaders; and community leaders) there was acceptance of both pictures and video (Ketaloya 2015). In the resultant videos, women expressed their views both within their family and in other social groupings.

Activities

It was anticipated that after a basic hands-on training that there would be members or sectors of the community that would take the opportunity to take pictures and video. Although some pictures were taken to support reports to the government on the situation of the severe El Nino drought of 2015 at the time, little other audio-visual material of environmental relevance was taken. We attribute this reluctance to a general lack of confidence on the part of communities' members in the use of the technology, not a lack of desire to tell their stories. An alternative approach was needed.

Video Filming and Production

In response to a lack of spontaneous videos, ECA and CI staff, as well as a freelance journalist went into communities to take video footage of story-telling (Fig. 1). On Ware, the community was made aware that videos on story-telling about changes in their environment from the past and of current issues would be filmed. On one Sunday when the community was relaxing in natural groupings, story-telling videos

Fig. 1 Videographer Paul Maolai taking video on Kwaraiwa Island

were shot. Apart from initial introduction and request for prior consent, participants expressed their stories without prompting so that topics of conversation were driven by themselves without undue outside influence. This method was also used on Koyagaugau, whilst on Kwaraiwa and Tubetube a combination of this method and of leaders bringing people together was used.

Story-telling was spoken in the local Bwanabwana language as people felt more able to express themselves in their first language. To facilitate editing and the opportunity of these stories to be shared more widely, these video shorts were translated into English. Mathew Jaymes, a journalist from Kwaraiwa Island went through all the video footage, transcribing what was said. The main stories were then extracted, placed in a logical flow and subtitles added. By having Bwanabwana language remain, as spoken, is important not only within the four communities, but also for other nearby communities who trade with these islanders and understand much of what is said. Where appropriate some scenery shots (commonly known in video production as B-roll) where then added. This was then edited. At the end of 2016 after a years work, local video products from the islands of Ware; Tubetube and Kwaraiwa (Engineer Group) and Koyagaugau, were shown to these communities.

Videos were first shown to those who participated in them to gauge their response. Their comments were noted and their informed consent was given prior to the videos being shown to the wider community. Videos were projected onto a white sheet 'screen' with external speakers for sound. This was shown in an open 'village square' at Ware, in a church building at Koyagaugau, a primary school classroom at Kwaraiwa and an elementary school classroom at Tubetube.

On Ware, a community of 1200, they watched carefully as they saw their own people in the video, especially the elderly speaking about the best times of their lives growing up, their lifestyle, the abundance of natural resources, food security and respect to nature's beauty in the years gone by. Only to see all these rapidly disappear from population increase, unsustainable management of use of marine resources, disrespect to traditional rules and climate change impacts (Aigoma and Mitchell 2017).

Often highlighted by the Ware community was the impact and changes caused by tropical cyclone Justin in March 1997, that came with strong winds, rain and a sea surge that took away a lined grove of fruit and nut trees along the foreshore and rubbled the fringing reefs. This was seen as the beginning of ongoing hardship by the people of Ware. It was an event that left immediate visual impact and a loss of the coastal fringe of the village (Fig. 2).

Post cyclone Justin (there have been several cyclones since, the most recent being tropical cyclone Ita in 2014) the causes of the changes mentioned above however tend often to be bundled together in peoples' stories.

Ware Island in 1979 had a resident population of 510 (340 people/km^2 of arable land on Ware) with a further 25% living elsewhere. In 2017, the resident population had grown to near 1200 (Elima Peter pers. comm. 2017) (800 people/km^2 of arable land on Ware). Within this time, the fallow of the subsistence garden reduced from 5–15 years in 1994 (Hide et al. 1994) to 2–3 years in 2005 (Foale 2005). This decrease in fallow is exacerbated by resultant soil erosion and loss of soil organic matter from burning and exposure to high intensity rainfall events. This was further exacerbated by water stress from drought such as the El Nino of 2015, and insect infestation of already weakened crops. The capability of the land to support the local population under current subsistence practice has been surpassed in this period. The marine resources especially near-shore have also altered and been heavily over utilised. Based on surveys of these resources, reef fishes for subsistence appear to be declining in size and abundance and if this trend continues will

Fig. 2 Ware Island looking east along the ridge with the main village on the right of the picture

become depleted (Wangunu 2013). The purchasing power of people to buy food on Ware is low with an annual household income of PGK750 ± 250 (Gwabu et al. 2013).

Video Response

During the community viewing of the videos, there was a bit of fun and laughter, but everyone became thoughtful in their realisation of what was expressed. There was a buzz of discussion around the entire island community the next day as it had raised so much to think about, based on what they themselves had said. The main point expressed within the video that people were drawn to on Ware was the erosion of and lack of 'Yakasisi', or respect in English (Aigoma 2017). The aspects of respect are further explored here.

Respect for Elders' Guidance and Decisions Based on Their Understanding of Changes They Have Coped with Over Their Lifetimes

> People had great respect in the past. What I see today is that the respect from the past is not here any more. Former Female Public Servant, Ware Island.[1]

> Such disrespect and reckless attitudes has played a negative impact on life on Ware Island.[2]

> In the past, when our elders tambu the plants we saw good results. Village man, Ware Island.[3]

> We taught our children to have respect. I believe if we follow those old time rules set by the old people we will make Ware Island a better place to live. Former Cooperative Trader, Ware Island.[4]

In the process of story-telling, especially by elders, the next generations were hearing and gaining an appreciation, first hand on how experiential learning led to customary management and the setting of rules. Elders, by having a span of memories over a relatively long time frame in the community and expressing changes they have seen and heard of, led others to reflect on their recent experiences and what the future may hold. The video has also become a valuable

---

[1]Former Female Public Servant. (2016) Quotation taken from Community Video, Ware Island subtitles.

[2]Elder. (2016) Quotation taken from Community Video, Ware Island subtitles.

[3]Village Man. (2016) Quotation taken from Community Video, Ware Island subtitles.

[4]Former Cooperative Trader. (2016) Quotation taken from Community Video, Ware Island subtitles.

audio-visual record, a snapshot of this point in time before the elders pass away with what they expressed.

Respect for Their Customary Ecological Knowledge

A customary management practice within these communities, which is ecologically beneficial, is to close an area of reef to fishing, known as 'bwagai' (tambu). This 'bwagai' is represented by standing a stick on the reef, as sign which people respected in the past, however today this respect is eroding.

> It is up to us to respect the rules we put so in return it will benefit us. Village man, Koyagaugau.[5]

The practice of 'bwagai' and the customary management of the community's reefs in the past, was described by elders of Ware, where rotational harvesting of marine resources was a formerly enforced in order to allow time for harvested areas to recover. Now all reefs are open to fishing with a resultant depletion of resources accentuated by the 'tragedy of the commons', the individual drive to collect before someone else does.

Respect for the Environment

> I also see people don't have respect for the reefs, people are reckless using the reefs. Elder, Ware Island.[6]

Respect of the natural environment (Fig. 3), especially within the sea was found to be important to islanders. Also on the land it was expressed that people should not burn the grassland carelessly, killing local wildlife, vegetation and 'killing' soil nutrients (Aigoma 2017).

> Respect is shown by how the coconut is consumed. A person can't just cut open the coconut and scatter the rubbish everywhere, we taught our children to have respect. Elder Kumpauli Clan, Ware Island.[7]

'Yakasisi', respect in its many forms was seen as a foundational attitude upon which to re-establish eroding customs, customary management and in the implementation of rules and actions in order to cope with change. Identifying and disaggregating the causes of change in the environment, including those due to

[5]Village Man. (2016) Quotation taken from Community Video, Koyagaugau Island subtitles.
[6]Elder. (2016) Quotation taken from Community Video, Ware Island subtitles.
[7]Elder Kumpauli Clan. (2016) Quotation taken from Community Video, Ware Island subtitles.

Fig. 3 Village woman, Koyagaugau Island expressing herself in story-telling video [Village Woman (2016) Quotation taken from community video, Koyagaugau Island subtitles]

changing climate, will assist in determining a series of actions to either mitigate these changes or improve the environment.

Next Steps

After the showing of videos to the community on Ware, many individuals downloaded these onto their smart phones opening up another option for disseminating the videos that needs due consideration.

All four island communities gave their support for their videos to be shown to other communities in Milne Bay Province and the rest of Papua New Guinea where communities are facing similar problems along with other Pacific Island countries as well. Many of the communities in Milne Bay are remote and logistically difficult to reach with regular extension support, therefore materials such as these that are intuitive and easy to understand are an important alternative.

It is anticipated that a management guidance package, Spreading-the-Reach of *Community-based Marine Management Toolkit* that includes the videos, exercises based on the video, a step-by-step cartoon guided interactive workbook on marine resource management and factsheets that support this, could serve this purpose. Eco Custodian Advocates has developed such a toolkit in the Milne Bay Province of PNG, based on its work with these island communities. The dispersal of the toolkit as a combination of electronic and hard copy resource materials, could be through schools where there is existing education and extension capacity.

Conclusion

In conclusion, community-driven participatory videos have been an important tool to raise awareness, stimulate discussion and action for resource management and as a way for local villagers' to express their perceptions of climate change and other impacts upon their environment. The videos have highlighted ways in which communiites can take steps to sustainably manage their resources especially by revitalizing respect for elders, customary ecological knowledge and the environment. We believe the tool of participatory videos can continue to be an imporant approach in facilitating sound resource management and its application can and should be expanded in PNG and beyond.

If through 'Yakasisi' natural resources are in a healthier state as a result of respecting the environment and restored management, it is anticipated that these subsistence altered ecosystems, upon which the communities depend, will be more resilient to the shocks of climate change.

Acknowledgements The authors are grateful to Paul Maolai a freelance journalist for filming, Whitney Anderson of Conservation International for putting the videos together, Mathew Jaymes, a radio announcer from the National Broadcasting Commission, Alotau for video translation from Bwanabwana language into English.

We are also grateful to Whitney Anderson and Scott Atkinson of the Marine team, CI Asia-Pacific Field Division in reviewing and commenting on the manuscript.

This work was supported with funding from the USAID Pacific-American Climate Fund (Grant# PACAM-14-0004) and an Australian Government Grant, through the Department of Foreign Affairs and Trade.

References

Aigoma, G. (2017). (unpublished), *Report on fieldwork to Ware, Kwaraiwa, Koyagaugau, Anagusa and Tubetube islands, December 2016*. Alotau, PNG: Eco Custodian Advocates.

Aigoma, G. & Mitchell, D. K. (2017). (unpublished), Spreading the Reach of community based marine management. *ECA Newsletter*, *1*, 4–5. July-December 2016.

Foale, S. (2005). *Sharks, sea slugs and skirmishes: Managing marine and agricultural resources on small, overpopulated islands in Milne Bay, PNG*. RMAP Program, Research School of Pacific and Asian Studies: The Australian National University.

Gwabu, C., Krishna, R., Sirabis, W., Yawanies, B., Ernest, J., Patiken, A. & Kapoila. J. (2013). (unpublished), *Farming Systems Survey for Ware, Nuakata and Kiriwina Islands* (Report to Conservation International). Alotau, PNG: National Agriculture Research Institute, PNG.

Hide, R. L., et al. (1994). Milne Bay Province: Text summaries, maps, code lists and village identification. *Agricultural systems of Papua New Guinea* (Working Paper No. 6). Canberra, Department of Human Geography: Australian National University.

Ketaloya, S. (2015). (unpublished), *Report on Gender Survey Kwaraiwa and Dawson [Koyagaugau] island communities. September 2015* (Report to Conservation International). Alotau, PNG.

Mitchell, D. K. & Aigoma, G. A. (2017). Behind the lens: First-hand images and videos collected by communities that document the impact of climate change in Milne Bay PNG. In Leal, W. L. (ed.), *Climate change adaptation in pacific countries: Fostering resilience and improving the quality of life* 18: 303–308. Berlin: Springer.
USAID PACAM. (2015). Pacific-American Climate Fund, Conservation International: Spreading the Reach of Community-Based Marine Management in Milne Bay. Fact Sheet, USAID, Manila.
Wangunu, N. (2013). (unpublished), *Report on Ware Island community managed marine area (CMMA) work.* Alotau, PNG: Conservation International.
Women's Leader. (2016). Quotation taken from Community Video, Ware Island subtitles

Limits and Barriers to Transformation: A Case Study of April Ridge Relocation Initiative, East Honiara, Solomon Islands

Michael Otoara Ha'apio, Keith Morrison, Ricardo Gonzalez, Morgan Wairiu and Elisabeth Holland

Abstract Increasing vulnerability to extreme environmental events (EEEs), exacerbated by climate change, is making adaptation inevitable for rural communities in Small Islands Developing states (SIDs), including the Pacific Island Countries and Territories (PICTs). Particularly the communities' located along the coastal areas that are experiencing sea level rise and coastal erosions, storm surges and flooding. Governments and development agencies across the Pacific have begun to implement adaptation policies to climate change at the community level to build resilience. This paper reports what limits and barriers rural household face for long-term adaptation, using community relocation from Mataniko Riverside to April Ridge, East Honiara, Solomon Islands, as a case study. Two hundred forty six (246) families were affected by the flash flood of Mataniko Riverside in April 2014. The Solomon Island government offered flood victims plots of land in an area safe from flooding. As of July 2015, the date of the study, the relocation process had been stalled, with flood victims still waiting for the promised plots of land. Questionnaires, oral interviews and focus group discussions with flood victims identified vulnerability, flood prone area and changing weather patterns as major limits, and government failures and the socioeconomic reality of these households as major barriers to adaptation. The study determined government failures to include a complicated land tenure system, absence of infrastructure development at the new site, inconsistent commitment to ensure completion of the land transfer to the settlers, and the lack of access to credit. Socioeconomic attributes including insufficient income, lack of formal education and skills, and consequential limited livelihood alternatives, also act as crucial barriers. The research findings indicate

M. O. Ha'apio (✉) · M. Wairiu · E. Holland
Pacific Centre for Environment and Sustainable Development,
The University of the South Pacific, Suva, Fiji
e-mail: mhaapio@gmail.com

K. Morrison
Sustainable Community Development Research Institute, Canterbury, New Zealand

R. Gonzalez
Department of Forest Sciences, Faculty of Agricultural and Forest Sciences,
Universidad de La Frontera, Temuco, Chile

W. Leal Filho (ed.), *Climate Change Impacts and Adaptation Strategies for Coastal Communities*, Climate Change Management,
https://doi.org/10.1007/978-3-319-70703-7_24

the need to design a relocation policy that addresses the limits and barriers identified here, specifically the land tenure system, and the financial support available to facilitate the relocation process.

Introduction

Adaptation to climate change impacts is inevitable for rural communities across the Solomon Islands and the Pacific region, due to the increasing intensity and frequency of extreme climate events. Coastal communities are particularly vulnerable. While rural communities have been practicing forms of adaptations for decades (Agrawal and Gibson 1999), adaptation methods and practices at these coastal villages and communities are often inadequately coordinated and implemented. Despite the fact that rural village communities have been trying to become more resilient to negative impacts of such climate related events, they also face constraints to achieving longer term adaptation. In this paper the study use the adaptation model of Islam et al. (2014), which addresses the limits and barriers that reduce people's ability to identify, assess and manage adaptation risks in an effective way, and the natural, technological, socioeconomic and institutional factors that affect personal well-being. The model implies that authorities must ensure the cost-effective allocation of resources to minimize the identified limits and barriers of the process, in order for rural communities to benefit from the implementation of long-term adaptation models. Failure to focus on these limits and barriers will lead to ineffective adaptation at the community level. The research team has adapted the model to avoid ambiguity that arises in terminology, to more explicitly consider the various scales at which the adaptation process operates, as well as to include a more nuanced understanding of what drives the adaptation process.

About 85% of the population of Solomon Islands live in rural households and rely on subsistence agriculture, forestry and marine resources (Albert et al. 2012). The high dependency on natural resources has strained productivity, which eventually leads to illegal fishing techniques and methods (Ha'apio et al. 2014). Over the past decades, people from the rural villages have increasingly migrated to Honiara, the capital city, looking for employment and better sources of income (Kabutaulaka 2001). An increasing number of domestic migrants have settled on squatter settlements (Bennett 2002), which currently comprises 35% of the city's population (HTCPB 2015). Squatter settlements have been rapidly spilling over into adjacent customary land, where some settlements are reportedly growing at over 9% per annum (SINS 2012). According to Keen and McNeil (2016), most of squatter settlements lack basic services such as water and electricity, road access, drainage and sanitation, and consequently are ill-equipped to deal with severe weather events. In addition, these quasi-legal settlers are residing on lands with no tenure security.

During a flash flood in 2014, the homes of 248 families, most of them squatters who had illegally settled in flood prone areas, were destroyed at Mataniko Riverside (Keen and McNeil 2016). In their effort to rehabilitate these families, the government promised to relocate them to a new site, now called April Ridge, near Gilbert Camp, Aekafo and Lau Valley, in East Honiara. The purpose of the relocation was to ensure that the settlers had access to land, which could be used to re-build their homes on, and their livelihoods. During 2014, the government selected, surveyed and subdivided April Ridge land into individual plots to be transferred to the affected families. In spite of the initial quick government response, the relocation process has not been executed as planned. This has caused the affected families to move and settle illegally at the new site, without acquiring legal titles.

The rationale behind this paper is to inform readers that coastal villages and communities are facing the increasing intensity and frequency of disaster events due climate change. In order to adapt to these increasing climatic events, policy makers should design a strategy that enable vulnerable communities especially those in coastal areas to resettle at safer locations. This paper therefore, analyses the relocation process of Mataniko Riverside community in Honiara, Solomon Islands and why it is stalled. It analyses the limits and barriers to long term adaptation and recommends alternate solutions for coastal communities and villages to adopt in the future adaptation programs.

Limits and Barriers to Long Term Adaptation

Barriers to adaptation can prevent the development and implementation of adaptations programs (Adger et al. 2009a, b). Due to the presence of barriers, high adaptive capacity does not necessarily translate into successful adaptation (O'Brien et al. 2006). Barriers to adaptation arise due to certain factors in adaptation programs, such as the nature of the systems involved or the larger context within which the people and systems operate (Moser and Ekstrom 2010). Adaptation programs are, however, not the only scale at which adaptation occurs. It also occurs at a smaller, single project scale and a larger, political scale that includes national policymaking. Barriers and limits to adaptation programs can be overcome by clarifying the adaptation processes that occur simultaneously at these multiple scales (Morrison and Singh 2009).

According to Huang et al. (2011), adaptation to climate change is an adjustment in natural or human systems in response to actual or expected climatic stimuli or their effects, which moderates harm or exploits beneficial opportunities. It is well known, however, that local adaptation measures are often reactive and short-term (Bohle et al. 1994), which can limit the scope for adaptation and become barriers for long-term adaptation. Morrison (2016) has clarified that short-term reactive maladaptation occurs due to confusion over the purpose for adaptation, or, to use the above definition by Huang et al. (2011), confusion over what is meant by moderating harm and beneficial opportunities. The latest depends on the scale at which adaptation is being considered. What is, however, common across scales is the "horizon" of the over-riding purpose sought to be achieved (Morrison 2016). Clarity about the "horizon" provides reforming endeavours with coherence of adaptation processes across scales. With this clarity, it is possible to see how limits and barriers to local adaptation measures emerge at multiple spatial and temporal scales (Adger et al. 2005). Coherent sets of adaptation measures at multiple spatial and temporal scales are necessary; otherwise maladaptation occurs (Morrison 2016). There can, however, be multiple successful adaptation options at each scale. They are what "human scale development" theory has defined as the multiple possible "satisfiers" to ensure that essential human needs are met (Max-Neef et al. 1991; Ekins and Max-Neef 1992; Morrison and Singh 2009).

The distinction between needs and "satisfiers" is important for understanding adaptation and resilience. This is because, even though there is a fixed set of universal needs, "satisfiers" are continually developed. The more adaptation options ("satisfiers") there are, the greater flexibility and hence adaptive capacity (Rappaport 1979). Therefore, a key function of policy and planning is to ensure innovation for providing multiple sets of "satisfiers", that are coherent across multiple scales of the adaptation process.

Failure to develop at least one set of coherent "satisfiers" across all scales results in maladaptation. It arises when particular "satisfiers" are confused with essential needs, and so are achieved at all costs, which results in a reduction of adaptive capacity and flexibility. Such mistakes produce hard, non-negotiable reactions at a specific scale that are liable to create constraints at other scales.

Transition management research has modelled the multiple scales of adaptation (Geels 2002; Foxon et al. 2009), defining three scales of adaptation through ecological metaphors. "Niche" refers to local innovative projects; "regimes" to programs institutionally mandated by policies; and "landscapes" to cultural norms and political processes (Morrison and Singh 2009). Whereas transition management research models a purely bottom-up process whereby local innovative ideas and practices emerge in a "niche" and eventually transform "regimes" to modify "landscapes", innovation occurs at the regime and landscape scales but is shaped by innovative personnel through creative dialogue with collaborators operating at different scales. This is how we see a process of transformation potentially guided by the same "horizon" that can provide long-term societal goals and integrates adaptation across all scales. The study argue that clarity about the nested multi-scaler adaptive processes could help dialogue between personnel working at different scales, for example through the use of ecological metaphors provided by transition management research.

Limits and barriers to climate change adaptation at all scales usually increase the cost of adaptation, especially through the cost of planning and implementation (Klein et al. 2014). Therefore, seeking coherent sets of "satisfiers" can increase the efficiency of adaptation. Most importantly, however, it is necessary to avoid limits to adaptation, as limiting constraints are what lead to outright maladaptation rather than only inefficiency. Limits are the insurmountable factors that constrain effective adaptation to climate change (McCarthy 2001). These limits are faced when thresholds associated with social and/or natural systems are exceeded (IPCC 2012). They can, however be avoided altogether through innovative development of multiple sets of possible 'satisfiers,' which can be both varied and synergistic. On the other hand, there are also conditions or factors that render adaptation to climate change difficult without creating maladaptation. These are defined as barriers to adaptation, which are often mutable (Adger et al. 2009a, b) and can be overcome with rigorous effort, creative management, change of thinking, prioritization, and related shifts in resources, land uses and institutions (Moser and Ekstrom 2010). The process of overcoming barriers is therefore a process of innovatively developing multiple sets of "synergistic satisfiers" to increase the efficiency and flexibility of adaptation. Therefore the same process makes it possible to overcome both limits and barriers to adaptation.

Case Study Sites

The study covers two sites: the Mataniko Riverside, a flood prone area along the Mataniko River in the urban area of Honiara, and the April Ridge site, a peri-urban grassland located eastward Honiara that is free from flood risk. Mataniko area has from 13 to 23 days of rain per year (WWCI 2016). According to Ogo et al. (1987), the average rainfall in Honiara, historically 3000 mm per year, is increasing annually. More frequent rainfall has been recorded during November to April, while higher rainfall drops during May to October (WWCI 2016).

The flood prone area of Mataniko River side was occupied by informal squatters who have relatively low income, poor infrastructure and are highly depend on the informal economic sector for their livelihoods. Although the newly allocated land area at April Ridge is sloping and not well terraced, it is more fertile than the flood prone area at Mataniko Riverside. After the flood, the government's intension was to relocate about 246 families, that were identified as flood's victims whose homes and food gardens were lost and, and with it their livelihoods. The Ministry of Lands, Housing and Survey surveyed the land in preparation for the formal allocation and transfer of lots to respective families. This would have provided the victims the ability to participate in agriculture such as farming, poultry or piggery for selling at various market outlets in the city. Given the better location and proximity to reach city market outlets, settlers would obtain relatively larger profit margins compared to village producers. Therefore, their ability to deal with limits and barriers to adaptation slightly increased at this site.

Data Gathering

The study adopted a mixed method approach for data collection, employing qualitative and quantitative techniques. The primary respondents were mainly heads of households that were affected by the flash floods of Mataniko Riverside in 2014. Primary data were collected by a survey at the household level, interviews with key informants, and focus group discussions (FGD). A total of 8 focus group discussions were made. For the survey process, we randomly selected 150 households from a total of 246, and 82 responded to the surveys. Secondary source data was collected from several ministries and a literature review.

The field work was conducted by the principal researcher of this project during two visits, from 19th October to 18th November 2015, and 25th January to 13th March 2016.

Methods

Before the household surveys and interviews were applied, the principal researcher explained the objective of the research project to the respondents, in order to ensure that they answered the questions genuinely.

Semi-structured questionnaires were carried out before the individual survey. The interviews were made to experts and also household members other than the household head, such as spouses, sons and daughters and relatives, in order to verify the household head respondents' views on their socio-economic attributes. For example, expert opinions from the government and the Honiara city council were intended to inform the effectiveness of the planning and implementation of the relocation initiative. These engagements were critical to ascertain the types of limits and barriers the government officers and city representatives perceived in regards to the relocation process.

Household surveys focused on the rising socioeconomic attributes of the 60 randomly selected households. It aimed to determine the income range, expenses and any monetary surplus, which would be available for possible adaptation investment. Each questionnaire contained four sections with a total of 85 variables. Questions covered topics such as the family's sources and level of income, the number of family members, how they judge their farm operations in the previous years compared to the current year, the types of commercial activity in which participants were involved, the constraints they faced in their adaptation program" and income sources, among others.

Oral interviews aimed at obtaining information from flood victims regarding their experience, in particular challenges they faced to relocate.

Focus group discussions were intended to determine the participants' agreement or disagreement about the relocation strategy, and how the national and local governments and relevant stakeholders had planned and implemented the initiative.

Results and Discussion

The research team has analysed data for emergent themes, which were then categorized according to the five categories of limits and barriers to adaptation defined by Islam et al. (2014), plus the added a category of land tenure system

Table 1 Limits and barriers to adaptation at April Ridge, East Honiara, Solomon Islands

Category of limits and barriers	Themes from respondents
Natural limits	Frequent rain and flooding in the area. Steep sloping land
Technological	No construction of hard structure such as dykes, barriers or blockade, detached breakwaters to prevent floods into the community
Economics	Very low income and no access to credit facilities. Lack of access to markets
Social	Lack of education, skills and livelihood alternatives. Lack of strong cooperation amongst households
Formal institutional	Lack of re allocation policy Lack of re allocation plan (e.g. a plan including the proper urban development of the new site, including road construction, health services and schools) Lack of coordination among ministries (bureaucracy and lack of political will, e.g., delay in allocation of land) Lack of micro credit government sponsored institution Lack of law enforcement (e.g. to control illegal settlement and illegal urban development in flood prone areas)
Further institutional category: land tenure system	There is no policy in place for relocation and state land could only acquire by normal land acquisition process

(See Table 1). To maintain participant anonymity, we assigned numbers to respondents from 3 to 60. The exceptions are interview numbers 1 and 2, which were assigned to the two village chiefs, and therefore are inherently identifiable.

Natural Limits

The study has identified increased flooding risk and changing weather patterns as the main natural limits. Because of the high rain frequency and volume, villagers surrounding the river always expect floods. The respondents are reluctant to build permanent homes in fear that their homes will be destroyed by a flood. According to Respondent 3, "we are just living temporarily at this site because the rain might fall heavily again and cause a flood to destroy our homes."

April Ridge area: Almost a third of the 10.72 ha of the identified land is sloping. The area has a lower flooding risk than the Mataniko riverside, but also lacks any pre-existing infrastructure such as electricity, water and roads, resulting in serious limitations to long term adaptation. One of the respondents confirmed that, although the land may be fertile, the lack of appropriate urban development makes it vulnerable to flooding, soil erosion and waterlogging during rainy seasons. Chief Michael Fa'abona said "the lower grounds may be suitable for farming, but given the lack of a drainage system during rainy season it collects pool of water which is not favourable for cropping."

Technological Barriers

Mataniko Riverside—The residents, especially those with homes built illegally, expressed concern that there are no barriers along the river bank to stop or divert floods from reaching their houses. These barriers may consist of dykes, channel systems, tree plantations along the riverside, or any other innovative technique to decrease the risks from flooding.

April Ridge—Respondents at this community highlighted the need to build proper roads and drainage systems. Chief Michael Fa'abona said that, "there is no proper road to this site and we have to carry our building materials and other goods long distance to this site".

Oral interviews—We also found that some respondents have lost confidence in weather forecast by Meteorology Office. According to respondents 5 "I lost confidence since they are imprecise and inconsistence in the accuracy of their forecasting". To achieve longer term adaptation, the population should be educated about the probabilistic nature of the forecasting system, and how to recognize weather patterns for themselves. Simultaneously, the Meteorology Office must adopt the latest technology in weather forecasting, improve coverage of meteorological stations, especially in key sensitive points to sentinel in advance catastrophic events, and employ skilled personnel in satellite technology. The lack of adequate forecasting technology is counter-productive and maladaptive.

Economic Barriers

Mataniko Riverside—Economic barriers are more pertinent in Mataniko Riverside than in April Ridge because Mataniko Riverside dwellers do not have enough land space for gardening. Traditionally, these villagers depended on subsistence agriculture (gardening) on small plots of land for their main source of income. Now, because of frequent flooding and scarcity of land, they have turned to selling general goods in informal markets, and selling handcrafts. Some have also turned to fishing. Chief John Toki of Mataniko Village said "the settlers nevertheless consider fishing a risky activity due to cyclones, and most of them do not want to continue to fish. Some, however, consider they have no option but to continue in it to support their livelihood". A number of barriers prevent them from replacing fishing with work in other sectors. Key respondent interviews identified low income, lack of access to credit to invest in alternative livelihood activities, scarcity of land and land infertility as key barriers.

April Ridge—Chief Michael Fa'abona stated that, "people of this area express that their hope and opportunities were quashed by lack of access to capital from the formal banking institution because of low income". Respondents mentioned that access to credit would enable them to diversify their activities. As Respondent Number 9 said, "We are poor and do not have sufficient access to credit and therefore we cannot increase our capacity to sell extra to meet the city dwellers' demand for agricultural produce".

Interviews showed that neither the Ministry of Agriculture and Livestock (MAL) or Ministry of Commerce, Industries, Labour and Immigration (MCILI) provide financial assistance to these settlers. Instead, banks are the main lenders to the people within the country. However, banks can only provide loans to customers that have assets (land, bank savings), which a majority of settlers could not provide if they submit a loan proposal. A senior Business Industrial officer within the MCILI confirmed that, in the early 90s, loan facilities for settlers were much better than because of the Development Bank of Solomon Islands (DBSI) (now defunct) used to lend agricultural loans to villagers.

Without access to credit from banks, other sources have had to be relied upon. Respondent 13 said, "A few of the villagers had received financial assistance from their member of parliament, but these were not enough to even start up or sustain the agriculture business". Also, in a few cases, financial constraint has forced farmers and villagers to borrow money from community members at interest rate ranging from 20 to 30%. According to the Village Chief of April Valley, "when we borrow money from other community members, we have to repay the principal plus interest. If we do not pay on the promised date, the interest accumulates and often finds ourselves in this cycle of debt".

Lastly, the Chief Planning officer within the MLHS insisted that "the flood victims must pay normal land acquisition fees before their allocated plot of land could be transferred to them". According to Keen and McNeil (2016), the cost of the allocated plots range from SBD$30,000 to $70,000 (Equivalent to USD$3800 to $8900). This was corroborated by interviews with the April Ridge Chief Michael Fa'abona. Respondents found, however, that the charge is impossible for them to pay. Respondent Number 16 from April said "we are poor and how could we pay the hefty fees? "Chief Michael Fa'abona of April Ridge summed up the difficult situation: "If the government (MLHS) is serious about the relocation of its citizens, it should waive some of the charges to the participants to ensure resilience of the communities in the longer term".

In sum, settlers recognize the need for diversity, but there was lack of willingness by institutions and government to facilitate it through access to credit. Even though community members appeared to be aware of what was needed for their long-term resilience, they were constrained from doing so by institutional limitations.

Social Barriers

Mataniko Riverside—Social barriers are also more pertinent in Mataniko Riverside than April Ridge. Long term adaptation to climate change and extreme environmental events are constrained by lack of knowledge and skills for alternative livelihoods, and limited by the availability of alternative livelihood activities. As was expressed by Respondent 17, "due to low levels of education, villagers struggle to obtain formal employment in the city. Most are settlers who decided to reside in town instead of returning back to their respective villages/communities to utilize their farming and fishing skills. Further, they do not have relevant knowledge and skills for formal employment".

It was explained during an oral history interviews from Mataniko River side that villagers are illiterate and not qualified to get jobs in the formal sector; "we do not have any other skills other than farming or fishing to change our professions" said the Respondent 7. Most of the flood victims claimed lack of higher education because most of them were displaced employees of the Solomon Islands Plantation Limited (SIPL) during the ethnic tension. They used to work as cheap labourers in the plantation because of lack of education and skills.

April Ridge—Chief Michael Fa'abona, "I was one of the few educated persons at our community to attend secondary and tertiary school and we must encourage better education if we want to improve our standard of living". Now, some of the children in the settlements are also progressing up to high school education level. Chief Fa'abona continued to say that "only by obtaining higher education will help us to indirectly benefit through gaining employment that will alleviate our families from poverty and assist us to build resilience".

At April Ridge, there is a privately established school that provides early childhood education (ECE) for the residents 'children. However, after completing their early childhood education, children must travel into the city to attend primary and secondary schools. According to Chief Michael Fa'abona, the government should consider investing in primary and secondary schools at the community if it is serious about building long-term adaptation at the community level. Community members intuit that improving their long-term resilience requires increasing their adaptation options. They were seeking to expand their adaptation options by developing the opportunities that formal education can provide, without losing the resilience that subsistence skills provide. They were actively seeking to develop multiple satisfiers, which increase the flexibility of their livelihood (Rappaport 1979).

Formal Institutional Barriers

Lack of institutional capacity was found to be a widespread cause of multiple constraints to the adaptation process. The main institutional barriers identified were the lack of an urban development plan for the flood prone area, the lack of a relocation policy programme, poor land tenure system and poor infrastructure development by the government.

Mataniko Riverside—Respondents expressed that, since the majority of the people have moved to April Ridge and other parts of the country, the relevant authorities should commence transforming the river side into a proper residential community of the town. Chief John Toki said, "Now, since the area is not that heavily populated as before, the responsible authority should commence with proper planning, construction of barriers against future floods and converting the area into urban housing zone area". The respondents claim that the Honiara city council has now create a formal town development plan, with one of its objectives to stop residents erecting homes less than 100 m close to the river bank (Honiara Local planning scheme, 2015[1]). However, as Respondent 15 expressed it, "It requires strong institution to enforce such important development legislation, to ensure people adhere to not rebuilding their houses close to the riverside." Now many houses are built less than 20 m away from the riverside and thus will continue to be exposed to floods.

April Ridge—Almost all interviewees blamed the MLHS for delaying the transfer of land titles to the flash flood victims. According to Chief Michael Fa'abona, the flood victims will never rebuild their livelihoods if this process is not completed. "These people need land to start building their homes and to participate in agribusiness or other small businesses". He continued to say that, "the government has secured the land with the intention to rehabilitate and assist the flood victims, therefore the land should transfer to the victims without delay".

Insight into the institutional limitations was provided by Respondent 20. They expressed that the government, donor aid partners and implementing agencies should shift their focus from a project to a programme focus when designing and implementing their adaptation initiatives. From a functional perspective, programs generally require policies to mandate them, whereas projects can be developed in an ad-hoc manner. It was pointed out by respondent No. 20 that previous adaptation activities implemented by the COM were also project oriented and focused, and when the duration of the project lapsed there was no continuity or ownership of the project by the beneficiaries, causing most of the adaptation projects to fail once the duration of the project expired. Respondent 20 further explained that, after the flood, donor aid partners provided funding through relevant authorities for projects to assist the flood victims relocate from the affected site, but only for about 3 months after the flood. Because of the land tenure complication, the relocation has now taken more than two years, and funds are no longer available to complete it. Respondent 20 continued by saying, "If the government takes ownership of the relocation process as a program rather than as a project, it would help overcome the issue of a project setting a precedent". To make this change would require addressing formal institutional barriers brought by lack of policies.

Another insight into the situation was provided by several respondents who pointed out that the relocation of Matakino Riverside was promised by the government immediately prior to the national general election of 2014. While politicians appeared to enthusiastically initiate adaptation measures, a portion of government proactivity may be attributed to election agendas and not reflect the

[1]The Local Planning Scheme is prepared for Honiara City Council by the Physical Planning Division of the Ministry of Lands, Housing and Survey and supported by the Secretariat of Pacific Community (SPC).

actual resource capacities. Nevertheless, the politicians did initiate an option that initially resonated well with community expectations, fulfilling the criteria for coordination between local and government efforts. However, the reforms lacked the institutional capacity to be fully implemented.

Land Tenure System

According to the Chairperson of the Flood Victims Committee (FVC), the plots of land are supposed to be transferred to the victims free of charge. This statement was contrary to the Ministry of Lands Housing and Survey officials, who stated that land must be paid for by the flood victims. It is unclear what the source of the confusion was. Whatever the actual reason for the confusion, it delayed the relocation. As explained by the Senior Land officer in the government, "these plots of land are crown land, and therefore although priority is preserved for the flood victims, they have to go through the normal process of tender for the land'. The process ceased to be vigorously pursued after the subdivision in 2015.

The failure to transfer titles appears to be due to the government having not yet decided on a policy for allocation of the land to the flood victims. According to Flood Victim Committee Chairperson the government has already subdivided 268 20 m × 20 m plots of land into urban development plots at April Ridge, which is more than adequate for flood victims to complete their recovery and commence the rehabilitation of their livelihoods. The process only stopped because government intervened and changed the process. A Senior MECDM officer cross-validated that that the lack of reliable government relocation policy prevented the MLHS from transferring the land and relocating flood victims. As such, "the government is reluctant to set precedents for future relocation initiatives across the country" as most communities are vulnerable to some form of extreme event and climate change. Besides, since announcing the relocation process, most victims aggressively see the opportunity for them to own land in the capital and settle at the areas even before government had approved the details for relocation. Therefore a purely institutional hurdle appears to have become a major barrier if not limit to the success of the relocation.

Besides, the government did not construct a proper road between Honiara City and the April Ridge community. Residents have to walk for 40–50 min from the main road to reach the allocated plots of land. Chief Michael Fa'abona commented: "before the government allocates plots of land they should ensure proper roads are constructed to ease with development at the site". This prevents the April Ridge residents from developing livelihood options that would otherwise be available to them and oversight by the government". Moreover, because there is neither a health center in the community, it is difficult for the sick, weak and children to seek medical assistance. "The closest health center takes us almost 2 h walk to reach" said the village Chief. Interviews with relevant institutions were not able to provide any reasons for the oversights in planning by the government.

Once again, the community appears to be well aware of what facilities would enable them to take best advantage of the relocation, but it seems that there was lack of coordination between government institutions in developing a coherent relocation plan.

Cultural Barriers

Human adaptation to climate change is a heterogeneous process influenced by natural limits and barriers, economic, technology, social, and formal institutions, (Nielsen and Reenberg 2010), land tenure systems, and also cultural perspectives. Relevant literature now increasingly acknowledges that factors such as class, gender and culture play a large role when choosing or rejecting adaptation strategies at the local scale (Denton 2002). For example, when making decisions of whether to move from Mataniko Riverside to April Ridge, men more frequently made the decision to move compared to their spouses. Some women expressed that there was not proper water, sanitation and health facility at the new site, but submitted to their husband's decision and agreed to move. According to a female respondent, "if only our male counterparts could listen to some of our concerns then we would not have experienced some of the difficulties as encountered during our relocation process".

Conclusion

The main limits to adaptation identified in this case study were: vulnerability wrought by a flood prone area, changing weather patterns and the varying topography. The corresponding barriers were poverty and the lack of institutional clarity about land tenure, in particular no government policy on relocation. The poverty that is further related to the lack of access to credit, lack of formal education, lack of subsistence skills. The government's failure to provide infrastructure development at the new site, limited alternative livelihood sources, unfavourable financial support by local and national governments, no commitment to ensure completion of land transfer to the settlers.

While the residents of these two sites participate in their own adaptation strategies, they are also forced to cope with the limits and barriers in their endeavour to adapt. The natural and economic limits are similar for the two communities, while technological, social and formal institutional barriers are more contextual to each study site. The limits and barriers are also interrelated and at times overlapped and combined to constrain adaptation.

If the government is serious about long-term transformation of these people, it should prioritize the building of infrastructures such as roads, electricity, telephone services, water, schools and health centres for residents before embarking on a relocation initiative. Given the interrelated nature and combined influence of these identified barriers, overcoming them is complex and requires planned and calculated adaptation strategies. The full extent of the range of adaptation options ("satisfiers") to fulfil the needs of a community has to be appreciated in the planning phase, or else the resilience of communities can be further threatened, rather than enhanced. The planned adaptation should include zoning and relocation of communities to areas that would minimize the impact of climate change risks on the communities. In addition, both internal and external factors pose barriers to adaptation at these sites, and some barriers are reinforced by others. To overcome these barriers, planned adaptation should occur at multiple scales through dialogue of personnel between sectors and with the community.

There is a need for personnel in the various government ministries to show innovation, to provide a diversity of options for multiple solutions by proposing new policies where necessary, and also to seek coherence of policies across sectors through dialogue with personnel from other ministries and sectors. There appears to be a need for greater innovation and pro-activity to be shown by public servants.

There is also a need for institutional reform to allow access to less expensive credit and to improve enforcement of by-laws to avoid the creation of vulnerable livelihoods in flood zones and coastal communities. Designers and implementers of adaptation must also shift their strategy focus from projects to programs. This will assist in providing a sense of continuity, and thus achieve long-term adaptation by helping overcome institutional inertia against adaptation.

Finally, pro-actively helping settlers build their capacity to create alternative livelihood activities would help diversify their incomes, and therefore position them better to overcome identified barriers and limits. In particular, the findings indicate a need for further study into the determinants and implications of the limits and adhering to the land tenure system, particularly when government owned land is involved. When there are both customary and formal land tenure systems co-existing in the socio-cultural and political "landscape", it is necessary to suggest alternate options that governments could consider when designing a relocation policy.

Policy Implications

The findings of this study provide an opportunity to make several specific policy recommendations that may be internationally applicable.

1. Relocation is vital for long-term resilience of communities, especially the coastal communities threatened by natural limits to adaptation. While it may be an easy concept to implement in the rural areas where there is a common land tenure system, it becomes very complex in semi-urban centers where land ownership and tenure security is in question. Therefore, explicit policies on relocation are necessary to identify vulnerable communities and enable a smooth relocation process.

2. Cost is one of the main identified barriers preventing effective adaptation and such responsible government's ministries should investigate ways of establishing micro-finance schemes for community residents, to enable them to start up and expand their businesses to facilitate adaptation. Reforms could include establishment of a scheme guaranteed by the government for rural dwellers to obtain loans from commercial banks with lower interest.
3. Finally, further in-depth studies are needed in order to move towards an improved characterization of the multi-scale temporal and spatial process of adaptation and to identify the most suitable means to overcome the limits and barriers at the coastal community level.

Acknowledgements The authors of the paper would thank the University of the South Pacific and Solomon Islands Government through the Ministry of Education for co-funding one of the trips to the Solomon Islands.

References

Adger, W. N., Arnell, N. W., & Tompkins, E. L. (2005). Successful adaptation to climate change across scales. *Global Environmental Change, 15*(2), 77–86. University of East Anglia, UK.

Adger, W. N., Dessai, S., Goulden, M., Hulme, M., Lorenzoni, I., Nelson, D. R., et al. (2009a). Are there social limits to adaptation to climate change? *Climatic Change, 93*(3–4), 335–354. Norwich, UK: University of East Anglia.

Adger, W. N, Lorenzoni, I., & O'Brien, K. L. (2009b). Adapting to climate change: Thresholds, values, governance. Cambridge: Cambridge University Press.

Agrawal, A., & Gibson, C. C. (1999). Enchantment and disenchantment: The role of community in natural resource conservation. *World Development, 27*(4), 629–649. Bloomington, USA: Indiana University.

Albert, J. A.,Trinidad, A., Cabral, R., & Boso, D. (2012). Economic value of coral reefs in the Solomon Islands. Case study findings from coral trade and non-coral trade communities. Honiara, Solomon Islands, World Fish. www.coraltriangleinititaivenet/knowledge-hub/document.

Bennett, J. (2002). *Roots of conflict in Solomon Islands-though much is taken, much abides: Legacies of tradition and colonialism*. Australia: Australian National University.

Bohle, H. G., Downing, T. E., & Watts, M. J. (1994). Climate change and social vulnerability: Toward a sociology and geography of food insecurity. *Global Environmental Change, 4*(1), 37–48.

Denton, F. (2002). Climate change vulnerability, impacts, and adaptation: Why does gender matter? *Gender & Development, 10*(2), 10–20.

Ekins, P., & Max-Neef, M. (Eds.). (1992). Real-life economics: Understanding wealth creation. London.

Foxon, T., Reed, M., & Stringer, L. (2009). Governing long-term socio-ecological change: What can resilience and transitions approaches learn from each other? *Environmental Policy and Governance, 19,* 3–20.

Geels, F. W. (2002). Technological transitions as evolutionary reconfiguration processes: a multi-level perspective and a case study. *Research Policy, 31,* 1257–1274.

Ha'apio, M., Filho, W. L., Gonzalez, R., Holland, E., & Wairiu, M. (2014). Mapping the economic costs and benefits of Coral Triangle Initiative (CTI) and Mangrove Rehabilitation Projects

(MRP) in Solomon Islands: A study of two MPAs and one MRP. *International Journal of Sustainable Development & World Ecology, 21*(5), 414–421.

Huang, C., Vaneckova, P., Wang, X., FitzGerald, G., Guo, Y., & Tong, S. (2011). Constraints and barriers to public health adaptation to climate change: a review of the literature. *American Journal of Preventive Medicine, 40*(2), 183–190.

HTCPB (Honiara Town and Country Planning Board). (2015). Honiara, Local Planning Scheme 2015, Ministry of lands Housing and Survey. Honiara.

IPCC. (2012). Managing the risks of extreme events and disasters to advance climate change adaptation. In: C. B. Field, V. Barros, T. F. Stocker, D. Qin, D. J. Dokken, K. L. Ebi, et al. (Eds.), *A special report of working groups I and II of the intergovernmental panel on climate change*. Cambridge: Cambridge University Press.

Islam, M. M., Sallu, S., Hubacek, K., & Paavola, J. (2014). Limits and barriers to adaptation to climate variability and change in Bangladeshi coastal fishing communities. *Marine Policy, 43,* 208–216.

Kabutaulaka, T. T. (2001). *Beyond ethnicity: The political economy of the Guadalcanal crisis in Solomon Islands*. Australia: Australian National University.

Keen, M., & McNeil, A. (2016). *After the floods: urban displacement, lessons from Solomon Islands*. Honiara, Solomon Islands.

Klein, J. T., Midgley, G. F., Preston, B. L., Alam, M., Berkhout, F. G. H., Dow, K., et al. (2014). In: C. B. Field, V. R. Barros, D. J. Dokken, K. J. Mach, M. D. Mastrandrea, T. E. Bilir, et al. (Eds.), *Climate change 2014: Impacts, adaptation, and vulnerability. Part A: global and sectoral aspects* (Vol. 1, pp. 899–943). United States: Cambridge University Press.

Max-Neef, M. A., Elizalde, A., & Hopenhayn, M. (1991). *Human scale development: conception, application and further reflections*. New York: The Apex Press.

McCarthy, J. J. (2001). *Climate change 2001: impacts, adaptation, and vulnerability: Contribution of working group II to the third assessment report of the intergovernmental panel on climate change*. Cambridge University Press.

Morrison, K. D. (2016). The role of traditional knowledge to frame understanding of migration as adaptation to the "slow disaster" of sea level rise in the South Pacific. In K. Sudmeier-Rieux, M. Jaboyedoff, M. Fernandez, I. Penna & J. C. Gaillard (Eds.), *Identifying emerging issues in disaster risk reduction, migration, climate change and sustainable development* (pp. 249–246). Springer.

Morrison, K. D., & Singh, S. J. (2009). Adaptation and indigenous knowledge as a bridge to sustainability. In Priscila Lopes & Alpina Begossi (Eds.), *Current trends in human ecology* (pp. 125–155). Newcastle upon Tyne: Cambridge Scholars Publishing.

Moser, S. C., & Ekstrom, J. A. (2010). A framework to diagnose barriers to climate change adaptation. In *Proceedings of the national academy of sciences* (Vol. 107(51), 4705–4709). Berkeley, CA: University of California.

Nielsen, J. O., & Reenberg, A. (2010). Cultural barriers to climate change adaptation: A case study from Northern Burkina Faso. *Global Environmental Change, 20*(1), 142–152. Copenhagen K, Denmark.

O'Brien, K., Eriksen, S., Sygna, L., & Naess, L. O. (2006). Questioning complacency: climate change impacts, vulnerability, and adaptation in Norway. *AMBIO: A Journal of the Human Environment, 35*(2), 50–56. Springer on behalf of Royal Swedish Academy of Sciences.

Ogo, T., Inomata, N., & Yamamoto, Y. (1987). Plant species used in Papua New Guinea and Solomon Islands. *Japanese Journal of Tropical Agriculture, 31*(1), 16–27, Japan.

Rappaport, R. A. (1979). *Ecology, meaning, and religion* (p. 174). Richmond, CA: North Atlantic Books.

SINS (Solomon Islands National Statistical Office). (2012). *2009 Population & Housing Census: Report on Migration and Urbanisation*. Honiara: Solomon Islands Government.

WWCI (World Weather and Climate Information). (2016). Honiara, Solomon Islands. Retrieved on September 12, 2016 from: https://weather-and-climate.com/average-monthly-Rainfall-Temperature-Sunshine,Honiara,Solomon-Islands.

Impacts of Climate Change in Coastal Areas: Lessons Learned and Experiences

Walter Leal Filho

Abstract This conclusions paper summarises some of the main development and trends related to climate change in coastal areas and outlines future research needs.

Introduction

Sixty percent of the world's cities with a population of over 5 million are located within 100 km of the coast (Nicholls et al. 2007). Coastal areas are among the world's most diverse and productive environments (McLean et al. 2001), which currently are experiencing growing exploitation pressures (Nicholls and Lowe 2004). Coastal systems, including estuaries, are among the most vulnerable areas to many impacts of climate change (U.S. EPA 2009; Camarsa et al. 2012) that could affect coasts in various ways (Zikra et al. 2015). Coastal areas are now affected by increasing flooding, accelerated erosion, water scarcity and droughts, water pollution, habitat destruction, loss of biodiversity, and seawater intrusion (Nicholls and Lowe 2004; Camarsa et al. 2012). These events have environmental, economic and societal impacts, and are already altering the livelihoods of coastal communities (Camarsa et al. 2012).

Impacts of Climate Change in Coastal Areas: An Overview

As documented in various chapters on this book, the impacts of climate change may be classified into 3 main areas: social impacts, economic impacts and environmental impacts. Due to their relevance, they will be outlined in turn.

W. Leal Filho (✉)
School of the Science and the Environment, Manchester Metropolitan University, Chester Street, Manchester, UK
e-mail: w.leal@mmu.ac.uk

W. Leal Filho (ed.), *Climate Change Impacts and Adaptation Strategies for Coastal Communities*, Climate Change Management,
https://doi.org/10.1007/978-3-319-70703-7_25

1. ***Social Impacts***: Even though it is often overlooked, damages associated with or caused by climate change in coastal areas have strong social impacts. In particular, they may lead to displacements of proportions of older residents and transient populations, who may need to move away from the coasts to become less vulnerable. The process may also lead to higher level of unemployment levels and high seasonality of work, up to physical isolation due to poor transport links.

Other negative effects as listed in the literature (e.g. McLean et al. 2001) which may be mentioned are:

- increased loss of property and coastal habitats;
- loss of renewable and subsistence resources;
- loss of tourism, recreation, and transportation functions;
- loss of non-monetary cultural resources and values;
- decline in soil and water quality.

However, it is important to recognize that the aforementioned impacts are largely associated with human settlements and developed areas. In general, most unaltered and naturally functioning shorelines are capable of adapting or responding to sea-level rise (Boesch et al. 2000).

2. ***Economic Impacts***: Climate change also threats key economic activities in coastal areas and leads to loss of a related share of GDP. For instance, in Europe the economic value of coastal areas within 500 m from the seas accounts between €500–1000 billion. According to the European Commission, the annual costs of taking mitigating and adapting measures are estimated at around €6 billion by the year 2020 (European Commission 2016). The economic impact of tropical storms and hurricanes might reach billions of dollars in insured losses alone (Boesch et al. 2000).
3. ***Environmental Impacts***: This area, which is the focus of this paper, is probably the one characterised by a greater degree of complexity and will therefore be described at a greater level of detail.

Coasts are sensitive to sea level rise (Zikra et al. 2015), which is considered to be one of the most significant threats to shoreline systems (Boesch et al. 2000). About 70% of the coastlines worldwide are projected to experience a sea level change within ± 20% of the global mean (Core Writing Team et al. 2014). The Pacific region is among the most vulnerable ones (Leal Filho 2017).

Increasing sea level leads to a number of impacts. These are:

(a) coastline retreat
(b) flooding
(c) swamping of low land on coastal plains
(d) coastal erosion
(e) underground seawater intrusion and
(f) destruction of the seawall and drainage systems (Cai et al. 2009; European Commission 2016).

In particular, overtopping of seawater results in the salinization of the surficial and sub-surficial zones of inland coastal areas (Yang et al. 2015). The induced rate of erosion is about two orders of magnitude greater than the rate of sea level rise. As beaches are lost, fixed structures nearby are increasingly exposed to the direct impact of storm waves, and will ultimately be damaged or destroyed (Zhang et al. 2004).

The erosion of the dune systems makes it more difficult for the water to recede back to sea through the drainage system (Gray et al. 2014). Increased anthropogenic activities (e.g. excessive pumping or intensive building close to the shore) in combination with climate change affect coastal groundwater systems globally (Oude Essink et al. 2010). Seawater intrusion caused by sea level rise diminishes an availability of fresh groundwater resources in coastal areas (Pham and Lee 2015; Tam et al. 2016).

A special mention should be made to **coastal erosion**, a matter of major concern to many countries, and subject to many climate change adaptation efforts (Leal Filho 2015). The already-severe coastal erosion problems seen in many parts of the world, are expected to exacerbate in the 21st century (Zhang et al. 2004). Coastal erosion threatens habitats and wildlife and has dramatic effects on coastal development and infrastructure (Boesch et al. 2000; European Commission 2016). Erosion events might also be a result of storms and extreme events (Boesch et al. 2000), and sediment budget reduction (Cai et al. 2009). In addition, chronic coastal erosion brought about by large wave events or changes in wave direction that shift coastal sand and sediments (Australian Government, Department of Climate Change 2009). Increased coastal erosion can reduce water clarity, limiting the depth at which sea grasses can grow (Paice and Chambers 2016a).

In terms of **losses of habitat and biodiversity**, some evidences are also available. The potential impacts of invasive species in response to climate change are relevant for all coastal ecosystems (Hellmann et al. 2008). Invasive plants and animals are a serious threat through predation and interference with food webs, competition for habitat and resources, and direct health effects on native plants and animals. Moreover, some native species may also become invasive, if they are able to tolerate high temperatures (Paice and Chambers 2016a). Rising global seas temperature and altered salinity stress temperature-sensitive organisms such as corals and causes their death or morbidity. It also makes estuary and near shore habitats inhospitable to species with narrow temperature tolerances (Agardy et al. 2005).

Changes of precipitation patterns are also a matter which deserves mention. Climate change affects rainfall amounts and their seasonal patterns (Paice and Chambers 2016a) as well as the frequency, severity and positions of cyclones and other storms (Australian Government, Department of Climate Change 2009). Altered frequencies and intensities of precipitation, and increased water temperatures impact the health of the coastlines (U.S. EPA 2009), exacerbate coastal erosion and flooding events, influence and damage many vital ecosystem services provided by coastal ecosystems (Paice and Chambers 2016b). The rising acidity of the ocean have significant impacts on coastal environments and marine ecosystems (Zikra et al. 2015) by limiting growth of carbonate-dependent organisms, including corals and shellfish (Paice and Chambers 2016a).

Impacts in Some Countries

Depending on a geographical location, coastal areas experience different direct and/ or indirect impacts of climate change. Below are examples of climate change impacts in different coastal areas.

Indonesia is the world's largest archipelagic state with over 81,000 km of coastline that is very vulnerable to climate change. Increasing trend of sea level rise, warmer ocean temperature and an increase of waves height are among significant climate change impacts in the country (Zikra et al. 2015).

South China experiences increasing occurrence and intensity of storm surges, aggravating saltwater intrusion, accelerating coastal erosion, exacerbated problems in urban drainage and flood control, inundating lowland areas, degrading mangroves and coral reef ecosystem (Cai et al. 2009; Du et al. 2013).

In **Senegal** about 25% of the shoreline is deemed to be at high risk of erosion and shoreline loss, notably, in the deltas and estuaries of the three main rivers. It is estimated that by 2080, as a result of a sea level rise, 75% of the coastline could become at high risk of erosion that will be accelerated by sand extraction and beach-top urbanization. It is also expected that by 2080 two thirds of the coastal area could be facing a high risk of submersion (World Bank 2013).

Climate change brings significant change to **Australia**'s coastal zone as well. It increases risks to settlements, industries, and natural ecosystems. Rising sea levels will have a large impact on many coastal environments such as beaches (e.g. beach loss), estuaries, coral reefs, wetlands (e.g. salinisation) and low-lying islands (e.g. inundation) (Australian Government, Department of Climate Change 2009).

Many coastal areas in **Europe** are already confronted with the effects of climate change, i.e., flooding, erosion, saline intrusion, and the loss of natural ecosystems such as wetlands (Camarsa et al. 2012). Despite the fact that each European coastal member state is exposed to different impacts of climate change, the following trends can be observed. For instance, (Directorate-General for Maritime Affairs and Fisheries European Commission 2009):

- Along the **Baltic** coastline, the overall vulnerability to coastal flooding and erosion due to sea level rise is expected to be low. Most climate change impacts are projected for marine species, because migration from the semi-enclosed Baltic Sea will be difficult when the sea surface temperature rises.
- The main climate risk in the Atlantic marine basin and the North Sea basin is flooding due to the sea level rise and storm surges. Southern countries could become more exposed to freshwater shortage in the future due to prolonged and more intense periods of drought.
- The coastline of the Mediterranean marine basin is highly exposed to erosion, large areas are affected by saltwater intrusion and dry periods projected to increase in length and frequency.

About 53% of the total **U.S.** population live on the 17% of land in the coastal zone, and this number is growing. Atlantic and Gulf coastlines are especially vulnerable to long-term sea-level rise and an increase in the frequency of storm surges or hurricanes (Boesch et al. 2000). Moser et al. (2014) compiled the regional threats from climate change, for instance:

(a) **Pacific Northwest**: The substantial global sea level rise is regionally moderated by the continuing uplift of land. Coastal storm surges are expected to be higher due to increase in sea level alone, and more intense persistent storm tracks will increase coastal flooding from inland runoff.
(b) **California**: Coastal storm surges are expected to be higher due to increases in sea level alone, and more intense persistent storm tracks will increase coastal flooding risks from inland runoff. There is also a growing risk of coastal erosion, temporary flooding, and permanent inundation.
(c) **Alaska**: Current and projected increases in Alaska's ocean temperatures and changes in ocean chemistry are expected to alter the distribution and productivity of Alaska's marine fisheries.
(d) **Gulf Coast**: Hurricanes, land subsidence, sea level rise, and erosion already pose great risks to the area, causing among others permanent land loss. Coastal inland and water temperatures are expected to rise; coastal inland areas are expected to become drier. There is still uncertainty about future frequency and intensity of Gulf Mexico hurricanes, but sea level rise will increase storm surges (Moser et al. 2014).

Adaptation and Mitigation Responses in Coastal Areas

Adaptation and mitigation measures responses to the impacts of climate change in coastal areas are expected to vary strongly among and within regions and countries. According to the European Commission, the integrated coastal management approach builds the foundations for sustainable coastal management and development, supporting socio-economic development, biodiversity and ecosystem services (European Commission 2016). It includes all steps: information collection, planning (in its broadest sense), decision making, management and monitoring of implementation (Cai et al. 2009). In addition to the integrated approach, coastal adaptation options based on local community participation, disaster risk reduction and ecosystems-based approaches (Core Writing Team et al. 2014).

Structures as seawalls, breakwaters, jetties, groins, and offshore dikes beach nourishment, man-made dunes, planting of mangroves and rise grass, and coastal shelter belts are used to control coastal erosion (Cai et al. 2009) and sand fences' (made of dry willow branches or wicker) for dune systems (Camarsa et al. 2012).

Catchment management activities that improve drainage design (e.g. flood retention basins, retrofitting of existing artificial drainage networks) can protect coastal stability during flooding, and may be important in areas where the frequency of intense rainfall events is predicted to increase (Paice and Chambers 2016a). U.S. EPA divides adaptation options for coastal areas into the following categories (U.S. EPA 2009):

- Maintain/restore wetlands: focus on facilitating wetland migration through changes in legislation and regulations, and prohibitions on shoreline hardening;
- Maintain sediment transport: include either trapping sediment that would otherwise migrate or reintroducing sediment into systems;
- Preserve coastal land/development (including infrastructure);
- Maintain shorelines utilizing "soft" measures: aim to develop living shorelines through beach nourishment, planting dune grasses, marsh creation, and planting submerged aquatic vegetation;
- Maintain shorelines utilizing "hard" measures: e.g., constructing bulkheads, seawalls, revetments, and breakwaters, or reinforcing dikes and headlands;
- Invasive species management;
- Preserve habitat for vulnerable species;
- Maintain water quality and
- Maintain water availability.

This list of measures is by no means comprehensive and their degree of applicability is likely to differ among countries, but it does illustrate the wide range of areas where action may be taken.

Conclusions

Global climate change affects the physical, biological, and biogeochemical characteristics of the coasts, modifying their ecological structure, functions, and provided goods and services (McLean et al. 2001). The major impacts are summarized in sea level rise, storm surge and erosion that are aggravated by unsustainable management (Elsharouny 2016). These threat key economic activities, lead to loss of development, destruction of cultural heritage and competition between stakeholders for resources (Camarsa et al. 2012). Both long-term planning and updated protection measures are required to eliminate the adverse impacts of the hazard and guarantee sustained and harmonious development of the coastal economy (Cai et al. 2009).

But in order to yield sustainable benefits, technical measures need to be complemented by fostering the capacity to prepare for climate risks and recover from specific events, and by the development of targeted support for those most vulnerable, especially—but not only—in developing countries.

References

Agardy, T., Alder, J., Dayton, P., Curran, S., Kitchingman, A., Wilson, M., Catenazzi, A., Restrepo, J., Birkeland, C., Blaber, S., Saifullah, S., Branch, G., Dee Boersma, Nixon, S., Dugan, P., Davidson, N., Vorosmarty, C. (2005). Costal Systems (Chap. 19). In *Ecosystems and human well-being: current state and trends*. Washington, Covelo, London: IslandPress. Available at: http://www.millenniumassessment.org/documents/document.288.aspx.pdf.

Australian Government, Department of Climate Change. (2009). Climate change risks to Australia's Coast. Commonwealth of Australia. Available at: http://www.environment.gov.au/system/files/resources/fa553e97-2ead-47bb-ac80-c12adffea944/files/cc-risks-full-report.pdf.

Boesch, D.F., Field, J.C., & Scavia, D. (Eds.) (2000). The potential consequences of climate variability and change on coastal areas and marine resources: Report of the Coastal Areas and Marine Resources Sector Team, U.S. National Assessment of the Potential Consequences of Climate Variability and Change, U.S. Global Change Research Program, NOAA Coastal Ocean Program Decision Analysis Series. Silver Spring, MD: NOAA/National Centers for Coastal Ocean Science. Available at: http://aquaticcommons.org/id/eprint/14637.

Cai, F., Su, X., Liu, J., Li, B., & Lei, G. 2009. Coastal erosion in China under the condition of global climate change and measures for its prevention. *Progress in Natural Science, 19*, 415–426. Available at: http://dx.doi.org/10.1016/j.pnsc.2008.05.034.

Camarsa, G., Silva, J.P., Toland, J., Eldridge, J., Hudson, T., Jones, W., O'Hara, E., Thorpe, E., & Thévignot, C. (2012). LIFE and coastal management. European Commission. Available at: http://ec.europa.eu/environment/life/publications/lifepublications/lifefocus/documents/coastal.pdf.

Core Writing Team, Pachauri, R.K., & Meyer, L.A. (2014). Climate change 2014: Synthesis report. Contribution of working groups I, II and III to the fifth assessment report of the intergovernmental panel on climate change. IPCC. Available at: http://ipcc.ch/pdf/assessment-report/ar5/syr/SYR_AR5_FINAL_full_wcover.pdf.

Directorate-General for Maritime Affairs and Fisheries European Commission. (2009). The economics of climate change adaptation in EU coastal areas—Summary report. Available at: https://ec.europa.eu/maritimeaffairs/sites/maritimeaffairs/files/docs/body/executive_summary_en.pdf.

Du, Y.-D., Cheng, X.-H., Wang, X.-W., Al, H., Duan, H.-L., He, J., & Wu, X.-X. (2013). A review of assessment and adaptation strategy to climate change impacts on the coastal areas in South China. *Advances in climate change research, 4*, 201–207. Available at: http://dx.doi.org/10.3724/SP.J.1248.2013.201.

Elsharouny, M.R.M.M. (2016). Planning coastal areas and waterfronts for adaptation to climate change in developing countries. *Procedia Environmental Sciences, 34*, 348–359. Available at: https://doi.org/10.1016/j.proenv.2016.04.031.

European Commission. (2016). The challenge of climate change to the European coastal areas. Available at: http://ec.europa.eu/environment/iczm/state_coast.htm. Accessed 10 June 2017.

Gray, S.R.J., Gagnon, A.S., Gray, S.A., O'Dwyer, B., O'Mahony, C., Muir, D., Devoy, R.J.N., Falaleeva, M., & Gault, J. (2014). Are coastal managers detecting the problem? Assessing stakeholder perception of climate vulnerability using Fuzzy Cognitive Mapping. *Ocean & Coastal Management, 94*, 74–89. Available at: http://dx.doi.org/10.1016/j.ocecoaman.2013.11.008.

Hellmann, J.J., Byers, J.E., Bierwagen, B.G., & Dukes, J.S. (2008). Five potential consequences of climate change for invasive species. *Conservation Biology, 22*, 534–543. Available at: https://doi.org/10.1111/j.1523-1739.2008.00951.x.

Leal Filho, W. (Ed.). (2015). *Handbook of climate change adaptation*. Berlin: Springer.

Leal Filho, W. (Ed.). (2017). *Climate change adaptation in pacific countries: Fostering resilience and improving the quality of life*. Berlin: Springer.

McLean, R.F., Tsyban, A., Burkett, V., Codignott, J.O., Forbes, D.L., Mimura, N., Beamish, R.J., & Ittekkot, V. (2001). Coastal zones and marine ecosystems. In *Climate change 2001: Impacts, adaptation, and vulnerability* (pp. 343–379). Cambridge, UK: Cambridge University Press. Available at: http://www.ipcc.ch/ipccreports/tar/wg2/pdf/wg2TARchap6.pdf.

Moser, S.C., Davidson, M.A., Kirshen, P., Mulvaney, P., Murley, J.F., Neumann, J.E., Petes, L., & Reed, D. (2014). Coastal zone development and ecosystems (Chap. 25). In *Climate change impacts in the United States: The third national climate assessment* (pp. 579–618). U.S. Global Change Research Program. Available at: https://doi.org/10.7930/J0MS3QNW.

Nicholls, R.J., & Lowe, J.A. (2004). Benefits of mitigation of climate change for coastal areas. *Global Environmental Change, 14*, 229–244. Available at: https://doi.org/10.1016/j.gloenvcha.2004.04.005.

Nicholls, R.J., Wong, P.P., Burkett, V., Codignotto, J., Hay, J., McLean, R., Ragoonaden, S., & Woodroffe, C.D. (2007). Coastal systems and low-lying areas, In *Climate Change 2007: Impacts, adaptation and vulnerability* (pp. 315–356). Contribution of working group II to the Fourth Assessment Report of the Intergovernmental Panel on Climate Change. Cambridge, UK: Cambridge University Press. Available at: https://www.ipcc.ch/pdf/assessment-report/ar4/wg2/ar4-wg2-chapter6.pdf.

Oude Essink, G.H.P., van Baaren, E.S., & de Louw, P.G.B. (2010). Effects of climate change on coastal groundwater systems: A modeling study in the Netherlands. *Water resources research, 46*, 1–16. Available at: https://doi.org/10.1029/2009WR008719.

Paice, R., & Chambers, J. (2016a). *Climate change adaptation planning for protection of coastal ecosystems*. Coast Adapt Information Manual 10. National Climate Change Adaptation Research Facility. Available at: https://www.nccarf.edu.au/sites/default/files/tool_downloads/IM_10_coastal%20ecosystems_uploaded.pdf.

Paice, R., & Chambers, J. (2016b). *Climate change impacts on coastal ecosystems* (No. Impact Sheet 8). Coast Adapt. Available at: https://coastadapt.com.au/sites/default/files/factsheets/T312_9_Coastal_Ecosystems.pdf.

Pham, H.V., & Lee, S.-I. (2015). Assessment of seawater intrusion potential from sea-level rise and groundwater extraction in a coastal aquifer. *Desalination and Water Treatment, 53*, 2324–2338. Available at: http://dx.doi.org/10.1080/19443994.2014.971617.

Tam, V.T., Batelaan, O., & Beyen, I. (2016). Impact assessment of climate change on a coastal groundwater system, Central Vietnam. *Environmental Earth Sciences, 78*, 908–1012. Available at: https://doi.org/10.1007/s12665-016-5718-y.

U.S. EPA. (2009). *Synthesis of adaptation options for coastal areas*. U.S. Environmental Protection Agency, Climate Ready Estuaries Program. Available at: https://www.epa.gov/sites/production/files/2014-04/documents/cre_synthesis_1-09.pdf.

World Bank. (2013). *Economic and spatial study of the vulnerability and adaptation to climate change of coastal areas in Senegal*. Available at: https://openknowledge.worldbank.org/bitstream/handle/10986/16986/837830WP0P12030Box0382112B00PUBLIC0.pdf?sequence=1.

Yang, J., Graf, T., & Ptak, T. (2015). Impact of climate change on freshwater resources in a heterogeneous coastal aquifer of Bremerhaven, Germany: A three-dimensional modelling study. *Journal of Contaminant Hydrology, 177–178*, 107–121. Available at: http://dx.doi.org/10.1016/j.jconhyd.2015.03.014.

Zhang, K., Douglas, B.C., & Leatherman, S.P. (2004). Global warming and coastal erosion. *Climatic Change, 64*, 41–58. Available at: https://doi.org/10.1023/B:CLIM.0000024690.32682.48.

Zikra, M., Suntoyo, Lukijanto (2015). Climate change impacts on Indonesian coastal areas. *Procedia Earth and Planetary Science, 14*, 57–63. Available at: https://doi.org/10.1016/j.proeps.2015.07.085.

MIX
Papier aus verantwortungsvollen Quellen
Paper from responsible sources
FSC® C105338

Printed by Books on Demand, Germany